Mathematical
Surveys
and
Monographs

Volume 175

Algebraic Design Theory

Warwick de Launey

Dane Flannery

American Mathematical Society
Providence, Rhode Island

2010 *Mathematics Subject Classification.* Primary 05-02, 05Bxx, 05E18, 16B99, 20Dxx; Secondary 05-04, 15A24, 16S99, 20B20, 20J06.

For additional information and updates on this book, visit
www.ams.org/bookpages/surv-175

Library of Congress Cataloging-in-Publication Data

De Launey, Warwick, 1958–
    Algebraic design theory / Warwick De Launey, Dane Flannery.
        p. cm. — (Mathematical surveys and monographs ; v. 175)
    Includes bibliographical references and index.
    ISBN 978-0-8218-4496-0 (alk. paper)
    1. Combinatorial designs and configurations. I. Flannery, D. L. (Dane Laurence), 1965–
II. Title

QA166.25.D43 2011
511'.6–dc23   2011014837

To Scott Godfrey MD, Richard Lam MD, and Mark Scholz MD

— Warwick de Launey

To my parents, Lois and Ivan

— Dane Flannery

Contents

Preface

Over the past several decades, algebra has become increasingly important in combinatorial design theory. The flow of ideas has for the most part been from algebra to design theory. Moreover, despite our successes, fundamental algebraic questions in design theory remain open. It seems that new or more sophisticated ideas and techniques will be needed to make progress on these questions. In the meantime, design theory is a fertile source of problems that are ideal for spurring the development of algorithms in the active field of computational algebra.

We hope that this book will encourage the investigation, by researchers at all levels, of the algebraic questions posed by design theory. To this end, we provide a large selection of the algebraic objects and applications to be found in design theory. We also isolate a small number of problems that we think are important.

This book is a technical work that takes an unusually abstract approach. While the approach is non-standard, it offers uniformity and enables us to highlight the principal themes in such a way that they can be studied for their own sake, rather than as a means to an end in special cases.

Everything begins with the following notion of orthogonality. Fix an integer $b > 1$, and a non-empty set $\mathcal{A}$ (an 'alphabet') excluding zero. Let Λ be a set (an 'orthogonality set') of $2 \times b$ arrays whose non-zero entries come from $\mathcal{A}$. Much of design theory is concerned with instances of the question

> When does there exist a $v \times b$ array D such that every $2 \times b$ subarray of D is in Λ?

If D exists, then we say that its rows are *pairwise Λ-orthogonal*. Since essentially combinatorial constraints are being placed on pairs of distinct rows, and because of antecedents in the design of experiments, we call D a *pairwise combinatorial design*, or $\mathrm{PCD}(v, \Lambda)$ for short. Chapter 2 describes families of widely-studied pairwise combinatorial designs. These designs are of interest in diverse fields including electrical engineering, statistical analysis, and finite geometry.

This book develops a theory of *square* pairwise combinatorial designs, i.e., those with $v = b$. For such designs we use the abbreviated notation $\mathrm{PCD}(\Lambda)$. Each of the principal design-theoretic themes finds expression. The 'ambient rings' introduced in Chapter 5 allow the free interplay of these themes: orthogonality, equivalence, transposability, composition, transference, the proliferation of inequivalent designs, the automorphism group, and links to group ring (norm) equations.

We pay particular attention to designs that possess a type of *regular* group action. The acting group has a certain central subgroup Z, and the corresponding 2-cocycles with coefficients in Z have a significant influence on properties of the design. Such a design is said to be *cocyclic*. This book contains a general theory for cocyclic pairwise combinatorial designs, plus many case studies. Along the way, we encounter numerous classical designs and other well-known mathematical objects.

This is a book of ideas. It is our opinion that design theory is still—even now—in its infancy. Thus, at this stage, ideas are more valuable than a compendium of our present state of knowledge (which will keep growing rapidly beyond the confines of a single volume). We have aimed to stimulate a creative reader rather than to be encyclopedic.

With respect to cocyclic designs, the chief omissions from our book are Noboru Ito's work on Hadamard groups; and work by Kathy Horadam, her colleagues, and her students.

Our book covers some of Ito's results, but from a different perspective. Starting in the 1980s, Ito produced a sequence of papers identifying regular group actions on the expanded design of a Hadamard matrix. We are content to refer the reader to those papers.

The first author, together with Horadam, founded the theory of cocyclic designs in the early 1990s. Horadam and her school have since published many results focusing on Hadamard, complex Hadamard, and generalized Hadamard matrices. That material is covered in Horadam's engaging book [87]. There one will find topics such as shift equivalence of cocycles, equivalence classes of relative difference sets, and the connection between generalized Hadamard matrices and presemifields, that are not in this book.

We have tried to make the book as accessible as possible; we especially hope that our treatment of the new ideas is welcoming and open-ended. Proofs are given for nearly all results outside of the 'algebraic primer' chapter and the chapter on Paley matrices. The book also contains a wealth of examples and case studies which should persuade the reader that the concepts involved are worthy of pursuit.

Acknowledgments. We are indebted to K. T. Arasu, Robert Craigen, Kathy Horadam, Hadi Kharaghani, S. L. Ma, Michael J. Smith, and Richard M. Stafford, whose collaborations with the first author form the basis of several chapters and sections.

We received useful advice and feedback from Joe Buhler, Alla Detinko, John Dillon, Al Hales, Kathy Horadam, Bill Kantor, Padraig Ó Catháin, Dick Stafford, Tobias Rossmann, and Jennifer Seberry. We are grateful to everyone for their help.

Many thanks are due as well to Sergei Gelfand and Christine Thivierge of the American Mathematical Society, who guided us toward publication.

Finally, we thank Science Foundation Ireland for financial assistance from the Research Frontiers Programme and Mathematics Initiative 2007 (grants 08/RFP/MTH1331 and 07/MI/007).

On November 8, 2010, Warwick de Launey passed away after a long illness.

This book represents Warwick's vision for Design Theory, gained from his years of experience and achievement in the subject. It was my privilege to share in the struggle to bring this vision to a wider audience.

The support of Warwick's wife, Ione Rummery, was constant throughout our writing of the book, and is deeply appreciated.

Warwick has dedicated the book to his doctors. Their care gave him the time he needed to complete his vision.

Dane Flannery
March 27, 2011

CHAPTER 1

Overview

1.1. What is a combinatorial design?

An experimental scientist, statistician, or engineer often studies variables that depend on several factors. Usually the experimenter has specific goals. She may wish to eliminate (as far as possible) the effect of one factor, or to gauge the effects of certain factors on the response variable. An experimental design is a schedule for a series of measurements that efficiently meets these needs. For example, it may be necessary to measure the weights of samples using a beam balance, or to compare the yields of strains of a crop. In the former case, the estimates are most accurate if the objects are weighed in combinations prescribed by a 'weighing matrix'. In the latter case, one can use a 'balanced incomplete block design'.

These and other efficient designs obey simple combinatorial rules determined by the setting. For example, a weighing matrix $W(n, k)$ is an $n \times n$ matrix of 0s, 1s, and -1s such that

- each row and column contains exactly k non-zero entries,
- each pair of distinct rows has inner product equal to zero.

Also, a balanced incomplete block design $BIBD(v, b, r, k, \lambda)$ is a $v \times b$ matrix of 0s and 1s such that

- each row contains exactly r 1s,
- each column contains exactly k 1s,
- each pair of distinct rows has inner product equal to λ.

The realization that an optimal experimental design corresponds to a schedule satisfying combinatorial rules, and that these designs may be used to save time and money, led to the development of combinatorial design theory: the field of mathematics concerned with finite objects, called combinatorial designs, obeying specified combinatorial rules. The field grows apace. It now has major interactions with coding theory, the study of sequences with autocorrelation properties, extremal graph theory, and finite geometry. In this book, we focus on *pairwise* combinatorial designs, each of which can be displayed as a square array whose rows taken pairwise obey a combinatorial rule. These seem to be the most pervasive sort of design.

1.2. What is Algebraic Design Theory?

This book is about the emerging area that we call *algebraic design theory*: the application of algebra and algebraic modes of reasoning in design theory.

The unresolved problems for pairwise combinatorial designs are not so easily categorized. It seems that these problems are algebraic, and some of them may be classed as a modern kind of Diophantine problem. In the situations where a design corresponds to a solution of a group ring (or norm) equation, this judgment is obviously correct; however, current algebraic techniques have not succeeded in answering vital questions. Thus, we sometimes fall back on constructions that supply us with examples and insights, but bypass the issue of extending current techniques to the point where they can answer our questions.

1.3. What is in this book?

Chapters 3–16 and Chapter 20 establish a general abstract framework for pairwise combinatorial designs. Chapters 2, 17, 18, 19, 21, 22, and 23 are case studies.

1.3.1. Theory.

Algebraic essentials. Chapter 3 collects together basic algebraic definitions and results. Parts of this chapter could be skipped by a reader with sufficient algebraic background. More algebra is filled in as we proceed through the book.

Orthogonality. Chapter 2 introduces the notions of orthogonality set Λ and pairwise combinatorial design $\mathrm{PCD}(\Lambda)$. We give many examples of familiar designs that are $\mathrm{PCD}(\Lambda)$s. Then Chapter 4 treats orthogonality in design theory at greater length. A natural and important problem is to determine, for each orthogonality set Λ, the maximum number of rows that are pairwise Λ-orthogonal. After discussing this problem, we show how the lattice of orthogonality sets with a given *alphabet* $\mathcal{A}$ may be studied using dual maps λ and δ between orthogonality sets and *design sets*. We further show how Λ determines a range of Λ-*equivalence* operations. These make up the group $\Pi_\Lambda = \langle \Pi_\Lambda^{\mathrm{row}}, \Pi_\Lambda^{\mathrm{col}} \rangle$ of row and column equivalence operations, and the group Φ_Λ incorporating Π_Λ and the global equivalence operations. We calculate Π_Λ and Φ_Λ for all the designs from Chapter 2.

Ambient rings. Chapter 5 constructs several rings for the necessary matrix algebra with designs. Let Λ be an orthogonality set with alphabet $\mathcal{A}$. At the very least, an *ambient ring* for Λ is just a ring containing $\mathcal{A}$. More productively, it is an involutory ring that also contains a *row group* $R \cong \Pi_\Lambda^{\mathrm{row}}$ and a *column group* $C \cong \Pi_\Lambda^{\mathrm{col}}$ in its group of units. This kind of ambient ring has a *matrix algebra model for Λ-equivalence*. Section 5.2 constructs such a ring for any Λ. Later, in Chapter 13, we show that a *central group* $Z \cong \Pi_\Lambda^{\mathrm{row}} \cap \Pi_\Lambda^{\mathrm{col}}$ may always be included in $R \cap C$. This extra structure is needed to model cocyclic development.

In Chapter 6 we consider Λ-orthogonality as a 'Gram Property': a $(0, \mathcal{A})$-array D is a $\mathrm{PCD}(\Lambda)$ if and only if the Gram matrix DD^* over an ambient ring $\mathcal{R}$ with involution $*$ lies in a prescribed set $\mathrm{Gram}_{\mathcal{R}}(\Lambda)$. Thus, although we start from a definition of orthogonality shorn of algebra, we can in fact think of each $\mathrm{PCD}(\Lambda)$ as a solution to a matrix ring equation.

Transposability. When does Λ-orthogonality, a pairwise condition on the rows of an array, impose a pairwise condition on the array's columns? In Chapter 7 we prove a theorem that answers this question for orthogonality sets like those in Chapter 2, as well as the new case of generalized weighing matrices over a non-abelian group. We also describe transposable orthogonality sets outside the scope of our theorem.

Composition and transference. Chapter 8 discusses the construction of new designs from other (generally smaller) designs. This is an old and recurring theme. The chapter divides into two sections: one on *composition*, the other on *transference*. These ideas have some overlap, but it is convenient to have both at our disposal.

Designs may be composed by means of a *substitution scheme*, which consists of a template array and a set of *plug-in matrices* that satisfy relations determined by the ambient ring for the template array. Since there is more than one choice of ring, a single template array can be employed in different compositions.

Transference (a term of our coinage) is a vehicle for another recurring theme in design theory. Here, existence of one kind of design implies the existence of another kind of design. The connections can be quite unexpected, and, to our eyes, exhibit no discernible common pattern. In Chapter 8 we give examples of transference. These depend on special properties of the alphabet.

The automorphism group. Let $\mathcal{R}$ be any ambient ring with row group $R \cong \Pi_\Lambda^{\mathrm{row}}$ and column group $C \cong \Pi_\Lambda^{\mathrm{col}}$. The ordered pair (P, Q) of monomial matrices $P \in \mathrm{Mon}(v, R)$ and $Q \in \mathrm{Mon}(v, C)$ is an *automorphism* of a PCD(Λ), D, if

$$PDQ^* = D.$$

Then $\mathrm{Aut}(D)$ consists of all automorphisms of D. This group is independent of the choice of ambient ring $\mathcal{R}$.

In Chapter 9 we show that Λ-orthogonality bounds the size of $\mathrm{Aut}(D)$. As an example, we find the automorphism groups for a class of generalized Hadamard matrices. We show that, in this case, the bound is nearly sharp. We then find the automorphism groups of some familiar orthogonal designs. Chapter 9 also presents a simple depth-first backtrack algorithm for computing $\mathrm{Aut}(D)$.

Expanded and associated designs. Chapter 9 introduces the *expanded design* and *associated design* of a pairwise combinatorial design. The associated design of a balanced generalized weighing matrix is a group divisible design. If D is a PCD(Λ) with ambient ring $\mathcal{R}$, then the expanded design of D is the block matrix

$$\mathcal{E}(D) = [rDc]_{r \in R, c \in C}$$

(the multiplication being done over $\mathcal{R}$). The expanded design $\mathcal{E}(D)$ is used to compute the automorphism group of D, and it figures prominently in the theory of cocyclic development.

Group-developed arrays. Chapter 10 is about a host of concepts: regular actions on square arrays, group development, associates, and group ring equations. The $(0, \mathcal{A})$-array A is group-developed modulo a group G if its rows and columns can be labeled with the elements of G so that, for some map $h : G \rightarrow \{0\} \cup \mathcal{A}$,

$$A = [h(xy)]_{x, y \in G}.$$

This is equivalent to G acting regularly on A. For each such h we get an element

$$\sum_{x \in G} h(x)x$$

of the group ring $\mathcal{R}[G]$ of G over an ambient ring $\mathcal{R}$, called a *G-associate* of A.

Associates and group ring equations. In Chapter 10 we show how each G-associate corresponds to a solution of a group ring equation with $(0, \mathcal{A})$-coefficients. The particulars of the group ring equation depend on the orthogonality set Λ. So the enumeration of group-developed PCD(Λ)s is equivalent to solving a group ring equation. The connection is akin to that between pairwise combinatorial designs and the Gram Property.

Associates and regular subgroups. We also show that the G-associates of an array correspond to the conjugacy classes of regular subgroups in the automorphism group of the array. We thereby gain a practical method to find all G-associates of a PCD(Λ).

Associates and difference sets. Finally, in Chapter 10 we see how the G-associates of designs give rise to (relative) difference sets.

Origins of cocyclic development. Chapter 11 elucidates the origins of cocyclic design theory. We give two derivations, both of which begin with the notion of an f-developed array

$$[f(x,y)(g(xy))]_{x,y \in G},$$

where $f : G \times G \to \mathrm{Sym}(\mathcal{A} \cup \{0\})$ is a map fixing 0. The first derivation is in the context of higher-dimensional designs. The second derivation springs from the fact that a group-developed array $A = [g(xy)]_{x,y \in G}$ is equivalent, via row and column permutations, to the array $[g(axy)]_{x,y \in G}$ obtained by developing the ath row of A for any a. If one requires that an f-developed array is Λ-equivalent to the array obtained by 'f-developing' the ath row of A, then f must be a 2-cocycle.

Cocycles. Let G and Z be groups, where Z is abelian. A map $f : G \times G \to Z$ such that

$$f(a,b)f(ab,c) = f(b,c)f(a,bc) \qquad \forall\, a,b,c \in G$$

is a *2-cocycle*. Chapter 12 contains an elementary exposition of 2-cocycles. The discussion revolves around central short exact sequences

$$1 \longrightarrow Z \overset{\iota}{\longrightarrow} E \overset{\pi}{\longrightarrow} G \longrightarrow 1.$$

Each map (apart from the first) is a group homomorphism whose kernel is the image of the preceding homomorphism, and ι maps into the center of the extension group E. We define a cocycle $f_{\iota,\tau} : G \times G \to Z$ for each 'transversal map' $\tau : G \to E$; moreover, every cocycle $f : G \times G \to Z$ arises in this way. Chapter 12 discusses cocycles of product groups, cocycles calculated by collection within a polycyclic group, and monomial representations from a cocycle.

Cocyclic pairwise combinatorial designs. Chapter 13 formulates the theory of cocyclic designs in terms of matrix algebra.

Let D be a PCD(Λ), where Λ has ambient ring $\mathcal{R}$ containing the row group R, column group C, and central group Z. Then D is *cocyclic* if for some cocycle $f : G \times G \to Z$ there are monomial matrices $P \in \mathrm{Mon}(v, R)$, $Q \in \mathrm{Mon}(v, C)$ and a map $g : G \to \{0\} \cup \mathcal{A}$ such that

$$D = P[f(x,y)g(xy)]_{x,y \in G} Q$$

over $\mathcal{R}$. We say that f is a *cocycle of* D with *indexing group* G, and that f is a Λ-*cocycle*. Any extension group for f is an *extension group* of D. Chapter 13 asks

1.3.1. QUESTION. *Given a* PCD(Λ), *D, what are all the cocycles of D?*

1.3.2. QUESTION. *Given an orthogonality set Λ and a cocycle $f : G \times G \to Z$, is there a* PCD(Λ) *with cocycle f?*

Four approaches to cocyclic designs are proposed.

1.3.3. For a given orthogonality set Λ,

   (1) study the cocycles of known highly-structured PCD(Λ)s;
   (2) determine Λ-cocycles via the extension group;
   (3) determine Λ-cocycles via the indexing group;
   (4) use composition to prove existence of PCD(Λ)s.

All these approaches are taken in the book.

Centrally regular actions, cocyclic associates, and group ring equations. Chapters 14 and 15 set out more theory of cocyclic designs. We describe relationships between cocyclic development, actions on the expanded design, cocyclic associates, and group ring equations. Taken together, Chapters 14 and 15 are analogous to Chapter 10.

Chapter 15 contains deeper material: an application of character theory to the existence question for circulant complex Hadamard matrices, and Ito's striking results on allowable extension groups of cocyclic Hadamard matrices (in line with part (2) of 1.3.3).

Cocyclic development tables. A *development table* displays all the ways in which a cocyclic PCD(Λ) with indexing group G may be f-developed for some cocycle $f : G \times G \to Z$. Chapter 20 explains how to compute development tables when G is solvable.

The theory for familiar classes of designs. Chapter 16 is a 'bridging' chapter. It refreshes some of the preceding theory in the book, with particular regard to (complex) Hadamard matrices, balanced weighing matrices, and orthogonal designs. Theoretical results needed for the case studies are here.

1.3.2. Practice.

Many pairwise combinatorial designs. In Chapter 2 we prove basic results about familiar PCD(Λ)s, and state what is done in the book concerning those designs.

Paley matrices. The automorphism groups and all the regular actions for the Paley conference and Hadamard matrices are described in Chapter 17. This case study is extremely rich, yet it is tractable enough that we can answer Question 1.3.1 and carry out part (1) of the agenda 1.3.3.

A large family of cocyclic Hadamard matrices. Chapter 18 is a nice example of part (4) of 1.3.3. Beginning with Paley matrices and applying plug-in techniques, we obtain a large family of cocyclic Hadamard matrices, and thus many maximal-sized relative difference sets with central forbidden subgroup of size 2.

Substitution schemes. Chapter 19 considers the cocyclic Hadamard matrices with cocycle $f : G \times G \to \langle -1 \rangle$ obtained from a central short exact sequence

$$1 \longrightarrow \langle -1 \rangle \longrightarrow E = L \ltimes N \xrightarrow{\pi} G = K \ltimes N \longrightarrow 1,$$

where $|K| = 4$, π is the identity on N, and $\pi(L) = K$ (this includes the atomic case $|G| = 4p$, $p > 3$ prime). All such Hadamard matrices are defined by a substitution scheme that, among other things, generalizes the Williamson construction. This study is an instance of part (3) of 1.3.3.

Cocyclic Hadamard matrices and elementary abelian groups. A primary aim in the study of cocyclic Hadamard matrices is to answer Question 1.3.2 for a given group G. For most G this problem breaks up into a practicable algebraic component and a difficult combinatorial component. However, if G is an elementary abelian 2-group then we show in Chapter 21 that the problem can be solved algebraically: here, nearly every cocycle is a cocycle of a Sylvester Hadamard matrix. Chapter 21 is a good example of parts (2) and (3) of the agenda 1.3.3.

Systems of cocyclic orthogonal designs. Chapter 22 gives a complete algebraic solution of the problem of classifying *concordant* systems $\{D_1, \ldots, D_r\}$ of cocyclic orthogonal designs, where in each D_i every relevant indeterminate appears exactly once in each row and column. The approach is via the extension group, and there is a strong reliance on Chapter 21.

Asymptotic existence of cocyclic Hadamard matrices. In the final chapter we present a proof of the best known (to date) asymptotic existence result for cocyclic Hadamard matrices. Our proof combines knowledge about the existence of complex complementary sequences with ideas from Chapters 21 and 22.

CHAPTER 2

Many Kinds of Pairwise Combinatorial Designs

This chapter introduces the fundamental concepts of orthogonality set Λ and pairwise combinatorial design $\mathrm{PCD}(\Lambda)$. We will see that many familiar designs are $\mathrm{PCD}(\Lambda)$s. The orthogonality set Λ in these cases is almost always *transposable*: D is a $\mathrm{PCD}(\Lambda)$ if and only if the transpose $D^\top$ is a $\mathrm{PCD}(\Lambda)$. When this is so, we give a justification of transposability.

Throughout the chapter we occasionally use terminology from design theory and algebra without definition. That terminology will be defined later. We adopt the conventions that i denotes a square root of -1, and if X is a matrix with entries from the field $\mathbb{C}$ of complex numbers, then X^* is the complex conjugate transpose of X. For a matrix X whose non-zero entries all lie in a group, X^* denotes the transpose of the matrix obtained from X by inverting its non-zero entries. The $v \times v$ identity matrix and the $v \times v$ matrix of 1s are denoted I_v, J_v respectively. For natural numbers a and b, $b \neq 0$, we write $\lfloor a/b \rfloor$ for the integer part of a/b, and $\{a/b\} = a/b - \lfloor a/b \rfloor$ for its fractional part.

The computer algebra system MAGMA [14] was used to generate many of the examples in this chapter; these were output in a canonical form based on maximal lexicographic ordering.

2.1. Orthogonality sets

Our first definition is the most important one in the book.

2.1.1. DEFINITION. Let $\mathcal{A}$ be a non-empty finite set (an 'alphabet') such that $0 \notin \mathcal{A}$. Let Λ be a set of $2 \times b$ $(0, \mathcal{A})$-arrays that is closed under row and column permutations; i.e., if

$$\begin{bmatrix} x_1 & x_2 & \cdots & x_b \\ y_1 & y_2 & \cdots & y_b \end{bmatrix}$$

is in Λ, then, for any permutation π of $\{1, 2, \ldots, b\}$,

$$\begin{bmatrix} y_1 & y_2 & \cdots & y_b \\ x_1 & x_2 & \cdots & x_b \end{bmatrix} \quad \text{and} \quad \begin{bmatrix} x_{\pi(1)} & x_{\pi(2)} & \cdots & x_{\pi(b)} \\ y_{\pi(1)} & y_{\pi(2)} & \cdots & y_{\pi(b)} \end{bmatrix}$$

are in Λ. Suppose also that no array in Λ has a repeated row. Then we call Λ an *orthogonality set*. We say that $(x_1, x_2, \ldots, x_b)$ and $(y_1, y_2, \ldots, y_b)$ are Λ-*orthogonal* if

$$\begin{bmatrix} x_1 & x_2 & \cdots & x_b \\ y_1 & y_2 & \cdots & y_b \end{bmatrix} \in \Lambda.$$

2.1.2. REMARK. To begin with, 0 in Definition 2.1.1 is simply a special element not contained in the alphabet $\mathcal{A}$; but eventually we will define it as the zero of an 'ambient' ring for Λ.

We are interested in arrays whose rows are pairwise Λ-orthogonal.

2.1.3. DEFINITION. Let Λ be an orthogonality set of $2 \times b$ $(0, \mathcal{A})$-arrays. A *pairwise combinatorial design* $\mathrm{PCD}(v, \Lambda)$ is a $v \times b$ array D such that all pairs of distinct rows of D are Λ-orthogonal. If $v = b$, then we just write $\mathrm{PCD}(\Lambda)$.

Definitions 2.1.1 and 2.1.3 are entirely combinatorial. They are also very broad, covering many different kinds of orthogonality and combinatorial designs.

2.1.4. EXAMPLE. A Latin square of order n is an $n \times n$ array with entries from $\mathcal{A} = \{1, \dots, n\}$ such that each element of $\mathcal{A}$ appears exactly once in each row and column. In this case Λ is the set of $2 \times n$ $\mathcal{A}$-arrays with no repeated element in any row or column.

2.1.5. EXAMPLE. A binary error-correcting block code of length b, with v codewords and distance d, is equivalent to a $v \times b$ $(0, 1)$-array where every two distinct rows differ in at least d columns. The alphabet is $\{1\}$, and the orthogonality set is all $2 \times b$ $(0, 1)$-arrays with at least d columns containing different entries.

2.1.6. EXAMPLE. A set of v mutually orthogonal Latin squares of order n corresponds to a $(v + 2) \times n^2$ array of entries from $\mathcal{A} = \{1, \dots, n\}$ such that every pair of rows contains each ordered pair of elements of $\mathcal{A}$ exactly once. Here Λ is the orbit of

$$
\begin{bmatrix}
1 & 1 & \cdots & 1 & 2 & 2 & \cdots & 2 & \cdots & n & n & \cdots & n \\
1 & 2 & \cdots & n & 1 & 2 & \cdots & n & \cdots & 1 & 2 & \cdots & n
\end{bmatrix}
$$

under row and column permutations.

The following are mutually orthogonal Latin squares for $n = 5$.

1 2 3 4 5	1 2 3 4 5	1 2 3 4 5	1 2 3 4 5
2 3 4 5 1	3 4 5 1 2	4 5 1 2 3	5 1 2 3 4
3 4 5 1 2	5 1 2 3 4	2 3 4 5 1	4 5 1 2 3
4 5 1 2 3	2 3 4 5 1	5 1 2 3 4	3 4 5 1 2
5 1 2 3 4	4 5 1 2 3	3 4 5 1 2	2 3 4 5 1

We now form a 6×25 array A. The ith row of A for $1 \leq i \leq 4$ is the concatenation of rows $1, 2, 3, 4, 5$ (in that order) of the ith Latin square. Adding the canonical rows

```
1 2 3 4 5 1 2 3 4 5 1 2 3 4 5 1 2 3 4 5 1 2 3 4 5
1 1 1 1 1 2 2 2 2 2 3 3 3 3 3 4 4 4 4 4 5 5 5 5 5
```

yields A as follows.

$$(2.1)$$

```
1 2 3 4 5 2 3 4 5 1 3 4 5 1 2 4 5 1 2 3 5 1 2 3 4
1 2 3 4 5 3 4 5 1 2 5 1 2 3 4 2 3 4 5 1 4 5 1 2 3
1 2 3 4 5 4 5 1 2 3 2 3 4 5 1 5 1 2 3 4 3 4 5 1 2
1 2 3 4 5 5 1 2 3 4 4 5 1 2 3 3 4 5 1 2 2 3 4 5 1
1 2 3 4 5 1 2 3 4 5 1 2 3 4 5 1 2 3 4 5 1 2 3 4 5
1 1 1 1 1 2 2 2 2 2 3 3 3 3 3 4 4 4 4 4 5 5 5 5 5
```

Notice that orthogonality with the last two rows ensures that each of the other rows is the concatenation of the rows of a Latin square. Furthermore, rows i and $j \neq i$ being orthogonal for all $i, j \in \{1, 2, 3, 4\}$ ensures that the Latin squares are mutually orthogonal.

2.1.7. EXAMPLE. An orthogonal array of strength 2 and index λ is equivalent to a $v \times \lambda n^2$ array of entries from $\mathcal{A} = \{1, \ldots, n\}$ such that, for all $a, b \in \mathcal{A}$, every pair of distinct rows contains exactly λ columns equal to $[a, b]^\top$. The 6×25 array (2.1) is an orthogonal array of strength 2 and index 1.

2.1.8. EXAMPLE. An orthogonal array of strength 3 and index 1 is equivalent to a $v \times n^3$ array of entries from $\mathcal{A} = \{1, \ldots, n\}$ such that, for all $a, b, c \in \mathcal{A}$, every triple of distinct rows contains exactly one column equal to $[a, b, c]^\top$.

In Example 2.1.8, the combinatorial nature of the array depends on how triples rather than pairs of rows behave. Even here the notion of an orthogonality set is pertinent, since every orthogonal array of strength 3 is an orthogonal array of strength 2.

This book deals with *square* pairwise combinatorial designs. We discuss some familiar PCD(Λ)s in the subsequent sections.

2.2. Symmetric balanced incomplete block designs

2.2.1. DEFINITION. Let v, k, λ be integers, where $v > k > \lambda \geq 0$. A *symmetric balanced incomplete block design* SBIBD(v, k, λ) is a set of v subsets of size k, called *blocks*, of a fixed set of v elements, called *varieties*, such that each unordered pair of distinct varieties is in exactly λ blocks.

Symmetric balanced incomplete block designs are among the earliest kinds of designs to have been studied. Of paramount concern is the existence question: for which v, k, λ does there exist an SBIBD(v, k, λ)? Infinite families of these designs are rare. Most of the families are constructed using other designs in this chapter; see [**92, 93, 94, 95, 96, 103, 104, 110**].

We now recast the definition of SBIBD(v, k, λ). An *incidence matrix* for the design is a $v \times v$ $(0, 1)$-matrix $A = [a_{ij}]$ such that $a_{ij} = 1$ if and only if variety i is contained in block j.

2.2.2. EXAMPLE. We display incidence matrices for an SBIBD($11, 5, 2$) and an SBIBD($13, 4, 1$).

$$
\begin{bmatrix}
1 & 1 & 1 & 1 & 1 & 0 & 0 & 0 & 0 & 0 & 0 \\
1 & 1 & 0 & 0 & 0 & 1 & 1 & 1 & 0 & 0 & 0 \\
1 & 0 & 1 & 0 & 0 & 1 & 0 & 0 & 1 & 1 & 0 \\
1 & 0 & 0 & 1 & 0 & 0 & 1 & 0 & 1 & 0 & 1 \\
1 & 0 & 0 & 0 & 1 & 0 & 0 & 1 & 0 & 1 & 1 \\
0 & 1 & 1 & 0 & 0 & 0 & 1 & 0 & 0 & 1 & 1 \\
0 & 1 & 0 & 1 & 0 & 0 & 0 & 1 & 1 & 1 & 0 \\
0 & 1 & 0 & 0 & 1 & 1 & 0 & 0 & 1 & 0 & 1 \\
0 & 0 & 1 & 1 & 0 & 1 & 0 & 1 & 0 & 0 & 1 \\
0 & 0 & 1 & 0 & 1 & 0 & 1 & 1 & 1 & 0 & 0 \\
0 & 0 & 0 & 1 & 1 & 1 & 1 & 0 & 0 & 1 & 0
\end{bmatrix}
$$

$$
\begin{bmatrix}
1 & 1 & 1 & 1 & 0 & 0 & 0 & 0 & 0 & 0 & 0 & 0 & 0 \\
1 & 0 & 0 & 0 & 1 & 1 & 1 & 0 & 0 & 0 & 0 & 0 & 0 \\
1 & 0 & 0 & 0 & 0 & 0 & 0 & 1 & 1 & 1 & 0 & 0 & 0 \\
1 & 0 & 0 & 0 & 0 & 0 & 0 & 0 & 0 & 0 & 1 & 1 & 1 \\
0 & 1 & 0 & 0 & 1 & 0 & 0 & 1 & 0 & 0 & 1 & 0 & 0 \\
0 & 1 & 0 & 0 & 0 & 1 & 0 & 0 & 1 & 0 & 0 & 1 & 0 \\
0 & 1 & 0 & 0 & 0 & 0 & 1 & 0 & 0 & 1 & 0 & 0 & 1 \\
0 & 0 & 1 & 0 & 1 & 0 & 0 & 0 & 1 & 0 & 0 & 0 & 1 \\
0 & 0 & 1 & 0 & 0 & 1 & 0 & 0 & 0 & 1 & 1 & 0 & 0 \\
0 & 0 & 1 & 0 & 0 & 0 & 1 & 1 & 0 & 0 & 0 & 1 & 0 \\
0 & 0 & 0 & 1 & 1 & 0 & 0 & 0 & 0 & 1 & 0 & 1 & 0 \\
0 & 0 & 0 & 1 & 0 & 1 & 0 & 1 & 0 & 0 & 0 & 0 & 1 \\
0 & 0 & 0 & 1 & 0 & 0 & 1 & 0 & 1 & 0 & 1 & 0 & 0
\end{bmatrix}
$$

Every column in the first design has five 1s and every pair of rows has precisely two columns of 1s. Similarly, every column in the second design has four 1s, and every pair of rows has precisely one column of 1s. Note also that every row of the first design contains five 1s, and every row of the second design contains four 1s. We show below that this regularity is not a coincidence.

Since each block contains exactly k varieties,

(2.2)
$$J_v A = k J_v.$$

We now prove that

$$(2.3) \qquad AA^\top = kI_v + \lambda(J_v - I_v).$$

The inner product of any two distinct rows of A is λ, so we only need to show that every row of A contains exactly k 1s. Let r_i denote the number of columns of A containing a 1 in the ith row. Let B be the $v \times r_i$ submatrix of A obtained by removing all columns not containing 1 in the ith row of A. Since the ith row of B consists of r_i 1s and the other rows each contain λ 1s, B contains exactly $r_i + \lambda(v - 1)$ 1s. Each column of B has exactly k 1s, so $r_i k = r_i + \lambda(v - 1)$. Thus $r_i = \lambda(v - 1)/(k - 1)$. Now counting the number of 1s in A first by row and then by column, we get $vr_i = vk$. This proves (2.3), and additionally

$$(2.4) \qquad \lambda(v - 1) = k(k - 1).$$

Since this is a necessary condition for there to be an SBIBD(v, k, λ), we will suppose in the remainder of this section that (2.4) holds.

2.2.3. DEFINITION. Let $v > k > \lambda \geq 0$ be integers satisfying equation (2.4), and let $\Lambda_{\mathrm{SBIBD}(v,k,\lambda)}$ be the set of $2 \times v$ $(0,1)$-matrices X such that

$$XX^\top = \begin{bmatrix} k & \lambda \\ \lambda & k \end{bmatrix}.$$

2.2.4. REMARK. $\Lambda_{\mathrm{SBIBD}(v,k,\lambda)}$ is non-empty by definition, and is the orbit under row and column permutations of the matrix

$$\begin{bmatrix} 1 & \cdots & 1 & 1 & \cdots & 1 & 0 & \cdots & 0 & 0 & \cdots & 0 \\ 1 & \cdots & 1 & 0 & \cdots & 0 & 1 & \cdots & 1 & 0 & \cdots & 0 \end{bmatrix},$$

with λ columns $[1,1]^\top$, $k - \lambda$ columns $[1,0]^\top$, $k - \lambda$ columns $[0,1]^\top$, and $v - 2k + \lambda$ columns $[0,0]^\top$; see the next lemma.

2.2.5. LEMMA. *For $v > k > \lambda \geq 0$ satisfying (2.4), $v \geq 2k - \lambda$.*

PROOF. Let x and $y \neq x$ be non-negative integers such that $xy \geq k(k - 1)$. Then $(x+y)^2 - (x-y)^2 \geq 4k(k-1)$, and so $(x+y)^2 - (2k-1)^2 \geq (x-y)^2 - 1 \geq 0$. Thus $x + y \geq 2k - 1$. So by (2.4), $\lambda + v - 1 \geq 2k - 1$, i.e., $v \geq 2k - \lambda$. □

2.2.6. THEOREM. *A is an incidence matrix of an SBIBD(v, k, λ) if and only if it is a PCD$(\Lambda_{\mathrm{SBIBD}(v,k,\lambda)})$.*

PROOF. Since an incidence matrix A of any SBIBD(v, k, λ) satisfies (2.3), we have the forward implication. Assuming (2.3) and (2.4), we verify that A has every column sum equal to k.

Let k_i denote the number of 1s in the ith column. Then

$$(2.5) \qquad \sum_{i=1}^{v} k_i(k_i - 1) = \lambda v(v - 1), \qquad \sum_{i=1}^{v} k_i = vk.$$

Using (2.4) and (2.5) we get $\sum_{i=1}^{v} k_i^2 = vk(k - 1) + \sum_{i=1}^{v} k_i = vk^2$, and then

$$\sum_{i=1}^{v} (k_i - k)^2 = \sum_{i=1}^{v} k_i^2 + vk^2 - 2k \sum_{i=1}^{v} k_i = vk^2 + vk^2 - 2k(vk) = 0.$$

Thus $k_i = k$ for all i. □

In particular, if A is a PCD$(\Lambda_{\mathrm{SBIBD}(v,k,\lambda)})$, then (2.2) holds.

2.2.7. THEOREM. $A^\top$ *is a* $\mathrm{PCD}(\Lambda_{\mathrm{SBIBD}(v,k,\lambda)})$ *if and only if A is too.*

PROOF. Suppose that A is a $\mathrm{PCD}(\Lambda_{\mathrm{SBIBD}(v,k,\lambda)})$. Then by (2.2) and (2.3) we have

$$(2.6) \qquad J_v A = A J_v.$$

Moreover, $\det(AA^\top) = (k-\lambda)^{v-1}(k+(v-1)\lambda) = k^2(k-\lambda)^{v-1} \neq 0$. Therefore A is invertible. By (2.6),

$$A^\top A = A^{-1} A A^\top A = A^{-1}(kI_v + \lambda(J_v - I_v))A = kI_v + \lambda(J_v - I_v)$$

as required. $\qquad\qquad\qquad\qquad\qquad\qquad\qquad\qquad\qquad\qquad\qquad\square$

2.2.8. COROLLARY. $\Lambda_{\mathrm{SBIBD}(v,k,\lambda)}$ *is transposable.*

2.3. Hadamard matrices

2.3.1. DEFINITION. A *Hadamard matrix of order n* is an $n \times n$ $(1,-1)$-matrix H such that

$$(2.7) \qquad HH^\top = nI_n.$$

2.3.2. EXAMPLE. Writing $-$ for -1, here is a Hadamard matrix of order 12:

$$
\begin{bmatrix}
1 & 1 & 1 & 1 & 1 & 1 & 1 & 1 & 1 & 1 & 1 & 1 \\
1 & 1 & 1 & 1 & 1 & 1 & - & - & - & - & - & - \\
1 & 1 & 1 & - & - & - & 1 & 1 & 1 & - & - & - \\
1 & 1 & - & 1 & - & - & 1 & - & - & 1 & 1 & - \\
1 & 1 & - & - & 1 & - & - & 1 & - & 1 & - & 1 \\
1 & 1 & - & - & - & 1 & - & - & 1 & - & 1 & 1 \\
1 & - & 1 & 1 & - & - & - & 1 & - & - & 1 & 1 \\
1 & - & 1 & - & 1 & - & - & - & 1 & 1 & 1 & - \\
1 & - & 1 & - & - & 1 & 1 & - & - & 1 & - & 1 \\
1 & - & - & 1 & 1 & - & 1 & - & 1 & - & - & 1 \\
1 & - & - & 1 & - & 1 & - & 1 & 1 & 1 & - & - \\
1 & - & - & - & 1 & 1 & 1 & 1 & - & - & 1 & - \\
\end{bmatrix}.
$$

2.3.3. DEFINITION. Let $\Lambda_{\mathrm{H}(n)}$ denote the set of $2 \times n$ $(1,-1)$-matrices X such that $XX^\top = nI_2$.

2.3.4. REMARKS.

  (1) $\Lambda_{\mathrm{H}(n)}$ is non-empty if and only if n is even.
  (2) $\Lambda_{\mathrm{H}(n)}$ consists of two orbits under row and column permutations if $n = 2$, and $(\lfloor \frac{n}{4} \rfloor + 1)(\frac{n}{2} + 1)$ orbits if $n \geq 4$.

2.3.5. THEOREM. *An $n \times n$ $(1,-1)$-matrix is a Hadamard matrix if and only if it is a $\mathrm{PCD}(\Lambda_{\mathrm{H}(n)})$.*

2.3.6. THEOREM. $\Lambda_{\mathrm{H}(n)}$ *is transposable.*

PROOF. Certainly (2.7) implies that H is invertible over the rational field $\mathbb{Q}$, with $H^{-1} = \frac{1}{n}H^\top$. So $H^\top H = nI_n$. $\qquad\qquad\qquad\qquad\qquad\square$

Hadamard matrices were first studied by Sylvester [149] and Hadamard [80] in the 19th century, and over the years have become the subject of vigorous research activity. Applications occur, for example, in electrical engineering (circuit design) and statistics (experimental design). Horadam's book [87] is an excellent reference for the use of these matrices in signal and data processing. Overall, the literature on Hadamard matrices is immense.

Hadamard matrices are a source of block designs. Normalize any Hadamard matrix H of order $4t$, and remove the first row and column to obtain the matrix A; then $(A + J)/2$ is an incidence matrix of an SBIBD$(4t - 1, 2t - 1, t - 1)$. If H has constant row and column sums then the same trick of replacing -1s with zeros applied directly to H results in another symmetric balanced incomplete block design.

We note the following widely-known result.

2.3.7. THEOREM. *A Hadamard matrix H with constant row and column sums has square order.*

PROOF. Let n be the order of H, and let s be the constant row and column sum. Then $n J_n = J_n H H^\top = s J_n H^\top = s^2 J_n$, so $n = s^2$. □

One may easily show that a Hadamard matrix of order $n > 2$ can exist only if 4 divides n. The famous Hadamard Conjecture states that this condition on the order is sufficient for existence.

1. CONJECTURE. *There is a Hadamard matrix of order $4t$ for all $t \geq 1$.*

This book gives a comprehensive account of Hadamard matrices characterized by a special regular group action: *cocyclic* Hadamard matrices. The Sylvester and Paley matrices are discussed in detail. A result on the asymptotic existence of cocyclic Hadamard matrices is proved. We describe most of the direct constructions known for cocyclic Hadamard matrices. Finally, a large family of cocyclic Hadamard matrices is given, which in turn provides a large class of central relative difference sets. This material is drawn from [**43, 47, 51, 54, 55, 56, 57, 65**].

2.4. Weighing matrices

2.4.1. DEFINITION. Let $n \geq k \geq 1$. A *weighing matrix* $\mathrm{W}(n, k)$ *of order n and weight k* is an $n \times n$ $(0, \pm 1)$-matrix W such that

$$W W^\top = k I_n.$$

2.4.2. EXAMPLE. We display a $\mathrm{W}(10, 8)$ below.

$$\begin{bmatrix}
1 & 1 & 1 & 1 & 1 & 1 & 1 & 1 & 0 & 0 \\
1 & 1 & 1 & 1 & - & - & - & - & 0 & 0 \\
1 & - & 1 & - & 1 & - & 0 & 0 & 1 & 1 \\
1 & - & - & 1 & - & 1 & 0 & 0 & 1 & 1 \\
1 & 1 & - & - & 0 & 0 & 1 & - & - & 1 \\
1 & 1 & - & - & 0 & 0 & - & 1 & 1 & - \\
1 & - & 0 & 0 & 1 & 1 & - & - & - & - \\
1 & - & 0 & 0 & - & - & 1 & 1 & - & - \\
0 & 0 & 1 & - & - & 1 & - & 1 & - & 1 \\
0 & 0 & 1 & - & - & 1 & 1 & - & 1 & -
\end{bmatrix}$$

2.4.3. DEFINITION. Let $\Lambda_{\mathrm{W}(n,k)}$ be the set of $2 \times n$ $(0, \pm 1)$-matrices X such that $X X^\top = k I_2$.

2.4.4. REMARK. $\Lambda_{\mathrm{W}(n,k)}$ is non-empty if and only if $n \geq k + \{k/2\}$.

2.4.5. THEOREM. *An $n \times n$ $(0, \pm 1)$-matrix is a $\mathrm{W}(n, k)$ if and only if it is a* PCD$(\Lambda_{\mathrm{W}(n,k)})$.

2.4.6. THEOREM. *The orthogonality set $\Lambda_{\mathrm{W}(n,k)}$ is transposable.*

PROOF. Cf. the proof of Theorem 2.3.6; a $\mathrm{W}(n, k)$ is invertible over $\mathbb{Q}$. □

Weighing matrices take their name from an application noted by Hotelling ([**89**]) in which items are weighed using a chemical balance. The idea is that the percentage error can be reduced by making measurements with some of the items placed on the left-hand pan and some on the right-hand pan. Hotelling's paper was the first of many on the use of combinatorial designs to improve the precision of laboratory measurements. A schedule for n weighings of n objects can be expressed in the form of an $n \times n$ $(0, \pm 1)$-matrix W. Hotelling showed that for maximum efficiency, $WW^\top$ should be a scalar matrix kI_n for large k. When $n \equiv 0 \pmod 4$, the best possible design is a Hadamard matrix; and when $n \equiv 2 \pmod 4$, the best weight is $k = n - 1$. A weighing matrix of order n and weight $n - 1$ is a *conference matrix*. Several constructions are known for these designs. The most extensive family was found by Gilman [**75**] and Paley [**128**].

2.4.7. EXAMPLE. The (symmetric) conference matrix

$$\begin{bmatrix}
0 & 1 & 1 & 1 & 1 & 1 & 1 & - & - & - \\
1 & 0 & 1 & 1 & 1 & - & - & - & 1 & 1 \\
1 & 1 & 0 & 1 & - & 1 & - & 1 & - & 1 \\
1 & 1 & 1 & 0 & - & - & 1 & 1 & 1 & - \\
1 & 1 & - & - & 0 & 1 & 1 & - & 1 & 1 \\
1 & - & 1 & - & 1 & 0 & 1 & 1 & - & 1 \\
1 & - & - & 1 & 1 & 1 & 0 & 1 & 1 & - \\
- & - & 1 & 1 & - & 1 & 1 & 0 & 1 & 1 \\
- & 1 & - & 1 & 1 & - & 1 & 1 & 0 & 1 \\
- & 1 & 1 & - & 1 & 1 & - & 1 & 1 & 0
\end{bmatrix}$$

is the optimal design for weighing ten objects, and in particular is an improvement on the design of Example 2.4.2.

A *Paley* conference matrix is cocyclic. In Chapter 17 we discuss this weighing matrix together with the two types of Paley Hadamard matrices (which are also cocyclic). We describe their automorphism groups and the associated regular actions.

2.5. Balanced weighing matrices

2.5.1. DEFINITION. A weighing matrix $W = [w_{ij}]$ of order v, such that $[w_{ij}^2]$ is an incidence matrix of an SBIBD(v, k, λ), is called a *balanced weighing matrix* BW(v, k, λ).

2.5.2. REMARK. A conference matrix is a balanced weighing matrix.

2.5.3. EXAMPLE. We display a BW$(7, 4, 2)$ and a BW$(8, 7, 6)$.

$$\begin{bmatrix}
1 & 1 & 1 & 1 & 0 & 0 & 0 \\
1 & - & 0 & 0 & 1 & 1 & 0 \\
1 & 0 & - & 0 & - & 0 & 1 \\
1 & 0 & 0 & - & 0 & - & - \\
0 & 1 & - & 0 & 0 & 1 & - \\
0 & 1 & 0 & - & 1 & 0 & 1 \\
0 & 0 & 1 & - & - & 1 & 0
\end{bmatrix}
\qquad
\begin{bmatrix}
0 & 1 & 1 & 1 & 1 & 1 & 1 & 1 \\
1 & 0 & - & - & 1 & - & 1 & 1 \\
1 & 1 & 0 & 1 & - & - & - & 1 \\
1 & 1 & - & 0 & - & 1 & 1 & - \\
1 & - & 1 & 1 & 0 & - & 1 & - \\
1 & 1 & 1 & - & 1 & 0 & - & - \\
1 & - & 1 & - & - & 1 & 0 & 1 \\
1 & - & - & 1 & 1 & 1 & - & 0
\end{bmatrix}$$

The second design is a conference matrix of order 8.

2.5.4. DEFINITION. Let $v > k > \lambda \geq 0$ be integers satisfying (2.4), with λ even. Let Λ be the set of $(0, \pm 1)$-matrices $[x_{ij}]$ such that $[x_{ij}^2] \in \Lambda_{\text{SBIBD}(v,k,\lambda)}$. Then define $\Lambda_{\text{BW}(v,k,\lambda)} = \Lambda_{\text{W}(v,k)} \cap \Lambda$.

2.5.5. REMARK. $\Lambda_{\mathrm{BW}(v,k,\lambda)}$ is non-empty (assuming the conditions on v, k, λ in Definition 2.5.4 are met).

2.5.6. THEOREM. *A $v \times v$ $(0, \pm 1)$-matrix is a $\mathrm{BW}(v, k, \lambda)$ if and only if it is a $\mathrm{PCD}(\Lambda_{\mathrm{BW}(v,k,\lambda)})$.*

PROOF. A $\mathrm{BW}(v, k, \lambda)$ is a $\mathrm{W}(v, k)$ whose rows are pairwise Λ-orthogonal (for Λ as in Definition 2.5.4). Consequently, by Theorem 2.4.5, a $\mathrm{BW}(v, k, \lambda)$ is a $v \times v$ matrix whose rows are $\Lambda_{\mathrm{W}(v,k)}$-orthogonal and Λ-orthogonal. □

2.5.7. THEOREM. *$\Lambda_{\mathrm{BW}(v,k,\lambda)}$ is transposable.*

PROOF. By Corollary 2.2.8, Λ is transposable. Also, $\Lambda_{\mathrm{W}(v,k)}$ is transposable by Theorem 2.4.6; and clearly the intersection of transposable orthogonality sets is transposable. □

Balanced weighing matrices were first studied by Mullin [**122**]. That paper and a flurry of others [**9, 10, 123, 124, 136**] exhibited infinite families and connections with symmetric balanced incomplete block designs. In this book, we show that some of these families are cocyclic. We further show that any cocyclic balanced weighing matrix is equivalent to a relative difference set with normal forbidden subgroup of order 2.

2.6. Orthogonal designs

2.6.1. DEFINITION. Let $n, a_1, \ldots, a_r$ be positive integers such that $n \geq \sum_{i=1}^{r} a_i$. An *orthogonal design* $\mathrm{OD}(n; a_1, \ldots, a_r)$ of order n and type $(a_1, \ldots, a_r)$ is an $n \times n$ matrix D with entries from a set $\{0, \pm x_1, \ldots, \pm x_r\}$, where the x_is are commuting indeterminates, such that

$$DD^{\top} = (a_1 x_1^2 + \cdots + a_r x_r^2) I_n.$$

2.6.2. EXAMPLE. An $\mathrm{OD}(12; 3, 3, 3, 3)$ is shown below.

$$\begin{bmatrix}
A & A & A & B & -B & C & -C & -D & B & C & -D & -D \\
A & -A & B & -A & -B & -D & D & -C & -B & -D & -C & -C \\
A & -B & -A & A & -D & D & -B & B & -C & -D & C & -C \\
B & A & -A & -A & D & D & D & C & C & -B & -B & -C \\
B & -D & D & D & A & A & A & C & -C & B & -C & B \\
B & C & -D & D & A & -A & C & -A & -D & C & B & -B \\
D & -C & B & -B & A & -C & -A & A & B & C & D & -D \\
-C & -D & -C & -D & C & A & -A & -A & -D & B & -B & -B \\
D & -C & -B & -B & -B & C & C & -D & A & A & A & D \\
-D & -B & C & C & B & B & -D & B & -D & A & -A & D & -A \\
C & -B & -C & C & D & -B & -D & -B & A & -D & -A & A \\
-C & -D & -D & C & -C & -B & B & B & D & A & -A & -A
\end{bmatrix}$$

This is the Baumert-Hall array of order 12.

There are still many open questions and conjectures for orthogonal designs, such as the following.

2. CONJECTURE. *For any odd positive integer t, there exists an $\mathrm{OD}(4t; t^4)$.*

2.6.3. DEFINITION. Let $\Lambda_{\mathrm{OD}(n;a_1,\ldots,a_r)}$ be the set of $2 \times n$ $(0, \pm x_1, \ldots, \pm x_r)$-matrices X such that $XX^{\top} = (a_1 x_1^2 + \cdots + a_r x_r^2) I_2$.

2.6.4. THEOREM. *An $n \times n$ $(0, \pm x_1, \ldots, \pm x_r)$-matrix is an $\mathrm{OD}(n; a_1, \ldots, a_r)$ if and only if it is a $\mathrm{PCD}(\Lambda_{\mathrm{OD}(n;a_1,\ldots,a_r)})$.*

2.6.5. LEMMA. *Let* $\alpha = \sum_{i=1}^{r} a_i$. *Then* $\Lambda_{\mathrm{OD}(n;a_1,\ldots,a_r)}$ *is non-empty if and only if* $n \geq \alpha + \{\alpha/2\}$.

PROOF. Let $X \in \Lambda_{\mathrm{OD}(n;a_1,\ldots,a_r)}$. As the number of non-zero entries in a row of X, α is not greater than n. Suppose that $\alpha = n$ is odd. Replacing each x_i in X by 1 produces an element of $\Lambda_{\mathrm{H}(n)}$. But then n is even (see Remark 2.3.4). This contradiction proves one direction.

To prove the other direction, first note that if all the frequencies a_i are 1, then

$$\begin{bmatrix} x_1 & x_2 & x_3 & x_4 & \cdots & x_{\alpha-1} & x_\alpha & 0 & \cdots & 0 \\ x_2 & -x_1 & x_4 & -x_3 & \cdots & x_\alpha & -x_{\alpha-1} & 0 & \cdots & 0 \end{bmatrix}$$

is in $\Lambda_{\mathrm{OD}(n;a_1,\ldots,a_r)}$ if α is even, and

$$\begin{bmatrix} x_1 & x_2 & x_3 & x_4 & \cdots & x_{\alpha-2} & x_{\alpha-1} & x_\alpha & 0 & 0 & \cdots & 0 \\ x_2 & -x_1 & x_4 & -x_3 & \cdots & x_{\alpha-1} & -x_{\alpha-2} & 0 & x_\alpha & 0 & \cdots & 0 \end{bmatrix}$$

is in $\Lambda_{\mathrm{OD}(n;a_1,\ldots,a_r)}$ if α is odd. To complete the proof we do an induction on the number m of frequencies a_i which exceed 1. We have just shown sufficiency for $m = 0$; so assume that $m \geq 1$ and sufficiency for $l < m$. Let $a_k \geq 2$ and set $t = \lfloor a_k/2 \rfloor$. By induction there is a $2 \times (n-2t)$ $(0, \pm x_1, \ldots, \pm x_r)$-matrix Y such that $YY^\top = ((\sum_{i \neq k} a_i x_i^2) + (a_k - 2t)x_k^2)I_2$. Then

$$\left[\begin{array}{cccc|c} x_k & x_k & \cdots & x_k & x_k \\ x_k & -x_k & \cdots & x_k & -x_k \end{array} \; \right| \; Y \; \right] \in \Lambda_{\mathrm{OD}(n;a_1,\ldots,a_r)},$$

and we are done. $\qquad\square$

The next lemma will be used in Chapter 4 to determine groups of equivalence operations for $\Lambda_{\mathrm{OD}(n;a_1,\ldots,a_r)}$.

2.6.6. LEMMA. *If* $a_i \leq a_j$ *for some* i, j, *then* $\Lambda_{\mathrm{OD}(n;a_1,\ldots,a_r)}$ *contains an array of the form*

$$X = \left[\begin{array}{cccccc|c} x_i & \cdots & x_i & x_j & \cdots & x_j \\ x_j & \cdots & x_j & -x_i & \cdots & -x_i \end{array} \; \right| \; Y \; \right]$$

where

$$YY^\top = \left((a_j - a_i)x_j^2 + \sum_{k \neq i,j} a_k x_k^2 \right) I_2.$$

PROOF. Put $\alpha = \sum_{k=1}^{r} a_k$. By Lemma 2.6.5, $n \geq \alpha + 1$ if α is odd, whereas $n \geq \alpha$ if α is even. Since $\alpha' = a_j - a_i + \sum_{k \neq i,j} a_k = \alpha - 2a_i$ is even if and only if α is, we have that $n - 2a_i \geq \alpha' + 1$ if α' is odd, and $n - 2a_i \geq \alpha'$ if α' is even. So Y as stated exists: $Y \in \Lambda_{\mathrm{OD}(n-2a_i;a_1,\ldots,a_{i-1},a_{i+1},\ldots,a_{j-1},a_j-a_i,a_{j+1},\ldots,a_r)}$. $\qquad\square$

An orthogonal design D gives rise to weighing matrices in different ways. The simplest of these is to set all indeterminates in D equal to 1. Another way begins by writing $D = \sum_{i=1}^{r} x_i W_i$ for $(0, \pm 1)$-matrices W_i. Then

$$\sum_i W_i W_i^\top x_i^2 + \sum_{i<j} (W_i W_j^\top + W_j W_i^\top)x_i x_j = DD^\top = (a_1 x_1^2 + \cdots + a_r x_r^2)I_n.$$

So

(2.8) $$W_i W_j^\top = -W_j W_i^\top, \qquad i \neq j$$

and

(2.9) $$W_i W_i^\top = a_i I_n, \qquad 1 \leq i \leq r.$$

Hence an $OD(n; a_1, \ldots, a_r)$ corresponds to r disjoint pairwise anti-amicable weighing matrices. (Compatible matrices A, B are said to be *amicable* if $AB^\top = BA^\top$; they are *anti-amicable* if $AB^\top = -BA^\top$.)

2.6.7. THEOREM. $\Lambda_{OD(n; a_1, \ldots, a_r)}$ *is transposable.*

PROOF. By Theorem 2.4.6, $W_i^\top$ is a $W(n, a_i)$ if and only if W_i is. Moreover,

$$a_i W_j^\top W_i = W_i^\top W_i W_j^\top W_i = -W_i^\top W_j W_i^\top W_i = -a_i W_i^\top W_j$$

if $i \neq j$. So the $W_i^\top$ satisfy (2.8) and (2.9) if and only if the W_i do. Therefore $D^\top$ is an $OD(n; a_1, \ldots, a_r)$ if and only if D is.   □

The following does not seem to have been noted before.

2.6.8. THEOREM. *Let* $r \geq 2$. *There is no* $OD(n; a_1, \ldots, a_r)$ *with constant row and column sums.*

PROOF. Suppose that D is an $OD(n; a_1, \ldots, a_r)$ with constant row and column sum $\sum_i b_i x_i$. Then

$$\sum_i a_i x_i^2 \cdot J_n = J_n D D^\top = \left(\sum_i b_i x_i \right)^2 J_n,$$

so that $\sum_i a_i x_i^2 = \left(\sum_i b_i x_i \right)^2$. Putting $x_j = 0$ for all $j \neq i$ yields $a_i = b_i^2$, and then $\sum_{i \neq j} b_i b_j x_i x_j = 0$. Thus, only one of the b_i can be non-zero; but $r > 1$.   □

2.6.9. COROLLARY. *There exist no disjoint, anti-amicable weighing matrices with constant row and column sums.*

Orthogonal designs were introduced by Geramita, Geramita, and Seberry in [**70**], as an aid to constructing Hadamard and weighing matrices. These designs were vital in the proof of Seberry's theorem [**156**] on the asymptotic existence of Hadamard matrices. That result and subsequent improvements yield large classes of other designs. See the book [**71**] for an introduction to the construction methods, and [**25, 69, 84, 109**] for related ideas and later progress.

Construction of orthogonal designs is tied to the problem of finding small sets of sequences with good combined correlation properties [**18**]. Another application is in the area of space-time block codes [**150**]. The algebraic properties of orthogonal designs have exercised several authors [**63, 68, 159, 160**]; the connection between amicable orthogonal designs and quasi-Clifford algebras described by Gastineau-Hills [**68**] especially warrants further inquiry.

Any cocyclic orthogonal design corresponds to a collection of relative difference sets in the same group, having the same forbidden normal subgroup of order 2. We develop a satisfactory algebraic theory for cocyclic orthogonal designs of type $(1, \ldots, 1)$. In Chapter 23, we use a pair of these designs to prove an asymptotic existence result for cocyclic Hadamard matrices.

2.7. Complex Hadamard matrices

2.7.1. DEFINITION. An $n \times n$ matrix C of fourth roots of unity such that

$$CC^* = nI_n$$

is a *complex Hadamard matrix of order* n.

2.7.2. EXAMPLE. We display complex Hadamard matrices of orders 6 and 10 (in additive notation; $0, 1, 2, 3$ stand for $1, i, -1, -i$ respectively):

$$
\begin{bmatrix}
0 & 0 & 0 & 0 & 0 & 0 \\
0 & 0 & 1 & 2 & 2 & 3 \\
0 & 1 & 3 & 0 & 2 & 2 \\
0 & 2 & 1 & 3 & 0 & 2 \\
0 & 2 & 3 & 2 & 1 & 0 \\
0 & 3 & 2 & 1 & 3 & 1
\end{bmatrix},
\quad
\begin{bmatrix}
0 & 0 & 0 & 0 & 0 & 0 & 0 & 0 & 0 & 0 \\
0 & 0 & 0 & 0 & 1 & 2 & 2 & 2 & 2 & 3 \\
0 & 0 & 1 & 2 & 3 & 0 & 0 & 2 & 2 & 2 \\
0 & 0 & 2 & 1 & 3 & 2 & 2 & 0 & 0 & 2 \\
0 & 1 & 3 & 3 & 2 & 1 & 3 & 1 & 3 & 1 \\
0 & 2 & 0 & 2 & 1 & 0 & 2 & 3 & 0 & 2 \\
0 & 2 & 0 & 2 & 3 & 2 & 1 & 0 & 2 & 0 \\
0 & 2 & 2 & 0 & 1 & 3 & 0 & 0 & 2 & 2 \\
0 & 2 & 2 & 0 & 3 & 0 & 2 & 2 & 1 & 0 \\
0 & 3 & 2 & 2 & 1 & 2 & 0 & 2 & 0 & 0
\end{bmatrix}.
$$

2.7.3. DEFINITION. Denote the set of all $2 \times n$ $(\pm 1, \pm i)$-matrices C such that $CC^* = nI_2$ by $\Lambda_{\mathrm{CH}(n)}$.

2.7.4. REMARK. $\Lambda_{\mathrm{CH}(n)}$ is non-empty if and only if n is even.

2.7.5. THEOREM. *A complex Hadamard matrix of order n is a* $\mathrm{PCD}(\Lambda_{\mathrm{CH}(n)})$.

2.7.6. THEOREM. $\Lambda_{\mathrm{CH}(n)}$ *is transposable.*

PROOF. Let C be a complex Hadamard matrix. Then $C^{-1} = \frac{1}{n}C^*$, so $C^*C = nI_n$. Now $C^\top = \overline{C^*}$, and $(C^\top)^* = \overline{C}$. So $C^\top(C^\top)^* = \overline{C^*C} = nI_n$. $\square$

Turyn wrote the seminal paper [**152**] on complex Hadamard matrices. Several other papers on the subject are [**23, 24, 53, 69, 111, 151, 153**].

3. CONJECTURE (Turyn). *There is a complex Hadamard matrix of every order $2t \geq 2$.*

We show that Turyn's conjecture implies the Hadamard Conjecture. The proof contains the first example in this book of 'transference', whereby matrices derived from one kind of pairwise combinatorial design are used to make another kind of pairwise combinatorial design.

2.7.7. THEOREM. *If there exists a complex Hadamard matrix of order $2t$, then there exist an $\mathrm{OD}(4t; 2t, 2t)$ and a Hadamard matrix of order $4t$.*

PROOF. Write the complex Hadamard matrix C as $A + iB$, where A, B are $2t \times 2t$ $(0, \pm 1)$-matrices. Then

$$2tI_{2t} = CC^* = (A + iB)(A^\top - iB^\top) = AA^\top + BB^\top + i(BA^\top - AB^\top).$$

Equating real and imaginary parts gives $AA^\top + BB^\top = 2tI_{2t}$ and $AB^\top = BA^\top$. Therefore

$$x_1 \begin{bmatrix} A & B \\ -B & A \end{bmatrix} + x_2 \begin{bmatrix} B & -A \\ A & B \end{bmatrix}$$

is an $\mathrm{OD}(4t; 2t, 2t)$. Putting $x_1 = x_2 = 1$, we get that

$$\begin{bmatrix} A + B & -A + B \\ A - B & A + B \end{bmatrix}$$

is a Hadamard matrix of order $4t$. $\square$

A *regular complex Hadamard matrix* is a complex Hadamard matrix which has all row and column sums equal. As with Hadamard matrices, there is a number-theoretic constraint on the existence of regular complex Hadamard matrices.

2.7.8. THEOREM. *If there is a regular complex Hadamard matrix of order n, then n is a sum of two squares.*

PROOF. Let C be a regular complex Hadamard matrix of order n, say $J_n C = C J_n = (a+ib)J_n$. Then $n J_n = J_n C C^* = (a^2+b^2)J_n$, so $n = a^2+b^2$.   □

Certain complex Hadamard matrices are group-developed modulo $C_s \times C_t$ for various s and t. They are equivalent to $s \times t$ arrays of fourth roots of unity with perfect periodic autocorrelation properties—see [2]. In this book, we will show how character theory implies non-existence results for circulant complex Hadamard matrices; and conjecture that $1, 2, 4, 8, 16$ are the only admissible orders for these cocyclic designs.

2.8. Complex generalized Hadamard matrices

2.8.1. DEFINITION. An $n \times n$ matrix B of complex kth roots of unity such that

$$BB^* = nI_n$$

is called a *complex generalized Hadamard matrix* $\mathrm{CGH}(n;k)$.

2.8.2. EXAMPLE. The two inequivalent $\mathrm{CGH}(7;6)$s are given below (an entry a stands for ω^a, where ω is a primitive sixth root of unity).

$$
\begin{bmatrix}
0 & 0 & 0 & 0 & 0 & 0 & 0 \\
0 & 0 & 0 & 2 & 3 & 3 & 4 \\
0 & 1 & 3 & 5 & 0 & 3 & 3 \\
0 & 2 & 5 & 4 & 3 & 0 & 2 \\
0 & 3 & 4 & 2 & 5 & 2 & 0 \\
0 & 4 & 2 & 0 & 3 & 4 & 1 \\
0 & 4 & 2 & 3 & 1 & 0 & 4
\end{bmatrix}
\qquad
\begin{bmatrix}
0 & 0 & 0 & 0 & 0 & 0 & 0 \\
0 & 0 & 0 & 2 & 3 & 3 & 4 \\
0 & 1 & 4 & 4 & 0 & 3 & 2 \\
0 & 2 & 3 & 0 & 4 & 1 & 4 \\
0 & 3 & 5 & 3 & 3 & 0 & 1 \\
0 & 4 & 2 & 2 & 0 & 5 & 3 \\
0 & 4 & 2 & 5 & 2 & 3 & 0
\end{bmatrix}
$$

2.8.3. DEFINITION. Let $\Lambda_{\mathrm{CGH}(n;k)}$ be the set of $2 \times n$ matrices X of kth roots of unity such that $XX^* = nI_2$.

We denote the set of natural numbers (i.e., non-negative integers) by $\mathbb{N}$.

2.8.4. THEOREM. *Let $p_1, \ldots, p_r$ be the primes dividing $k > 1$. Then $\Lambda_{\mathrm{CGH}(n;k)}$ is non-empty if and only if $n = a_1 p_1 + \cdots + a_r p_r$ for some $a_1, \ldots, a_r \in \mathbb{N}$.*

PROOF. The orthogonality set $\Lambda_{\mathrm{CGH}(n;k)}$ is non-empty precisely when there are kth roots of unity $\omega_1, \ldots, \omega_n$ such that $\sum_{i=1}^{n} \omega_i = 0$. By the main theorem of [113], this is equivalent to n being an $\mathbb{N}$-linear combination of the prime divisors of k.   □

The following lemma is needed later in this chapter.

2.8.5. LEMMA. *Let p be a prime, and let ω be a primitive pth root of unity. Then $\sum_{i=0}^{n} a_i \omega^i = 0$ for some $n < p$ and $a_0, \ldots, a_n \in \mathbb{N}$ not all zero if and only if $n = p - 1$ and $a_0 = \cdots = a_n$.*

PROOF. Let $f(x) = \sum_{i=0}^{n} a_i x^i \in \mathbb{Z}[x]$. The minimal polynomial of ω over $\mathbb{Q}$ is the pth cyclotomic polynomial $h(x) = 1 + x + \cdots + x^{p-1}$. Since $h(x)$ divides every polynomial in $\mathbb{Z}[x]$ with ω as a root, and $f(x)$ has degree no greater than $p - 1$, we see that if $f(\omega) = 0$ then $f(x)$ must be an integer multiple of $h(x)$.   □

2.8.6. THEOREM. *An $n \times n$ array of kth roots of unity is a $\mathrm{CGH}(n;k)$ if and only if it is a $\mathrm{PCD}(\Lambda_{\mathrm{CGH}(n;k)})$.*

2.8.7. THEOREM. $\Lambda_{\mathrm{CGH}(n;k)}$ *is transposable.*

PROOF. Cf. the proof of Theorem 2.7.6.   □

Complex generalized Hadamard matrices were introduced by Butson [**16**]. The papers [**17, 146**] relate some of these matrices to relative difference sets, maximum-length linear-recurring sequences, error-correcting codes, and orthogonal arrays of strength 2. A great deal has been written about sequences of roots of unity with perfect periodic autocorrelation properties; they correspond to circulant complex generalized Hadamard matrices. However, it seems that a systematic investigation of complex generalized Hadamard matrices has not yet been undertaken.

2.9. Complex generalized weighing matrices

2.9.1. DEFINITION. A *complex generalized weighing matrix* $\mathrm{CGW}(v,k;m)$ is a $v \times v$ array W over $\mathbb{C}$ whose non-zero entries are all mth roots of unity, such that

$$WW^* = kI_v.$$

2.9.2. EXAMPLE. We display a $\mathrm{CGW}(10,8;12)$ and a $\mathrm{CGW}(15,12;6)$, where ζ and γ respectively denote a primitive complex twelfth root of unity and a primitive complex cube root of unity.

$$
\begin{bmatrix}
0 & \zeta^3 & \zeta^3 & \zeta^4 & \zeta^4 & 0 & 1 & \zeta^7 & \zeta^7 & 1 \\
\zeta^4 & 1 & \zeta^4 & 0 & \zeta^3 & \zeta^7 & \zeta^3 & 1 & 0 & \zeta^7 \\
1 & \zeta^4 & 0 & \zeta^7 & \zeta^3 & \zeta^3 & \zeta^7 & 1 & \zeta^4 & 0 \\
\zeta^4 & 0 & \zeta^7 & 1 & \zeta^4 & 1 & 0 & \zeta^7 & \zeta^3 & \zeta^3 \\
\zeta^7 & \zeta^7 & \zeta^3 & 1 & 0 & \zeta^4 & \zeta^4 & 0 & \zeta^3 & 1 \\
1 & 0 & \zeta^4 & \zeta^3 & \zeta^7 & \zeta^3 & 0 & \zeta^4 & 1 & \zeta^7 \\
\zeta^7 & \zeta^3 & \zeta^7 & 0 & 1 & \zeta^4 & 1 & \zeta^3 & 0 & \zeta^4 \\
\zeta^4 & \zeta^4 & 1 & \zeta^3 & 0 & \zeta^7 & \zeta^7 & 0 & 1 & \zeta^3 \\
\zeta^3 & \zeta^7 & 0 & \zeta^4 & 1 & 1 & \zeta^4 & \zeta^4 & \zeta^7 & 0 \\
0 & 1 & 1 & \zeta^7 & \zeta^7 & 0 & \zeta^3 & \zeta^3 & \zeta^4 & \zeta^4
\end{bmatrix}
\quad
\begin{bmatrix}
\gamma^2 & \gamma^2 & \gamma & -\gamma & 0 & -\gamma^2 & -\gamma & 0 & 0 & \gamma^2 & \gamma^2 & -\gamma^2 \\
0 & \gamma^2 & 0 & \gamma^2 & \gamma^2 & 0 & -\gamma^2 & -\gamma & \gamma^2 & -\gamma & \gamma & \gamma^2 \\
-\gamma^2 & \gamma^2 & -\gamma & 0 & -\gamma & \gamma^2 & -\gamma & \gamma^2 & -\gamma^2 & 0 & \gamma^2 & 0 \\
-\gamma & -\gamma^2 & -\gamma^2 & \gamma^2 & 0 & \gamma^2 & \gamma^2 & 0 & 0 & -\gamma & -\gamma & \gamma^2 \\
\gamma^2 & 0 & \gamma^2 & -\gamma^2 & -\gamma & 0 & \gamma^2 & -\gamma^2 & -\gamma^2 & 0 & \gamma & 0 \\
0 & -\gamma & 0 & -\gamma^2 & -\gamma^2 & 0 & \gamma^2 & \gamma^2 & -\gamma & \gamma^2 & -\gamma^2 & -\gamma \\
-\gamma & -\gamma^2 & \gamma^2 & 0 & \gamma^2 & -\gamma^2 & \gamma^2 & -\gamma & \gamma^2 & 0 & -\gamma & 0 \\
-\gamma^2 & 0 & -\gamma & -\gamma & -\gamma & \gamma^2 & 0 & -\gamma^2 & \gamma^2 & \gamma^2 & 0 & -\gamma^2 \\
\gamma & 0 & \gamma^2 & \gamma^2 & -\gamma^2 & 0 & -\gamma & -\gamma^2 & \gamma^2 & -\gamma & 0 & \gamma^2 \\
0 & \gamma^2 & 0 & -\gamma & \gamma & 0 & -\gamma & -\gamma^2 & \gamma^2 & -\gamma^2 & \gamma^2 & \gamma^2 \\
\gamma^2 & -\gamma & \gamma^2 & -\gamma^2 & 0 & -\gamma & -\gamma^2 & 0 & 0 & \gamma^2 & \gamma^2 & -\gamma \\
\gamma^2 & \gamma & -\gamma^2 & 0 & -\gamma^2 & -\gamma & -\gamma^2 & \gamma^2 & -\gamma & 0 & \gamma^2 & 0
\end{bmatrix}
$$

2.9.3. REMARK. A $\mathrm{CGW}(v,1;m)$ is a $v \times v$ monomial matrix whose non-zero entries are mth roots of unity.

2.9.4. DEFINITION. Let $v \geq k > 0$ and $m > 0$. Let $\mathcal{A}$ be the set of all complex mth roots of unity, and let $\Lambda_{\mathrm{CGW}(v,k;m)}$ be the set of $2 \times v$ $(0,\mathcal{A})$-matrices X such that $XX^* = kI_2$.

2.9.5. REMARK. $\Lambda_{\mathrm{CGW}(v,k;m)}$ is non-empty if $v \geq 2k - \alpha$ and $k \geq \alpha$ where $\alpha = \sum_{p|m \text{ prime}} a_p p$ for some $a_p \in \mathbb{N}$.

2.9.6. THEOREM. *A $v \times v$ $(0,\mathcal{A})$-matrix is a $\mathrm{CGW}(v,k;m)$ if and only if it is a* $\mathrm{PCD}(\Lambda_{\mathrm{CGW}(v,k;m)})$.

2.9.7. THEOREM. $\Lambda_{\mathrm{CGW}(v,k;m)}$ *is transposable.*

Complex generalized weighing matrices were considered by Berman [**9, 10**]. Apart from special cases such as complex Hadamard matrices, and prime m, these designs have received little scrutiny.

2.10. Generalized Hadamard matrices over groups

Let G be a finite non-trivial group, and denote by $\mathbb{Z}[G]$ the group ring of G over the integers $\mathbb{Z}$. If $S \subseteq G$ then we also write S for the element $\sum_{x \in S} x$ of $\mathbb{Z}[G]$.

2.10.1. DEFINITION. Let n be a positive integer divisible by $|G|$. A *generalized Hadamard matrix* $\mathrm{GH}(n; G)$ of order n over G is an $n \times n$ matrix H with entries from G, such that, over $\mathbb{Z}[G]$,

$$(2.10) \qquad HH^* = nI_n + \frac{n}{|G|}G(J_n - I_n).$$

2.10.2. EXAMPLE. The following are generalized Hadamard matrices over the additive group $\mathbb{Z}_3$ of integers modulo 3:

$$
\begin{bmatrix}
0 & 0 & 0 & 0 & 0 & 0 \\
0 & 0 & 1 & 1 & 2 & 2 \\
0 & 1 & 0 & 2 & 1 & 2 \\
0 & 1 & 2 & 0 & 2 & 1 \\
0 & 2 & 1 & 2 & 0 & 1 \\
0 & 2 & 2 & 1 & 1 & 0
\end{bmatrix},
\begin{bmatrix}
0 & 0 & 0 & 0 & 0 & 0 & 0 & 0 & 0 & 0 & 0 & 0 \\
0 & 0 & 0 & 0 & 1 & 1 & 1 & 1 & 2 & 2 & 2 & 2 \\
0 & 0 & 0 & 1 & 0 & 2 & 2 & 2 & 1 & 1 & 1 & 2 \\
0 & 0 & 1 & 2 & 2 & 0 & 1 & 2 & 0 & 1 & 2 & 1 \\
0 & 1 & 0 & 2 & 2 & 1 & 2 & 0 & 2 & 0 & 1 & 1 \\
0 & 1 & 2 & 0 & 1 & 2 & 0 & 2 & 0 & 2 & 1 & 1 \\
0 & 1 & 2 & 1 & 2 & 0 & 0 & 1 & 2 & 1 & 0 & 2 \\
0 & 1 & 2 & 2 & 0 & 2 & 1 & 1 & 1 & 0 & 2 & 0 \\
0 & 2 & 1 & 0 & 2 & 0 & 2 & 1 & 1 & 2 & 1 & 0 \\
0 & 2 & 1 & 1 & 0 & 2 & 1 & 0 & 2 & 2 & 0 & 1 \\
0 & 2 & 1 & 2 & 1 & 1 & 0 & 2 & 1 & 0 & 0 & 2 \\
0 & 2 & 2 & 1 & 1 & 1 & 2 & 0 & 0 & 1 & 2 & 0
\end{bmatrix}.
$$

2.10.3. EXAMPLE. We display a $\mathrm{GH}(4; C_2^2)$ and a $\mathrm{GH}(16; C_4)$.

$$
\begin{bmatrix}
1 & 1 & 1 & 1 \\
1 & a & b & ab \\
1 & ab & a & b \\
1 & b & ab & a
\end{bmatrix}
$$

$$
\begin{bmatrix}
1 & 1 & 1 & 1 & 1 & c & c^2 & c^3 & 1 & c^3 & c & c^2 & 1 & c^2 & c^3 & c \\
1 & 1 & 1 & 1 & c & 1 & c^3 & c^2 & c & 1 & c^3 & c & c^2 & 1 & c & c^3 \\
1 & 1 & 1 & 1 & c^2 & c^3 & 1 & c & c & c^2 & 1 & c^3 & c^3 & c & 1 & c^2 \\
1 & 1 & 1 & 1 & c^3 & c^2 & c & 1 & c^2 & c & c^3 & 1 & c & c^3 & c^2 & 1 \\
1 & c & c^2 & c^3 & 1 & 1 & 1 & 1 & c^2 & c^3 & c & 1 & c^3 & c & c^2 \\
c & 1 & c^3 & c^2 & 1 & 1 & 1 & 1 & c^2 & 1 & c & c^3 & c & 1 & c^3 & c \\
c^2 & c^3 & 1 & c & 1 & 1 & 1 & 1 & c^3 & c & 1 & c^2 & c & c^2 & 1 & c^3 \\
c^3 & c^2 & c & 1 & 1 & 1 & 1 & 1 & c & c^3 & c^2 & 1 & c^2 & c & c^3 & 1 \\
1 & c^3 & c & c^2 & 1 & c^2 & c^3 & c & 1 & 1 & 1 & 1 & c & c^2 & c^3 \\
c & 1 & c^3 & c & c^2 & 1 & c & c^3 & 1 & 1 & 1 & 1 & c & 1 & c^3 & c^2 \\
c & c^2 & 1 & c^3 & c^3 & c & 1 & c^2 & 1 & 1 & 1 & 1 & c^2 & c^3 & 1 & c \\
c^2 & c & c^3 & 1 & c & c^3 & c^2 & 1 & 1 & 1 & 1 & 1 & c^3 & c^2 & c & 1 \\
1 & c^2 & c^3 & c & 1 & c^3 & c & c^2 & 1 & c^2 & c^3 & c & 1 & 1 & 1 & 1 \\
c^2 & 1 & c & c^3 & c & 1 & c^3 & c & c & 1 & c^3 & c^2 & 1 & 1 & 1 & 1 \\
c^3 & c & 1 & c^2 & c & c^2 & 1 & c^3 & c^2 & c^3 & 1 & c & 1 & 1 & 1 & 1 \\
c & c^3 & c^2 & 1 & c^2 & c & c^3 & 1 & c^3 & c^2 & c & 1 & 1 & 1 & 1 & 1
\end{bmatrix}
$$

The second design is a special case of a general construction for $\mathrm{GH}(p^{2t}; G)$ where G is any p-group.

There are many unanswered questions about generalized Hadamard matrices over groups. For example, the following conjecture is open.

4. CONJECTURE. *A $\mathrm{GH}(n; G)$ can exist only if G is a p-group.*

If we work modulo the ideal $\mathbb{Z}G = \{zG \mid z \in \mathbb{Z}\}$ of $\mathbb{Z}[G]$, then (2.10) becomes

$$(2.11) \qquad HH^* = nI_n,$$

which is like the defining equation (2.7) for Hadamard matrices. It is easily seen that existence of a Hadamard matrix of order n is equivalent to existence of a generalized Hadamard matrix of order n over the cyclic group of order 2. Indeed, if ϕ is an injective homomorphism from $\mathbb{Z}_k$ into the multiplicative group of $\mathbb{C}$, and $[h_{ij}]$ is a $\mathrm{GH}(n; \mathbb{Z}_k)$, then $[\phi(h_{ij})]$ is a $\mathrm{CGH}(n; k)$. By Lemma 2.8.5, this process may be reversed when k is prime.

2.10.4. DEFINITION. Let G be a finite group, and let $\Lambda_{\mathrm{GH}(n; G)}$ be the set of $2 \times n$ arrays X of elements of G such that

$$XX^* = nI_2 + \frac{n}{|G|}G(J_2 - I_2).$$

2.10.5. REMARK. $\Lambda_{\mathrm{GH}(n;G)}$ is non-empty if and only if $|G|$ divides n.

2.10.6. THEOREM. *An $n \times n$ array of elements of G is a $\mathrm{GH}(n;G)$ if and only if it is a $\mathrm{PCD}(\Lambda_{\mathrm{GH}(n;G)})$.*

We now look at properties of the transpose $H^\top$ of a $\mathrm{GH}(n;G)$, H. The following operations on H preserve (2.10):

- pre/postmultiplication of every element in a row by a fixed element of G;
- postmultiplication of every element in a column by a fixed element of G.

(These are examples of *local equivalence operations*. In Chapter 4, we will show that each orthogonality set determines a group of local equivalence operations.)

When G is non-abelian, premultiplying every element in a column by a fixed element of G may not preserve (2.10). To see this, pick any two different rows a and b of H, and choose a column i which contributes a non-central element x to the inner product of row a of H with row b of H^*. Then choose $y \in G$ such that $yxy^{-1} \neq x$. Premultiply every element in column i by y. The inner product of row a of H and row b of H^* will now be $yxy^{-1} - x + (nG/|G|)$. Hence

2.10.7. THEOREM ([**22**]). *Let G be non-abelian. Suppose that H is a $\mathrm{GH}(n;G)$ whose transpose is also a $\mathrm{GH}(n;G)$. Then there exists a $\mathrm{GH}(n;G)$ whose transpose is not a $\mathrm{GH}(n;G)$.*

On the other hand, when G is abelian, the transpose of a $\mathrm{GH}(n;G)$ is also a $\mathrm{GH}(n;G)$. We give three proofs of this. In the first two proofs, we aim to show that H^* and H commute. Then $H^*H = nI_n \pmod{\mathbb{Z}G}$, and applying the inversion automorphism $x \mapsto x^{-1}$ of G entrywise we get $H^\top(H^\top)^* = nI_n$ as required. Note that if H^* commutes with H modulo $\mathbb{Z}G$, then H^* is a $\mathrm{GH}(n;G)$ regardless of whether or not G is abelian; but inversion is a homomorphism precisely because G is abelian.

We start with the most elementary proof. The quotient ring $\mathcal{R} = \mathbb{Q}[G]/\mathbb{Q}G$ is not an integral domain, so we cannot form its field of fractions. However, because $\mathcal{R}$ is commutative, the determinant of an $\mathcal{R}$-matrix is well-defined. The determinant of H is a unit (it divides the determinant n^n of $nI_n = HH^*$). Thus we may invert H in the usual way by calculating cofactors. Then $H^{-1} = \frac{1}{n}H^*$ by (2.11), and so $H^*H = HH^*$.

Next, we use the fact that two group ring elements which evaluate under each character to the same complex number are equal. If χ_0 is the principal character then $\chi_0(H)\chi_0(H^*) = nJ_n = \chi_0(H^*)\chi_0(H)$. Applying a non-principal character χ to (2.10), we have $\chi(H)\chi(H^*) = nI_n = \chi(H^*)\chi(H)$. Therefore $H^*H = HH^*$.

Finally, we mention an idea due to Jungnickel. There is a well-known result that the dual of a square resolvable transversal design is a square resolvable transversal design with the same parameters [**105**]. Moreover, it is well-known that a generalized Hadamard matrix H is equivalent to a class regular transversal design. Under this equivalence, the dual corresponds to H^*. It follows that H^* is a $\mathrm{GH}(n;G)$. Note: this proves that H^* is a $\mathrm{GH}(n;G)$ even when G is non-abelian. Since A^* is a $\mathrm{PCD}(\Lambda_{\mathrm{GH}(n;G)})$ if and only if A is too, we say that $\Lambda_{\mathrm{GH}(n;G)}$ is a *conjugate transposable* orthogonality set.

We summarize the above deliberations as a theorem.

2.10.8. THEOREM. *Let G be a group of order dividing n.*

(1) $\Lambda_{\mathrm{GH}(n;G)}$ *is conjugate transposable.*

(2) $\Lambda_{\mathrm{GH}(n;G)}$ *is transposable if and only if* G *is abelian.*

Generalized Hadamard matrices were introduced as early as 1978 [**62**, **105**, **131**, **142**]. They feature in recursive constructions for balanced incomplete block designs, affine resolvable block designs, transversal designs, and difference matrices. They are used to construct optimal error-correcting block codes [**35**, **36**, **115**], and large sets of mutually orthogonal F-squares [**37**]. They also have applications in finite geometry [**28**, **117**] and graph theory [**31**, **58**, **144**]. Constructions are given in [**16**, **17**, **29**, **36**, **38**, **39**, **45**, **46**, **62**, **73**, **105**, **143**, **148**]. Some non-existence results appear in [**15**, **33**, **34**], and classification results appear in [**28**, **32**, **117**]. Generalized Hadamard matrices are known to exist over any group of prime power order [**39**]. None have been found for other groups (cf. Conjecture 4). See [**87**], and the surveys [**19**, **35**], for more information.

2.11. Balanced generalized weighing matrices

2.11.1. DEFINITION. Let G be a finite group. A $v \times v$ matrix W with entries from $\{0\} \cup G$ such that

$$WW^* = kI_v + \frac{\lambda}{|G|}G(J_v - I_v)$$

over $\mathbb{Z}[G]$ is a *balanced generalized weighing matrix* $\mathrm{BGW}(v, k, \lambda; G)$.

2.11.2. EXAMPLE. We display a $\mathrm{BGW}(13, 9, 6; S_3)$ and a $\mathrm{BGW}(15, 7, 3; C_3)$. Here $S_3 = \langle a, b \mid a^2 = b^3 = (ab)^2 = 1 \rangle$ is the symmetric group of degree 3, and ω is a generator for the cyclic group of order 3.

$$
\begin{bmatrix}
ab^2 & a & a & ab & 1 & ab^2 & 1 & ab & 1 & 0 & 0 & 0 & 0 \\
a & 1 & a & 1 & ab^2 & ab & 0 & 0 & 0 & 1 & ab^2 & ab & 0 \\
1 & ab & ab^2 & 1 & 0 & 0 & a & ab & 0 & ab^2 & a & 0 & 1 \\
a & ab^2 & 1 & 0 & ab & 0 & ab^2 & 0 & 1 & 0 & a & 1 & ab \\
1 & 1 & 0 & ab^2 & 0 & ab^2 & 0 & a & ab & 0 & ab & 1 & a \\
1 & ab^2 & 0 & 0 & a & a & 1 & 1 & 0 & ab & 0 & ab & ab^2 \\
ab & 0 & 1 & 1 & 1 & 0 & 0 & ab^2 & ab^2 & ab & 0 & a & a \\
ab^2 & 0 & ab & 0 & 0 & ab & a & 1 & ab^2 & a & 1 & 1 & 0 \\
ab & 0 & 0 & ab^2 & ab & 1 & 1 & 0 & a & a & ab^2 & 0 & 1 \\
0 & a & ab^2 & 0 & a & 1 & 0 & ab^2 & ab & 1 & 1 & 0 & ab \\
0 & 1 & ab & a & 0 & 1 & ab & 0 & 1 & ab^2 & 0 & a & ab^2 \\
0 & ab & 0 & a & 1 & 0 & ab^2 & 1 & a & 1 & ab & ab^2 & 0 \\
0 & 0 & 1 & ab & ab^2 & a & ab & a & 0 & 0 & 1 & ab^2 & 1
\end{bmatrix}
$$

$$
\left[
\begin{array}{c|ccccccc|ccccccc}
0 & 1 & 1 & 1 & 1 & 1 & 1 & 1 & 0 & 0 & 0 & 0 & 0 & 0 & 0 \\
\hline
1 & 1 & \omega & 0 & \omega^2 & 0 & 0 & 0 & 1 & \omega^2 & 0 & \omega & 0 & 0 & 0 \\
1 & \omega & 0 & \omega^2 & 0 & 0 & 0 & 1 & \omega^2 & 0 & \omega & 0 & 0 & 0 & 1 \\
1 & 0 & \omega^2 & 0 & 0 & 0 & 1 & \omega & 0 & \omega & 0 & 0 & 0 & 1 & \omega^2 \\
1 & \omega^2 & 0 & 0 & 0 & 1 & \omega & 0 & \omega & 0 & 0 & 0 & 1 & \omega^2 & 0 \\
1 & 0 & 0 & 0 & 1 & \omega & 0 & \omega^2 & 0 & 0 & 0 & 1 & \omega^2 & 0 & \omega \\
1 & 0 & 0 & 1 & \omega & 0 & \omega^2 & 0 & 0 & 0 & 1 & \omega^2 & 0 & \omega & 0 \\
1 & 0 & 1 & \omega & 0 & \omega^2 & 0 & 0 & 0 & 1 & \omega^2 & 0 & \omega & 0 & 0 \\
\hline
0 & 1 & \omega^2 & 0 & \omega & 0 & 0 & 0 & 0 & 0 & \omega^2 & 0 & \omega^2 & \omega^2 & 1 \\
0 & \omega^2 & 0 & \omega & 0 & 0 & 0 & 1 & 0 & \omega^2 & 0 & \omega^2 & \omega^2 & 1 & 0 \\
0 & 0 & \omega & 0 & 0 & 0 & 1 & \omega^2 & \omega^2 & 0 & \omega^2 & \omega^2 & 1 & 0 & 0 \\
0 & \omega & 0 & 0 & 0 & 1 & \omega^2 & 0 & 0 & \omega^2 & \omega^2 & 1 & 0 & 0 & \omega^2 \\
0 & 0 & 0 & 0 & 1 & \omega^2 & 0 & \omega & \omega^2 & \omega^2 & 1 & 0 & 0 & \omega^2 & 0 \\
0 & 0 & 0 & 1 & \omega^2 & 0 & \omega & 0 & \omega^2 & 1 & 0 & 0 & \omega^2 & 0 & \omega^2 \\
0 & 0 & 1 & \omega^2 & 0 & \omega & 0 & 0 & 1 & 0 & 0 & \omega^2 & 0 & \omega^2 & \omega^2
\end{array}
\right]
$$

2.11.3. EXAMPLE. The BGW$(10, 9, 8; C_4)$ below derives from an example of Berman [10, p. 1025].

$$\begin{bmatrix} 0 & \alpha & \alpha & 1 & 1 & \alpha^3 & \alpha^3 & \alpha^2 & \alpha & \alpha^3 \\ 1 & \alpha^3 & 1 & 0 & 1 & \alpha^3 & 1 & 1 & \alpha^2 & \alpha^2 \\ \alpha & \alpha^3 & 0 & 1 & \alpha^2 & \alpha^2 & 1 & \alpha^3 & 1 & \alpha^3 \\ \alpha^3 & \alpha & \alpha & \alpha^3 & 0 & \alpha^2 & 1 & 1 & \alpha^3 & 1 \\ 1 & 1 & \alpha^3 & 1 & 1 & 0 & \alpha^2 & \alpha^3 & \alpha^3 & 1 \\ \alpha^2 & 0 & 1 & 1 & \alpha^3 & 1 & 1 & \alpha^2 & \alpha^3 & \alpha \\ \alpha^3 & 1 & \alpha^2 & 1 & \alpha^2 & 1 & \alpha^3 & 1 & 0 & \alpha^2 \\ 1 & \alpha^3 & \alpha & \alpha^2 & \alpha^3 & 1 & \alpha^3 & 0 & 1 & 1 \\ \alpha^2 & \alpha^2 & \alpha^3 & \alpha^3 & 1 & 1 & 0 & 1 & 1 & \alpha^3 \\ 1 & \alpha & 1 & \alpha^3 & \alpha^2 & 1 & \alpha & \alpha^3 & \alpha & 0 \end{bmatrix}$$

This is a generalized conference matrix.

2.11.4. DEFINITION. Let $\Lambda_{\mathrm{BGW}(v,k,\lambda;G)}$ denote the set of $2 \times v$ $(0, G)$-matrices X such that

$$XX^* = kI_2 + \frac{\lambda}{|G|} G(J_2 - I_2)$$

over $\mathbb{Z}[G]$.

Now let Λ be the set of $2 \times v$ $(0, G)$-matrices $X = [x_{ij}]$ such that $Y = [x_{ij}^{|G|}]$ satisfies

$$YY^\top = \begin{bmatrix} k & \lambda \\ \lambda & k \end{bmatrix}.$$

Then $\Lambda_{\mathrm{BGW}(v,k,\lambda;G)}$ is the subset of Λ whose elements X satisfy $XX^* = kI_2$, viewed as a matrix equation over the quotient ring $\mathbb{Z}[G]/\mathbb{Z}G$.

2.11.5. THEOREM. $\Lambda_{\mathrm{BGW}(v,k,\lambda;G)}$ *is non-empty if and only if* $|G|$ *divides* λ *and* v, k, λ *satisfy* (2.4).

2.11.6. THEOREM. *A* $(0, G)$*-matrix is a* BGW$(v, k, \lambda; G)$ *if and only if it is a* PCD$(\Lambda_{\mathrm{BGW}(v,k,\lambda;G)})$.

The following theorem will be proved in Chapter 7, where we give a unified approach to all the transposability results of this chapter.

2.11.7. THEOREM.
  (1) $\Lambda_{\mathrm{BGW}(v,k,\lambda;G)}$ *is conjugate transposable.*
  (2) $\Lambda_{\mathrm{BGW}(v,k,\lambda;G)}$ *is transposable if and only if* G *is abelian.*

Balanced generalized weighing matrices have been studied by de Launey [36], Rajkundlia [131, 132], and Seberry [142]. Ionin [96] used these matrices to obtain symmetric balanced incomplete block designs (see Section 2.2). For a cameo in graph theory, see [30]. Constructions are given in [6, 10, 36, 38, 39, 72, 74, 116, 131, 132].

2.12. Generalized weighing matrices

We can have generalized weighing matrices that are not balanced.

2.12.1. DEFINITION. Let G be a finite group. A $v \times v$ matrix W with entries from $\{0\} \cup G$ such that

$$WW^* = kI_v$$

over $\mathbb{Z}[G]/\mathbb{Z}G$ is a *generalized weighing matrix* GW$(v, k; G)$.

2.12.2. EXAMPLE. We display a $\mathrm{GW}(12, 9; C_3)$ that is not balanced.

$$\begin{bmatrix}
0 & \omega & 1 & \omega & 1 & 1 & 0 & 1 & 0 & 1 & \omega^2 & \omega^2 \\
1 & \omega & \omega^2 & 0 & 0 & \omega & 1 & 1 & 1 & \omega^2 & 1 & 0 \\
1 & 1 & 0 & \omega^2 & 1 & 0 & \omega^2 & 0 & \omega & 1 & 1 & \omega \\
\omega^2 & 0 & 1 & 1 & 1 & \omega & 1 & \omega^2 & \omega & 0 & 0 & 1 \\
0 & \omega^2 & 1 & 1 & \omega^2 & 1 & 0 & 1 & 0 & \omega & 1 & \omega \\
1 & \omega^2 & \omega & 0 & 0 & 1 & 1 & \omega^2 & 1 & 1 & \omega & 0 \\
1 & 0 & \omega^2 & 1 & 1 & 1 & \omega & \omega & \omega^2 & 0 & 0 & 1 \\
\omega^2 & 1 & 0 & 1 & \omega & 0 & \omega & 0 & 1 & 1 & 1 & \omega^2 \\
0 & 1 & 1 & \omega^2 & \omega & 1 & 0 & 1 & 0 & \omega^2 & \omega & 1 \\
1 & 1 & 1 & 0 & 0 & \omega^2 & 1 & \omega & 1 & \omega & \omega^2 & 0 \\
\omega & 1 & 0 & \omega & \omega^2 & 0 & 1 & 0 & \omega^2 & 1 & 1 & 1
\end{bmatrix}$$

2.12.3. REMARK. By Lemma 2.8.5, a $\mathrm{CGW}(v, k; m)$ for prime m is equivalent to a generalized weighing matrix over the cyclic group of order m.

2.12.4. DEFINITION. Let $\Lambda_{\mathrm{GW}(v,k;G)}$ denote the set of $2 \times v$ $(0, G)$-matrices X such that $XX^* = kI_2$ modulo $\mathbb{Z}G$.

2.12.5. THEOREM. *A $v \times v$ $(0, G)$-array is a $\mathrm{GW}(v, k; G)$ if and only if it is a $\mathrm{PCD}(\Lambda_{\mathrm{GW}(v,k;G)})$.*

2.12.6. THEOREM. $\Lambda_{\mathrm{GW}(v,k;G)}$ *is non-empty if and only if $v \geq k + |G|\{k/|G|\}$.*

PROOF. Suppose that $X \in \Lambda_{\mathrm{GW}(v,k;G)}$, and rearrange the columns of X until all k non-zero entries in the first row are left-most. In the first k columns of X we must then have exactly $t|G|$ non-zero entries in the second row, for some t. There are $k - t|G| \geq |G|\{k/|G|\}$ remaining non-zero entries in the second row. Hence $v \geq k + (k - t|G|) \geq k + |G|\{k/|G|\}$. Conversely, if $v \geq k + |G|\{k/|G|\}$, then we form the matrix with first row having the first k entries equal to 1, and remaining entries equal to 0, and second row beginning with the elements of G listed $t = \lfloor k/|G| \rfloor$ times, then $k - t|G|$ entries equal to 0, then $\{k/|G|\}|G|$ entries equal to 1, and finally $v - k - \{k/|G|\}|G|$ zeros. $\square$

When the group is G is abelian (respectively, non-abelian), $W^\top$ (respectively, W^*) is a $\mathrm{GW}(v, k; G)$ if and only if W is. We prove all this in Chapter 7. Some papers on generalized weighing matrices are [**10, 34, 121, 135**].

2.13. Summary

Figure 2.1 depicts inclusions between the types of designs in this chapter.

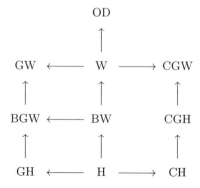

FIGURE 2.1. PCD inclusion diagram

The inclusion A → B means that the designs of type A form a subclass of the designs of type B. Thus any weighing matrix is an orthogonal design (with one indeterminate), and we deduce that a Hadamard matrix is an orthogonal design from the arrows H → BW → W → OD. This diagram may be enlarged. We could, for example, insert a vertex COD, for *complex orthogonal designs*, at the upper right, with arrows from OD and CGW.

For ease of reference later on in the book, we now recap the definitions of the various orthogonality sets.

2.13.1. DEFINITION. Let G be a finite group, and let $x_1, \ldots, x_r$ be commuting indeterminates.

(1) $\Lambda_{\mathrm{SBIBD}(v,k,\lambda)}$ is the set of $2 \times v$ $(0,1)$-matrices X such that $XX^\top = kI_2 + \lambda(J_2 - I_2)$.

(2) $\Lambda_{\mathrm{H}(n)}$ is the set of $2 \times n$ $(1,-1)$-matrices X such that $XX^\top = nI_2$.

(3) $\Lambda_{\mathrm{W}(n,k)}$ is the set of $2 \times n$ $(0,\pm 1)$-matrices X such that $XX^\top = kI_2$.

(4) $\Lambda_{\mathrm{BW}(n,k,\lambda)}$ is the set of $2 \times n$ $(0,\pm 1)$-matrices $X = [x_{ij}]$ in $\Lambda_{\mathrm{W}(n,k)}$ such that $[x_{ij}^2] \in \Lambda_{\mathrm{SBIBD}(n,k,\lambda)}$.

(5) $\Lambda_{\mathrm{OD}(n;a_1,\ldots,a_r)}$ is the set of $2 \times n$ $(0,\pm x_1, \ldots, \pm x_r)$-matrices X such that $XX^\top = (a_1 x_1^2 + \cdots + a_r x_r^2)I_2$.

(6) $\Lambda_{\mathrm{CH}(n)}$ is the set of $2 \times n$ $(\pm 1, \pm i)$-matrices X such that $XX^* = nI_2$.

(7) $\Lambda_{\mathrm{CGH}(n;k)}$ is the set of $2 \times n$ matrices X whose entries are complex kth roots of unity and such that $XX^* = nI_2$.

(8) $\Lambda_{\mathrm{CGW}(v,k;n)}$ is the set of $2 \times v$ arrays X with each entry either zero or a complex nth root of unity, such that $XX^* = kI_2$.

(9) $\Lambda_{\mathrm{GH}(n;G)}$ is the set of $2 \times n$ matrices X with entries from G such that $XX^* = nI_2 + \frac{n}{|G|}G(J_2 - I_2)$.

(10) $\Lambda_{\mathrm{GW}(v,k;G)}$ is the set of $2 \times v$ $(0,G)$-matrices X such that $XX^* = kI_2$ over $\mathbb{Z}[G]/\mathbb{Z}G$.

(11) $\Lambda_{\mathrm{BGW}(v,k,\lambda;G)}$ is the set of $2 \times v$ $(0,G)$-matrices X such that $XX^* = kI_2 + \frac{\lambda}{|G|}G(J_2 - I_2)$.

As we have shown, Definitions 2.1.3 and 2.13.1 cover all the designs of this chapter; that is, D is a design of type T if and only if D is a $\mathrm{PCD}(\Lambda_\mathsf{T})$.

CHAPTER 3

A Primer for Algebraic Design Theory

This chapter is a digest of some algebraic essentials. The first few sections are on groups, monoids, and actions. Section 3.4 is on ring theory. Miscellaneous topics in matrix algebra and linear groups are then collected in Sections 3.5 and 3.6. The final section, on matrix representations, is a prerequisite for later chapters, such as Chapter 10.

General references for this chapter are [**97**, **134**]. We omit proofs of standard results.

3.1. Groups

3.1.1. Definition. Let G be a set on which a binary operation is defined. Denote the image of $(a, b) \in G \times G$ under the operation by ab. Then G is a *group* with identity element $1 \in G$ if for all $a, b, c \in G$ the following hold.

- *Associativity*: $a(bc) = (ab)c$.
- *Identity*: $1a = a = a1$.
- *Invertibility*: there is an inverse $a^{-1} \in G$ such that $aa^{-1} = 1 = a^{-1}a$.

The *trivial group*, denoted 1, consists only of its identity. The group G is *abelian* if for all $a, b \in G$ we have

- *Commutativity*: $ab = ba$.

The group operation is called multiplication, or addition if G is abelian. In the latter case ab is written $a + b$, the identity is 0, and the inverse of $a \in G$ is $-a$.

3.1.2. Order. The *order* $|G|$ of a group G is the cardinality of its element set. The order $|g|$ of $g \in G$ is defined to be ∞ if there is no non-zero integer s such that $g^s = 1$. (Here g^s denotes the s-fold product of g with itself, if s is positive; $g^0 = 1$; and if s is negative then g^s denotes the $|s|$-fold product of g^{-1} with itself.) Otherwise $|g|$ is defined to be the smallest positive integer s such that $g^s = 1$. Finite order elements are also referred to as *torsion* elements. Note that $|g|$ is the order of the *cyclic group* $\langle g \rangle$ consisting of all powers g^s, $s \in \mathbb{Z}$. An element of order 2 (hence self-inverse) is an *involution*. The *exponent* of a finite group G is the least positive integer e such that $g^e = 1$ for all $g \in G$. The exponent of G divides $|G|$.

If $|G|$ is a power of a prime p then G is called a *p-group*. Groups of order p are cyclic, and those of order p^2 are abelian.

3.1.3. Subgroups. A non-empty subset A of a group G is a *subgroup* of G if $ab^{-1} \in A$ for all $a, b \in A$. That is, A is a group under the restriction to A of the operation of G. We write $A \leq G$, or $A < G$ if $A \neq G$ is a *proper* subgroup.

Given non-empty subsets A, B of G, define $AB = \{ab \mid a \in A, b \in B\}$. If A or B is a singleton $\{x\}$ then we can write xB or Ax for AB. Note that $A \leq G$ if and only if $1 \in A$ and $aA = A$ for all $a \in A$. When A is finite, it is a subgroup of G if and only if $AA = A$.

Let $A \leq G$. The subset xA of G is called a *(left) coset* of A. The cosets of A partition G into disjoint subsets, each of size $|A|$. Therefore

3.1.1. THEOREM (Lagrange). *If A is a subgroup of a finite group G then $|A|$ divides $|G|$.*

The *index* of $A \leq G$ in G, denoted $|G : A|$, is the number of cosets of A. So $|G : A| = |G|/|A|$ if G is finite. A *transversal of A in G* is an irredundant and complete set T of representatives for the cosets. That is, if $g \in G$ then $gA = tA$ for a unique element t of T.

The following theorem is the most important result on existence of subgroups.

3.1.2. THEOREM (Sylow). *Let G be a finite group, of order $p^a q$, where p is prime and $\gcd(p, q) = 1$. Then G contains a subgroup of order p^a.*

Such a subgroup of largest possible p-power order in G is a *Sylow p-subgroup*. In fact G contains subgroups of every order p^b, $1 \leq b \leq a$.

3.1.4. Normal subgroups. For elements a, x of a group G, $a^x = x^{-1}ax$ is the *conjugate of a by x*. The conjugate $A^x = x^{-1}Ax$ of $A \leq G$ is also a subgroup of G.

3.1.3. THEOREM. *The Sylow p-subgroups of a finite group are all conjugate to each other.*

If $A \leq G$ and $A^x = A$ for all $x \in G$ then A is a *normal* subgroup; the usual notation is $A \trianglelefteq G$. The subset AB of G is a subgroup if either subgroup A or B is normal. If $A \trianglelefteq G$ and $B \trianglelefteq G$ then AB and $A \cap B$ are normal in G too. Any subgroup of index 2 has to be normal.

The trivial subgroup of a group G and G itself are normal. A non-trivial group with no proper non-trivial normal subgroups is *simple*.

3.1.5. Normalizers. Let $A, B \leq G$. If $A^x = A$ then we say that x *normalizes* A. The *normalizer of A in B* is the set of elements of B that normalize A; this is a subgroup of G, denoted $N_B(A)$.

3.1.4. LEMMA. *A subgroup A of G has precisely $|G : N_G(A)|$ conjugates.*

3.1.5. LEMMA. *Let $|G| = p^a q$, where p is prime and $\gcd(p, q) = 1$. Denote the number of Sylow p-subgroups of G by n_p. Then $n_p \equiv 1 \pmod{p}$ and $n_p | q$.*

If a group has a unique Sylow p-subgroup then that subgroup is normal.

3.1.6. COROLLARY. *Let p be a prime, and let n be a positive integer less than p. Then every group of order np contains a normal subgroup of order p.*

3.1.6. Center and centralizers. Elements a, b of a group G *commute* if $ab = ba$. The *center* $Z(G)$ of G is the set of elements that commute with every element. This is a normal subgroup of G. An element or subgroup of G is said to be *central* if it is contained in $Z(G)$. A normal subgroup of order 2 is necessarily central in G. The *centralizer* $C_G(H)$ of $H \leq G$ is the set of elements of G that commute with every element of H. This set forms a subgroup of G. Note that $C_G(H) \trianglelefteq N_G(H)$, and $C_G(G) = Z(G)$.

3.1.7. Quotient groups. If $N \trianglelefteq G$ then $xNyN = xyN^yN = xyN$ for all $x, y \in G$. This defines a multiplication on the set of cosets of N in G. Indeed, that set is a group, the *quotient* (or *factor*) *group* G/N, of order $|G : N|$. If A, B are normal in C, and $A \leq B$, then B/A is normal in C/A.

3.1.8. Homomorphisms. Let K and L be groups. A *homomorphism* from K to L is a map $\phi : K \to L$ such that $\phi(k_1 k_2) = \phi(k_1)\phi(k_2)$ for all $k_1, k_2 \in K$. The normal subgroup $\ker \phi = \{k \in K \mid \phi(k) = 1\}$ of K is the *kernel* of ϕ. The homomorphism ϕ is an *embedding*, i.e., is injective, if $\ker \phi$ is trivial. If ϕ is a bijection then K and L are said to be *isomorphic* (by the isomorphism ϕ), and we write $K \cong L$. More generally, the image $\text{im}(\phi) = \phi(K) = \{\phi(k) \mid k \in K\}$ of ϕ is a subgroup of L isomorphic to $K/\ker \phi$. If ϕ is surjective (onto), then L is called a *homomorphic image* of K. Any quotient group K/H of K is a homomorphic image of K, under the natural projection homomorphism that maps $k \in K$ to the coset $kH \in K/H$.

There are several standard isomorphism theorems for groups (one has been mentioned already, i.e., $K/\ker \phi \cong \text{im}(\phi)$). Let C be a group with subgroup A and normal subgroup B. Then $AB/B \cong A/(A \cap B)$. If also $A \leq B$ and $A \trianglelefteq C$ then $(C/A)/(B/A) \cong C/B$.

An isomorphism $K \to K$ is an *automorphism*. The set of automorphisms of K is a group $\text{Aut}(K)$ under composition of functions. The *inner* automorphisms $y \mapsto y^x$ are defined for each $x \in K$; these form a normal subgroup of $\text{Aut}(K)$ that is isomorphic to $K/Z(K)$.

A subgroup $H \leq K$ such that $\phi(H) = H$ for all $\phi \in \text{Aut}(K)$ is *characteristic*. Characteristic subgroups are normal. Note that if H is characteristic in L, and L is normal in K, then H is normal in K.

3.1.9. Finitely generated free groups. Given a finite set $X = \{x_1, \ldots, x_r\}$, we may form the *free group on* X. The non-trivial elements of this group are the finite length words $x_{m_1}^{e_1} x_{m_2}^{e_2} \cdots x_{m_k}^{e_k}$ where each e_i is ± 1. We assume that words are *reduced*, i.e., they do not contain a subword of the form xx^{-1} or $x^{-1}x$. Taking the identity to be the empty word, the group operation is concatenation of words, together with any necessary cancellations of an element and its inverse to obtain a reduced word.

The *free abelian group* on X is the abelian group whose elements are the words $x_1^{e_1} x_2^{e_2} \cdots x_r^{e_r}$, $e_i \in \mathbb{Z}$, governed by the multiplication

$$(x_1^{e_1} x_2^{e_2} \cdots x_r^{e_r})(x_1^{f_1} x_2^{f_2} \cdots x_r^{f_r}) = x_1^{e_1+f_1} x_2^{e_2+f_2} \cdots x_r^{e_r+f_r}.$$

The *rank* of a free (or free abelian) group is the size $|X|$ of its generating set.

If F is the free group on X and G is a group then we define a homomorphism $\phi : F \to G$ merely by specifying a set map $\sigma : X \to G$. Then the homomorphism is defined on all of F recursively: $\phi(xy) = \phi(x)\phi(y)$ for all $x, y \in F$, and $\phi(x) = \sigma(x)$ if $x \in X$.

3.1.10. Finitely generated groups. For any set X of elements of a group G, define $\langle X \rangle$ to be the set of finite length products of elements from $\{x, x^{-1} \mid x \in X\}$; this is a subgroup of G. We call X a *generating set* of $\langle X \rangle$, the elements of X are *generators* of the subgroup, and the subgroup is *generated in G by (the elements of)* X. A group G is *finitely generated* if it has a finite generating set, i.e., $G = \langle X \rangle$ for some finite subset X.

3.1.11. Normal closure. Let Y be a set of elements of a group G. The *normal closure* of Y in G is the subgroup generated by $\{x^{-1}yx \mid x \in G, y \in Y\}$. A normal closure is a normal subgroup. If $Y \subseteq N \trianglelefteq G$, then N will contain the normal closure of Y in G; thus, the normal closure is the smallest normal subgroup of G containing Y.

3.1.12. Presentations. Suppose that X is a finite generating set of a group G. Let F be the free group on X. The map that sends $x \in F$ to $x \in G$ defines a homomorphism $\phi : F \to G$; hence G is a homomorphic image of F. If $\ker \phi$ has a finite generating set Y in F (which will be the case, for example, if G is finite) then we write

$$(3.1) \qquad G = \langle x : x \in X \mid y = 1 : y \in Y \rangle,$$

or just $G = \langle X \mid Y \rangle$. The right-hand side of (3.1) is a *finite presentation* for G. Elements y of Y are called *relators*; the equality $y = 1$ is a *relation*. Notice that $\ker \phi$ is the normal closure of Y in F.

Conversely, if Y is any finite non-empty subset of the free group F on X, then we may consider the group G with presentation $\langle X \mid Y \rangle$. Here G is isomorphic to F/R where R is the normal closure of Y in F.

3.1.13. Finitely presented abelian groups. We write $G = \mathrm{Ab}\langle X \mid Y \rangle$ to mean that all generators of G (elements of the finite generating set X) commute, subject to relations specified by Y.

Finitely generated abelian groups are well-understood as direct products. We discuss group products in more detail later. For the moment, note that if A and B are groups then the set $\{(a, b) \mid a \in A, b \in B\}$ forms a group $A \times B$, the direct product of A and B, under 'componentwise' operations, e.g., $(a, b)(c, d) = (ac, bd)$. We also write $C = A \times B$ if A, B are normal subgroups of C such that $C = AB$ and $A \cap B = 1$. The direct product notation $A \times B$ is replaced by the direct sum notation $A \oplus B$ if we think of A and B as additive abelian groups.

Let A be a finitely generated abelian group. The set H of finite order elements of A is a finite subgroup of A, the *torsion subgroup* or *torsion part* of A. Now a finitely generated free abelian group of rank n is isomorphic to the direct sum $\mathbb{Z}^n$ of n copies of $\mathbb{Z}$ (the infinite cyclic group, i.e., free abelian group of rank 1). It may be shown that $A = N \oplus H$ where N is free abelian. Whereas H is unique, N is not: we can have $A = N_1 \oplus H = N_2 \oplus H$ but $N_1 \neq N_2$. However, $N_1 \cong N_2$, so that the rank of N is an invariant of A. The torsion subgroup H may be written uniquely in *torsion-invariant form*

$$H \cong \mathbb{Z}_{d_1} \oplus \cdots \oplus \mathbb{Z}_{d_n}$$

where $d_1 \mid \cdots \mid d_n$ (here, $\mathbb{Z}_d$ denotes the integers modulo d, thought of as an additive cyclic group of order d). The finite abelian group H may also be written in *primary-invariant form*

$$H \cong \mathbb{Z}_{e_1} \oplus \cdots \oplus \mathbb{Z}_{e_n},$$

where the e_i are prime powers and $e_1 \leq \cdots \leq e_n$.

3.1.14. Homsets. Let G and C be groups, where C is abelian. The set of all homomorphisms from G to C forms an abelian group $\mathrm{Hom}(G, C)$ under 'pointwise' addition: if $\phi_1, \phi_2 \in \mathrm{Hom}(G, C)$ then define $(\phi_1 + \phi_2)(g) = \phi_1(g) + \phi_2(g)$ for all $g \in G$.

We note some properties of the Hom operator. Firstly, Hom is bi-additive:

$$\operatorname{Hom}(G_1 \times G_2, C) \cong \operatorname{Hom}(G_1, C) \oplus \operatorname{Hom}(G_2, C)$$

and

$$\operatorname{Hom}(G, C_1 \oplus C_2) \cong \operatorname{Hom}(G, C_1) \oplus \operatorname{Hom}(G, C_2).$$

For example, restricting each homomorphism $G_1 \times G_2 \to C$ to G_1 and to G_2 gives homomorphisms $G_1 \to C$ and $G_2 \to C$. Conversely, a homomorphism $\psi_i : G_i \to C$ extends to $\psi \in \operatorname{Hom}(G_1 \times G_2, C)$ by setting ψ_i on G_j ($j \neq i$) to be trivial. Similar remarks apply to restriction and extension in the other variable of Hom.

Secondly, let $e_1, \dots, e_n \in \mathbb{N}$, and suppose that A is a finitely generated abelian group $\operatorname{Ab}\langle a_1, \dots, a_n \mid e_i a_i = 0 \rangle$. Each element of $\operatorname{Hom}(A, C)$ corresponds to an assignment $a_i \mapsto c_i \in C$ such that $e_i c_i = 0$, $1 \leq i \leq n$. If A is finite of exponent e, and e divides d, then $\operatorname{Hom}(A, \mathbb{Z}_d) \cong A$.

3.1.15. Linear algebra over abelian groups. Let C be a finitely generated abelian group, and let $M = [m_{ij}]$ be an $r \times n$ matrix over $\mathbb{Z}$. To solve the system

$$(3.2) \qquad\qquad \sum_{j=1}^{n} m_{ij} x_j = 0, \qquad 1 \leq i \leq r$$

of equations over C, we consider the group

$$G = \langle a_1, \dots, a_n \mid \sum_{j=1}^{n} m_{ij} a_j = 0, 1 \leq i \leq r \rangle.$$

If $(x_1, \dots, x_n) = (c_1, \dots, c_n)$ is a solution of (3.2), then the map $\phi : \{a_1, \dots, a_n\} \to C$ such that $\phi(a_i) = c_i$ extends to a homomorphism from G to C. Conversely, if $\phi : G \to C$ is a homomorphism, then setting $x_j = \phi(a_j)$ for $1 \leq j \leq n$ produces a solution of (3.2). So there is a bijection between the solution set of (3.2) and $\operatorname{Hom}(G, C)$.

3.1.16. Solvable groups. The *commutator* of elements a, b of a group G is $[a, b] = a^{-1} b^{-1} a b$. Note that a and b commute if and only if $[a, b] = 1$. If $A, B \subseteq G$ then $[A, B]$ denotes the subgroup of G generated by all commutators $[a, b]$, $a \in A$, $b \in B$. Since $[a, b]^{-1} = [b, a]$ we have $[B, A] = [A, B]$.

The *derived subgroup* of G is $G' = [G, G]$. This is the smallest normal subgroup of G with abelian quotient; $G' = 1$ if and only if G is abelian.

Set $G^{(2)} = [G', G']$ and $G^{(i)} = [G^{(i-1)}, G^{(i-1)}]$ for $i \geq 3$. The subgroups $G^{(i)}$ are normal in G, and the quotients $G^{(i)}/G^{(i+1)}$ are abelian. If $G^{(n)} = 1$ for some n then G is *solvable*. The class of solvable groups includes the abelian groups, and all groups of prime-power order. It also includes all odd order groups (by the 'Feit-Thompson theorem'). Subgroups and homomorphic images of solvable groups are solvable.

3.1.17. Nilpotent groups. A finite group G that is the direct product of its Sylow p-subgroups is *nilpotent*. In particular, G has a unique Sylow p-subgroup for each prime p dividing $|G|$, and finite p-groups are themselves nilpotent. A nilpotent group is solvable. Each nilpotent group $G \neq 1$ has non-trivial center $Z(G)$, and any normal subgroup $N \neq 1$ of G has non-trivial intersection with $Z(G)$.

An *elementary abelian p-group* is an abelian group of exponent dividing p; equivalently, it is a direct product of cyclic groups of order p. The number of factors in the direct product is the *rank* of the elementary abelian group. An elementary abelian p-group of rank r viewed additively is a vector space of dimension r over $\mathbb{Z}_p$, the integers modulo p.

3.1.18. The Frattini subgroup. The *Frattini subgroup* $\mathrm{Frat}(G)$ of a finite group G is the intersection of its maximal subgroups (the proper subgroups of G not properly contained in any proper subgroup). Clearly $\mathrm{Frat}(G)$ is characteristic, so any automorphism of G induces an automorphism of $G/\mathrm{Frat}(G)$. Suppose that G is a finite p-group. Then $\mathrm{Frat}(G) = G'P$ where P is the subgroup of G generated by the pth powers of all elements of G. Consequently $G/\mathrm{Frat}(G)$ is an elementary abelian p-group. In this case, if $|G/\mathrm{Frat}(G)| = p^r$ then r is the minimum possible size of a generating set for G.

3.1.19. Group products. Design theory contains many instances of two or more designs being combined in a product operation to obtain a larger design. The larger design often inherits a group of symmetries that is a product of groups associated to the smaller designs.

We say that a group G is a *product* of its subgroups K and L if $G = KL$. Let $C = K \cap L$. Choose a transversal S for the left cosets kC of C in K, and a transversal T for the right cosets $C\ell$ of C in L. Then $\{kc\ell \mid k \in S, \ell \in T, c \in C\}$ is the set of distinct elements of G. Thus $|G| = |K : C||L : C||C| = |K||L|/|C|$ if G is finite.

When $C = 1$, we write $G = K \cdot L$ and say that G *factors*. If $C = 1$ and $L \trianglelefteq G$, then G is a *semidirect product* $K \ltimes L$ (or $L \rtimes K$). The semidirect product $G = K \ltimes L$ is also known as a *split extension* (G *splits* over L), and K is a *complement* of L in G. If $C = 1$ and $[K, L] = 1$ then G is the familiar direct product; or direct sum if both K and L are abelian. We can use the same notation $K \times L$ or $K \oplus L$ for this 'internal' direct product as we use for the 'external' direct product (i.e., the set of ordered pairs $\{(k, \ell) \mid k \in K, \ell \in L\}$ with componentwise operations), since the two groups are canonically isomorphic.

The condition $[K, L] = 1$ implies that K and L are both normal subgroups of $G = KL$, and furthermore that $C \leq Z(G)$. If $[K, L] = 1$ then we call G the *central product* $K \curlyvee_C L$ of K and L. Thus, a central product is a product of two subgroups that centralize each other. The subgroup intersection can be non-trivial (a trivial intersection gives a direct product), but it must be central.

The product defined by the single condition $C \leq Z(G)$ on C is written $G = K \diamond_C L$. When $C = \langle z \rangle$ is cyclic, we use the notation $K \diamond_z L$. The subscript spells out the intersection of the factors; we may write $K \diamond L$ for $K \diamond_C L$ and $K \curlyvee L$ for $K \curlyvee_C L$.

We can attach meaning to our product notation when the groups are considered abstractly. For example, given groups R_1, R_2, C and R_3, we write $R_3 = R_1 \diamond_C R_2$ to indicate that R_3 contains isomorphic copies K and L of R_1 and R_2, respectively, such that $R_3 = KL$ and $K \cap L$ is a central subgroup of R_3 isomorphic to C. The notation can be quite subtle. The equation $R_3 = R_1 \diamond_C R_2$ could mean that R_3 contains isomorphic copies K of R_1 and L of R_2 such that $R_3 = KL$ and $K \cap L$ is some *prescribed* subgroup C of R_3. If $R \cong S \diamond T$ and $U \cong S \diamond T$, then R and U need not be isomorphic.

3.1.7. EXAMPLE. The presentation $\langle a, b, z \mid a^2 = b^2 = z, z^2 = 1, b^a = bz \rangle$ defines a group G isomorphic to the quaternion group Q_8 of order 8. Let $K = \langle a \rangle$ and $L = \langle b \rangle$. Then $G = K \diamond_z L$; so we have $\mathrm{Q}_8 \cong \mathbb{Z}_4 \diamond \mathbb{Z}_4$. But $\mathbb{Z}_2 \times \mathbb{Z}_4 \cong \mathbb{Z}_4 \diamond \mathbb{Z}_4$ too.

3.1.8. PROPOSITION. *Let K_1, K_2, L_1, L_2, C be finite groups.*

(1) *If $K = K_1 \diamond_C K_2$ and $L = L_1 \cdot L_2$ then*

$$K \times L = (K_1 \times L_1) \diamond_C (K_2 \times L_2).$$

(2) *If $K = K_1 \diamond_C K_2$, $L = L_1 \diamond_C L_2$, and $K \cap L = C$, then*

$$K \curlyvee_C L = (K_1 \curlyvee_C L_1) \diamond_C (K_2 \curlyvee_C L_2).$$

PROOF. We prove (2) only. By hypothesis, $K = K_1 K_2$, $L = L_1 L_2$, $K_1 \cap K_2 = L_1 \cap L_2 = K \cap L = C$ is central in K and L, and $[K, L] = 1$. Then $K_1 \cap L_1 = C$ and $K_1 \curlyvee_C L_1$ is a subgroup of $K \curlyvee_C L$. Similarly $K_2 \curlyvee_C L_2 \le K \curlyvee_C L$. Now each element of $K \curlyvee L$ may be written as $k_1 k_2 \ell_1 \ell_2 = k_1 \ell_1 k_2 \ell_2$ for $k_i \in K_i$ and $\ell_i \in L_i$. Thus $K \curlyvee L = (K_1 \curlyvee L_1)(K_2 \curlyvee L_2)$. If $k_1 \ell_1 = k_2 \ell_2$ then $k_2^{-1} k_1 = \ell_2 \ell_1^{-1} \in K \cap L = C$, so $k_1 \in k_2 C \subseteq K_2$, i.e., $k_1 \in K_1 \cap K_2 = C$. Similarly, $\ell_1 \in C$. This shows that $(K_1 \curlyvee L_1) \cap (K_2 \curlyvee L_2) = C$, proving the claim in full. $\square$

We end the subsection with a definition that utilizes an 'external' semidirect product.

3.1.9. DEFINITION. Let G be a group. The *holomorph* of G, $\mathrm{Hol}(G)$, is the group with element set $\{(g, \alpha) \mid g \in G, \alpha \in \mathrm{Aut}(G)\}$ and multiplication defined by $(g, \alpha)(h, \beta) = (g\alpha(h), \alpha\beta)$. We have $\mathrm{Hol}(G) \cong G \rtimes \mathrm{Aut}(G)$.

3.1.20. Group factorization. There is a simple numerical condition for a finite group to split over one of its normal subgroups.

3.1.10. THEOREM (Schur-Zassenhaus). *Suppose that G is a finite group with a normal subgroup N of order n, such that n and $m = |G : N|$ are coprime. Then G has subgroups of order m, and $G = N \rtimes H$ for any subgroup H of order m.*

PROOF. See [**134**, 9.1.2, p. 253]. $\square$

Using the Schur-Zassenhaus theorem, we can show that if G has square-free order $p_1 \cdots p_m$, p_i prime, then G factors into a product of cyclic subgroups:

$$(3.3) \qquad G \cong \mathbb{Z}_{p_m} \ltimes (\mathbb{Z}_{p_{m-1}} \ltimes (\cdots \ltimes (\mathbb{Z}_{p_2} \ltimes \mathbb{Z}_{p_1}) \cdots)).$$

Any group G of square-free order is solvable, by a theorem of Hölder, Burnside, and Zassenhaus (see [**134**, 10.1.10, p. 290]); in fact G is *metacyclic*, i.e., it has a normal cyclic subgroup with cyclic quotient. Thus (3.3) is a special case of the following.

3.1.11. THEOREM. *If G is a finite solvable group whose exponent is square-free, then G can be factorized as a product of cyclic subgroups.*

PROOF. We induct on $|G|$. Clearly the theorem is true for cyclic G. Suppose that G is non-cyclic and the theorem is true for all groups of order less than $|G|$. Since G is solvable, it has a proper normal subgroup N such that G/N is cyclic. The exponent of G/N divides the exponent of G, so is square-free. Therefore, we can assume that $|G : N| = q$, q prime. Select an element g of $G \setminus N$, so that q is the maximum power of q dividing $|g|$. Say $|g| = qm$ where q and m are coprime. Then $g' = g^m$ has order q and is not in N: if it were, then for integers x, y such that $qx + my = 1$ we would have $g = (g^q)^x (g^m)^y \in N$. Therefore $G = \langle g' \rangle \ltimes N$. By induction, N is factorizable into a product of cyclic subgroups. Hence so too is our group G. $\square$

3.1.21. Ubiquitous families of groups. Certain groups are ubiquitous in the book. In this subsection we standardize notation for these groups.

The symbol C_n is reserved for a cyclic group of order n; as we have been doing, this will be written additively as $\mathbb{Z}_n$.

For any group A, we sometimes use the shorthand A^r to denote the direct product $A \times \cdots \times A$ with r factors; thus C_p^r denotes the elementary abelian p-group of rank r.

We use the notation E_n for $C_{p_1}^{e_1} \times \cdots \times C_{p_r}^{e_r}$, where $n = p_1^{e_1} \cdots p_r^{e_r}$ is the prime factorization of n. So if $q = p^r$ where p is prime, then E_q is the elementary abelian group of order q.

The dihedral group D_{2n} of order $2n$ is given by

$$\langle a, b \mid a^2 = b^n = 1, \, b^a = b^{-1} \rangle.$$

This group has a splitting cyclic subgroup $\langle b \rangle$ of index 2. If $n > 2$ is even then D_{2n} has center $\langle b^{n/2} \rangle$ of order 2.

The quaternion group Q_8 has elements $\{\pm 1, \pm i, \pm j, \pm k\}$ and relations $i^2 = j^2 = k^2 = -1$, $(-1)^2 = 1$, $ij = k$ (cf. Example 3.1.7). The generalized quaternion 2-group Q_{2^n} of order 2^n, $n \geq 3$, has presentation

$$\langle a, b \mid a^{2^{n-1}} = 1, \, b^2 = a^{2^{n-2}}, \, a^b = a^{-1} \rangle.$$

Like D_{2^n}, Q_{2^n} is metacyclic (it has a cyclic subgroup of index 2), and its center is of order 2; however, Q_{2^n} does not split over a cyclic subgroup of index 2—unlike D_{2^n}. We can define more generally the *dicyclic group*

$$Q_{4t} = \langle a, b \mid a^{2t} = 1, \, b^2 = a^t, \, a^b = a^{-1} \rangle$$

of order $4t$, for any t. If 2^{n-2} is the largest power of 2 dividing t then Q_{4t} contains Q_{2^n} as Sylow 2-subgroup.

The quaternion and dihedral groups of order 8 are examples of *extraspecial* groups. An extraspecial p-group, for p prime, is one whose derived subgroup and center are equal and of order p. Such a group has exponent p or p^2. Indeed, if G is an extraspecial p-group then $\mathrm{Frat}(G) = Z(G)$ and $G/Z(G)$ is an elementary abelian p-group. Each non-abelian group of order p^3 is extraspecial. In general, an extraspecial p-group is a central product of non-abelian groups of order p^3. Since $Q_8 \curlyvee Q_8 \cong D_8 \curlyvee D_8$ but $D_8 \curlyvee Q_8 \not\cong D_8 \curlyvee D_8$, an extraspecial 2-group is isomorphic either to $D_8 \curlyvee \cdots \curlyvee D_8$ or to $D_8 \curlyvee \cdots \curlyvee D_8 \curlyvee Q_8$ (with the right number of factors; the center of each factor is amalgamated). The latter two types of extraspecial group at a given order cannot be isomorphic, because they have different numbers of elements of order 4.

3.2. Monoids

A monoid is a generalization of a group.

3.2.1. DEFINITION. Let M be a set equipped with a binary operation $M \times M \to M$, where we write ab for the image of (a, b) under the operation. Then M is a *monoid* if

- $a(bc) = (ab)c$ for all $a, b, c \in M$,
- there is an element $1 \in M$ such that $1a = a = a1$ for all $a \in M$.

The monoid M is *commutative* if $ab = ba$ for all $a, b \in M$.

3.2.1. Generators and relations for monoids. Just like a group, a monoid can be defined by a presentation, i.e., by a generating set and a set of monoid relations.

3.2.2. EXAMPLE.

(1) The free monoid generated by elements a, b consists of all finite length words $a^{m_1}b^{n_1} \cdots a^{m_r}b^{n_r}$ where $m_i, n_i \in \mathbb{N}$, with concatenation of words as the operation. The empty word is the identity.
(2) The free commutative monoid generated by a, b consists of all words $a^m b^n$ $(m, n \in \mathbb{N})$.
(3) The monoid generated by a, b subject to the relation $ab = b$ has elements $\{1, a^m, b^n, b^n a^m \mid m, n > 0\}$.

3.2.2. Submonoids. A subset A of a monoid M is a *submonoid* if A contains the identity element 1 of M and $ab \in A$ for all $a, b \in A$. If M is an infinite group then it may have submonoids that are not subgroups (whereas a submonoid of a finite group is definitely a subgroup). The order of a submonoid of a finite monoid M does not have to divide the order of M.

3.2.3. EXAMPLE. The monoid generated by a, b subject to the relations $ba = ab = b$, $a^2 = 1$, $b^2 = b$ has just three elements: $1, a, b$. However $\{1, a\}$ and $\{1, b\}$ are submonoids of order 2.

3.2.3. Monoids embedded in groups. Sometimes, we can embed a monoid in a group by adding inverses.

3.2.4. EXAMPLE. The monoids $(\mathbb{Z} \setminus \{0\}, \times)$ and $(\mathbb{N}, +)$ extend to the groups $(\mathbb{Q} \setminus \{0\}, \times)$ and $(\mathbb{Z}, +)$ respectively. The monoid of Example 3.2.3 is not embeddable in any group.

A commutative monoid can be embedded in a group if and only if it possesses a cancellation law, i.e., $ab = ac \Rightarrow b = c$.

3.2.4. Homomorphisms. Let M and N be monoids. A map $\phi : M \to N$ is a *monoid homomorphism* if $\phi(1)$ is the identity in N, and $\phi(ab) = \phi(a)\phi(b)$ for all $a, b \in M$. A monoid homomorphism is an *isomorphism* if it is bijective.

3.2.5. EXAMPLE.

(1) The free monoid generated by a is isomorphic to $(\mathbb{N}, +)$.
(2) The monoid M of Example 3.2.3 is isomorphic to the submonoid $\{0, \pm 1\}$ of $(\mathbb{Z}, \times)$. The only homomorphism $\phi : M \to G$ where G is a group is the trivial one.

3.3. Group actions

A core part of this book deals with groups acting on designs. For instance, Chapter 9 defines the automorphism group of a $\mathrm{PCD}(\Lambda)$, which acts on the design in a natural way.

3.3.1. Preliminary definitions. The set of all permutations of a non-empty set Ω forms a group under function composition, the symmetric group $\mathrm{Sym}(\Omega)$. A group G *acts* on Ω if there is a homomorphism $\pi : x \mapsto \pi_x$ of G into $\mathrm{Sym}(\Omega)$. The homomorphism π (a 'permutation representation' of G) defines a left *action* of G

on Ω, which is the map $G \times \Omega \to \Omega$ given by $(x, a) \mapsto \pi_x(a)$. We write $\pi_x(a)$ as xa and say that x *moves a to xa*. This action is 'associative':

$$x(ya) = \pi_x(\pi_y(a)) = (\pi_x\pi_y)(a) = \pi_{xy}(a) = (xy)a.$$

Alternatively, we could define a right action of G on Ω by $ax = \pi_x^{-1}(a)$. In this book we mostly use left actions; for one exception, see the comments below on multiplying permutations.

An element x of G *fixes* (or *stabilizes*) $a \in \Omega$ if $xa = a$. If x fixes every element of Ω then it *acts trivially*. The set of such elements in G is called the *kernel of the action*; this normal subgroup of G is the kernel of the corresponding permutation representation. If the action has trivial kernel—so that the only element of G fixing every element of Ω is the identity—then we say that the action is *faithful*, and that G is a *permutation group*. If G is a permutation group on Ω then we sometimes do not distinguish it from its isomorphic image in $\mathrm{Sym}(\Omega)$.

When Ω is finite, $G \leq \mathrm{Sym}(\Omega)$ is finite. We assume that Ω is finite for the rest of the section, but much of what we say is valid for infinite Ω too. If Ω has size n then $\mathrm{Sym}(\Omega)$ is also denoted $\mathrm{Sym}(n)$, and G is a permutation group of *degree n*.

From the viewpoint of permutation representations, the names of the elements of Ω are irrelevant. So we may as well take $\Omega = \{1, \ldots, n\}$. A *cycle* in $\mathrm{Sym}(n)$ is a permutation of the form $(i_1, \ldots, i_k)$, in 'cycle notation', i.e., $i_1 \mapsto i_2 \mapsto \cdots \mapsto i_k \mapsto i_1$ where the i_j are distinct. Each element of $\mathrm{Sym}(\Omega)$ is a product of disjoint cycles. When multiplying permutations α, β in cycle notation, we compose from left to right; so $\alpha\beta$ means the function α applied first, then β. This presupposes a right action on the underlying set.

The elements of $\mathrm{Sym}(n)$ that can be written as a product of an even number of cycles of length 2 form a (normal) subgroup of index 2, the *alternating group* $\mathrm{Alt}(n)$.

3.3.2. Wreath product. Let G and H be groups, where $G \leq \mathrm{Sym}(n)$. Then the *wreath product* $H \wr G$ is the semidirect product $H^n \rtimes G$, where

$$g(h_1, h_2, \ldots, h_n)g^{-1} = (h_{g^{-1}1}, h_{g^{-1}2}, \ldots, h_{g^{-1}n}), \qquad g \in G, \; h_i \in H.$$

The group of signed permutation matrices is a wreath product: here $H = \langle -1 \rangle \cong C_2$, so that H^n is the group of all $(1, -1)$-diagonal matrices, and $G \cong \mathrm{Sym}(n)$ is the group of all $n \times n$ permutation matrices.

3.3.3. Permutation isomorphism. Suppose that groups G_1 and G_2 act on the sets Ω_1 and Ω_2, respectively. Suppose further that there are a bijection $f : \Omega_1 \to \Omega_2$ and an isomorphism $\phi : G_1 \to G_2$ such that $f(xa) = \phi(x)f(a)$ for all $x \in G_1$, $a \in \Omega_1$. Then the action of G_1 on Ω_1 is said to be *permutation isomorphic* to the action of G_2 on Ω_2. In the case that $\Omega_1 = \Omega_2$ and G_1, G_2 are subgroups of $\mathrm{Sym}(\Omega_1)$, the natural actions of G_1 and G_2 on Ω_1 are permutation isomorphic if and only if G_1 and G_2 are conjugate in $\mathrm{Sym}(\Omega_1)$.

3.3.4. Orbits and stabilizers. Each action of G on Ω defines an equivalence relation on Ω: a and b are equivalent if and only if $xa = b$ for some $x \in G$. The equivalence class $Ga = \{xa \mid x \in G\}$ is an *orbit* (the orbit of a under G).

Let Δ be a non-empty subset of Ω. The *pointwise stabilizer* in G of Δ, G_Δ, is the set of elements of G which fix every element of Δ. If $\Delta = \{a\}$ then we write G_a for the stabilizer of Δ. The kernel of the action of G on Ω is precisely $\cap_{a \in \Omega} G_a$.

The *(setwise) stabilizer* in G of Δ is the set of elements that fix Δ setwise, i.e., move each element of Δ to some other element of Δ. Both kinds of stabilizers are subgroups of G. Notice that the setwise stabilizer of any orbit is the whole of G.

The map $xa \mapsto xG_a$ defines a bijection between the orbit Ga and the set of cosets of the stabilizer G_a in G. Hence $|Ga| = |G : G_a|$, and so the size of an orbit divides $|G|$ if G is finite.

3.3.5. Transitive and regular actions. An action of G on Ω is *transitive* if the only orbit is Ω itself. Equivalently, G acts transitively on Ω if for each pair $a, b \in \Omega$ there is an element x of G that moves a to b.

The group G is said to act *semiregularly* on Ω if $G_a = 1$ for all $a \in \Omega$. If G is semiregular then the kernel $\cap_{a \in \Omega} G_a$ of the action is trivial: G acts faithfully on Ω. An action that is transitive and semiregular is *regular*. Equivalently, G acts regularly if there is a single orbit Ω, of size $|G|$. An abelian group acting faithfully and transitively must act regularly.

Up to permutation isomorphism, any group G has a unique regular action. To prove this, first note that G acts regularly on the set of its own elements by multiplication in G, via the representation $\sigma : G \to \operatorname{Sym}(G)$ defined by $\sigma_x(a) = xa$. Then suppose that we have another regular action of G on a set Ψ. Fix an arbitrary element a_0 of Ψ. By regularity, each element a of Ψ is xa_0 for a unique $x \in G$, so we can define a bijection f from Ψ to G by $xa_0 \mapsto x$. It is then easy to check that $f(ya) = yf(a)$ for all $y \in G$ and $a \in \Psi$.

The regular action of G on the set of its elements, as above, is by multiplication on the left. Similarly we have a right regular action, via the (faithful) representation $\tau : G \to \operatorname{Sym}(G)$ defined by $\tau_x(a) = ax^{-1}$. While they are permutation isomorphic, in general these actions give distinct injective homomorphisms of G into $\operatorname{Sym}(G)$. Note, however, that $\sigma_x \tau_y(a) = xay^{-1} = \tau_y \sigma_x(a)$ for all $a \in G$. Here is a stronger statement.

3.3.1. LEMMA. $C_{\operatorname{Sym}(G)}(\sigma(G)) = \tau(G)$ *and* $C_{\operatorname{Sym}(G)}(\tau(G)) = \sigma(G)$.

PROOF. We prove the first claim only. Suppose that $\pi \in \operatorname{Sym}(G)$ and $\pi\sigma_x(g) = \sigma_x\pi(g)$ for all $x, g \in G$. Taking $g = 1$ we get $\pi(x) = x\pi(1)$. Therefore π acts by multiplication on the right by a fixed element of G, i.e., $\pi \in \tau(G)$. $\qquad\square$

We have an accompanying result for normalizers.

3.3.2. LEMMA. $N_{\operatorname{Sym}(G)}(\sigma(G)) = \tau(G) \rtimes \operatorname{Aut}(G)$ *and* $N_{\operatorname{Sym}(G)}(\tau(G)) = \sigma(G) \rtimes$ $\operatorname{Aut}(G)$. *That is, both normalizers coincide, and are isomorphic to* $\operatorname{Hol}(G)$.

PROOF. Set $N_1 = N_{\operatorname{Sym}(G)}(\sigma(G))$ and $N_2 = N_{\operatorname{Sym}(G)}(\tau(G))$. Certainly $\tau(G) \leq N_1$ and $\sigma(G) \leq N_2$. If $x \in G$ and $\eta \in \operatorname{Aut}(G)$ then $\eta^{-1}\sigma_x\eta = \sigma_{\eta^{-1}(x)}$ and $\eta^{-1}\tau_x\eta = \tau_{\eta^{-1}(x)}$; therefore $\operatorname{Aut}(G)$ normalizes each of $\sigma(G)$ and $\tau(G)$ in $\operatorname{Sym}(G)$. It follows that $\tau(G)\operatorname{Aut}(G) \leq N_1$ and $\sigma(G)\operatorname{Aut}(G) \leq N_2$.

For $\pi \in N_1$ define $\eta : G \to G$ by $\eta(x) = \sigma^{-1}(\pi^{-1}\sigma_x\pi)$. Then $\eta \in \operatorname{Aut}(G)$, and $\pi\eta = \tau_{\pi(1)^{-1}}$. Thus $N_1 \leq \tau(G)\operatorname{Aut}(G)$. Similarly $N_2 \leq \sigma(G)\operatorname{Aut}(G)$.

If σ_x or τ_x is a homomorphism on G, then $x = 1$. We conclude that $N_1 = \tau(G) \rtimes \operatorname{Aut}(G)$ and $N_2 = \sigma(G) \rtimes \operatorname{Aut}(G)$. Also, since $\sigma(G) \leq N_1$ and $\tau(G) \leq N_2$, we have $N_1 = N_2$. $\qquad\square$

We now describe the subgroup of $\operatorname{Sym}(G)$ generated by the images of the left and right regular representations of G.

3.3.3. LEMMA. $\sigma(G) \cap \tau(G) = \sigma(Z(G)) = \tau(Z(G))$.

PROOF. Suppose that $\sigma_x = \tau_y$. Then $xa = ay^{-1}$ for all $a \in G$. Taking $a = 1$ yields $y^{-1} = x$, and thus $xa = ax$ for all $a \in G$. So $x \in Z(G)$, and the restriction of σ (or τ) to $Z(G)$ is an isomorphism onto $\sigma(G) \cap \tau(G)$. □

3.3.4. THEOREM. $\langle \sigma(G), \tau(G) \rangle \leq \mathrm{Sym}(G)$ *is the central product* $\sigma(G) \curlyvee_W \tau(G)$ *where* $W = \sigma(Z(G)) = \tau(Z(G)) \cong Z(G)$.

PROOF. Since $\sigma(G)$ and $\tau(G)$ commute by Lemma 3.3.1, they generate a central product. The amalgamated subgroup is as stated by Lemma 3.3.3. □

3.3.5. REMARK. If G is abelian then $\sigma(G)\tau(G) = \sigma(G) = \tau(G) \cong G$.

3.3.6. Essential uniqueness of regular actions. In the previous subsection we saw that there is a single regular action of a finite group G, up to permutation isomorphism. We expand on this idea in this subsection.

Permutation representations π and τ of G in $\mathrm{Sym}(\Omega)$ are *similar* if there exists $f \in \mathrm{Sym}(\Omega)$ such that $f^{-1}\pi_x f = \tau_x$ for all $x \in G$. (As we see from the definitions, similarity is a permutation isomorphism between two actions of the same group on the same set.) Similar permutation representations of G in $\mathrm{Sym}(\Omega)$ have conjugate images, but the converse need not be true.

3.3.6. LEMMA. *Let* π, τ *be faithful representations of* G *in* $\mathrm{Sym}(\Omega)$. *Then* $\pi(G)$ *and* $\tau(G)$ *are conjugate in* $\mathrm{Sym}(\Omega)$ *if and only if* π *and* $\tau \circ \varepsilon$ *are similar for some* $\varepsilon \in \mathrm{Aut}(G)$.

PROOF. Only one direction needs proof. Suppose that $\pi(G) = f^{-1}\tau(G)f$ for some $f \in \mathrm{Sym}(\Omega)$, so that given $x \in G$ there is a unique $y \in G$ such that $f^{-1}\tau_x f = \pi_y$. Then $\varepsilon : x \mapsto y$ is an automorphism of G, and $\pi, \tau \circ \varepsilon^{-1}$ are similar. □

3.3.7. LEMMA. *Any two faithful regular permutation representations of* G *in* $\mathrm{Sym}(\Omega)$ *are similar.*

PROOF. Let π, τ be isomorphisms of G onto regular subgroups of $\mathrm{Sym}(\Omega)$. By Subsection 3.3.5, we know that there are bijections f and $\bar{f}$ such that $f(\pi_x(a)) = xf(a)$ and $\bar{f}(\tau_x(a)) = x\bar{f}(a)$ for all $x \in G$, $a \in \Omega$. Now $h = \bar{f}^{-1}f \in \mathrm{Sym}(\Omega)$ and $h^{-1}\tau_x h = \pi_x$, so π and τ are similar. □

3.3.8. COROLLARY. *Regular subgroups of* $\mathrm{Sym}(\Omega)$ *are conjugate if and only if they are isomorphic.*

3.3.7. Induced actions. Suppose that G acts on Ω. This one group action gives rise to several *induced actions*. Perhaps the simplest way to obtain an induced action is by restriction to a subgroup of G. There is also an induced action of G on each of its orbits. This induced action is transitive, so that if G acts semiregularly on Ω, then it is regular on orbits.

Let $N \trianglelefteq G$. Then $g(Na) = N(ga)$ for all $g \in G$ and $a \in \Omega$, implying that G acts on the set Φ of N-orbits in Ω. This induced action has kernel containing N, and so G/N acts on Φ.

There is an induced action of G on $\{(a_1, \ldots, a_k) \mid a_i \in \Omega, 1 \leq i \leq k\}$, defined by $x(a_1, \ldots, a_k) = (xa_1, \ldots, xa_k)$. The original action of G on Ω is *k-transitive* if this action on the set of k-tuples with distinct components is transitive.

3.3.8. Primitivity. Let $\Pi = \{A_1, \ldots, A_r\}$ be a partition of Ω, $|\Omega| > 1$, and suppose that a group G acts non-trivially on Ω. Call the sets A_i *components* of the partition. The action of G on Ω *preserves* Π if G permutes the elements of Π; that is, for every i and $x \in G$, $xA_i = \{xa \mid a \in A_i\}$ is some component A_{j_i}. So we have an induced action of G on Π. Any action on Ω preserves the two *trivial partitions*: the partition with just one component Ω, and the partition with $|\Omega|$ components, each containing a single element of Ω. An action of G on Ω is *primitive* if it preserves no non-trivial partitions; otherwise, it is *imprimitive*. A non-trivial partition preserved by G is called a *system of imprimitivity* (for G).

3.4. Rings

3.4.1. Bare necessities. A ring $\mathcal{R}$ is an additive abelian group with identity 0 which is endowed with an associative multiplication that distributes over the addition. The ring is *commutative* if $rs = sr$ for all $r, s \in \mathcal{R}$. Each of our rings will have a multiplicative identity $1 \neq 0$. A *subring* of $\mathcal{R}$ is a subset $\mathcal{S}$ that is a ring in its own right under the addition and multiplication of $\mathcal{R}$ (possibly $1_{\mathcal{S}} \neq 1_{\mathcal{R}}$).

Elements r and s of $\mathcal{R}$ such that $rs = sr = 1$ are *units*. The element s is the *multiplicative inverse* of r, and we write $s = r^{-1}$. The set of units of $\mathcal{R}$ is a group (the *group of units* of $\mathcal{R}$) under the ring multiplication.

A non-zero element r of $\mathcal{R}$ such that $rs = 0$ or $sr = 0$ for some non-zero $s \in \mathcal{R}$ is a *zero divisor*. An *integral domain* is a commutative ring with no zero divisors. A *field* is a commutative ring $\mathbb{F}$ whose group of units is $\mathbb{F}^{\times} := \mathbb{F} \setminus \{0\}$. So a field is an integral domain. If an integral domain is finite, then it is a field. The *field of fractions* of an integral domain $\mathcal{R}$ is obtained by adjoining to $\mathcal{R}$ the formal inverses r^{-1} for each non-zero $r \in \mathcal{R}$ that is not a unit.

The *characteristic* of a ring $\mathcal{R}$ is the smallest positive integer n such that $n \cdot 1 = 1 + \cdots + 1$ (n times) is zero in $\mathcal{R}$, if such an integer exists; otherwise $\mathcal{R}$ has characteristic zero. If an integral domain has positive characteristic n then n is a prime.

Suppose that $\mathcal{R}$ is a finite-dimensional vector space over a field $\mathbb{F}$. If there is a multiplication defined on $\mathcal{R}$, such that $\mathcal{R}$ is a ring under this multiplication and vector addition, and if further $c(ab) = (ca)b = a(cb)$ for all $a, b \in \mathcal{R}$ and $c \in \mathbb{F}$, then $\mathcal{R}$ is an $\mathbb{F}$-*algebra*. Note that $\mathbb{F}$ is embedded in the algebra $\mathcal{R}$, via $c \mapsto c1_{\mathcal{R}}$.

3.4.2. Group rings and monoid rings. The *group ring* $\mathcal{R}[G]$ of a group G over a ring $\mathcal{R}$ consists of the finite formal sums $\sum_{k \in G} a_k k$, where the coefficients a_k are elements of $\mathcal{R}$, with addition and multiplication defined by

$$\sum_k a_k k + \sum_k b_k k = \sum_k (a_k + b_k)k$$

and

$$\sum_k a_k k \cdot \sum_k b_k k = \sum_k \left(\sum_{k_1 k_2 = k} a_{k_1} b_{k_2} \right) k.$$

The ring $\mathcal{R}[G]$ is commutative if and only if $\mathcal{R}$ and G are both commutative.

If M is a monoid then the definition of the monoid ring $\mathcal{R}[M]$ is the same as above, with G replaced everywhere by M.

3.4.3. Polynomial rings. Let $\mathcal{R}$ be a ring, and let $X = \{x_1, \ldots, x_r\}$ be a set of commuting indeterminates. For $e = (e_1, \ldots, e_r) \in \mathbb{N}^r$, denote $x_1^{e_1} \cdots x_r^{e_r}$ by x^e. The ring $\mathcal{R}[X]$ of polynomials in the indeterminates $\{x_1, \ldots, x_r\}$ over $\mathcal{R}$ is the set of finite sums $\sum_e a_e x^e$, where the coefficients a_e are elements of $\mathcal{R}$. Addition and multiplication are given by

$$\sum_e a_e x^e + \sum_e b_e x^e = \sum_e (a_e + b_e) x^e$$

and

$$\sum_e a_e x^e \cdot \sum_f b_f x^f = \sum_d \left(\sum_{e+f=d} a_e b_f \right) x^d.$$

This ring is commutative if and only if $\mathcal{R}$ is commutative.

3.4.4. Ideals, homomorphisms, and quotients. An *ideal* $\mathcal{I}$ of a ring $\mathcal{R}$ is an additive subgroup of $\mathcal{R}$ such that $ra, ar \in \mathcal{I}$ for all $a \in \mathcal{I}$ and $r \in \mathcal{R}$. Given an ideal $\mathcal{I}$ of $\mathcal{R}$, we may define the *quotient ring* $\mathcal{R}/\mathcal{I}$ whose elements are the additive cosets $x + \mathcal{I} := \{x\} + \mathcal{I}$ of $\mathcal{I}$ in $\mathcal{R}$, and which is governed by the operations

$$(x + \mathcal{I})(y + \mathcal{I}) = xy + \mathcal{I}, \qquad (x + \mathcal{I}) + (y + \mathcal{I}) = x + y + \mathcal{I}.$$

Ideals are the kernels of ring homomorphisms. A map ϕ from $\mathcal{R}$ to a ring $\mathcal{S}$ is a *ring homomorphism* if $\phi(ab) = \phi(a)\phi(b)$ and $\phi(a + b) = \phi(a) + \phi(b)$ for all $a, b \in \mathcal{R}$. The *kernel* $\ker \phi$ of ϕ is the set of elements in $\mathcal{R}$ that are mapped to 0 in $\mathcal{S}$; this is an ideal of $\mathcal{R}$. The image $\phi(\mathcal{R}) = \{\phi(r) \mid r \in \mathcal{R}\}$ is a subring of $\mathcal{S}$. The map ϕ is an isomorphism from $\mathcal{R}$ to $\mathcal{S}$ if and only if $\ker \phi = 0$ and $\phi(\mathcal{R}) = \mathcal{S}$ (in which case we say that $\mathcal{R}$ and $\mathcal{S}$ are isomorphic, and write $\mathcal{R} \cong \mathcal{S}$, just as for groups). Conversely, if $\mathcal{I}$ is an ideal of $\mathcal{R}$, then $r \mapsto r + \mathcal{I}$ defines a surjective ring homomorphism from $\mathcal{R}$ to $\mathcal{R}/\mathcal{I}$ with kernel $\mathcal{I}$. For any homomorphism $\phi : \mathcal{R} \to \mathcal{S}$ we have $\mathcal{R}/\ker \phi \cong \phi(\mathcal{R})$.

Let U be a subset of a ring $\mathcal{R}$. The *ideal generated by U* consists of all two-sided linear combinations $\sum_{u \in U} r_u u t_u$ over $\mathcal{R}$. This is the smallest ideal of $\mathcal{R}$ containing U. The set U of ring elements furnishes a set of relations $\{u = 0 \mid u \in U\}$ in $\mathcal{R}$, and carrying out operations in $\mathcal{R}$ modulo these relations is equivalent to working in the quotient ring $\mathcal{R}/\mathcal{I}$, where $\mathcal{I}$ is the ideal generated by U.

If there is a finite subset U of the ring $\mathcal{R}$ such that every element of $\mathcal{R}$ is a finite sum of elements of the form $\pm u_1 \cdots u_k$, $u_i \in U$, then $\mathcal{R}$ is *finitely generated* (by U). Any commutative finitely generated ring is a homomorphic image of a polynomial ring over $\mathbb{Z}$.

An integral domain is always a subring of a field, namely its field of fractions. On the other hand, if $\mathcal{I}$ is a *maximal ideal* of the commutative ring $\mathcal{R}$ (the only ideal that properly contains $\mathcal{I}$ is $\mathcal{R}$) then $\mathcal{R}/\mathcal{I}$ is a field. More generally, an ideal $\mathcal{I} \neq \mathcal{R}$ is *prime* if and only if $\mathcal{R}/\mathcal{I}$ is an integral domain.

3.4.5. Finite fields. The size of a finite field must be a power of a prime p. Conversely, for each prime p and all $r \geq 1$ there is a unique field of size p^r, up to isomorphism.

The field of size p^r is called the *Galois field* of that size, denoted $\mathrm{GF}(p^r)$. The *prime subfield* of $\mathrm{GF}(p^r)$ is $\mathrm{GF}(p) \cong \mathbb{Z}_p = \mathbb{Z}/p\mathbb{Z}$, the ring of integers modulo p. With the field addition as vector addition, and multiplication by elements of $\mathrm{GF}(p)$ in $\mathrm{GF}(p^r)$ as scalar multiplication, $\mathrm{GF}(p^r)$ is a vector space of dimension r over its prime subfield.

The multiplicative group $\mathrm{GF}(p^r)^\times$ is cyclic. Any generator ω of this group is called a *primitive element*. In fact there is an irreducible polynomial $f(x) \in \mathbb{Z}_p[x]$ of degree r with ω as a root. The ideal $\langle f(x) \rangle$ of $\mathbb{Z}_p[x]$ generated by $f(x)$ is maximal, and $\mathrm{GF}(p^r) \cong \mathbb{Z}_p[x]/\langle f(x) \rangle$.

For each positive integer s dividing r, $\mathrm{GF}(p^r)$ has a unique subfield isomorphic to $\mathrm{GF}(p^s)$. We say that $\mathrm{GF}(p^r)$ is an *extension* of $\mathrm{GF}(p^s)$, whose *degree* is r/s.

The only field automorphism of $\mathbb{Z}_p$ is the identity. A basic result in Galois theory is that the (Galois) group of all field automorphisms of $\mathrm{GF}(p^r)$ is cyclic of order r. A generator for this group is the 'Frobenius map' $a \mapsto a^p$. Note that (by Fermat's little theorem) the Frobenius map is indeed the identity on the subfield $\mathbb{Z}_p$ of $\mathrm{GF}(p^r)$.

3.5. Matrices

3.5.1. Mat$(n, \mathcal{R})$. The set of all $n \times n$ matrices over a ring $\mathcal{R}$ forms a ring $\mathrm{Mat}(n, \mathcal{R})$ under matrix addition and multiplication, whose 1 is the identity matrix I_n. If $\mathcal{R}$ is commutative and $M \in \mathrm{Mat}(n, \mathcal{R})$ then the *cofactor matrix N* for M satisfies $NM = MN = \det(M)I_n$. Thus M is a unit of $\mathrm{Mat}(n, \mathcal{R})$ if and only if $\det(M)$ is a unit of $\mathcal{R}$.

3.5.2. Products. Let $A, B \in \mathrm{Mat}(n, \mathcal{R})$ where $A = [a_{ij}]$ and $B = [b_{ij}]$. The *Hadamard product $A \wedge B$* is an entrywise product: the matrix with (i,j)th entry $a_{ij}b_{ij}$. If A is $m \times m$ and B is $n \times n$, then the *Kronecker product $A \otimes B$* is the $mn \times mn$ matrix which, considered as an $m \times m$ block matrix, has (i,j)th block $a_{ij}B$. Lexicographically indexing rows by the pairs (i, k), and columns by the pairs (j, l), $1 \le i, j \le m$, $1 \le k, l \le n$, the entry in row (i, k) and column (j, l) of $A \otimes B$ is $a_{ij}b_{kl}$. Some authors define Kronecker product the other way around, i.e., define $A \otimes B$ to be the $m \times m$ block matrix whose (i,j)th block is Ab_{ij}. This is $B \otimes A$ in our notation, and $A \otimes B$, $B \otimes A$ are equivalent (via row and column permutations) to each other. The proofs of the following facts are routine.

3.5.1. LEMMA. *Let $A, \bar{A} \in \mathrm{Mat}(m, \mathcal{R})$ and $B, \bar{B}, X \in \mathrm{Mat}(n, \mathcal{R})$ where $X \ne 0$.*
  (i) *If all entries of B commute with all entries of $\bar{A}$ then $(A \otimes B)(\bar{A} \otimes \bar{B}) = A\bar{A} \otimes B\bar{B}$.*
  (ii) *$A \otimes B + \bar{A} \otimes B = (A + \bar{A}) \otimes B$.*
Suppose that no entry of X is a zero divisor. Then
  (iii) *$A \otimes X = 0$ if and only if $A = 0$,*
  (iv) *$A \otimes X = \bar{A} \otimes X$ if and only if $A = \bar{A}$.*

3.5.3. Perm(n). Using the δ-notation
$$\delta_b^a = \left\{ \begin{array}{ll} 1 & \text{if } a = b \\ 0 & \text{otherwise,} \end{array} \right.$$
each $\pi \in \mathrm{Sym}(n)$ has a matching permutation matrix
$$P_\pi = [\delta_{\pi(j)}^i]_{1 \le i, j \le n}.$$
The set $\mathrm{Perm}(n) = \{P_\pi \mid \pi \in \mathrm{Sym}(n)\}$ of permutation matrices in $\mathrm{Mat}(n, \mathcal{R})$ is a group under matrix multiplication, and $\pi \mapsto P_\pi$ defines an isomorphism of $\mathrm{Sym}(n)$ onto $\mathrm{Perm}(n)$. We can regard $\mathrm{Perm}(n)$ as a permutation group, acting on the set of $(0, 1)$-vectors each of which has a single non-zero entry. Multiplication of a matrix A by a compatible permutation matrix shuffles the rows or columns of A.

3.5.2. LEMMA. *Let A be an $n \times n$ matrix.*
  (i) *Row $\pi(i)$ of $P_\pi A$ is row i of A.*
  (ii) *Column $\pi(j)$ of $AP_\pi^\top$ is column j of A.*

PROOF. Part (i) follows from $[\delta^i_{\pi(k)}]_{ik}[a_{kj}]_{kj} = [a_{\pi^{-1}(i),j}]_{ij}$, and part (ii) from $[a_{ik}]_{ik}[\delta^k_{\pi(j)}]^\top_{kj} = [a_{ik}]_{ik}[\delta^j_{\pi(k)}]_{kj} = [a_{i,\pi^{-1}(j)}]_{ij}$. $\square$

That is, premultiplication of A by P_π permutes the rows of A as π permutes $\{1,\ldots,n\}$ (moves row i to row $\pi(i)$), and postmultiplication of A by $P^\top_\pi$ permutes the columns of A as π permutes $\{1,\ldots,n\}$ (moves column j to column $\pi(j)$).

3.5.4. Mon(n,C). A square matrix is *monomial* if it has exactly one non-zero entry in each row and each column. Let C be a finite group. An $n \times n$ monomial matrix M with non-zero entries from C may be written as $[g(i)\delta^i_{\pi(j)}]_{1 \le i,j \le n}$ for some permutation π of $\{1,\ldots,n\}$ and map $g : \{1,\ldots,n\} \to C$. Thus M is the product of a diagonal matrix $\mathrm{diag}(g(1),\ldots,g(n))$ and a permutation matrix P_π. The set of $n \times n$ monomial matrices over C forms a group, $\mathrm{Mon}(n,C)$. Although we locate $\mathrm{Mon}(n,C)$ in $\mathrm{Mat}(n,\mathcal{R}[C])$, the only non-zero elements of $\mathcal{R}[C]$ involved when multiplying matrices in $\mathrm{Mon}(n,C)$ are elements of C. Note that $\mathrm{Mon}(n,C) = \mathrm{Diag}(n,C) \rtimes \mathrm{Perm}(n)$ where $\mathrm{Diag}(n,C)$ is the subgroup of all diagonal matrices. Thus $\mathrm{Mon}(n,C) \cong C \wr \mathrm{Sym}(n)$. The 'base group' of this wreath product is the n-fold direct product $\mathrm{Diag}(n,C) \cong C^n$.

3.5.3. LEMMA. *The following hold.*
  (1) *If $M \in \mathrm{Mon}(n,C)$ then M^{-1} is the matrix M^* obtained by inverting each non-zero entry of M and taking the transpose.*
  (2) *For $n > 2$, or $n = 2$ and $|C| > 1$, the center of $\mathrm{Mon}(n,C)$ is the group of scalar matrices $\{cI_n \mid c \in Z(C)\} \cong Z(C)$.*

PROOF.
1. Let $M = DP$ where $D \in \mathrm{Diag}(n,C)$ has ith main diagonal entry $g(i)$, and $P = [\delta^i_{\pi(j)}]_{ij} \in \mathrm{Perm}(n)$. Then

$$M^{-1} = P^{-1}D^{-1} = P^\top D^{-1} = [\delta^j_{\pi(i)}]_{ij}D^{-1} = [\delta^j_{\pi(i)}g(j)^{-1}]_{ij} = M^*.$$

2. Let μ be the canonical projection homomorphism $\mathrm{Mon}(n,C) \to \mathrm{Perm}(n)$ defined by $\mu : DP \mapsto P$. If DP is in the center Z of $\mathrm{Mon}(n,C)$, then $P = \mu(DP)$ is in the center of $\mathrm{Perm}(n) \cong \mathrm{Sym}(n)$, which is trivial if $n > 2$. So $Z \le \mathrm{Diag}(n,C)$. Furthermore, $\mathrm{Perm}(n)$ acts on $\mathrm{Diag}(n,C)$ by conjugation: D^P is a diagonal matrix with the same diagonal entries as D's, but permuted according to how the element of $\mathrm{Sym}(n)$ represented by P permutes diagonal indices. Since this is a transitive action, D must be a scalar matrix. Verifying the claim for $n = 2$ and $|C| > 1$ is an easy exercise. $\square$

3.5.5. The Grammian. If $M = [m_{ij}]_{ij}$ is a matrix over a ring $\mathcal{R}$ and $\alpha : \mathcal{R} \to \mathcal{R}$ is a map, denote $[\alpha(m_{ij})]_{ij}$ by $\alpha(M)$. We call the matrix product $M\alpha(M)^\top$ a *Grammian* (or *Gram matrix*). Suppose that α is an anti-automorphism of $\mathcal{R}$ such that α^2 is the identity; i.e., $\alpha(r_1 + r_2) = \alpha(r_1) + \alpha(r_2)$, $\alpha(r_1 r_2) = \alpha(r_2)\alpha(r_1)$, and $\alpha(\alpha(r_1)) = r_1$ for all r_1, r_2. In this case, $\mathcal{R}$ is an *involutory ring* (with *involution* α). Note that if α is the identity map then a ring $\mathcal{R}$ with involution α is necessarily commutative.

3.6. Linear and related groups

3.6.1. General linear groups. The *general linear group* $\mathrm{GL}(n,\mathcal{R})$ is the group of units of $\mathrm{Mat}(n,\mathcal{R})$. A subgroup of $\mathrm{GL}(n,\mathcal{R})$ for some n and $\mathcal{R}$ is called a *linear group* (of degree n, over $\mathcal{R}$). If $\mathbb{F}$ is a finite field, say of size q, then it is customary to write $\mathrm{GL}(n,q)$ for $\mathrm{GL}(n,\mathbb{F})$. We can identify $\mathrm{GL}(n,p)$ with the automorphism group of the elementary abelian p-group of rank n.

3.6.2. Subgroups, quotients, and semilinearity. Let $\mathbb{F}$ be a field. The elements of $\mathrm{GL}(n,\mathbb{F})$ with determinant 1 form the *special linear group* $\mathrm{SL}(n,\mathbb{F})$. Since $\mathrm{SL}(n,\mathbb{F})$ is the kernel of the determinant map $\mathrm{GL}(n,\mathbb{F}) \to \mathbb{F}^\times$, $|\mathrm{SL}(n,q)| = |\mathrm{GL}(n,q)|/(q-1)$. (We follow the tradition of replacing '$\mathbb{F}$' in the notation by the size of $\mathbb{F}$, if it is finite.) Indeed, $\mathrm{GL}(n,\mathbb{F}) = \mathrm{SL}(n,\mathbb{F}) \rtimes T$, where T is all diagonal matrices $(1,\ldots,1,t)$, $t \in \mathbb{F}^\times$.

Let V be an n-dimensional vector space over $\mathbb{F}$. A map $f : V \to V$ is a *semilinear transformation* if there is an automorphism λ of $\mathbb{F}$ such that $f(u+v) = f(u)+f(v)$ and $f(au) = \lambda(a)f(u)$ for all $u,v \in V$ and $a \in \mathbb{F}$. The *general semilinear group* $\Gamma\mathrm{L}(V)$ is the set of invertible semilinear transformations of V, under function composition. We also use the notation $\Gamma\mathrm{L}(n,\mathbb{F})$. Clearly $\Gamma\mathrm{L}(V)$ contains $\mathrm{GL}(V)$, the group of invertible $\mathbb{F}$-linear transformations $V \to V$. In standard fashion, each choice of basis of V yields an isomorphism of $\mathrm{GL}(V)$ onto $\mathrm{GL}(n,\mathbb{F})$. With respect to the chosen basis, a semilinear transformation is a map of the form $v \mapsto x\lambda(v)$ for $v \in V$, where $x \in \mathrm{GL}(n,\mathbb{F})$ and the field automorphism λ is applied entrywise. Thus $\lambda x \lambda^{-1} = \lambda(x)$, and it follows that $\Gamma\mathrm{L}(n,\mathbb{F}) \cong \mathrm{GL}(n,\mathbb{F}) \rtimes \mathrm{Aut}(\mathbb{F})$. If $\mathbb{F}$ has size $q = p^r$ then $\mathrm{Aut}(\mathbb{F})$ is cyclic of order r, generated by the Frobenius p-powering map. Hence $|\Gamma\mathrm{L}(n,q)| = r|\mathrm{GL}(n,q)|$.

The prefix 'P' denotes the quotient obtained from a linear group by factoring out all scalars that it contains. Thus, the *projective general linear group* $\mathrm{PGL}(n,\mathbb{F})$ is the central quotient of $\mathrm{GL}(n,\mathbb{F})$. The groups $\mathrm{PSL}(n,q)$ are simple if $n > 2$ or $q > 3$.

3.6.3. Irreducibility. Let V be an n-dimensional vector space over a field $\mathbb{F}$. Any subgroup of $\mathrm{GL}(V)$ acts in the natural way faithfully on V, so embeds into $\mathrm{Sym}(V)$. A subgroup G of $\mathrm{GL}(V)$ or $\mathrm{GL}(n,\mathbb{F})$ is *irreducible* if the only subspaces of V stabilized setwise by G are the zero subspace and V itself; otherwise G is *reducible*.

Suppose that $\mathbb{F}$ has size q, and let A be an abelian subgroup of $\mathrm{GL}(n,q)$. Denote by $\langle A \rangle_\mathbb{F}$ the (commutative) 'enveloping algebra' of A. That is, $\langle A \rangle_\mathbb{F}$ is the smallest subalgebra of the full matrix $\mathbb{F}$-algebra $\mathrm{Mat}(n,q)$ containing A; its elements are all $\mathbb{F}$-linear combinations of the elements of A. If A is irreducible then $\langle A \rangle_\mathbb{F}$ is a field. More can be said.

3.6.1. PROPOSITION. *The abelian subgroup A of $\mathrm{GL}(n,q)$ is irreducible if and only if $\langle A \rangle_\mathbb{F}$ is a field extension of degree n over $\mathbb{F}$. If A is irreducible then it is cyclic, and $\langle A \rangle_\mathbb{F}^\times$ acts regularly on the non-zero vectors of V.*

There always exist elements of $\mathrm{GL}(n,q)$ of order $q^n - 1$, and if g is any such element then $\langle g \rangle$ is irreducible. A cyclic subgroup of $\mathrm{GL}(n,q)$ of order $q^n - 1$ is called a *Singer cycle*. To see why Singer cycles exist, consider the degree n field extension of $\mathbb{F}$ as the $\mathbb{F}$-vector space V, and choose a primitive element ω of V. Then the map $s : v \mapsto \omega v$ is $\mathbb{F}$-linear, invertible, and of order $q^n - 1$. In matrix terms, a Singer cycle S may be constructed as follows. As before let ω be a primitive element of the degree n extension of $\mathbb{F}$. The diagonal matrix $\mathrm{diag}(\omega,\omega^q,\ldots,\omega^{q^{n-1}})$ has order $q^n - 1$ and trace in $\mathbb{F}$, so is conjugate to an element s of $\mathrm{GL}(n,\mathbb{F})$; then $\langle s \rangle$ is a Singer cycle in $\mathrm{GL}(n,q)$.

3.6.2. EXAMPLE. If ω is a primitive element of the degree 2 extension of $\mathbb{F}$ then

$$\begin{bmatrix} 0 & -\omega^{q+1} \\ 1 & \omega + \omega^q \end{bmatrix}$$

generates a Singer cycle in $\mathrm{GL}(2, \mathbb{F})$.

3.7. Representations

A permutation representation of a group G specifies an action of G on a set. Analogously, a representation of G as a linear or matrix group specifies a certain kind of action of G on a vector space.

3.7.1. Matrix representations. Let G be a finite group and let $\mathcal{R}$ be a ring. A (*linear*) *representation of G over $\mathcal{R}$ of degree d* is a homomorphism of G into $\mathrm{GL}(d, \mathcal{R})$. The representation is *faithful* if it is injective. Let $\mathcal{R}^d$ denote the $\mathcal{R}$-*module* consisting of all d-tuples of elements of $\mathcal{R}$ under componentwise addition and multiplication by elements of $\mathcal{R}$. Each representation of G over $\mathcal{R}$ of degree d expresses G as a group of matrices that are symmetries of the $\mathcal{R}$-module $\mathcal{R}^d$. That is, G acts on $\mathcal{R}^d$ and preserves its module structure. (In fact $\mathcal{R}^d$ is a module for the group ring $\mathcal{R}[G]$: any degree d representation of G over $\mathcal{R}$ extends uniquely to a ring homomorphism of $\mathcal{R}[G]$ into $\mathrm{Mat}(d, \mathcal{R})$.) The symmetries of $\mathcal{R}^d$ are the invertible maps $\theta : \mathcal{R}^d \to \mathcal{R}^d$ such that $\theta(m+n) = \theta(m) + \theta(n)$ and $\theta(rm) = r\theta(m)$ for all $m, n \in \mathcal{R}^d$ and $r \in \mathcal{R}$.

One common example of an $\mathcal{R}$-module is a finite-dimensional vector space V over a field $\mathcal{R} = \mathbb{F}$. Let $\{v_1, \ldots, v_d\}$ be a basis for V, and for each $g \in \mathrm{GL}(V)$ and $1 \leq i, j \leq d$ define the scalars $a(g)_{ij}$ by $gv_j = \sum_i a(g)_{ij} v_i$; then the assignment $g \mapsto [a(g)_{ij}]_{1 \leq i,j \leq d}$ defines an isomorphism of $\mathrm{GL}(V)$ onto $\mathrm{GL}(d, \mathbb{F})$. Remember that if $|\mathbb{F}| = q$ is prime then additively V is an elementary abelian q-group (whose rank is the dimension d of V), and $\mathrm{GL}(V) \cong \mathrm{GL}(d, q) \cong \mathrm{Aut}(V)$.

Later in the book, we will see that an array with a regular group action may be decomposed into a sum of matrices obtained from (monomial) representations of the group. This will enable us to establish in a straightforward manner connections between cocyclic designs and regular group actions.

3.7.2. Indexing matrices by groups. To talk about group actions on a square array, we must index rows and columns by elements of a relevant group.

3.7.1. DEFINITION. Let $M \in \mathrm{Mat}(v, \mathcal{R})$, and let G be a group of order v. We say that M is *indexed* by G if, for some fixed total ordering $g_1 < \cdots < g_v$ of G, row i of M is indexed g_i and column j of M is indexed g_j. We write

$$M = [m_{g,h}]_{g,h \in G}$$

to mean that some total ordering of G is being used to index the matrix.

We can do algebra with matrices that are all indexed by the same group G under the same ordering of group elements. Then the indices are dummy variables; e.g.,

$$[m_{x,y}]_{x,y \in G}[n_{y,z}]_{y,z \in G} = [\textstyle\sum_{y \in G} m_{x,y} n_{y,z}]_{x,z \in G}.$$

Of course, the results of these calculations do not depend on the choice of total ordering of G.

3.7.3. Two regular matrix representations. A group G of order v has two faithful representations as a group of $v \times v$ permutation matrices, arising from the left and right regular actions of G on itself. Although different as homomorphisms, these representations are equivalent, i.e., similar in the sense of Subsection 3.3.6.

We assume that a particular ordering of G is used to index $v \times v$ matrices. For each $x \in G$ define the permutation matrices S_x, T_x by

$$S_x = [\delta^a_{xb}]_{a,b \in G}, \qquad T_x = [\delta^{ax}_b]_{a,b \in G}.$$

These matrices are transposes of each other when G is abelian. We have $S_x T_x = T_x S_x$ for all x, and

$$\sum_{x \in G} S_x = \sum_{x \in G} T_x = J_v$$

where J_v denotes the $v \times v$ all 1s matrix. The maps

$$\sigma : x \mapsto S_x, \qquad \tau : x \mapsto T_x$$

define isomorphisms from G onto the subgroups $\{S_x \mid x \in G\}$ and $\{T_x \mid x \in G\}$ of $\mathrm{Perm}(v)$. That is, σ and τ are faithful degree v representations of G over any ring. In a natural and obvious way they also provide faithful regular permutation representations of G.

3.7.2. LEMMA. *Let Ψ be the set $\{\alpha_c = [\delta^a_c]_{a \in G} \mid c \in G\}$ of $v \times 1$ $(0,1)$-vectors. Multiplying by S_x moves α_c to α_{xc}, and multiplying by T_x moves α_c to $\alpha_{cx^{-1}}$. Hence we obtain (permutation isomorphic) regular actions on Ψ by $\sigma(G)$ and $\tau(G)$.*

The next lemma shows the effect on arbitrary matrices of multiplication by S_x and T_x; cf. Lemma 3.5.2.

3.7.3. LEMMA. *Let M be a $v \times v$ matrix whose rows and columns are indexed by the elements of G under the same ordering as that chosen for S_x and T_x. Then row ax^{-1} of $T_x M$ is row a of M, and column xb of $M S_x^\top$ is column b of M.*

For any ring $\mathcal{R}$, σ and τ extend to ring homomorphisms $\mathcal{R}[G] \to \mathrm{Mat}(v, \mathcal{R})$. The following lemma identifies $\sigma(\mathcal{R}[G])$ and $\tau(\mathcal{R}[G])$ in $\mathrm{Mat}(v, \mathcal{R})$.

3.7.4. LEMMA. *Let $g : G \to \mathcal{R}$ be a map. Then*

(3.4) $$\sum_{x \in G} g(x) S_x = [g(ab^{-1})]_{a,b \in G},$$

(3.5) $$\sum_{x \in G} g(x) T_x = [g(a^{-1}b)]_{a,b \in G}.$$

PROOF. We have

$$\sum_{x \in G} g(x) S_x = \sum_{x \in G} g(x)[\delta^a_{xb}]_{a,b \in G}$$
$$= [\sum_{x \in G} g(x)\delta^a_{xb}]_{a,b \in G}$$
$$= [g(ab^{-1})]_{a,b \in G}.$$

A similar calculation proves (3.5). $\qquad \square$

The set of elements of a ring $\mathcal{R}$ that commute with every element of a subring $\mathcal{S}$ is a subring, the centralizer of $\mathcal{S}$ in $\mathcal{R}$. We show that the centralizer of $\sigma(\mathcal{R}[G])$ in $\mathrm{Mat}(v, \mathcal{R})$ is $\tau(\mathcal{R}[G])$, and the centralizer of $\tau(\mathcal{R}[G])$ is $\sigma(\mathcal{R}[G])$. (Recall that the images in $\mathrm{Sym}(v)$ of the permutation representations defining the left and right regular actions of G on itself centralize each other.)

3.7.5. THEOREM. *A matrix $M \in \mathrm{Mat}(v, \mathcal{R})$ commutes with T_x (respectively S_x) for all $x \in G$ if and only if $M = [g(ab^{-1})]_{a,b \in G}$ (respectively $M = [g(a^{-1}b)]_{a,b \in G}$) for some map $g : G \to \mathcal{R}$.*

PROOF. We write $M = [h(a,b)]_{a,b \in G}$, and define $g : G \to \mathcal{R}$ by $g(x) = h(x, 1)$. By Lemma 3.7.3,
$$T_x M T_x^{-1} = [h(ax, bx)]_{a,b \in G},$$
so that T_x commutes with M for all x if and only if $h(a,b) = h(ab^{-1}, 1) = g(ab^{-1})$ for all $a, b \in G$. □

3.7.6. COROLLARY. *If $\mathcal{R}$ is commutative, then $\sigma(\mathcal{R}[G])$ is the centralizer of $\tau(\mathcal{R}[G])$ in $\mathrm{Mat}(v, \mathcal{R})$, and vice versa.*

The above arguments carry over if the map $g : G \to \mathcal{R}$ is replaced by a map $G \to \mathrm{Mat}(m, \mathcal{R})$, and matrices are enlarged by means of the Kronecker product. One has the following higher-dimensional versions of (3.4) and (3.5):
$$\sum_{x \in G} (g(x) \otimes I_v)(I_m \otimes S_x) = \sum_{x \in G} (g(x) \otimes S_x) = [g(ab^{-1})]_{a,b \in G},$$
$$\sum_{x \in G} (g(x) \otimes I_v)(I_m \otimes T_x) = \sum_{x \in G} (g(x) \otimes T_x) = [g(a^{-1}b)]_{a,b \in G}.$$
A higher-dimensional version of Theorem 3.7.5 is then

3.7.7. THEOREM. *A matrix $M \in \mathrm{Mat}(mv, \mathcal{R})$ commutes with $I_m \otimes T_x$ for all $x \in G$ if and only if $M = \sum_{x \in G} (g(x) \otimes S_x)$ for some map $g : G \to \mathrm{Mat}(m, \mathcal{R})$.*

3.7.4. Regular matrix representations and group development. We now prove some results that will be needed in Chapter 10, where we characterize group development of arrays in terms of regular group actions on arrays.

Let G be a group of order v. Define $W_x = [\delta_x^{ab}]_{a,b \in G}$. Since
$$W_1^{-1} = W_1^{\top} = [\delta_1^{ba}]_{a,b \in G} = [\delta_1^{ab}]_{a,b \in G} = W_1,$$
W_1 is an involution. Next, notice that
$$W_x W_1 = [\delta_x^{ac}]_{a,c \in G}[\delta_{b^{-1}}^{c}]_{c,b \in G} = [\delta_x^{ab^{-1}}]_{a,b \in G} = [\delta_{xb}^{a}]_{a,b \in G} = S_x.$$
Similarly $W_1 W_x = T_x$. Further,

$$(3.6) \qquad W_1 S_x = W_1 W_x W_1 = T_x W_1.$$

The above equalities express the fact that $\sigma : x \mapsto S_x$ and $\tau : x \mapsto T_x$ are equivalent representations of G: $\sigma(x) = W_1^{-1} \tau(x) W_1$ for all $x \in G$. Now if $g : G \to \mathcal{R}$ is any map then
$$[g(ab)]_{a,b \in G} = \sum_{c \in G} g(c)[\delta_c^{ab}]_{a,b \in G} = \sum_{c \in G} g(c) W_c,$$
and so

$$(3.7) \qquad [g(ab)]_{a,b \in G} = \left(\sum_{c \in G} g(c) S_c\right) W_1 = W_1 \sum_{c \in G} g(c) T_c.$$

3.7.8. THEOREM. *Let $A \in \mathrm{Mat}(v, \mathcal{R})$. Then $T_x A S_x^{\top} = A$ for all $x \in G$ if and only if $A = [g(ab)]_{a,b \in G}$ for some map $g : G \to \mathcal{R}$.*

PROOF. By (3.6), $T_x A S_x^{\top} = A$ if and only if AW_1 centralizes T_x. Theorem 3.7.5 implies that this holds for all $x \in G$ if and only if $AW_1 = [g(ab^{-1})]_{a,b \in G}$ for some $g : G \to \mathcal{R}$. Then by (3.4) and (3.7),
$$[g(ab^{-1})]_{a,b \in G} W_1 = [g(ab)]_{a,b \in G},$$
so we are done. □

3.7.9. REMARK. Lemma 3.7.3 gives one of the directions in Theorem 3.7.8: if $M = [g(ab)]_{a,b \in G}$ then the entry in row a, column b of $T_x M S_x^\top$ is $g(ax.x^{-1}b) = g(ab)$.

The matrix $[g(ab)]_{a,b \in G}$ is said to be *group-developed*. Loosely speaking, a group-developed matrix is an image of a group multiplication table. We also call $[g(ab^{-1})]_{a,b \in G}$ and $[g(a^{-1}b)]_{a,b \in G}$ group-developed. For the same G and g, all such matrices are equivalent, up to row and column permutations.

The next lemma is used in Chapter 10, when we give a criterion to distinguish between all the ways in which an array may be group-developed.

3.7.10. LEMMA. *The normalizer of the subgroup*

$$K = \{(T_x, S_x) \mid x \in G\} \cong G$$

in $\mathrm{Perm}(v) \times \mathrm{Perm}(v)$ *is*

$$\{(S_x P_\alpha, T_y P_\alpha) \mid x, y \in G, \alpha \in \mathrm{Aut}(G)\}$$

where $P_\alpha = [\delta^a_{\alpha(b)}]_{a,b \in G}$.

PROOF. Note that W_1 centralizes P_α for any $\alpha \in \mathrm{Aut}(G)$.

Suppose that $(P, Q) \in \mathrm{Perm}(v) \times \mathrm{Perm}(v)$ normalizes K. Then $(T_x^P)^{W_1} = S_x^Q$, so that $P^{W_1} Q^{-1}$ centralizes every S_x. Therefore $P^{W_1} = T_z Q$ for some $z \in G$, by Lemma 3.3.1. Also, by Lemma 3.3.2, $P = S_x P_\alpha$ and $Q = T_y P_\beta$ for some $\alpha, \beta \in \mathrm{Aut}(G)$ and $x, y \in G$. Thus

$$P = (T_z Q)^{W_1} = S_z Q^{W_1} = S_z (T_y P_\beta)^{W_1} = S_{zy} P_\beta.$$

Since $P = S_x P_\alpha$ we have $P_\alpha = P_\beta$, and then $\alpha = \beta$. $\qquad\square$

CHAPTER 4

Orthogonality

In this and the next three chapters we develop various inter-related aspects of orthogonality in design theory.

4.1. How many rows can be pairwise Λ-orthogonal?

The question above may be stated more precisely as

4.1.1. QUESTION. *Given an orthogonality set Λ of $2 \times v$ arrays, what is the maximum number $v_{\max}(\Lambda)$ of Λ-orthogonal rows?*

Of course $v_{\max}(\Lambda) \leq (1 + |\mathcal{A}|)^v$, where $\mathcal{A}$ is the alphabet for Λ. For all the orthogonality sets Λ in Definition 2.13.1, $v_{\max}(\Lambda) \leq v$. In general $v_{\max}(\Lambda)$ can be much larger (even exponentially larger) than v; also $v_{\max}(\Lambda)$ can be much smaller than v. For example, suppose that two distinct v-dimensional $(0,1)$-rows are Λ-orthogonal if and only if their Hamming distance is not less than $d \leq v$. A set of $v_{\max}(\Lambda)$ mutually Λ-orthogonal rows is an error-correcting code with length v, minimum distance d, and maximum number $v_{\max}(\Lambda)$ of codewords. If $d = 1$ then $v_{\max}(\Lambda) = 2^v$; if $d = v$ then $v_{\max}(\Lambda) = 2$. This connection with error-correcting codes is enough to indicate the depth of Question 4.1.1.

The Hadamard Conjecture is equivalent to

$$v_{\max}(\Lambda_{H(4t)}) = 4t \qquad \forall\, t \geq 1.$$

There has been progress on bounding $v_{\max}(\Lambda_{H(4t)})$ from below. De Launey [42] used a result about writing any odd number as a sum of three primes that are close together to prove that $v_{\max}(\Lambda_{H(4t)})$ is at least approximately $4t/3$. Later de Launey and Gordon [48] showed that if certain short arithmetic sequences always contain a prime, then for each $\epsilon > 0$,

$$v_{\max}(\Lambda_{H(4t)}) > 2t - (4t)^{\frac{17}{22}+\epsilon}$$

for all sufficiently large t. Thus the above bound is conditional on the Extended Riemann Hypothesis (ERH). Graham and Shparlinski [78] noted that character sum estimates due to Iwaniec can be employed instead of the ERH to obtain the weaker bound

$$v_{\max}(\Lambda_{H(4t)}) > 2t - O(t^{\frac{113}{132}+o(1)}).$$

So for all sufficiently large t, we have about one half of a Hadamard matrix of order $4t$. In many situations, these bounds are quite useful. See, for example, the introduction to [42]; and also [49], which establishes the existence of high-distance error-correcting codes with rates falling just short of the limits implied by Plotkin's Bound.

Since knowing bounds on $v_{\max}(\Lambda)$ can be useful, and since there is no mature unified method for obtaining them, we propose the following research problem.

1. RESEARCH PROBLEM. *Find good upper and lower bounds on $v_{\max}(\Lambda)$.*

4.2. Non-trivial orthogonality sets

A non-empty orthogonality set Λ is *trivial* if every column of every array in Λ contains a zero. Non-triviality is a requirement in some of our later statements and proofs.

4.2.1. EXAMPLE. For $v = 2$ and $\mathcal{A} = \{1\}$, there are seven orbits of $2 \times v$ $(0, \mathcal{A})$-arrays under row and column permutations. The following is a complete set of orbit representatives:

$$\begin{bmatrix} 0 & 0 \\ 0 & 0 \end{bmatrix}, \quad \begin{bmatrix} 1 & 0 \\ 0 & 0 \end{bmatrix}, \quad \begin{bmatrix} 1 & 0 \\ 1 & 0 \end{bmatrix}, \quad \begin{bmatrix} 1 & 0 \\ 0 & 1 \end{bmatrix}, \quad \begin{bmatrix} 1 & 0 \\ 1 & 1 \end{bmatrix}, \quad \begin{bmatrix} 0 & 0 \\ 1 & 1 \end{bmatrix}, \quad \begin{bmatrix} 1 & 1 \\ 1 & 1 \end{bmatrix}.$$

An orthogonality set Λ for this v and $\mathcal{A}$ cannot include the first, third, or seventh orbit. If we insist on non-triviality then Λ must contain the fifth orbit. Thus, there are eight non-trivial orthogonality sets with $v = 2$ and $\mathcal{A} = \{1\}$.

The next example is a variation on the previous one.

4.2.2. EXAMPLE. Let $v = 3$ and $\mathcal{A} = \{1\}$. Omitting arrays with repeated rows, we get nine orbits of $2 \times v$ $(0, \mathcal{A})$-arrays under row and column permutations, with representatives

$$\begin{bmatrix} 0 & 1 & 1 \\ 1 & 1 & 1 \end{bmatrix}, \quad \begin{bmatrix} 0 & 0 & 1 \\ 1 & 1 & 1 \end{bmatrix}, \quad \begin{bmatrix} 1 & 0 & 0 \\ 1 & 1 & 0 \end{bmatrix}, \quad \begin{bmatrix} 1 & 1 & 0 \\ 1 & 0 & 1 \end{bmatrix}, \quad \begin{bmatrix} 1 & 0 & 0 \\ 0 & 1 & 1 \end{bmatrix}, \quad \begin{bmatrix} 1 & 0 & 0 \\ 0 & 1 & 0 \end{bmatrix},$$

$$\begin{bmatrix} 0 & 0 & 0 \\ 1 & 0 & 0 \end{bmatrix}, \quad \begin{bmatrix} 0 & 0 & 0 \\ 1 & 1 & 0 \end{bmatrix}, \quad \begin{bmatrix} 0 & 0 & 0 \\ 1 & 1 & 1 \end{bmatrix}.$$

To get a non-trivial orthogonality set, we must select at least one orbit from the first four. So there are 480 orthogonality sets with $v = 3$ and $\mathcal{A} = \{1\}$ that are non-trivial.

4.2.3. THEOREM. *The following orthogonality sets are non-empty and non-trivial.*

(1) $\Lambda_{\mathrm{SBIBD}(v,k,\lambda)}$, $v > k > \lambda \geq 1$ *(satisfying (2.4))*.
(2) $\Lambda_{\mathrm{H}(v)}$, $v \geq 2$ *even*.
(3) $\Lambda_{\mathrm{W}(v,k)}$, $k \geq 2$, $v \geq 2k - 2\lfloor k/2 \rfloor$.
(4) $\Lambda_{\mathrm{BW}(v,k,\lambda)}$, $v > k > \lambda \geq 1$, λ *even*.
(5) $\Lambda_{\mathrm{OD}(v;a_1,\ldots,a_r)}$, $v \geq \alpha + \{\alpha/2\}$ *where* $\alpha = \sum_{i=1}^{r} a_i \geq 2$.
(6) $\Lambda_{\mathrm{CH}(v)}$, $v \geq 2$ *even*.
(7) $\Lambda_{\mathrm{CGH}(v;m)}$, $v = \sum_{p|m \text{ prime}} a_p p \geq 2$ *where* $a_p \in \mathbb{N}$.
(8) $\Lambda_{\mathrm{CGW}(v,k;m)}$, $v \geq 2k - \alpha$ *where* $k \geq \alpha = \sum_{p|m \text{ prime}} a_p p \geq 2$, $a_p \in \mathbb{N}$.
(9) $\Lambda_{\mathrm{GH}(v;G)}$, $|G| \geq 2$ *dividing* v.
(10) $\Lambda_{\mathrm{BGW}(v,k,\lambda;G)}$, $v > k > \lambda \geq 1$ *satisfying (2.4)*, $|G|$ *dividing* λ.
(11) $\Lambda_{\mathrm{GW}(v,k;G)}$, $k \geq 2$, $v \geq 2k - |G|\lfloor k/|G| \rfloor$.

PROOF. The conditions for non-emptiness were given in Sections 2.2–2.12. Non-triviality is evident from the definition of each orthogonality set. $\qquad\square$

4.3. A big picture

This section introduces a further abstraction of orthogonality, by way of the notion of design set, and two mutually inverse maps between orthogonality sets and design sets.

4.3.1. The maps λ and δ. Let Λ be an orthogonality set with alphabet $\mathcal{A}$. We denote by $\delta(\Lambda)$ the (possibly empty) set of all $\mathrm{PCD}(\Lambda)$s.

4.3.1. LEMMA. *If $\Lambda_1 \subseteq \Lambda_2$ then $\delta(\Lambda_1) \subseteq \delta(\Lambda_2)$.*

4.3.2. DEFINITION. A set of $v \times v$ $(0, \mathcal{A})$-arrays with no repeated rows that is closed under row and column permutations is a *design set with alphabet $\mathcal{A}$.*

Since permuting the rows and columns of any $\mathrm{PCD}(\Lambda)$ produces another $\mathrm{PCD}(\Lambda)$, we have

4.3.3. LEMMA. *$\delta(\Lambda)$ is a design set with alphabet $\mathcal{A}$.*

Orthogonality set and design set are dual notions. For a design set Δ, let $\lambda(\Delta)$ be the set of all $2 \times v$ subarrays of the elements of Δ.

4.3.4. LEMMA. *If Δ is a design set then $\lambda(\Delta)$ is an orthogonality set.*

PROOF. Let X be a $2 \times v$ subarray of some $D \in \Delta$, and suppose that Y can be obtained from X by performing a sequence of row and column permutations. Applying the same operations to D yields $E \in \Delta$ such that Y is a $2 \times v$ subarray of E. So $\lambda(\Delta)$ is closed under row and column permutations. $\square$

Thus, the maps δ and λ allow us to pass from orthogonality sets to design sets and vice versa.

We have an analog of Lemma 4.3.1.

4.3.5. LEMMA. *$\Delta_1 \subseteq \Delta_2$ implies $\lambda(\Delta_1) \subseteq \lambda(\Delta_2)$.*

The map λ gives us direct access to the smallest orthogonality sets with non-empty design sets.

4.3.6. EXAMPLE. Carrying on with Example 4.2.2, for $v = 3$ and $\mathcal{A} = \{1\}$ we determine the smallest orthogonality sets $\lambda(\Delta)$ containing no array with a row of zeros. Each such set corresponds to an orbit Δ of a single 3×3 $(0, 1)$-array with no repeated rows and no zero rows. There are ten such orbits. Figure 4.1 shows the orbits, together with their corresponding smallest orthogonality sets.

There is a duality between λ and δ.

4.3.7. LEMMA. *$\delta(\Lambda) \supseteq \Delta$ if and only if $\lambda(\Delta) \subseteq \Lambda$.*

PROOF. Both premises are equivalent to "Every D in Δ is a $\mathrm{PCD}(\Lambda)$". $\square$

We rephrase Lemma 4.3.7.

4.3.8. THEOREM. *The following hold.*

(1) *$\delta(\Lambda)$ is the largest design set Δ such that every D in Δ is a $\mathrm{PCD}(\Lambda)$.*
(2) *$\lambda(\Delta)$ is the smallest orthogonality set Λ such that every D in Δ is a $\mathrm{PCD}(\Lambda)$.*

Theorem 4.3.8 prompts the next definition.

$$\Delta_0 = \left\{ \begin{bmatrix} 1 & 0 & 0 \\ 0 & 1 & 0 \\ 0 & 0 & 1 \end{bmatrix} \right\} \qquad \lambda(\Delta_0) = \left\{ \begin{bmatrix} 1 & 0 & 0 \\ 0 & 1 & 0 \end{bmatrix} \right\}$$

$$\Delta_1 = \left\{ \begin{bmatrix} 1 & 1 & 1 \\ 1 & 1 & 0 \\ 1 & 0 & 1 \end{bmatrix} \right\} \qquad \lambda(\Delta_1) = \left\{ \begin{bmatrix} 1 & 1 & 1 \\ 1 & 1 & 0 \end{bmatrix}, \begin{bmatrix} 1 & 1 & 0 \\ 1 & 0 & 1 \end{bmatrix} \right\}$$

$$\Delta_2 = \left\{ \begin{bmatrix} 1 & 1 & 1 \\ 1 & 1 & 0 \\ 1 & 0 & 0 \end{bmatrix} \right\} \qquad \lambda(\Delta_2) = \left\{ \begin{bmatrix} 1 & 1 & 1 \\ 1 & 1 & 0 \end{bmatrix}, \begin{bmatrix} 1 & 1 & 1 \\ 1 & 0 & 0 \end{bmatrix}, \begin{bmatrix} 1 & 1 & 0 \\ 1 & 0 & 0 \end{bmatrix} \right\}$$

$$\Delta_3 = \left\{ \begin{bmatrix} 1 & 1 & 1 \\ 1 & 1 & 0 \\ 0 & 0 & 1 \end{bmatrix} \right\} \qquad \lambda(\Delta_3) = \left\{ \begin{bmatrix} 1 & 1 & 1 \\ 1 & 1 & 0 \end{bmatrix}, \begin{bmatrix} 1 & 1 & 1 \\ 1 & 0 & 0 \end{bmatrix}, \begin{bmatrix} 1 & 1 & 0 \\ 0 & 0 & 1 \end{bmatrix} \right\}$$

$$\Delta_4 = \left\{ \begin{bmatrix} 1 & 1 & 1 \\ 1 & 0 & 0 \\ 0 & 1 & 0 \end{bmatrix} \right\} \qquad \lambda(\Delta_4) = \left\{ \begin{bmatrix} 1 & 1 & 1 \\ 1 & 0 & 0 \end{bmatrix}, \begin{bmatrix} 1 & 0 & 0 \\ 0 & 1 & 0 \end{bmatrix} \right\}$$

$$\Delta_5 = \left\{ \begin{bmatrix} 1 & 1 & 0 \\ 1 & 0 & 1 \\ 0 & 1 & 1 \end{bmatrix} \right\} \qquad \lambda(\Delta_5) = \left\{ \begin{bmatrix} 1 & 1 & 0 \\ 1 & 0 & 1 \end{bmatrix} \right\}$$

$$\Delta_6 = \left\{ \begin{bmatrix} 1 & 1 & 0 \\ 1 & 0 & 0 \\ 0 & 1 & 0 \end{bmatrix} \right\} \qquad \lambda(\Delta_6) = \left\{ \begin{bmatrix} 1 & 1 & 0 \\ 1 & 0 & 0 \end{bmatrix}, \begin{bmatrix} 1 & 0 & 0 \\ 0 & 1 & 0 \end{bmatrix} \right\}$$

$$\Delta_7 = \left\{ \begin{bmatrix} 1 & 1 & 0 \\ 1 & 0 & 0 \\ 0 & 0 & 1 \end{bmatrix} \right\} \qquad \lambda(\Delta_7) = \left\{ \begin{bmatrix} 1 & 1 & 0 \\ 1 & 0 & 0 \end{bmatrix}, \begin{bmatrix} 1 & 1 & 0 \\ 0 & 0 & 1 \end{bmatrix}, \begin{bmatrix} 1 & 0 & 0 \\ 0 & 1 & 0 \end{bmatrix} \right\}$$

$$\Delta_8 = \left\{ \begin{bmatrix} 1 & 1 & 0 \\ 1 & 0 & 1 \\ 1 & 0 & 0 \end{bmatrix} \right\} \qquad \lambda(\Delta_8) = \left\{ \begin{bmatrix} 1 & 1 & 0 \\ 1 & 0 & 1 \end{bmatrix}, \begin{bmatrix} 1 & 1 & 0 \\ 1 & 0 & 0 \end{bmatrix} \right\}$$

$$\Delta_9 = \left\{ \begin{bmatrix} 1 & 1 & 0 \\ 1 & 0 & 1 \\ 0 & 1 & 0 \end{bmatrix} \right\} \qquad \lambda(\Delta_9) = \left\{ \begin{bmatrix} 1 & 1 & 0 \\ 1 & 0 & 1 \end{bmatrix}, \begin{bmatrix} 1 & 1 & 0 \\ 1 & 0 & 0 \end{bmatrix}, \begin{bmatrix} 1 & 1 & 0 \\ 0 & 0 & 1 \end{bmatrix} \right\}$$

FIGURE 4.1. λ applied to minimal design sets

4.3.9. DEFINITION. An orthogonality set Λ is *irredundant* if $\Lambda = \lambda(\Delta)$ for some design set Δ. A design set Δ is *full* if there is an orthogonality set Λ such that $\Delta = \delta(\Lambda)$.

That is, λ maps design sets to irredundant orthogonality sets, and δ maps orthogonality sets to full design sets. We permit empty orthogonality sets to be irredundant, and empty design sets to be full.

4.3.10. EXAMPLE. For $v = 3$ and $\mathcal{A} = \{1\}$, every irredundant orthogonality set containing no arrays with a zero row is a union of some of the $\lambda(\Delta_i)$s as in Example 4.3.6. Even in this tiny case, we see that there are many irredundant orthogonality sets.

4.3.2. Functional relationships between λ and δ.

4.3.11. EXAMPLE. It is instructive to determine the images under δ of the orthogonality sets in Figure 4.1: see Figure 4.2.

In all cases $\delta\lambda(\Delta_i) \supseteq \Delta_i$, with equality only if $i = 0, 2, 3, 5$. It is easily checked that $\lambda\delta\lambda(\Delta_i) = \lambda(\Delta_i)$ for all i. So $\delta\lambda(\delta\lambda(\Delta_i)) = \delta\lambda(\Delta_i)$ for all i. Consequently, the sets $\delta\lambda(\Delta_i)$ are the minimal fixed points under $\delta\lambda$.

$$\delta\lambda(\Delta_0) = \left\{ \begin{bmatrix} 1 & 0 & 0 \\ 0 & 1 & 0 \\ 0 & 0 & 1 \end{bmatrix} \right\} \qquad\qquad \delta\lambda(\Delta_5) = \left\{ \begin{bmatrix} 1 & 1 & 0 \\ 1 & 0 & 1 \\ 0 & 1 & 1 \end{bmatrix} \right\}$$

$$\delta\lambda(\Delta_1) = \left\{ \begin{bmatrix} 1 & 1 & 1 \\ 1 & 1 & 0 \\ 1 & 0 & 1 \end{bmatrix}, \begin{bmatrix} 1 & 1 & 0 \\ 1 & 0 & 1 \\ 0 & 1 & 1 \end{bmatrix} \right\} \qquad \delta\lambda(\Delta_6) = \left\{ \begin{bmatrix} 1 & 1 & 0 \\ 1 & 0 & 0 \\ 0 & 1 & 0 \end{bmatrix}, \begin{bmatrix} 1 & 0 & 0 \\ 0 & 1 & 0 \\ 0 & 0 & 1 \end{bmatrix} \right\}$$

$$\delta\lambda(\Delta_2) = \left\{ \begin{bmatrix} 1 & 1 & 1 \\ 1 & 1 & 0 \\ 1 & 0 & 0 \end{bmatrix} \right\} \qquad\qquad \delta\lambda(\Delta_7) = \left\{ \begin{bmatrix} 1 & 1 & 0 \\ 1 & 0 & 0 \\ 0 & 0 & 1 \end{bmatrix}, \begin{bmatrix} 1 & 1 & 0 \\ 1 & 0 & 0 \\ 0 & 1 & 0 \end{bmatrix}, \begin{bmatrix} 1 & 0 & 0 \\ 0 & 1 & 0 \\ 0 & 0 & 1 \end{bmatrix} \right\}$$

$$\delta\lambda(\Delta_3) = \left\{ \begin{bmatrix} 1 & 1 & 1 \\ 1 & 1 & 0 \\ 0 & 0 & 1 \end{bmatrix} \right\} \qquad\qquad \delta\lambda(\Delta_8) = \left\{ \begin{bmatrix} 1 & 1 & 0 \\ 1 & 0 & 1 \\ 1 & 0 & 0 \end{bmatrix}, \begin{bmatrix} 1 & 1 & 0 \\ 1 & 0 & 1 \\ 0 & 1 & 1 \end{bmatrix} \right\}$$

$$\delta\lambda(\Delta_4) = \left\{ \begin{bmatrix} 1 & 1 & 1 \\ 1 & 0 & 0 \\ 1 & 0 & 0 \end{bmatrix}, \begin{bmatrix} 1 & 0 & 0 \\ 0 & 1 & 0 \\ 0 & 0 & 1 \end{bmatrix} \right\} \quad \delta\lambda(\Delta_9) = \left\{ \begin{bmatrix} 1 & 1 & 0 \\ 1 & 0 & 1 \\ 1 & 0 & 0 \end{bmatrix}, \begin{bmatrix} 1 & 1 & 0 \\ 1 & 0 & 1 \\ 0 & 1 & 0 \end{bmatrix}, \begin{bmatrix} 1 & 1 & 0 \\ 1 & 0 & 1 \\ 0 & 1 & 1 \end{bmatrix} \right\}$$

FIGURE 4.2. δ applied to the orthogonality sets in Figure 4.1

Example 4.3.11 suggests that $\delta\lambda$ is an idempotent operator: that is, $(\delta\lambda)^2 = \delta\lambda$. In fact we will show that both $\delta\lambda$ and $\lambda\delta$ are idempotent. The proof depends on the following result.

4.3.12. LEMMA. *Let Δ be a design set and let Λ be an orthogonality set. Then*

$$\text{(4.1a)} \qquad\qquad\qquad \lambda\delta(\Lambda) \subseteq \Lambda,$$

$$\text{(4.1b)} \qquad\qquad\qquad \delta\lambda(\Delta) \supseteq \Delta.$$

PROOF. These are obvious, but note that (4.1a) (respectively, (4.1b)) follows by taking $\Delta = \delta(\Lambda)$ (respectively, $\Lambda = \lambda(\Delta)$) in Lemma 4.3.7. $\qquad\square$

We note that $\lambda\delta$ preserves some properties of Λ, but not others.

4.3.13. THEOREM. *If no array in Λ has a zero row, then the same is true of $\lambda\delta(\Lambda)$.*

PROOF. Immediate from (4.1a). $\qquad\square$

4.3.14. EXAMPLE. Non-triviality is not preserved by $\lambda\delta$. Let

$$\Lambda = \left\{ \begin{bmatrix} 1 & 1 & 0 \\ 1 & 1 & 1 \end{bmatrix}, \begin{bmatrix} 1 & 0 & 0 \\ 0 & 1 & 0 \end{bmatrix} \right\}.$$

Then

$$\delta(\Lambda) = \left\{ \begin{bmatrix} 1 & 0 & 0 \\ 0 & 1 & 0 \\ 0 & 0 & 1 \end{bmatrix} \right\} \qquad \text{and} \qquad \lambda\delta(\Lambda) = \left\{ \begin{bmatrix} 1 & 0 & 0 \\ 0 & 1 & 0 \end{bmatrix} \right\}.$$

So Λ is non-trivial, whereas $\lambda\delta(\Lambda)$ is trivial.

The following theorem shows that irredundant orthogonality sets are fixed by $\lambda\delta$, and full design sets are fixed by $\delta\lambda$ (hence $\lambda\delta$ and $\delta\lambda$ are idempotent, as claimed).

4.3.15. THEOREM. *For all design sets Δ and orthogonality sets Λ,*

$$\text{(4.2a)} \qquad\qquad\qquad \lambda\delta\lambda(\Delta) = \lambda(\Delta),$$

$$\text{(4.2b)} \qquad\qquad\qquad \delta\lambda\delta(\Lambda) = \delta(\Lambda).$$

PROOF. Put $\Delta_1 = \Delta$, $\Lambda_1 = \lambda(\Delta)$, $\Delta_2 = \delta\lambda(\Delta)$, and $\Lambda_2 = \lambda\delta\lambda(\Delta)$. Then

$$\Lambda_2 = \lambda\delta(\Lambda_1) \subseteq \Lambda_1 \qquad \text{(by (4.1a))}$$
$$\Delta_2 = \delta\lambda(\Lambda_1) \supseteq \Delta_1 \qquad \text{(by (4.1b))}$$
$$\Lambda_2 = \lambda(\Delta_2) \supseteq \lambda(\Delta_1) = \Lambda_1 \qquad \text{(by Lemma 4.3.5).}$$

Hence $\Lambda_2 = \Lambda_1$, proving (4.2a). There is a dual proof of (4.2b). $\qquad\square$

For given v and $\mathcal{A}$, denote the lattice of irredundant orthogonality sets by $\mathcal{L}$, and denote the lattice of full design sets by $\mathcal{D}$. Equations (4.2a) and (4.2b) imply that the maps $\lambda : \mathcal{D} \to \mathcal{L}$ and $\delta : \mathcal{L} \to \mathcal{D}$ are inverses of each other. Without loss of generality, we may work solely within $\mathcal{L}$ and $\mathcal{D}$.

4.3.16. THEOREM. *The following hold.*
  (1) *Every orthogonality set Λ contains a unique irredundant orthogonality set $\Lambda' = \lambda\delta(\Lambda)$ such that D is a PCD(Λ) if and only if it is a PCD(Λ').*
  (2) *Every design set Δ is contained in a unique full design set $\Delta' = \delta\lambda(\Delta)$ such that the rows of D are $\lambda(\Delta)$-orthogonal if and only if $D \in \Delta'$.*

PROOF. We prove (1) only. By (4.1a), Λ certainly contains the irredundant orthogonality set Λ'. By (4.2b), $\delta(\Lambda) = \delta(\Lambda')$, i.e., D is a PCD(Λ) if and only if it is a PCD(Λ'). $\qquad\square$

The 'big picture' provided by $\mathcal{L}$, $\mathcal{D}$, λ, and δ contains much information about pairwise combinatorial designs for the nominated v and $\mathcal{A}$.

2. RESEARCH PROBLEM. *Further study the lattices $\mathcal{D}$ and $\mathcal{L}$.*

4.4. Equivalence

Permuting rows and columns are equivalence operations on PCD(Λ)s. We now describe other sorts of equivalence operations.

4.4.1. Local equivalence operations. We begin with 'local' operations, which change only some entries in an array.

Let $P_{\mathcal{A}}$ denote the group of permutations on $\{0\} \cup \mathcal{A}$ that fix 0 (the stabilizer in $\mathrm{Sym}(\{0\} \cup \mathcal{A})$ of 0). Then let $\Pi_{\Lambda}^{\mathrm{row}}$ denote the set of $\rho \in P_{\mathcal{A}}$ such that

$$\begin{bmatrix} x_1 & x_2 & \cdots & x_v \\ \rho(y_1) & \rho(y_2) & \cdots & \rho(y_v) \end{bmatrix} \in \Lambda \qquad \text{for all} \qquad \begin{bmatrix} x_1 & x_2 & \cdots & x_v \\ y_1 & y_2 & \cdots & y_v \end{bmatrix} \in \Lambda;$$

this is a subgroup of $P_{\mathcal{A}}$. The subgroup $\Pi_{\Lambda}^{\mathrm{col}}$ consists of those $\kappa \in P_{\mathcal{A}}$ such that

$$\begin{bmatrix} \kappa(x_1) & x_2 & \cdots & x_v \\ \kappa(y_1) & y_2 & \cdots & y_v \end{bmatrix} \in \Lambda \qquad \text{for all} \qquad \begin{bmatrix} x_1 & x_2 & \cdots & x_v \\ y_1 & y_2 & \cdots & y_v \end{bmatrix} \in \Lambda.$$

So, applying $\rho \in \Pi_{\Lambda}^{\mathrm{row}}$ to all the entries in a row of a PCD(Λ) produces a PCD(Λ). Similarly, applying $\kappa \in \Pi_{\Lambda}^{\mathrm{col}}$ to any column preserves Λ-orthogonality.

4.4.1. DEFINITION. Two $v \times v$ $(0, \mathcal{A})$-arrays X and Y are Λ-*equivalent* if Y can be obtained from X by a sequence of the following four types of *elementary row and column operations*:

  • interchanging two rows,
  • interchanging two columns,
  • replacing a row $[x_{ij}]_{1 \le j \le v}$ by the row $[\rho(x_{ij})]_{1 \le j \le v}$ for some $\rho \in \Pi_{\Lambda}^{\mathrm{row}}$,

- replacing a column $[x_{ij}]_{1 \le i \le v}$ by the column $[\kappa(x_{ij})]_{1 \le i \le v}$ for some $\kappa \in \Pi_\Lambda^{\mathrm{col}}$.

If X and Y are Λ-equivalent then we write $X \approx_\Lambda Y$.

Thus, if D is a PCD(Λ) and $E \approx_\Lambda D$, then E is a PCD(Λ).

4.4.2. NOTATION. The subgroup $\langle \Pi_\Lambda^{\mathrm{row}}, \Pi_\Lambda^{\mathrm{col}} \rangle$ of $P_\mathcal{A}$ generated by the elements of $\Pi_\Lambda^{\mathrm{row}}$ and $\Pi_\Lambda^{\mathrm{col}}$ is denoted Π_Λ.

4.4.3. EXAMPLE. When $\mathcal{A}$ is small, $\Pi_\Lambda^{\mathrm{row}}$ and $\Pi_\Lambda^{\mathrm{col}}$ are usually easy to calculate. For $\Lambda_{\mathrm{SBIBD}(v,k,\lambda)}$ we have $P_\mathcal{A} = \{\mathrm{id}\}$, and for $\Lambda_{\mathrm{H}(n)}$ we have $\mathcal{A} = \{\pm 1\}$ and $P_\mathcal{A} = \langle \text{multiplication by } -1 \rangle$. In each case, $\Pi_\Lambda^{\mathrm{row}} = \Pi_\Lambda^{\mathrm{col}} = \Pi_\Lambda$ is the full group $P_\mathcal{A}$.

We examine the structure of Π_Λ.

4.4.4. LEMMA. *Let $K = [\Pi_\Lambda^{\mathrm{row}}, \Pi_\Lambda^{\mathrm{col}}]$.*

(1) *For $X \in \Lambda$ and $\pi \in K$, if X' is the array obtained from X by replacing a single entry x_{ij} in X by $\pi(x_{ij})$, then $X' \in \Lambda$.*
(2) *K is a normal subgroup of both $\Pi_\Lambda^{\mathrm{row}}$ and $\Pi_\Lambda^{\mathrm{col}}$.*

PROOF.
1. To replace x_{ij} by $[\rho, \kappa](x_{ij})$ we apply κ to the jth column, then ρ to the ith row, then κ^{-1} to the jth column, and finally ρ^{-1} to the ith row. Each of these individual operations does not move the resulting array out of Λ, and their total effect changes at most one entry in X, namely x_{ij}. Since π is a product of elements of the form $[\rho, \kappa]$, this proves that $X' \in \Lambda$.

2. By part (1), $K \le \Pi_\Lambda^{\mathrm{row}} \cap \Pi_\Lambda^{\mathrm{col}}$. Since then $[\Pi_\Lambda^{\mathrm{row}}, K] \le [\Pi_\Lambda^{\mathrm{row}}, \Pi_\Lambda^{\mathrm{col}}] = K$ and $[K, \Pi_\Lambda^{\mathrm{col}}] \le [\Pi_\Lambda^{\mathrm{row}}, \Pi_\Lambda^{\mathrm{col}}] = K$, K is normal in both groups. $\square$

4.4.5. PROPOSITION. *If Λ is one of the non-trivial orthogonality sets defined in Definition 2.13.1, then $[\Pi_\Lambda^{\mathrm{row}}, \Pi_\Lambda^{\mathrm{col}}] = 1$. Hence $\Pi_\Lambda = \Pi_\Lambda^{\mathrm{row}} \curlyvee \Pi_\Lambda^{\mathrm{col}}$ for all such Λ.*

PROOF. Suppose that $\rho \in \Pi_\Lambda^{\mathrm{row}}$, $\kappa \in \Pi_\Lambda^{\mathrm{col}}$, and $[\rho, \kappa] \ne 1$. We check that each element of $\mathcal{A}$ is an entry x_{ij} of some $X \in \Lambda$, such that changing x_{ij} takes X outside Λ. But as we know from the proof of Lemma 4.4.4, the effect of $[\rho, \kappa]$ on X is to change at most one entry of X; namely, the intersection of the row and column to which ρ and κ are applied. Thus $[\rho, \kappa] = 1$. $\square$

As we now explain, we may assume that $K = 1$ in general. Let $\mathcal{A}'$ be the list of distinct K-orbits in $\mathcal{A}$. Replacing each non-zero entry in each element of Λ by its K-orbit results in an *induced orthogonality set* Λ' over $\{0\} \cup \mathcal{A}'$, such that $[\Pi_{\Lambda'}^{\mathrm{row}}, \Pi_{\Lambda'}^{\mathrm{col}}] = 1$. Consequently, if $K \ne 1$ then we can work with the smaller alphabet $\mathcal{A}'$ and orthogonality set Λ'.

Henceforth we assume that $[\Pi_\Lambda^{\mathrm{row}}, \Pi_\Lambda^{\mathrm{col}}] = 1$. With this condition in place, a sequence of Λ-equivalence operations amounts to premultiplication by a monomial matrix with non-zero entries in $\Pi_\Lambda^{\mathrm{row}}$, and postmultiplication by a monomial matrix with non-zero entries in $\Pi_\Lambda^{\mathrm{col}}$. Chapter 5 makes more of this fact.

4.4.2. Π_Λ, $\Pi_\Lambda^{\mathrm{row}}$, and $\Pi_\Lambda^{\mathrm{col}}$ for the familiar Λ. We now determine $\Pi_\Lambda^{\mathrm{row}}$, $\Pi_\Lambda^{\mathrm{col}}$, and Π_Λ for the orthogonality sets Λ in Definition 2.13.1. As in Subsection 3.3.5, for a finite group G we let σ, τ denote respectively the left and right regular permutation representations $G \to \mathrm{Sym}(G)$.

4.4.6. THEOREM. *Let Λ be any one of the non-trivial orthogonality sets in Definition 2.13.1 with $\mathcal{A}$ equal to a finite group G. Then $\Pi_\Lambda^{\mathrm{row}} = \sigma(G)$, $\Pi_\Lambda^{\mathrm{col}} = \tau(G)$, and $\Pi_\Lambda = \sigma(G) \curlyvee_W \tau(G)$, where $W = \sigma(Z(G)) = \tau(Z(G))$.*

PROOF. Certainly $\sigma(G) \leq \Pi_\Lambda^{\mathrm{row}}$ and $\tau(G) \leq \Pi_\Lambda^{\mathrm{col}}$. Then $\Pi_\Lambda^{\mathrm{row}}$ centralizes $\tau(G)$ by Proposition 4.4.5, so that $\Pi_\Lambda^{\mathrm{row}} \leq \sigma(G)$ by Lemma 3.3.1. Thus $\Pi_\Lambda^{\mathrm{row}} = \sigma(G)$. Similarly, $\Pi_\Lambda^{\mathrm{col}} = \tau(G)$. Therefore $\Pi_\Lambda = \sigma(G) \curlyvee_W \tau(G)$ by Theorem 3.3.4.   □

4.4.7. REMARK. In Theorem 4.4.6, $\Pi_\Lambda^{\mathrm{row}} = \Pi_\Lambda^{\mathrm{col}} = \sigma(G) = \tau(G) \cong G$ when G is abelian.

All but one of the orthogonality sets in Definition 2.13.1 has a finite group as its alphabet. The exception, $\Lambda_{\mathrm{OD}(n;a_1,\ldots,a_r)}$, is dealt with in the next subsection. We can be more explicit when the alphabet is a cyclic group.

4.4.8. THEOREM. *Let Λ be one of $\Lambda_{\mathrm{CH}(n)}$, $\Lambda_{\mathrm{CGH}(n;m)}$, or $\Lambda_{\mathrm{CGW}(v,k;m)}$. Let ζ_m be a primitive mth root of unity, with $m = 4$ if $\Lambda = \Lambda_{\mathrm{CH}(n)}$. Then $\Pi_\Lambda^{\mathrm{row}} = \Pi_\Lambda^{\mathrm{col}} = \langle\alpha\rangle$, where $\alpha(\zeta_m^k) = \zeta_m^{k+1}$ for $k = 0, 1, \ldots, m - 1$.*

PROOF. See Remark 4.4.7. If $G = \langle a \rangle$ then $\sigma(G) = \tau(G)$ is generated by the shift $a^i \mapsto a^{i+1}$.   □

4.4.3. Global equivalence operations. An orthogonality set Λ may have additional equivalence operations not in Π_Λ.

4.4.9. DEFINITION. Let Λ be an orthogonality set of $2 \times v$ $(0, \mathcal{A})$-arrays. Denote by Φ_Λ the set of all $\omega \in P_\mathcal{A}$ such that $[\omega(x_{ij})] \in \Lambda$ for all $[x_{ij}] \in \Lambda$. Then Φ_Λ is a group, containing $\Pi_\Lambda^{\mathrm{row}}$ and $\Pi_\Lambda^{\mathrm{col}}$ as subgroups. We call an element of $\Phi_\Lambda \setminus \Pi_\Lambda$ a *global equivalence operation* of Λ.

If $D = [x_{ij}]$ is a PCD(Λ) and $\omega \in \Phi_\Lambda$ then $\omega(D) = [\omega(x_{ij})]$ is also a PCD(Λ). While $\Pi_\Lambda \leq \Phi_\Lambda$, these two groups can be different.

4.4.10. THEOREM. *Let Λ be any one of the non-trivial orthogonality sets in Definition 2.13.1 with $\mathcal{A}$ equal to an abelian group G. Then $\Phi_\Lambda \cong \mathrm{Hol}(G)$.*

PROOF. Postmultiplication by a fixed $g \in G$ is an element of $\Pi_\Lambda^{\mathrm{col}}$. Since the multiset of column quotients of $A \in \Lambda$ contains each element of G exactly $v/|G|$ times, for any $x, y \in G$ there is some $A \in \Lambda$ in which $[x, y]^\top$ is a column. Hence $\phi(xg)\phi(yg)^{-1} = \phi(x)\phi(y)^{-1}$ for all $\phi \in \Phi_\Lambda$ and $x, y, g \in G$; so $\phi(x)^{-1}\phi(xg)$ is constant as x varies. This implies that $\alpha : G \to G$ defined by $\alpha(g) = \phi(1)^{-1}\phi(g)$ is a homomorphism. Indeed, since ϕ is a bijection, α is an automorphism. Then $\phi \mapsto (\phi(1), \alpha)$ defines an isomorphism $\Phi_\Lambda \to \mathrm{Hol}(G)$.   □

Next, we find Φ_Λ for orthogonal designs.

4.4.11. THEOREM. *Let $\Lambda = \Lambda_{\mathrm{OD}(v;a_1^{e_1},\ldots,a_m^{e_m})}$ where $a_i \neq a_j$ for $i \neq j$. Let x_{ij}, $j = 1, \ldots, e_i$, be the indeterminates which appear (up to negation) exactly a_i times in each row of an element of Λ. Write the alphabet $\mathcal{A}$ of Λ as*

$$\mathcal{A} = \bigcup_{i=1}^m \{\pm x \mid x \in \mathcal{A}_i\}$$

where $\mathcal{A}_i = \{x_{ij} \mid j = 1, \ldots, e_i\}$. Then

$$\Phi_\Lambda \cong (\mathrm{C}_2 \wr \mathrm{Sym}(\mathcal{A}_1)) \times \cdots \times (\mathrm{C}_2 \wr \mathrm{Sym}(\mathcal{A}_m)).$$

Proof. Let K be the group of all $\omega \in \operatorname{Sym}(\mathcal{A})$ such that

(i) ω preserves the partition $\bigcup_{x \in \mathcal{A}} \{x, -x\}$,
(ii) ω fixes each of the sets $\{\pm x \mid x \in \mathcal{A}_i\}$.

Then $K \cong (\mathrm{C}_2 \wr \operatorname{Sym}(\mathcal{A}_1)) \times \cdots \times (\mathrm{C}_2 \wr \operatorname{Sym}(\mathcal{A}_m))$. We show that $\Phi_\Lambda = K$.

Let $X \in \Lambda$ and denote by X' the array in Λ obtained by negating the first column of X. If $\omega \in \Phi_\Lambda$ then $\omega(X), \omega(X') \in \Lambda$. Furthermore,

$$(4.3) \qquad [\omega(X)\omega(X)^\top]_{11} = [XX^\top]_{11} = [\omega(X')\omega(X')^\top]_{11},$$

so that $\omega(x_1)^2 = \omega(-x_1)^2$ where x_1 is the upper leftmost entry of X. Since ω is a bijection, $\omega(-x_1) = -\omega(x_1)$. Then since each $x \in \mathcal{A}$ appears as the upper leftmost entry of some $X \in \Lambda$, ω satisfies (i).

Now

$$[XX^\top]_{11} = \sum_{i=1}^{m} a_i \sum_{x \in \mathcal{A}_i} x^2,$$

and for all $\omega \in \Phi_\Lambda$,

$$[\omega(X)\omega(X)^\top]_{11} = \sum_{i=1}^{m} a_i \sum_{x \in \mathcal{A}_i} \omega(x)^2,$$

by (i). Thus (ii) holds by (4.3), and so $\Phi_\Lambda \leq K$.

We now prove that $K \leq \Phi_\Lambda$. Let X be a $2 \times v$ $(0, \mathcal{A})$-array. We may write X uniquely in the form

$$X = \sum_{x \in \mathcal{A}} x U_x$$

where the U_x are $(0,1)$-matrices. Then $X \in \Lambda$ if and only if

$$(4.4) \quad U_x U_y^\top + U_y U_x^\top + U_{-x} U_{-y}^\top + U_{-y} U_{-x}^\top = U_x U_{-y}^\top + U_{-y} U_x^\top + U_{-x} U_y^\top + U_y U_{-x}^\top$$

for all $x, y \in \mathcal{A}$, $x \neq \pm y$, and

$$(4.5) \qquad U_x U_x^\top + U_{-x} U_{-x}^\top = a_i I_2$$

for all $x \in \mathcal{A}_i$, $1 \leq i \leq m$. Let $\omega \in K$; then

$$\omega(X) = \sum_{x \in \mathcal{A}} \omega(x) U_x = \sum_{x \in \mathcal{A}} x U_{\omega^{-1}(x)}.$$

Hence $\omega(X) \in \Lambda$ if and only if

$$(4.6) \quad U_{\omega^{-1}(x)} U_{\omega^{-1}(y)}^\top + U_{\omega^{-1}(y)} U_{\omega^{-1}(x)}^\top + U_{\omega^{-1}(-x)} U_{\omega^{-1}(-y)}^\top + U_{\omega^{-1}(-y)} U_{\omega^{-1}(-x)}^\top$$
$$= U_{\omega^{-1}(x)} U_{\omega^{-1}(-y)}^\top + U_{\omega^{-1}(-y)} U_{\omega^{-1}(x)}^\top + U_{\omega^{-1}(-x)} U_{\omega^{-1}(y)}^\top + U_{\omega^{-1}(y)} U_{\omega^{-1}(-x)}^\top$$

for all $x, y \in \mathcal{A}$, $x \neq \pm y$, and

$$(4.7) \qquad U_{\omega^{-1}(x)} U_{\omega^{-1}(x)}^\top + U_{\omega^{-1}(-x)} U_{\omega^{-1}(-x)}^\top = a_i I_2$$

for all $x \in \mathcal{A}_i$ and $1 \leq i \leq m$. By (i), (4.6) is (4.4) with $\omega^{-1}(x)$, $\omega^{-1}(y)$ in place of x, y respectively. Similarly (4.7) is equivalent to (4.5). Thus $\omega(X) \in \Lambda$, proving that $K \leq \Phi_\Lambda$. $\qquad \square$

Let $\Lambda = \Lambda_{\mathrm{OD}(v; a_1, \ldots, a_r)}$ be non-trivial. Also let $\mathcal{A} = \{\pm x_i \mid 1 \leq i \leq r\}$ where the indeterminate x_i occurs (counting negations) exactly a_i times in each row of every $X \in \Lambda$. Define $\alpha_0 \in \operatorname{Sym}(\mathcal{A})$ to be the involution that maps x_i to $-x_i$ for $i = 1, \ldots, r$.

4.4.12. THEOREM. $\Pi_\Lambda^{\mathrm{col}} = \langle \alpha_0 \rangle$.

PROOF. By Lemma 2.6.6, for each pair i, j $(i \neq j)$, Λ contains an array of the form

$$X = \left[\begin{array}{c} x_i \\ x_j \end{array} \middle| \; Y \; \right].$$

If $\kappa \in \Pi_\Lambda^{\mathrm{col}}$ then

$$X' = \left[\begin{array}{c} \kappa(x_i) \\ \kappa(x_j) \end{array} \middle| \; Y \; \right] \in \Lambda.$$

Therefore $\kappa(x_i)^2 = x_i^2$, $\kappa(x_j)^2 = x_j^2$, and $\kappa(x_i)\kappa(x_j) = x_i x_j$. It follows that $\kappa \in \langle \alpha_0 \rangle$. □

4.4.13. THEOREM. $\Pi_\Lambda^{\mathrm{row}} = \langle \alpha_0 \rangle$.

PROOF. Let $\omega \in \Pi_\Lambda^{\mathrm{row}}$. Since $\omega \in \Phi_\Lambda$, by Theorem 4.4.11 there is $\mu \in \mathrm{Sym}(r)$ such that $\{\omega(x_j), \omega(-x_j)\} = \{x_{\mu(j)}, -x_{\mu(j)}\}$. We first show that μ is the identity. Suppose not. Then there is a j such that $i := \mu(j) \neq j$ and $a_i \leq a_j$ (otherwise $a_j < a_{\mu(j)} < a_{\mu^2(j)} < \cdots < a_j$ for all j). Then by Lemma 2.6.6, Λ contains an array of the form

$$X = \left[\begin{array}{ccccccc} x_i & \cdots & x_i & x_j & \cdots & x_j \\ x_j & \cdots & x_j & -x_i & \cdots & -x_i \end{array} \middle| \; Y \; \right]$$

where Y has no entries equal to $\pm x_i$, and $[YY^\top]_{12} = 0$. Thus

$$X' = \left[\begin{array}{ccccccc} x_i & \cdots & x_i & x_j & \cdots & x_j \\ \omega(x_j) & \cdots & \omega(x_j) & \omega(-x_i) & \cdots & \omega(-x_i) \end{array} \middle| \; Y' \; \right] \in \Lambda$$

where Y' has no entries equal to $\pm x_i$ in its top row. But then for some $\epsilon = \pm 1$ and integers a_{kl},

$$[X'X'^\top]_{12} = \epsilon a_i x_i^2 + \sum_{(k,l) \neq (i,i)} a_{kl} x_k x_l.$$

This is a contradiction, because $[X'X'^\top]_{12} = 0$ and $\epsilon a_i \neq 0$. So μ is the identity, and $\omega(x_i) = \pm x_i$.

To complete the proof, we show that

$$\omega(x_i)/x_i = \omega(x_1)/x_1 \qquad \forall\, i.$$

Since ω fixes $\{\pm x_i\}$ setwise, $\omega(-x_i)/(-x_i) = \omega(x_i)/x_i$. If $a_i \geq a_j$ then

$$X = \left[\begin{array}{ccccccc} x_j & \cdots & x_j & x_i & \cdots & x_i \\ x_i & \cdots & x_i & -x_j & \cdots & -x_j \end{array} \middle| \; Y \; \right] \in \Lambda$$

where $[YY^\top]_{12} = 0$ and Y has no entries $\pm x_j$, by Lemma 2.6.6. So

$$X' = \left[\begin{array}{ccccccc} x_j & \cdots & x_j & x_i & \cdots & x_i \\ \omega(x_i) & \cdots & \omega(x_i) & \omega(-x_j) & \cdots & \omega(-x_j) \end{array} \middle| \; Y' \; \right] \in \Lambda$$

where

$$[X'X'^\top]_{12} = a_j (x_j \omega(x_i) + x_i \omega(-x_j)) + \sum_{k, l \neq j} a_{kl} x_k x_l.$$

Since $[X'X'^\top]_{12} = 0$ we must have $x_j \omega(x_i) + x_i \omega(-x_j) = 0$. Thus $\omega(x_i)/x_i = \omega(x_j)/x_j$ for all i, j, as required. □

4.4.14. EXAMPLE. We determine the inequivalent $OD(v; 1^v)$ for $v = 2, 4, 8$. Let

$$O_2 = \begin{bmatrix} A & B \\ B & -A \end{bmatrix},$$

$$O_4^{(1)} = \begin{bmatrix} A & B & C & D \\ B & -A & D & -C \\ C & -D & -A & B \\ D & C & -B & -A \end{bmatrix}, \qquad O_4^{(2)} = \begin{bmatrix} A & B & C & D \\ B & -A & -D & C \\ C & D & -A & -B \\ D & -C & B & -A \end{bmatrix},$$

$$O_8 = \begin{bmatrix} A & B & C & D & E & F & G & H \\ B & -A & D & -C & -F & E & -H & G \\ C & -D & -A & B & -G & H & E & -F \\ D & C & -B & -A & -H & -G & F & E \\ -E & -F & -G & -H & A & B & C & D \\ F & -E & -H & G & B & -A & -D & C \\ G & H & -E & -F & C & D & -A & -B \\ H & -G & F & -E & D & -C & B & -A \end{bmatrix}.$$

The elementary local and global equivalence operations are

(1) swapping a pair of rows,
(2) swapping a pair of columns,
(3) negating all entries in a row,
(4) negating all entries in a column,
(5) relabeling the indeterminates.

The first four operations are local, and the fifth is global.

Using operations (1)–(4), we may force the first row and column of an $OD(2; 1^2)$ to coincide with those of O_2. Then there is just one possibility (namely, $-A$) for the lower right entry.

Next we prove that, modulo local equivalence operations, $O_4^{(1)}$ and $O_4^{(2)}$ are the only $OD(4; 1^4)$s. Given any $OD(4; 1^4)$ with indeterminates A, B, C, D, consider first the layout of the indeterminates, ignoring signs for the moment. Since each indeterminate must appear exactly once in each row and column, we can swap columns and rows as needed to make the diagonal A, A, A, A. By orthogonality of rows, if the ith and jth entries of some column are X, Y respectively, then there is another column with ith and jth entries Y, X respectively. Thus, using the first two operations only, we get the following layout:

$$\begin{bmatrix} A & B & C & D \\ B & A & D & C \\ C & D & A & B \\ D & C & B & A \end{bmatrix}.$$

Now let us consider the deployment of signs. Negating rows and columns as needed, we suppose that the first row is $[A, B, C, D]$ and the first column is $[A, B, C, D]^\top$. But then the diagonal must be $A, -A, -A, -A$; otherwise, a pair of rows is not orthogonal. According to whether the last entry of the second row is negated, we find that the rest of the signs are forced. The two designs $O_4^{(1)}$ and $O_4^{(2)}$ result.

It is not possible to turn $O_4^{(2)}$ into $O_4^{(1)}$ without using the fifth equivalence operation. (This assertion is validated below.) However, relabeling C as D and D as C, and then exchanging the third and fourth columns and the third and fourth rows, turns the first array into the second. So there is just one $OD(4; 1^4)$ modulo local and global equivalence operations.

Any $OD(8; 1^8)$ may be put into the form

$$\left[\begin{array}{cc} D_1(A, B, C, D) & D_2(E, F, G, H) \\ D_3(E, F, G, H) & D_4(A, B, C, D) \end{array} \right]$$

where the D_i are $OD(4; 1^4)$s. By the previous paragraph, we assume that $D_1 = O_4^{(1)}$. Once this quadrant is set, we may no longer relabel, although we are still free to use local operations. Thus we can make D_4 equal to $O_4^{(1)}$ or $O_4^{(2)}$. After D_1 and D_4 are set, the signs in D_2 and D_3 are forced by orthogonality with the first row and column. Trying $D_4 = D_1$ quickly leads to a contradiction, whereas the other case leads to a solution. (This implies that the two $OD(4; 1^4)$s are not equivalent via local operations alone—if not, our two possibilities for D_4 would have either both succeeded or both failed.) Therefore O_8 is the unique $OD(8; 1^8)$ modulo local and global operations.

4.5. Matrices, arrays, and designs

Now we make some remarks on usage of the terms 'array', 'matrix', and 'design'. The word 'matrix' appears in nomenclature for many kinds of designs: Hadamard matrices, weighing matrices, and so on. Sometimes 'design' and 'array' are used. For example, we have orthogonal designs, and Williamson and Plotkin arrays. One reason for the blurring of distinction between the concepts of designs, arrays, and matrices is that existence of a design is often equivalent to existence of a matrix satisfying an equation over an implicit ring.

We believe that it is helpful to assign separate roles to each of these concepts. The immediate payoff is that we can begin with a purely combinatorial definition of orthogonality—Definition 2.1.1—without requiring an algebraic setting or inner product. Also, we will see that for any orthogonality set, there are *ambient* rings over which any pairwise combinatorial design of the prescribed type corresponds to a solution of a matrix equation.

Matrices. A matrix is an element of some matrix ring over a ring $\mathcal{R}$, so has no mathematical meaning unless $\mathcal{R}$ is specified. Two matrices are the same if and only if all of their matching entries agree. Except for standard usages (e.g., 'Hadamard matrices', etc.), when we speak of a 'matrix' we mean an element of a matrix ring.

Arrays. An array is a combinatorial object. It usually comes with an alphabet $\mathcal{A}$ from which its entries are drawn. Like matrices, two arrays are equal if and only if all matching entries agree. Any array can be regarded as a matrix over an ambient ring $\mathcal{R}$ by specifying an injection of $\mathcal{A}$ into $\mathcal{R}$. Algebraic arguments may then be used to establish combinatorial properties of the array.

Designs. A design is an equivalence class of arrays with certain combinatorial properties. For example, let D be a $PCD(\Lambda)$, and let E be obtained from D by permuting some rows and columns. Although E and D could be different arrays, E must be a $PCD(\Lambda)$. We regard D and E as being the same design. This generalizes the notion that an $SBIBD(v, k, \lambda)$ corresponds to its set of incidence matrices.

We may encode a permutation of the rows and columns of a $v \times v$ $(0, \mathcal{A})$-array X as a matrix product $P_\pi X P_\phi^{-1}$ over a ring containing $\mathcal{A}$, where $P_\pi, P_\phi \in \mathrm{Perm}(v)$. Thus $\mathrm{Perm}(v) \times \mathrm{Perm}(v)$ acts on the set of $v \times v$ $(0, \mathcal{A})$-arrays. If Y is in the same orbit as X then we write $Y \approx X$, and say that Y and X are *permutation equivalent*.

This defines an equivalence relation on the set of arrays. As we saw in Section 4.4, orthogonality sets have other equivalence operations. So a PCD(Λ) corresponds to a $\approx_\Lambda$-equivalence class of $(0, \mathcal{A})$-arrays.

CHAPTER 5

Modeling Λ-Equivalence

Matrix algebra is an important tool in the study of designs such as those in Chapter 2. However, as it stands, our combinatorial definition of orthogonality set lacks an algebraic context. In this chapter we show that any pairwise combinatorial design may be manipulated using matrix algebra over an *ambient ring*.

If we say nothing else, then an ambient ring $\mathcal{R}$ for a given orthogonality set Λ is simply a ring containing the alphabet $\mathcal{A}$ of Λ (of course, the extra symbol 0 will now become the zero of $\mathcal{R}$). To support a matrix algebra model for Λ-equivalence, an ambient ring $\mathcal{R}$ for Λ must also contain a *row group* $R \cong \Pi_\Lambda^{\text{row}}$ and a *column group* $C \cong \Pi_\Lambda^{\text{col}}$. The essential point is that application of any sequence of row and column operations to a $v \times v$ $(0, \mathcal{A})$-array D may be encoded as a matrix product PDQ over $\mathcal{R}$, where $P \in \text{Mon}(v, R)$ and $Q \in \text{Mon}(v, C)$.

After having amassed the necessary cohomological basics, in Chapter 13 we construct an ambient ring with a *central group* (in addition to the usual row and column groups). This more specialized ring is needed to deal with *cocyclic* designs.

5.1. A first look at the automorphism group

In this section we take our first (and fairly narrow) look at the automorphism group of a pairwise combinatorial design. This serves as motivation for the next section. Chapter 9 is a thorough treatment of automorphism groups of designs.

Let Λ be an orthogonality set of $2 \times v$ $(0, \mathcal{A})$-arrays, and let D be a PCD(Λ). There is some latitude in how to define the automorphism group $\text{Aut}(D)$ of D. However, the underlying principle is always the same: after deciding on a group $\mathcal{E}$ of Λ-equivalence operations, we define $\text{Aut}(D) = \text{Aut}_\mathcal{E}(D)$ to be the stabilizer of D in $\mathcal{E}$; that is, the subgroup of elements of $\mathcal{E}$ that fix D as an array.

Commonly, $\mathcal{E}$ is taken to be the group generated by the elementary row and column equivalence operations as set out in Definition 4.4.1. Assuming (without loss of generality) that $[\Pi_\Lambda^{\text{row}}, \Pi_\Lambda^{\text{col}}] = 1$, each element of $\mathcal{E}$ is a sequence of Λ-equivalence row operations followed by a sequence of Λ-equivalence column operations. So an element of $\mathcal{E}$ may be viewed as an ordered pair of $v \times v$ monomial matrices $P = [\rho_i \delta_{\pi(j)}^i]_{ij}$ and $Q = [\kappa_i \delta_{\phi(j)}^i]_{ij}$ where $\rho_i \in \Pi_\Lambda^{\text{row}}$, $\kappa_i \in \Pi_\Lambda^{\text{col}}$, and $\pi, \phi \in \text{Sym}(v)$. Let $D = [d_{ij}]_{ij}$. The set of pairs (P, Q) of such monomial matrices satisfying

(5.1) $\qquad [\sum_{k,l} \delta_{\pi(k)}^i \delta_{\phi(l)}^j \rho_i \kappa_j (d_{kl})]_{ij} = [\rho_i \kappa_j (d_{\pi^{-1}(i), \phi^{-1}(j)})] = [d_{ij}]$

forms a group.

5.1.1. DEFINITION. The *automorphism group* $\text{Aut}(D)$ of D is the set of pairs $([\rho_i \delta_{\pi(j)}^i], [\kappa_i \delta_{\phi(j)}^i])$ such that $\rho_i \in \Pi_\Lambda^{\text{row}}$, $\kappa_i \in \Pi_\Lambda^{\text{col}}$, $\pi, \phi \in \text{Sym}(v)$, and (5.1) holds. The elements of $\text{Aut}(D)$ are called *automorphisms* of D.

5.1.2. EXAMPLE. Let $G = \langle \omega \mid \omega^3 = 1 \rangle \cong C_3$. Recall that a 3×3 G-array D is a GH$(3; G)$ if and only if

$$DD^* = 3I_3 + (1 + \omega + \omega^2)(J_3 - I_3)$$

over the group ring $\mathbb{Z}[G]$, where the asterisk denotes complex conjugate transpose. We find the automorphism group of the following GH$(3; G)$:

$$D = \begin{bmatrix} 1 & 1 & 1 \\ 1 & \omega & \omega^2 \\ 1 & \omega^2 & \omega \end{bmatrix}.$$

Here the alphabet is $\mathcal{A} = \{1, \omega, \omega^2\}$, and $\Pi_\Lambda^{\mathrm{row}} = \Pi_\Lambda^{\mathrm{col}} = \langle \rho \rangle$ where $\rho \in \mathrm{Sym}(\mathcal{A})$ is the permutation that multiplies each element of $\mathcal{A}$ by ω. Let $R = \langle \omega \rangle$. By (5.1), an automorphism of D corresponds to a pair $(P, Q) \in \mathrm{Mon}(3, R) \times \mathrm{Mon}(3, R)$ such that $PDQ^* = D$. Certainly $\zeta = (\omega I_3, \omega I_3) \in \mathrm{Aut}(D)$. Write $P = P_1^* P_2$ and $Q = Q_1^* Q_2$ where P_1, Q_1 are diagonal matrices and P_2, Q_2 are permutation matrices. So $P_2 D Q_2^\top = P_1 D Q_1^*$. If $P_1 = \mathrm{diag}\,(\omega^{i_1}, \omega^{i_2}, \omega^{i_3})$ and $Q_1 = \mathrm{diag}\,(\omega^{j_1}, \omega^{j_2}, \omega^{j_3})$, then

$$P_1 D Q_1^* = \begin{bmatrix} \omega^{i_1 - j_1} & \omega^{i_1 - j_2} & \omega^{i_1 - j_3} \\ \omega^{i_2 - j_1} & \omega^{1 + i_2 - j_2} & \omega^{2 + i_2 - j_3} \\ \omega^{i_3 - j_1} & \omega^{2 + i_3 - j_2} & \omega^{1 + i_3 - j_3} \end{bmatrix} = \left[\omega^{i_a - j_b + (a-1)(b-1)} \right]_{a, b \in \{1, 2, 3\}}.$$

Since $P_2 D Q_2^\top$ has a row and column of 1s, $P_1 D Q_1^*$ does too. If the kth row and lth column are all 1s, then

$$j_1 = i_k, \qquad j_2 = i_k + k + 2, \qquad j_3 = i_k + 2k + 1,$$
$$i_1 = j_l, \qquad i_2 = j_l + 2l + 1, \qquad i_3 = j_l + l + 2$$

and

$$P_1 D Q_1^* = \omega^{j_l - i_k} \begin{bmatrix} 1 & \omega^{2k+1} & \omega^{k+2} \\ \omega^{2l+1} & \omega^{2k+2l} & \omega^{k+2l+2} \\ \omega^{l+2} & \omega^{2k+l+2} & \omega^{k+l+2} \end{bmatrix}.$$

Thus we get nine possibilities for $P_2 D Q_2^\top$, modulo $\langle \zeta \rangle$:

$$\begin{bmatrix} 1 & 1 & 1 \\ 1 & \omega & \omega^2 \\ 1 & \omega^2 & \omega \end{bmatrix}, \quad \begin{bmatrix} 1 & 1 & 1 \\ \omega^2 & 1 & \omega \\ \omega & 1 & \omega^2 \end{bmatrix}, \quad \begin{bmatrix} 1 & 1 & 1 \\ \omega & \omega^2 & 1 \\ \omega^2 & \omega & 1 \end{bmatrix},$$

$$\begin{bmatrix} 1 & \omega^2 & \omega \\ 1 & 1 & 1 \\ 1 & \omega & \omega^2 \end{bmatrix}, \quad \begin{bmatrix} \omega & 1 & \omega^2 \\ 1 & 1 & 1 \\ \omega^2 & 1 & \omega \end{bmatrix}, \quad \begin{bmatrix} \omega^2 & \omega & 1 \\ 1 & 1 & 1 \\ \omega & \omega^2 & 1 \end{bmatrix},$$

$$\begin{bmatrix} 1 & \omega & \omega^2 \\ 1 & \omega^2 & \omega \\ 1 & 1 & 1 \end{bmatrix}, \quad \begin{bmatrix} \omega^2 & 1 & \omega \\ \omega & 1 & \omega^2 \\ 1 & 1 & 1 \end{bmatrix}, \quad \begin{bmatrix} \omega & \omega^2 & 1 \\ \omega^2 & \omega & 1 \\ 1 & 1 & 1 \end{bmatrix}.$$

Therefore, since P_2 and Q_2 determine P_1 and Q_1 modulo $\langle \zeta \rangle$, each coset of $\langle \zeta \rangle$ in $\mathrm{Aut}(D)$ corresponds to a pair (P_2, Q_2) of permutation matrices that fixes the above set of nine matrices. For any permutation matrix T of order 3, $\langle (I_3, T), (T, I_3) \rangle$ acts transitively on the set of nine matrices. There are just two pairs (P_2, Q_2) that fix D:

$$(I_3, I_3) \qquad \text{and} \qquad \alpha = \left(\begin{bmatrix} 1 & 0 & 0 \\ 0 & 0 & 1 \\ 0 & 1 & 0 \end{bmatrix}, \begin{bmatrix} 1 & 0 & 0 \\ 0 & 0 & 1 \\ 0 & 1 & 0 \end{bmatrix} \right).$$

The entire automorphism group is generated by ζ, α,

$$\beta = \left(\begin{bmatrix} 0 & 0 & 1 \\ 1 & 0 & 0 \\ 0 & 1 & 0 \end{bmatrix}, \begin{bmatrix} 1 & 0 & 0 \\ 0 & \omega^2 & 0 \\ 0 & 0 & \omega \end{bmatrix} \right), \quad \text{and} \quad \gamma = \left(\begin{bmatrix} 1 & 0 & 0 \\ 0 & \omega^2 & 0 \\ 0 & 0 & \omega \end{bmatrix}, \begin{bmatrix} 0 & 1 & 0 \\ 0 & 0 & 1 \\ 1 & 0 & 0 \end{bmatrix} \right).$$

Direct calculation reveals that $\beta^\alpha = \beta^{-1}$, $\gamma^\alpha = \gamma^{-1}$, and $\beta^\gamma = \zeta\beta$; so $\mathrm{Aut}(D) \cong \langle \alpha \rangle \ltimes \langle \beta, \gamma \rangle$.

Example 5.1.2 illustrates how matrix algebra over an ambient ring provides a natural environment within which to model Λ-equivalence. In the next section, we show how to construct, for any given orthogonality set Λ, an ambient ring with a matrix model for Λ-equivalence.

5.2. Ambient rings with a model for Λ-equivalence

An ambient ring $\mathcal{R}$ for modeling Λ-equivalence must contain the alphabet $\mathcal{A}$ of Λ. It should also contain within its group of units an isomorphic copy R of $\Pi_\Lambda^{\mathrm{row}}$ and an isomorphic copy C of $\Pi_\Lambda^{\mathrm{col}}$.

Let $\rho : R \to \Pi_\Lambda^{\mathrm{row}}$ and $\kappa : C \to \Pi_\Lambda^{\mathrm{col}}$ be isomorphisms. For ease of reading, we write ρ_r for the image $\rho(r)$ of $r \in R$ under ρ, and κ_c for the image $\kappa(c)$ of $c \in C$ under κ. Suppose further that $*$ is an involution of $\mathcal{R}$ such that

$$r^* = r^{-1}, \quad c^* = c^{-1} \quad \forall\, r \in R,\, c \in C.$$

We use the following matrix notation.

5.2.1. NOTATION. If $X = [x_{ij}]_{ij}$ is a matrix over a ring with involution $*$, then X^* denotes the matrix $[x_{ji}^*]_{ij}$. That is, X^* is obtained by transposing X and applying $*$ to every entry.

Now we can proceed to the main problem of modeling Λ-equivalence operations. As noted earlier, since $[\Pi_\Lambda^{\mathrm{row}}, \Pi_\Lambda^{\mathrm{col}}] = 1$, every sequence of such operations can be encoded as an ordered pair of monomial matrices

$$([\rho_{r_i} \delta^i_{\pi(j)}], [\kappa_{c_i} \delta^i_{\phi(j)}]),$$

which maps a $(0, \mathcal{A})$-array $A = [a_{ij}]$ to

(5.2) $$([\rho_{r_i} \delta^i_{\pi(j)}], [\kappa_{c_i} \delta^i_{\phi(j)}])A := [\rho_{r_i} \kappa_{c_j} (a_{\pi^{-1}(i), \phi^{-1}(j)})]_{ij}.$$

Our goal is to model this operation as a matrix product PAQ^* over $\mathcal{R}$, where

$$P = [r_i \delta^i_{\pi(j)}] \in \mathrm{Mon}(v, R) \qquad \text{and} \qquad Q = [c_i \delta^i_{\phi(j)}] \in \mathrm{Mon}(v, C).$$

This will impose additional constraints on $\mathcal{R}$. To determine what is needed, we compare (5.2) with the following:

(5.3) $$(P, Q)A := PAQ^*$$
$$= [\textstyle\sum_{k,l} \delta^i_{\pi(k)} \delta^j_{\phi(l)} r_i a_{kl} c_j^{-1}]_{ij}$$
$$= [r_i a_{\pi^{-1}(i), \phi^{-1}(j)} c_j^{-1}]_{ij}.$$

Thus both calculations arrive at the same array if

(5.4) $$ra = \rho_r(a) \quad \text{and} \quad ac^{-1} = \kappa_c(a) \quad \forall\, a \in \mathcal{A},\, r \in R,\, c \in C.$$

Therefore, we would like (5.4) to hold. Notice that applying $*$ to (5.4) gives

(5.5) $$a^* r^{-1} = \rho_r(a)^* \quad \text{and} \quad ca^* = \kappa_c(a)^* \quad \forall\, a \in \mathcal{A},\, r \in R,\, c \in C.$$

5.2.2. REMARK. Since

$$(P_2, Q_2)((P_1, Q_1)A) = P_2(P_1 A Q_1^*)Q_2^* = (P_2 P_1)A(Q_2 Q_1)^* = (P_2 P_1, Q_2 Q_1)A,$$

(5.3) defines an action of $\operatorname{Mon}(v, R) \times \operatorname{Mon}(v, C)$ on the set of $v \times v$ $(0, \mathcal{A})$-arrays. This is an analog of the action of $\operatorname{Mon}(v, \Pi_\Lambda^{\mathrm{row}}) \times \operatorname{Mon}(v, \Pi_\Lambda^{\mathrm{col}})$ defined by (5.2), which relies crucially on the assumption that each row operation commutes with each column operation.

In summary, an ambient ring $\mathcal{R}$ supports a matrix model for Λ-equivalence operations if the following hold.

5.2.3. REQUIREMENTS FOR MODELING Λ-EQUIVALENCE
  (1) $\mathcal{R}$ is a ring with involution $*$ containing the alphabet $\mathcal{A}$, and its group of units contains a row group $R \cong \Pi_\Lambda^{\mathrm{row}}$ and column group $C \cong \Pi_\Lambda^{\mathrm{col}}$.
  (2) For all $r \in R$ and $c \in C$, $r^* = r^{-1}$ and $c^* = c^{-1}$.
  (3) There are isomorphisms $\rho : R \to \Pi_\Lambda^{\mathrm{row}}$ and $\kappa : C \to \Pi_\Lambda^{\mathrm{col}}$ such that $ra = \rho_r(a)$ and $ac^{-1} = \kappa_c(a)$ for all $a \in \mathcal{A}$, $r \in R$, and $c \in C$.

The next two results record what we have achieved so far.

5.2.4. LEMMA. *If $[d_{ij}]_{ij}$ is a $(0, \mathcal{A})$-array, then over any ring satisfying 5.2.3,*

$$[r_i \delta_{\pi(j)}^i][d_{ij}][c_i \delta_{\phi(j)}^i]^* = [\rho_{r_i} \kappa_{c_j}(d_{\pi^{-1}(i), \phi^{-1}(j)})].$$

5.2.5. THEOREM. *Let D be a $\operatorname{PCD}(\Lambda)$, and suppose that $\mathcal{R}$ satisfies 5.2.3. Then $([\rho_{r_i} \delta_{\pi(j)}^i], [\kappa_{c_i} \delta_{\phi(j)}^i])$ is an automorphism of D if and only if*

$$PDQ^* = D$$

over $\mathcal{R}$, where $(P, Q) = ([r_i \delta_{\pi(j)}^i], [c_i \delta_{\phi(j)}^i]) \in \operatorname{Mon}(v, R) \times \operatorname{Mon}(v, C)$.

We have shown that rings with the properties listed in 5.2.3 support a matrix algebra model for Λ-equivalence. It remains to prove that these rings always exist. We do so after an illuminating example.

5.2.6. EXAMPLE. Let $\Lambda = \Lambda_{\mathrm{GH}(v;G)}$. In this case $\mathcal{A} = G$, $\Pi_\Lambda^{\mathrm{row}} = \{\sigma_x \mid x \in G\}$, and $\Pi_\Lambda^{\mathrm{col}} = \{\tau_x \mid x \in G\}$, where $\sigma_x(g) = xg$ and $\tau_x(g) = gx^{-1}$ for $g \in G$. We may take $\mathcal{R} = \mathbb{Z}[G]$, $\mathcal{A} = R = C = G$, and define $\rho : R \to \Pi_\Lambda^{\mathrm{row}}$, $\kappa : C \to \Pi_\Lambda^{\mathrm{col}}$ by $\rho_x = \sigma_x$ and $\kappa_x = \tau_x$. The row group, column group, and alphabet of this orthogonality set coincide. The involution $*$ of $\mathcal{R}$ is inversion on G extended linearly to $\mathcal{R}$.

5.2.7. CONSTRUCTION. Let $\rho : R \to \Pi_\Lambda^{\mathrm{row}}$ and $\kappa : C \to \Pi_\Lambda^{\mathrm{col}}$ be isomorphisms. Suppose that R and C are given by the monoid presentations $\langle X_R \mid W_R \rangle$ and $\langle X_C \mid W_C \rangle$ respectively (that is, the words in the relation sets W_R, W_C do not involve inverses; this is valid because R and C are finite). Denote by $\mathcal{A}^*$ the set $\{a^* \mid a \in \mathcal{A}\}$. Let $M(\Lambda)$ be the monoid defined by the monoid presentation with generating set $X_R \cup X_C \cup \mathcal{A} \cup \mathcal{A}^*$ (disjoint union), and relations

$$W_R \cup W_C \quad \text{and} \quad \left. \begin{array}{c} ra = \rho_r(a), \ ac = \kappa_{c^{-1}}(a) \\ a^*r = \rho_{r^{-1}}(a)^*, \ ca^* = \kappa_c(a)^* \end{array} \right\} \forall r \in X_R, c \in X_C, a \in \mathcal{A}.$$

Define an anti-automorphism $*$ of $M(\Lambda)$ by

$$(a)^* = a^*, \ (a^*)^* = a, \ (r)^* = r^{-1}, \ (c)^* = c^{-1} \quad \forall r \in X_R, c \in X_C, a \in \mathcal{A}.$$

We extend $*$ to the monoid ring $\mathcal{R}(\Lambda) := \mathbb{Z}[M(\Lambda)]$ linearly.

5.2.8. THEOREM. $\mathcal{R}(\Lambda)$ *satisfies* 5.2.3.

PROOF. This is clear from the definitions. □

5.2.9. THEOREM. *The isomorphism class of* $M(\Lambda)$ *is independent of the choice of* ρ *and* κ.

PROOF. If $\rho' : R \to \Pi_\Lambda^{\mathrm{row}}$ and $\kappa' : C \to \Pi_\Lambda^{\mathrm{col}}$ are isomorphisms then $\rho' = \rho \circ \alpha$ and $\kappa' = \kappa \circ \beta$ where $\alpha = \rho^{-1}\rho' \in \mathrm{Aut}(R)$ and $\beta = \kappa^{-1}\kappa' \in \mathrm{Aut}(C)$. Let M be the monoid that results from replacing ρ by ρ' and κ by κ' in Construction 5.2.7. Then the map $M(\Lambda) \to M$ defined by $t \mapsto t$ for $t \in \mathcal{A} \cup \mathcal{A}^*$ and $r \mapsto \alpha^{-1}(r)$, $c \mapsto \beta^{-1}(c)$ for $r \in X_R$, $c \in X_C$ is an isomorphism. □

We can (and do) construct ambient rings satisfying 5.2.3 that differ from $\mathcal{R}(\Lambda)$. These rings involve further conditions tailored to particular Λ; indeed, they are homomorphic images of $\mathcal{R}(\Lambda)$. Thus $\mathcal{R}(\Lambda)$ is a sort of universal ambient ring.

We close this section with two results that cover the familiar orthogonality sets in Chapter 2.

5.2.10. THEOREM. *Suppose that* $\Pi_\Lambda^{\mathrm{row}}$ *and* $\Pi_\Lambda^{\mathrm{col}}$ *act regularly on* $\mathcal{A}$. *Fix* $a_0 \in \mathcal{A}$. *In the notation of Construction* 5.2.7, *there is an isomorphism* $\phi : C \to R$ *such that* $M(\Lambda)$ *is defined by the monoid presentation with generating set*

$$X_R \cup X_C \cup \{a_0, a_0^*\}$$

and relation set

$$W_R \cup W_C \cup \{\phi(c)a_0 = a_0 c,\ ca_0^* = a_0^* \phi(c) \mid c \in X_C\}.$$

PROOF. We have $\mathcal{A} = \{\rho_r(a_0) \mid r \in R\} = \{\kappa_c(a_0) \mid c \in C\}$. Hence there is a bijection $\phi : C \to R$ such that $a_0 c = \phi(c)a_0$. Then

$$\phi(cd)a_0 = a_0(cd) = (a_0 c)d$$
$$= (\phi(c)a_0)d = \phi(c)(a_0 d) = \phi(c)(\phi(d)a_0) = (\phi(c)\phi(d))a_0.$$

Since R acts regularly on $\mathcal{A}$, ϕ is therefore an isomorphism.

Now if $a \in \mathcal{A}$ then there exist (unique) $r \in R$ and $c \in C$ such that $\rho_r(a_0) = a$ and $\kappa_c(a_0) = a$. Thus, the generators a, a^* where $a \in \mathcal{A} \setminus \{a_0\}$ and the relations

$$ra = \rho_r(a),\ ac = \kappa_{c^{-1}}(a),\ a^*r = \rho_{r^{-1}}(a)^*,\ ca^* = \kappa_c(a)^*$$

may be deleted from the monoid presentation of $M(\Lambda)$. □

Theorem 5.2.10 generalizes Example 5.2.6. In Example 5.2.6, $R = C = \mathcal{A}$. We might wonder what happens if Construction 5.2.7 is modified by insisting only that $R = C$ in the case that $\Pi_\Lambda^{\mathrm{row}} = \Pi_\Lambda^{\mathrm{col}}$. The next theorem shows the result of this change when these groups are abelian.

5.2.11. THEOREM. *Suppose that* $\Pi_\Lambda^{\mathrm{row}} = \Pi_\Lambda^{\mathrm{col}}$ *is abelian. Let* B *be a complete and irredundant set of representatives for the orbits of* $\Pi_\Lambda^{\mathrm{row}}$ *in* $\mathcal{A}$. *Let* M *denote the monoid obtained by carrying out Construction* 5.2.7 *with* $R = C$. *Then* M *is defined by the monoid presentation with generating set*

$$X_C \cup B \cup B^*$$

and relation set

$$W_C \cup \{bc = cb,\ cb^* = b^*c \mid c \in X_C,\ b \in B\}.$$

PROOF. If $R = C$ then $\alpha := \kappa^{-1}\rho \in \mathrm{Aut}(C)$. Define the isomorphism $\kappa' : C \to \Pi_\Lambda^{\mathrm{col}}$ by $\kappa'_c = \kappa_{\alpha(c^{-1})}$, $c \in C$. Theorem 5.2.9 implies that we may replace κ by κ'. Then (5.4) and (5.5) become

$$(5.6) \qquad ca = \rho_c(a), \quad ac^{-1} = \kappa'_c(a), \quad a^*c^{-1} = \rho_c(a)^*, \quad ca^* = \kappa'_c(a)^*$$

for all $c \in C$ and $a \in \mathcal{A}$. Thus M is the monoid generated by $\mathcal{A} \cup \mathcal{A}^* \cup X_C$ subject to the relations (5.6). Since $\rho_c = \kappa'_{c^{-1}}$, we have

$$(5.7) \quad ca = \rho_c(a), \ ac = \rho_c(a), \ a^*c = \rho_{c^{-1}}(a)^*, \ ca^* = \rho_{c^{-1}}(a)^* \quad \forall\, c \in C, \ a \in \mathcal{A}.$$

Since C is abelian, (5.7) holds if (i) $bc = cb$, $cb^* = b^*c$ for all $c \in X_C$, $b \in B$; and (ii) $ca = \rho_c(a)$, $a^*c = \rho_{c^{-1}}(a)^*$ for all $c \in X_C$ and $a \in \mathcal{A}$. With $X_C \cup B \cup B^*$ as a generating set for M, the relations (ii) are superfluous. $\qquad\square$

5.3. Ambient rings for the familiar orthogonality sets

We determine $M(\Lambda)$, or a convenient homomorphic image, for the orthogonality sets Λ of Chapter 2. Then we give ambient rings for each Λ.

5.3.1. THEOREM. *Let Λ be one of the orthogonality sets in Definition 2.13.1.*

(1) *Except for $\Lambda = \Lambda_{\mathrm{OD}(n;a_1,\ldots,a_r)}$, $M(\Lambda)$ is as described in Theorem 5.2.10. In these cases, $\mathcal{A}$ is the element set of a finite group isomorphic to R and C, and R, C are both homomorphic images of $M(\Lambda)$.*

(2) *Let $\Lambda = \Lambda_{\mathrm{OD}(n;a_1,\ldots,a_r)}$, and let Y be a set of size r. Then the monoid M generated by c and $Y \cup Y^*$, subject to the relations*

$$c^2 = 1, \ cy = yc, \ cy^* = y^*c$$

for y ranging over Y, is an image of $M(\Lambda)$. The row and column groups are equal to $\langle c \rangle$.

(3) *Let ζ_m be a primitive mth root of unity, and let G be a finite group. Each of the following orthogonality sets has the indicated ambient ring $\mathcal{R}$, row group R, and column group C.*

(a) $\Lambda_{\mathrm{SBIBD}(v,k,\lambda)}$: $\mathcal{R} = \mathbb{Z}$, $R = C = \langle 1 \rangle$.
(b) $\Lambda_{\mathrm{H}(n)}$, $\Lambda_{\mathrm{W}(n,k)}$, $\Lambda_{\mathrm{BW}(n,k,\lambda)}$: $\mathcal{R} = \mathbb{Z}$, $R = C = \langle -1 \rangle$.
(c) $\Lambda_{\mathrm{OD}(n;a_1,\ldots,a_r)}$: $\mathcal{R} = \mathbb{Z}[x_1,\ldots,x_r]$, $R = C = \langle -1 \rangle$.
(d) $\Lambda_{\mathrm{CH}(n)}$: $\mathcal{R} = \mathbb{Z}[\zeta_4]$, $R = C = \langle \zeta_4 \rangle$.
(e) $\Lambda_{\mathrm{CGH}(v;m)}$, $\Lambda_{\mathrm{CGW}(v,k;m)}$: $\mathcal{R} = \mathbb{Z}[\zeta_m]$, $R = C = \langle \zeta_m \rangle$.
(f) $\Lambda_{\mathrm{GH}(n;G)}$, $\Lambda_{\mathrm{BGW}(v,k,\lambda;G)}$, $\Lambda_{\mathrm{GW}(v,k;G)}$: $\mathcal{R} = \mathbb{Z}[G]$, $R = C = G$.

The ring involution is trivial in (a)–(c), complex conjugation extended to $\mathcal{R}$ in (d)–(e), and the inversion map on G extended to $\mathcal{R}$ in (f).

PROOF.

1. Excluding $\Lambda_{\mathrm{OD}(n;a_1,\ldots,a_r)}$, Theorem 5.2.10 covers all the orthogonality sets in Definition 2.13.1, for which $\mathcal{A}$ is a finite group G and $\mathbb{Z}[G]$ is a homomorphic image of $\mathcal{R}(\Lambda)$.

2. This is a consequence of Theorem 5.2.11, using Theorems 4.4.12 and 4.4.13.

3. If $\mathcal{A}$ is a group G then part (1) applies. By Theorem 4.4.6 we can take $R = C = G$, and, as in Example 5.2.6, it is easy to see that $\mathbb{Z}[G]$ satisfies 5.2.3. Then (d)–(f) are immediate. In (a) and (b), $R = C = \mathcal{A} \subseteq \mathbb{Z}$, so $\mathcal{R}$ can be $\mathbb{Z}$. For (c), let $Y = \{y_1,\ldots,y_r\}$ and note that $\mathcal{R} = \mathbb{Z}[x_1,\ldots,x_r]$ is an image of $\mathbb{Z}[M]$ with

M as in part (2), under the homomorphism ψ defined by $\psi(y_i^*) = \psi(y_i) = x_i$ and $\psi(c) = -1$. $\qquad\square$

CHAPTER 6

The Grammian

The elements of each orthogonality set in Definition 2.13.1 are the solutions of a Gram matrix equation over an ambient ring. As we show in this chapter, to varying extents any orthogonality set can be so defined.

6.1. Orthogonality as a Grammian property

Let $\mathcal{R}$ be an ambient ring with involution $*$ for the orthogonality set Λ of $2 \times v$ $(0, \mathcal{A})$-arrays.

6.1.1. DEFINITION. Let $\Gamma_{\mathcal{R}}(\Lambda) = \{XX^* \mid X \in \Lambda\}$, i.e., $\Gamma_{\mathcal{R}}(\Lambda)$ is the set of Grammians over $\mathcal{R}$ of the elements of Λ. Suppose that a $2 \times v$ $(0, \mathcal{A})$-array Y with distinct rows is in Λ if and only if its Grammian YY^* over $\mathcal{R}$ is in $\Gamma_{\mathcal{R}}(\Lambda)$. Then we say that Λ has the *Gram Property over* $\mathcal{R}$.

Notice that Definition 6.1.1 does not call for the presence of a row group or column group in $\mathcal{R}$; it is only necessary that $\mathcal{R}$ allows us to define the Grammian of a $(0, \mathcal{A})$-array. So in this chapter, unless stated otherwise, an 'ambient ring' for Λ is simply an involutory ring containing $\mathcal{A}$.

If Λ has the Gram Property over $\mathcal{R}$, then $\mathrm{PCD}(\Lambda)$s are distinguished among $v \times v$ $(0, \mathcal{A})$-arrays by a Grammian condition.

6.1.2. PROPOSITION. *Denote the set* $\{DD^* \mid D \in \delta(\Lambda)\}$ *of Grammians of all* $\mathrm{PCD}(\Lambda)$s *by* $\mathrm{Gram}_{\mathcal{R}}(\Lambda)$. *If* Λ *has the Gram Property over* $\mathcal{R}$, *then for any* $v \times v$ $(0, \mathcal{A})$-array D *with distinct rows*,

$$D \text{ is a } \mathrm{PCD}(\Lambda) \quad \Longleftrightarrow \quad DD^* \in \mathrm{Gram}_{\mathcal{R}}(\Lambda).$$

6.1.3. EXAMPLE. The orthogonality set Λ for each of the following designs from Chapter 2 has the Gram Property over the indicated ring $\mathcal{R}$.

- $\mathrm{SBIBD}(v, k, \lambda)$: $\mathcal{R} = \mathbb{Z}$, $\Gamma_{\mathcal{R}}(\Lambda) = \{kI_2 + \lambda(J_2 - I_2)\}$.
- $\mathrm{H}(n)$: $\mathcal{R} = \mathbb{Z}$, $\Gamma_{\mathcal{R}}(\Lambda) = \{nI_2\}$.
- $\mathrm{W}(v, k)$: $\mathcal{R} = \mathbb{Z}$, $\Gamma_{\mathcal{R}}(\Lambda) = \{kI_2\}$.
- $\mathrm{BW}(v, k, \lambda)$: $\mathcal{R} = \mathbb{Z}[\mathrm{C}_2]$, $\Gamma_{\mathcal{R}}(\Lambda) = \{kI_2 + \frac{\lambda}{2}\mathrm{C}_2(J_2 - I_2)\}$.
- $\mathrm{OD}(n; a_1, \ldots, a_r)$: $\mathcal{R} = \mathbb{Z}[x_1, \ldots, x_r]$, $\Gamma_{\mathcal{R}}(\Lambda) = \{\sum_{i=1}^{r} a_i x_i^2 I_2\}$.
- $\mathrm{CH}(n)$: $\mathcal{R} = \mathbb{Z}[\mathrm{i}]$, $\Gamma_{\mathcal{R}}(\Lambda) = \{nI_2\}$.
- $\mathrm{CGH}(n; m)$: $\mathcal{R} = \mathbb{Z}[\zeta_m]$, $\Gamma_{\mathcal{R}}(\Lambda) = \{nI_2\}$.
- $\mathrm{CGW}(v, k; m)$: $\mathcal{R} = \mathbb{Z}[\zeta_m]$, $\Gamma_{\mathcal{R}}(\Lambda) = \{kI_2\}$.
- $\mathrm{GH}(n; H)$: $\mathcal{R} = \mathbb{Z}[H]/\mathbb{Z}H$, $\Gamma_{\mathcal{R}}(\Lambda) = \{nI_2\}$.
- $\mathrm{GW}(v, k; H)$: $\mathcal{R} = \mathbb{Z}[H]/\mathbb{Z}H$, $\Gamma_{\mathcal{R}}(\Lambda) = \{kI_2\}$.
- $\mathrm{BGW}(v, k, \lambda; H)$: $\mathcal{R} = \mathbb{Z}[H]$, $\Gamma_{\mathcal{R}}(\Lambda) = \{kI_2 + \frac{\lambda}{|H|}H(J_2 - I_2)\}$.

71

6.1.4. EXAMPLE. Every orthogonality set Λ with alphabet $\{1\} \subseteq \mathbb{Z}$ has the Gram Property over $\mathbb{Z}$. To see this, first note that each permutation equivalence class of $2 \times v$ $(0,1)$-arrays consists of elements with row sums r_1, r_2 and inner product λ, say. The orthogonality set Λ contains all such arrays (without repeated rows) if and only if $\begin{bmatrix} r_1 & \lambda \\ \lambda & r_2 \end{bmatrix}$ and $\begin{bmatrix} r_2 & \lambda \\ \lambda & r_1 \end{bmatrix}$ are in $\Gamma_{\mathbb{Z}}(\Lambda)$.

We now prove that every orthogonality set has the Gram Property over at least one ambient ring. Let $M_0 = M_0(\Lambda)$ be the (free) monoid generated by $\mathcal{A} \cup \mathcal{A}^* = \{a, a^* \mid a \in \mathcal{A}\}$. Define $*$ to be the anti-automorphism of M_0 such that $(a)^* = a^*$ and $(a^*)^* = a$ for all $a \in \mathcal{A}$. Then extend $*$ linearly to the monoid ring $\mathbb{Z}[M_0]$, so that $\mathbb{Z}[M_0]$ becomes a ring with involution $*$.

6.1.5. PROPOSITION. *Let X and Y be $2 \times v$ $(0, \mathcal{A})$-arrays. The following hold over $\mathbb{Z}[M_0]$.*

    (1) *$XX^* = YY^*$ if and only if $X = YQ$ for some $Q \in \mathrm{Perm}(v)$.*
    (2) *If $Y \in \Lambda$ and $XX^* = YY^*$, then $X \in \Lambda$.*

PROOF. The elements $[XX^*]_{12} = \sum_{i=1}^{v} x_{1i} x_{2i}^*$ and $[YY^*]_{12} = \sum_{i=1}^{v} y_{1i} y_{2i}^*$ of $\mathbb{Z}[M_0]$ can be equal only if $x_{1i} x_{2i}^* = y_{1,\pi(i)} y_{2,\pi(i)}^*$ for all i and some $\pi \in \mathrm{Sym}(v)$. Since M_0 is the free monoid on $\mathcal{A} \cup \mathcal{A}^*$, (1) is clear. Now (2) follows from (1). $\square$

Thus, Λ has the Gram Property over $\mathbb{Z}[M_0]$. In addition, $\mathbb{Z}[M_0]$ supports a matrix model for permutation equivalence (as does any ring containing $\mathcal{A}$). Ideally, Λ will have the Gram Property over an ambient ring that supports a matrix model for Λ-equivalence—which is the case for the orthogonality sets in Example 6.1.3. We say more about this point in the last section.

6.2. Non-degeneracy

We use the Grammian to define non-pathological orthogonality sets.

6.2.1. DEFINITION. Let Λ be an orthogonality set with alphabet $\mathcal{A}$ and ambient ring $\mathcal{R}$ containing a row group R and column group C (in the group of units of $\mathcal{R}$). For a $(0, \mathcal{A})$-array X, denote the set of diagonal entries and the set of off-diagonal entries of the Grammian of X over $\mathcal{R}$ by $\mathrm{wt}(X)$ and $\mathrm{df}(X)$ respectively. Then X is *non-degenerate over $\mathcal{R}$* if, for r_1, r_2, r_3, r_4 ranging over R, the sets

$$r_1 \mathrm{wt}(X) r_2 \qquad \text{and} \qquad r_3 \mathrm{df}(X) r_4$$

are all disjoint from each other. We say that Λ is *non-degenerate over $\mathcal{R}$* if every $\mathrm{PCD}(\Lambda)$ is non-degenerate.

This definition is needed in Chapters 9 and 14. It is also natural to ask that the off-diagonal entries of the Grammian of a design be somehow different from the diagonal entries.

6.2.2. THEOREM. *Each orthogonality set in Example 6.1.3 is non-degenerate over its stated ambient ring.*

PROOF. In all cases $\mathrm{Gram}_{\mathcal{R}}(\Lambda)$, if non-empty, is a singleton. We verify that Λ is non-degenerate by eye (leaving aside trivialities such as $k = \lambda$ or $H = \langle 1 \rangle$). $\square$

It is easy to find examples of degenerate orthogonality sets.

6.2.3. THEOREM. *No array in an irredundant non-degenerate orthogonality set* Λ *has a zero row.*

PROOF. If the claim were false, then the Grammian of some $\mathrm{PCD}(\Lambda)$ would have a row and a column of zeros. $\qquad\square$

6.2.4. EXAMPLE. The orthogonality set Λ with ambient ring $\mathbb{Z}$ formed as the union of the orbits of

$$\begin{bmatrix} 1 & 1 & 0 \\ 1 & 1 & 1 \end{bmatrix}, \quad \begin{bmatrix} 1 & 1 & 0 \\ 1 & 0 & 1 \end{bmatrix}$$

under row and column permutations is degenerate. For example,

$$A = \begin{bmatrix} 0 & 1 & 1 \\ 1 & 0 & 1 \\ 1 & 1 & 1 \end{bmatrix}$$

is a $\mathrm{PCD}(\Lambda)$ such that $\mathrm{wt}(A)$ and $\mathrm{df}(A)$ are not disjoint. There are 60 non-trivial orthogonality sets in Example 4.2.2 not containing an array with a zero row. Only two of these are irredundant and non-degenerate: they are unions of the orbits of

$$\begin{bmatrix} 1 & 0 & 0 \\ 0 & 1 & 0 \end{bmatrix}, \quad \begin{bmatrix} 1 & 1 & 0 \\ 1 & 0 & 1 \end{bmatrix}.$$

6.3. Gram completions and composition of orthogonality sets

This section will be needed in Chapter 8; and in Sections 10.8 and 15.8, where we compose cocyclic pairwise combinatorial designs via their associates.

6.3.1. DEFINITION. An orthogonality set Λ is *Gram complete* over an ambient ring $\mathcal{R}$ if there are subsets U, V of $\mathcal{R}$ such that Λ consists of all $2 \times v$ $(0, \mathcal{A})$-arrays X with distinct rows satisfying $[XX^*]_{11}, [XX^*]_{22} \in U$ and $[XX^*]_{12} \in V$.

If Λ is Gram complete then it has the Gram Property. We can always enlarge Λ (for fixed $\mathcal{R}$) to obtain a Gram complete orthogonality set.

6.3.2. NOTATION. Define $\mathrm{wt}(\Lambda) = \bigcup_{X \in \Lambda} \mathrm{wt}(X)$ and $\mathrm{df}(\Lambda) = \bigcup_{X \in \Lambda} \mathrm{df}(X)$.

6.3.3. PROPOSITION. *There is a unique smallest orthogonality set* $\overline{\Lambda}$ *containing* Λ *that is Gram complete over* $\mathcal{R}$.

PROOF. $\overline{\Lambda}$ is the set of $2 \times v$ $(0, \mathcal{A})$-arrays Y with distinct rows such that $[YY^*]_{11}, [YY^*]_{22} \in \mathrm{wt}(\Lambda)$ and $[YY^*]_{12} \in \mathrm{df}(\Lambda)$. $\qquad\square$

The orthogonality set $\overline{\Lambda}$ is the *Gram completion of* Λ over $\mathcal{R}$. Of course, Λ is Gram complete if and only if $\Lambda = \overline{\Lambda}$.

Next, we discuss ways to compose orthogonality sets. For $i = 1, 2$, let Λ_i be an orthogonality set of $2 \times v_i$ $(0, \mathcal{A}_i)$-arrays, $v_i \geq 2$, with ambient ring $\mathcal{R}_i$.

6.3.4. DEFINITION. Suppose that $\mathcal{R}$ is a ring containing the $\mathcal{R}_i$ as subrings. Define $\mathcal{A} = \{a_1 a_2 \mid a_1 \in \mathcal{A}_1, a_2 \in \mathcal{A}_2\} \subseteq \mathcal{R}$, excluding 0 if necessary. Let $\Lambda_1 \otimes \Lambda_2$ denote the closure under column permutations of the set of all $2 \times v_1 v_2$ subarrays with distinct rows of the arrays $X_1 \otimes X_2$ where $X_1 \in \Lambda_1$ and $X_2 \in \Lambda_2$. Then the orthogonality set $\Lambda_1 \otimes \Lambda_2$ with alphabet $\mathcal{A}$ is the *Kronecker product* of Λ_1 and Λ_2 over $\mathcal{R}$.

There may be several viable choices for the ring $\mathcal{R}$ in Definition 6.3.4. It is not compulsory that $\mathcal{R}$ has an involution.

We now present another, related way of composing orthogonality sets.

6.3.5. DEFINITION. Suppose that $\mathcal{R}$ is a ring with involution $*$ containing the $\mathcal{R}_i$ as subrings, and the restriction of $*$ to each $\mathcal{R}_i$ is the involution of that ring. Suppose further that $\mathcal{R}_1$ and $\mathcal{R}_2$ centralize each other. For $i = 1, 2$, let $\mathrm{wt}_i = \mathrm{wt}(\Lambda_i)$ and $\mathrm{df}_i = \mathrm{df}(\Lambda_i)$. Also let $\mathcal{A}$ be the subset $\{a_1 a_2 \mid a_1 \in \mathcal{A}_1, a_2 \in \mathcal{A}_2\}$ of $\mathcal{R}$, assuming that $0 \notin \mathcal{A}$. Let $\Lambda_1 \curlywedge \Lambda_2$ denote the set of all $2 \times v_1 v_2$ $(0, \mathcal{A})$-arrays X (with distinct rows) such that

$$[XX^*]_{11}, [XX^*]_{22} \in \mathrm{wt}_1 \mathrm{wt}_2 \quad \text{and} \quad [XX^*]_{12} \in \mathrm{wt}_1 \mathrm{df}_2 \cup \mathrm{df}_1 \mathrm{wt}_2 \cup \mathrm{df}_1 \mathrm{df}_2.$$

Then $\Lambda_1 \curlywedge \Lambda_2$ is an orthogonality set with ambient ring $\mathcal{R}$.

Again, existence of a suitable ring $\mathcal{R}$ for all the orthogonality sets involved is a prerequisite for the composition to be possible.

6.3.6. PROPOSITION. *Suppose that no array $X \otimes Y$ where $X \in \Lambda_1$ and $Y \in \Lambda_2$ has a repeated row. Then $\Lambda_1 \curlywedge \Lambda_2$ is the Gram completion of $\Lambda_1 \otimes \Lambda_2$ over $\mathcal{R}$ as in Definition 6.3.5.*

PROOF. Let $\Lambda = \Lambda_1 \otimes \Lambda_2$. For $i = 1, 2$ let $w_i \in \mathrm{wt}_i$ and $d_i \in \mathrm{df}_i$; so there exist $X_i, Y_i \in \Lambda_i$ such that $w_i = [X_i X_i^*]_{11}$ and $d_i = [Y_i Y_i^*]_{12}$. Since $(A \otimes B)(A \otimes B)^* = AA^* \otimes BB^*$ for $A \in \{X_1, Y_1\}$ and $B \in \{X_2, Y_2\}$, we have $w_1 w_2 \in \mathrm{wt}(\Lambda)$ and $w_1 d_2$, $d_1 w_2, d_1 d_2 \in \mathrm{df}(\Lambda)$. Thus

$$\mathrm{wt}(\Lambda) = \mathrm{wt}_1 \mathrm{wt}_2 \quad \text{and} \quad \mathrm{df}(\Lambda) = \mathrm{wt}_1 \mathrm{df}_2 \cup \mathrm{df}_1 \mathrm{wt}_2 \cup \mathrm{df}_1 \mathrm{df}_2,$$

so that $\overline{\Lambda} = \Lambda_1 \curlywedge \Lambda_2$.   □

6.4. The Gram Property and Λ-equivalence

The orthogonality sets in Example 6.1.3 all have the Gram Property over an ambient ring with a row group and a column group. In this section, we show how to approximate this conjunction for an arbitrary orthogonality set.

6.4.1. THEOREM. *Let Λ be an orthogonality set with alphabet $\mathcal{A}$. Then there are*

  (i) *an ambient ring $\mathcal{R}$ for Λ satisfying Requirements 5.2.3,*
  (ii) *an orthogonality set $\tilde{\Lambda}$ with alphabet $\tilde{\mathcal{A}}$ and ambient ring $\tilde{\mathcal{R}}$, and*
  (iii) *a surjective homomorphism ϕ of $\tilde{\mathcal{R}}$ onto $\mathcal{R}$,*

such that

  (iv) *$\tilde{\Lambda}$ consists of all $(0, \tilde{\mathcal{A}})$-arrays that are mapped into Λ by (the entrywise extension of) ϕ, and*
  (v) *$\tilde{\Lambda}$ has the Gram Property over $\tilde{\mathcal{R}}$.*

Thus, each $\mathrm{PCD}(\Lambda)$ lifts to a $\mathrm{PCD}(\tilde{\Lambda})$ that is a solution of a Gram matrix equation over $\tilde{\mathcal{R}}$. Moreover, the image $\mathcal{R}$ of $\tilde{\mathcal{R}}$ supports a matrix model for Λ-equivalence.

We begin the proof of Theorem 6.4.1. Let $\mathcal{T}$ be a complete irredundant set of representatives for the Π_Λ-orbits in $\mathcal{A}$. Define a monoid $\tilde{M}$ as follows. Select isomorphisms $\rho : R \to \Pi_\Lambda^{\mathrm{row}}$ and $\kappa : C \to \Pi_\Lambda^{\mathrm{col}}$, where the groups R and C are given by monoid presentations $R = \langle X_R \mid W_R \rangle$, $C = \langle X_C \mid W_C \rangle$. Then $\tilde{M}$ is given by the presentation with generating set $X_R \cup X_C \cup \{\tilde{t}, \tilde{t}^* \mid t \in \mathcal{T}\}$ and relations

(6.1) $$W_R \cup W_C \cup \{[r, c] = 1 \mid r \in X_R, c \in X_C\}.$$

Let $*$ be the anti-automorphism of $\tilde{M}$ such that $(x)^* = x^{-1}$ for $x \in X_R \cup X_C$, and $(\tilde{t})^* = \tilde{t}^*$, $(\tilde{t}^*)^* = \tilde{t}$ for $t \in \mathcal{T}$. Take the monoid presentation for $M(\Lambda)$ specified in Construction 5.2.7, and add all commutator relations $[r, c] = 1$ for $r \in X_R$ and $c \in X_C$: call the monoid with this new presentation M. Define $\phi : \tilde{M} \to M$ to be the monoid homomorphism such that

$$\phi(r) = r, \quad \phi(c) = c, \quad \phi(\tilde{t}) = t, \quad \phi(\tilde{t}^*) = t^*$$

for all $r \in X_R$, $c \in X_C$, and $t \in \mathcal{T}$. Then ϕ extends to a homomorphism from $\tilde{\mathcal{R}} = \mathbb{Z}[\tilde{M}]$ onto $\mathcal{R} = \mathbb{Z}[M]$. Let

$$\tilde{\mathcal{A}} = \{r\tilde{t}c \mid r \in R, c \in C, t \in \mathcal{T}\} \subseteq \tilde{M}.$$

Note that $\phi(\tilde{\mathcal{A}}) = \mathcal{A}$. Finally, put $\tilde{\Lambda} = \phi^{-1}(\Lambda)$.

Except for part (v), all of Theorem 6.4.1 is apparent from the definitions in the previous paragraph. Proving part (v) takes up the rest of this section.

6.4.2. LEMMA. *Let* $a_1, a_2, a_3, a_4 \in \tilde{\mathcal{A}}$. *In* $\tilde{M}$, $a_1 a_2^* = a_3 a_4^*$ *if and only if* $a_3 = a_1 c$ *and* $a_4 = a_2 c$ *for some* $c \in C$.

PROOF. Let G be the group generated by $X_R \cup X_C \cup \{\tilde{t}, \tilde{t}^* \mid t \in \mathcal{T}\}$, subject to the relations (6.1). So $\tilde{M} \subseteq G$. Suppose that $a_1 a_2^* = a_3 a_4^*$; then $a_3^{-1} a_1 a_2^* (a_4^*)^{-1} = 1$ in G. For $i = 1, 2, 3, 4$, we may write $a_i = r_i \tilde{t}_i c_i$ where $r_i \in R$, $c_i \in C$, and $t_i \in \mathcal{T}$. Hence

$$w = c_3^{-1} \tilde{t}_3^{-1} r_3^{-1} r_1 \tilde{t}_1 c_1 c_2^{-1} \tilde{t}_2^* r_2^{-1} r_4 (\tilde{t}_4^*)^{-1} c_4 = 1,$$

which implies that

$$\tilde{t}_3^{-1} \tilde{t}_1 \tilde{t}_2^* (\tilde{t}_4^*)^{-1} = 1 \qquad \text{and} \qquad c_3^{-1} r_3^{-1} r_1 c_1 c_2^{-1} r_2^{-1} r_4 c_4 = 1.$$

Therefore $\tilde{t}_3 = \tilde{t}_1$ and $\tilde{t}_4 = \tilde{t}_2$; further $r_3^{-1} r_1 = r_4^{-1} r_2 = r$ and $c_1 c_2^{-1} = c_3 c_4^{-1} = c$, say. So we have

$$\tilde{t}_1^{-1} r \tilde{t}_1 c \tilde{t}_2^* r^{-1} (\tilde{t}_2^*)^{-1} c^{-1} = c_3 w c_3^{-1} = 1,$$

which cannot be true in G if $r \neq 1$. The equalities $r_3 = r_1$ and $r_4 = r_2$ then yield the desired conclusion. $\square$

6.4.3. PROPOSITION. *Suppose that* Y *is a* $2 \times v$ $(0, \tilde{\mathcal{A}})$-*array such that* $YY^* = XX^*$ *for some* $X \in \tilde{\Lambda}$. *Then* $Y \in \tilde{\Lambda}$.

PROOF. Divide the set of columns of X into four subsets X_{ab}, $a, b \in \{0, 1\}$, where X_{ab} is all columns that have a zero in row 1 if and only if $a = 0$, and have a zero in row 2 if and only if $b = 0$. Define the sets Y_{ab} of columns of Y similarly. Since $[YY^*]_{12} = [XX^*]_{12}$, Y_{11} and X_{11} have the same cardinality. Then enforcing $[YY^*]_{11} = [XX^*]_{11}$ and $[YY^*]_{22} = [XX^*]_{22}$ in turn proves that $|Y_{ab}| = |X_{ab}|$ for all ab.

Each column in Y_{ab} is matched with a column in X_{ab} such that their respective contributions to YY^* and XX^* are the same. Say $[a_1, a_2]^\top \in Y_{11}$ and $[a_3, a_4]^\top \in X_{11}$ contribute $a_1 a_2^* = a_3 a_4^*$ to the respective equal sums $[YY^*]_{12}$ and $[XX^*]_{12}$. By Lemma 6.4.2, $a_3 = a_1 c$ and $a_4 = a_2 c$ for some $c \in C$. In this way, Lemma 6.4.2 also implies that each column in $Y_{10} \cup Y_{01}$ can be obtained from its matched column in

X via postmultiplying by a single element of C. That is, $Y = XQ$ for some $Q \in$ Mon(v, C). Hence $\phi(Y) = \phi(X)\phi(Q) \in \Lambda$, i.e., $Y \in \tilde{\Lambda}$ as required. $\qquad\square$

CHAPTER 7

Transposability

For many of the orthogonality sets Λ in Definition 2.13.1, the transpose of a PCD(Λ) is also a PCD(Λ). When this is not so, a more general condition holds: if D is a PCD(Λ), then the array obtained by applying a permutation α of the alphabet set to each non-zero entry in $D^\top$ is a PCD(Λ). So there are plenty of orthogonality sets Λ where Λ-orthogonality of the rows imposes a pairwise condition on the columns. In this chapter, we discuss this intriguing aspect of orthogonality.

7.1. The main problems

7.1.1. Transposability.

7.1.1. DEFINITION. An orthogonality set Λ is *transposable* if the transpose of every PCD(Λ) is also a PCD(Λ).

7.1.2. PROBLEM. *Decide whether a given orthogonality set is transposable.*

Problem 7.1.2 has a nice visualization when $|\mathcal{A}| = 1$. To determine whether a $v \times v$ $(0,1)$-array is a PCD(Λ), one may draw the corresponding bipartite row-column graph. Λ-orthogonality means that the degrees of (row) vertices on the left come from a restricted set. Furthermore, the number of common neighbors of any pair of vertices on the left comes from a restricted set that depends on the degrees of the two left-hand vertices. The transposability question asks whether, for every bipartite graph whose left-hand vertices have the degree and neighborhood properties, the right-hand vertices do too.

7.1.3. EXAMPLE. Let Λ be the orthogonality set of all $2 \times v$ $(0,1)$-arrays whose row sums are even, and the inner product of the two rows is odd. Hence, a $v \times v$ $(0,1)$-array D is a PCD(Λ) if and only if $DD^\top \equiv J + I \pmod 2$. For example,

$$X = \begin{bmatrix} 1 & 1 & 0 & 0 & 0 \\ 0 & 1 & 1 & 0 & 0 \\ 0 & 1 & 0 & 1 & 0 \\ 0 & 1 & 0 & 0 & 1 \\ 1 & 0 & 1 & 1 & 1 \end{bmatrix}$$

is a PCD(Λ) for $v = 5$. Its row-column graph is given in Figure 7.1. All vertices on the left have even degree, and the common neighborhood sets all contain just one vertex. The same is true for the right-hand vertices; so $X^\top$ is a PCD(Λ).

We show that Λ is transposable. Suppose that D is a PCD(Λ). If v is even then $(J + I)^2 \equiv I$, implying that $J + I$ is invertible modulo 2. But $DJ \equiv 0$; so D is not invertible modulo 2, and we cannot have $DD^\top \equiv J + I$. Consequently v must be odd. Then $(D + J)(D + J)^\top \equiv I$, so $(D + J)^\top (D + J) \equiv I$. Thus $D^\top D \equiv JD + D^\top J + J + I$. Comparing diagonals, we see that $D^\top D$ has zero diagonal modulo 2; i.e., the column sums of D are even. Thus $JD \equiv D^\top J \equiv 0$, and so $D^\top D \equiv J + I$: $D^\top$ is a PCD(Λ).

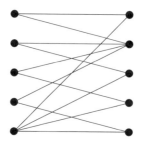

FIGURE 7.1. Example of a row-column graph

7.1.2. Self-duality. Some of the orthogonality sets in Definition 2.13.1 are not transposable. Yet Λ-orthogonality of rows imposes a pairwise condition on columns. For example, if D is a GH($n;G$) then $D^\top$ may not be a GH($n;G$), whereas the conjugate transpose D^*, obtained by transposing D and inverting each entry, is always a GH($n;G$). This sort of example motivates the next definition.

7.1.4. DEFINITION. Let Λ be an orthogonality set with alphabet $\mathcal{A}$. If there is a permutation α of $\{0\} \cup \mathcal{A}$ fixing 0 such that $\alpha(D)^\top = [\alpha(d_{ji})]_{ij}$ is a PCD(Λ) whenever $D = [d_{ij}]_{ij}$ is a PCD(Λ), then Λ is α-*transposable*, or *self-dual*.

Definition 7.1.4 generalizes transposability and conjugate transposability (in the latter case, $\mathcal{A}$ is the element set of a finite group, and α is inversion $x \mapsto x^{-1}$).

The main problem concerning self-duality is as follows.

7.1.5. PROBLEM. *Determine whether a given orthogonality set Λ is self-dual. Find the permutations α such that Λ is α-transposable.*

We will describe a matrix algebra approach to Problems 7.1.2 and 7.1.5 that provides answers for the orthogonality sets in Definition 2.13.1 (all are self-dual; see Theorem 7.3.5 below). Since our results have limited scope, we recommend the following project.

3. RESEARCH PROBLEM. *Develop techniques for solving Problems 7.1.2 and 7.1.5. Obtain conditions for self-duality (or transposability). Characterize the self-dual (or transposable) orthogonality sets.*

7.2. A functional approach to self-duality

The maps δ and λ from Section 4.3 can be used to frame the transposability problems.

Suppose that Λ is an orthogonality set with alphabet $\mathcal{A}$ such that $\delta(\Lambda)$ is non-empty, and every PCD(Λ) has distinct columns. Let $\top$ be the transpose operator on arrays. For $\alpha \in \mathrm{Sym}(\mathcal{A})$, let $\top_\alpha$ be the operator that maps a $(0, \mathcal{A})$-array to its transpose and then applies α to each non-zero entry. We write $\Lambda^\top$ for $\lambda \top \delta(\Lambda)$, and $\Lambda^{\top\alpha}$ for $\lambda \top_\alpha \delta(\Lambda)$. Thus

7.2.1. PROPOSITION. Λ *is α-transposable if and only if $\Lambda^{\top\alpha} \subseteq \Lambda$.*

7.2.2. PROPOSITION. $\top_\alpha : \Lambda \mapsto \Lambda^{\top\alpha}$ *maps orthogonality sets to irredundant orthogonality sets.*

PROOF. Since $\top_\alpha \delta(\Lambda)$ is a non-empty design set, the orthogonality set $\Lambda^{\top\alpha} = \lambda(\top_\alpha \delta(\Lambda))$ is irredundant. $\square$

In light of the above, it is natural to inspect the sequence $\Lambda, \Lambda^{\top\alpha}, \Lambda^{\top^2_\alpha}, \dots$ in the hope of finding k such that $\Lambda^{\top^{k+1}_\alpha} = \Lambda^{\top^k_\alpha}$. If this occurs then $\Lambda^{\top^k_\alpha}$ is α-transposable. Certainly $\Lambda^{\top^l_\alpha} = \Lambda^{\top^k_\alpha}$ for some l and $k < l$. However, l need not be equal to $k + 1$.

7.2.3. EXAMPLE. Figure 7.2 shows an iterative application of $\lambda\top\delta$. Note the eventual repetition: $\Lambda^{\top^4} = \Lambda^{\top^2}$. None of the orthogonality sets $\Lambda^{\top^i}$, $0 \le i \le 3$, is transposable.

$$\Lambda = \left\{ \begin{bmatrix} 1\ 0\ 0 \\ 0\ 1\ 0 \end{bmatrix}, \begin{bmatrix} 1\ 0\ 0 \\ 1\ 1\ 0 \end{bmatrix} \right\}$$

$$\delta(\Lambda) = \left\{ \begin{bmatrix} 1\ 0\ 0 \\ 0\ 1\ 0 \\ 0\ 0\ 1 \end{bmatrix}, \begin{bmatrix} 1\ 0\ 0 \\ 1\ 1\ 0 \\ 0\ 1\ 0 \end{bmatrix} \right\}$$

$$\top\delta(\Lambda) = \left\{ \begin{bmatrix} 1\ 0\ 0 \\ 0\ 1\ 0 \\ 0\ 0\ 1 \end{bmatrix}, \begin{bmatrix} 1\ 1\ 0 \\ 1\ 0\ 1 \\ 0\ 0\ 0 \end{bmatrix} \right\}$$

$$\lambda\top\delta(\Lambda) = \left\{ \begin{bmatrix} 1\ 0\ 0 \\ 0\ 1\ 0 \end{bmatrix}, \begin{bmatrix} 1\ 1\ 0 \\ 1\ 0\ 1 \end{bmatrix}, \begin{bmatrix} 1\ 1\ 0 \\ 0\ 0\ 0 \end{bmatrix} \right\}$$

$$\delta\lambda\top\delta(\Lambda) = \left\{ \begin{bmatrix} 1\ 0\ 0 \\ 0\ 1\ 0 \\ 0\ 0\ 1 \end{bmatrix}, \begin{bmatrix} 1\ 1\ 0 \\ 1\ 0\ 1 \\ 0\ 0\ 0 \end{bmatrix}, \begin{bmatrix} 1\ 1\ 0 \\ 1\ 0\ 1 \\ 0\ 1\ 1 \end{bmatrix} \right\}$$

$$\top\delta\lambda\top\delta(\Lambda) = \left\{ \begin{bmatrix} 1\ 0\ 0 \\ 0\ 1\ 0 \\ 0\ 0\ 1 \end{bmatrix}, \begin{bmatrix} 1\ 1\ 0 \\ 1\ 0\ 0 \\ 0\ 1\ 0 \end{bmatrix}, \begin{bmatrix} 1\ 1\ 0 \\ 1\ 0\ 1 \\ 0\ 1\ 1 \end{bmatrix} \right\}$$

$$(\lambda\top\delta)^2(\Lambda) = \left\{ \begin{bmatrix} 1\ 0\ 0 \\ 0\ 1\ 0 \end{bmatrix}, \begin{bmatrix} 1\ 1\ 0 \\ 1\ 0\ 1 \end{bmatrix}, \begin{bmatrix} 1\ 1\ 0 \\ 1\ 0\ 0 \end{bmatrix} \right\}$$

$$\delta(\lambda\top\delta)^2(\Lambda) = \left\{ \begin{bmatrix} 1\ 0\ 0 \\ 0\ 1\ 0 \\ 0\ 0\ 1 \end{bmatrix}, \begin{bmatrix} 1\ 1\ 0 \\ 1\ 0\ 0 \\ 0\ 1\ 0 \end{bmatrix}, \begin{bmatrix} 1\ 0\ 0 \\ 1\ 1\ 0 \\ 1\ 0\ 1 \end{bmatrix}, \begin{bmatrix} 1\ 1\ 0 \\ 1\ 0\ 1 \\ 0\ 1\ 1 \end{bmatrix} \right\}$$

$$\top\delta(\lambda\top\delta)^2(\Lambda) = \left\{ \begin{bmatrix} 1\ 0\ 0 \\ 0\ 1\ 0 \\ 0\ 0\ 1 \end{bmatrix}, \begin{bmatrix} 1\ 0\ 1 \\ 0\ 1\ 1 \\ 0\ 0\ 0 \end{bmatrix}, \begin{bmatrix} 1\ 1\ 1 \\ 0\ 1\ 0 \\ 0\ 0\ 1 \end{bmatrix}, \begin{bmatrix} 1\ 1\ 0 \\ 1\ 0\ 1 \\ 0\ 1\ 1 \end{bmatrix} \right\}$$

$$(\lambda\top\delta)^3(\Lambda) = \left\{ \begin{bmatrix} 1\ 0\ 0 \\ 0\ 1\ 0 \end{bmatrix}, \begin{bmatrix} 0\ 0\ 0 \\ 1\ 1\ 0 \end{bmatrix}, \begin{bmatrix} 1\ 1\ 0 \\ 1\ 0\ 1 \end{bmatrix}, \begin{bmatrix} 1\ 0\ 0 \\ 1\ 1\ 1 \end{bmatrix} \right\}$$

$$\delta(\lambda\top\delta)^3(\Lambda) = \left\{ \begin{bmatrix} 1\ 0\ 0 \\ 0\ 1\ 0 \\ 0\ 0\ 1 \end{bmatrix}, \begin{bmatrix} 1\ 0\ 1 \\ 0\ 1\ 1 \\ 0\ 0\ 0 \end{bmatrix}, \begin{bmatrix} 1\ 1\ 1 \\ 0\ 1\ 0 \\ 0\ 0\ 1 \end{bmatrix}, \begin{bmatrix} 1\ 1\ 0 \\ 1\ 0\ 1 \\ 0\ 1\ 1 \end{bmatrix} \right\}$$

$$\top\delta(\lambda\top\delta)^3(\Lambda) = \left\{ \begin{bmatrix} 1\ 0\ 0 \\ 0\ 1\ 0 \\ 0\ 0\ 1 \end{bmatrix}, \begin{bmatrix} 0\ 1\ 1 \\ 0\ 1\ 0 \\ 0\ 0\ 1 \end{bmatrix}, \begin{bmatrix} 1\ 1\ 0 \\ 1\ 0\ 1 \\ 1\ 0\ 0 \end{bmatrix}, \begin{bmatrix} 1\ 1\ 0 \\ 1\ 0\ 1 \\ 0\ 1\ 1 \end{bmatrix} \right\}$$

$$(\lambda\top\delta)^4(\Lambda) = \left\{ \begin{bmatrix} 1\ 0\ 0 \\ 0\ 1\ 0 \end{bmatrix}, \begin{bmatrix} 1\ 1\ 0 \\ 1\ 0\ 1 \end{bmatrix}, \begin{bmatrix} 1\ 1\ 0 \\ 1\ 0\ 0 \end{bmatrix} \right\}$$

FIGURE 7.2. Examples of applying the operator $\lambda\top\delta$

7.3. Conjugate equivalence operations

Let Λ be an orthogonality set with alphabet $\mathcal{A}$ such that $\delta(\Lambda) \neq \emptyset$. The group Π_Λ consists of the permutations of $\mathcal{A}$ that may be applied to the non-zero entries of a $\mathrm{PCD}(\Lambda)$ in any one of its rows or columns to obtain another $\mathrm{PCD}(\Lambda)$. Aside from these there are the global equivalence operations, which together with Π_Λ form the group Φ_Λ of permutations that may be applied to *every* entry of a $\mathrm{PCD}(\Lambda)$ to obtain a $\mathrm{PCD}(\Lambda)$. We show below that when Λ is self-dual, this group may be doubled in size by adding the permutations α for which Λ is α-transposable.

7.3.1. DEFINITION. A *conjugate equivalence operation* of Λ is a permutation α of $\{0\} \cup \mathcal{A}$ fixing 0 such that $\alpha(D^\top)$ is a $\mathrm{PCD}(\Lambda)$ whenever D is a $\mathrm{PCD}(\Lambda)$.

7.3.2. LEMMA. *If non-empty, the set of conjugate equivalence operations is a single coset $\alpha\Phi_\Lambda$ of Φ_Λ in $\mathrm{Sym}(\{0\} \cup \mathcal{A})$.*

PROOF. If D is a $\mathrm{PCD}(\Lambda)$ and α, β are conjugate equivalence operations then $\alpha\beta(D) = \alpha(\beta(D)^\top)^\top$ is a $\mathrm{PCD}(\Lambda)$. Thus $\alpha\beta \in \Phi_\Lambda$, and $\beta \in \alpha^{-1}\Phi_\Lambda = \alpha\Phi_\Lambda$. $\square$

7.3.3. DEFINITION. Let Ψ_Λ denote the group of permutations α such that $\alpha(D) \in \delta(\Lambda)$ for all $D \in \delta(\Lambda)$, or $\alpha(D)^\top \in \delta(\Lambda)$ for all $D \in \delta(\Lambda)$.

7.3.4. THEOREM. Ψ_Λ *consists of the conjugate equivalence operations and the elements of Φ_Λ. Moreover,*

(1) Φ_Λ *has index 2 in Ψ_Λ if and only if Λ is self-dual, but not transposable;*

(2) *if Λ is not self-dual, or is transposable, then $\Psi_\Lambda = \Phi_\Lambda$.*

PROOF. By Lemma 7.3.2, $|\Psi_\Lambda : \Phi_\Lambda| \leq 2$. So (1) implies (2). Since $|\Psi_\Lambda : \Phi_\Lambda| = 2$ if and only if Λ has non-identity conjugate equivalence operations (all in the same coset of Φ_Λ), and Λ is transposable if and only if the identity is a conjugate equivalence operation, (1) follows. $\square$

In general it seems to be difficult to determine whether an orthogonality set is self-dual. Nonetheless, we do know Ψ_Λ for the orthogonality sets Λ in Chapter 2.

7.3.5. THEOREM. *In cases 1–8 of Definition 2.13.1, and in cases 9–11 when G is abelian, the orthogonality sets are transposable. The orthogonality sets in cases 9–11 for non-abelian G are conjugate transposable but not transposable, and Ψ_Λ is obtained by adjoining to Φ_Λ the anti-automorphism $x \mapsto x^{-1}$ of G.*

Nearly all of Theorem 7.3.5 was proved in Section 2.1. We still have to prove that $\Lambda_{\mathrm{GW}(v,k;G)}$ and $\Lambda_{\mathrm{BGW}(v,k,\lambda;G)}$ are self-dual. This is done in the next section.

7.4. A matrix algebra approach to transposability and self-duality

7.4.1. LEMMA. *Let $\mathcal{R}$ be a commutative ring. Suppose that $AB = kI_v$ where $A, B \in \mathrm{Mat}(v, \mathcal{R})$ and $k \neq 0$ is not a zero divisor in $\mathcal{R}$. Then $BA = kI_v$.*

PROOF. Let C be the cofactor matrix for A, so $AC = CA = \det(A)I_v$. Hence $\det(A)B = CAB = kC$, and then

$$\det(A)AB = A\det(A)B = kAC = kCA = \det(A)BA,$$

implying that $\det(A)(AB - BA) = 0$. Since $\det(A)$ divides k^v, $\det(A)$ is not a zero divisor. Therefore $AB = BA$. $\square$

We turn now to the proof of Theorem 7.3.5. We select the weakest hypotheses necessary to derive some basic linear algebra facts that will finish the job.

Let Λ be an orthogonality set with alphabet $\mathcal{A}$. Suppose that $\mathcal{S}$ is a ring containing $\mathcal{A}$, and α is a permutation of $\{0\} \cup \mathcal{A}$ such that $\alpha(0) = 0$ and $\alpha^{-1} = \alpha$. Suppose further that $X \in \Lambda$ if and only if

$$(7.1) \qquad\qquad X\alpha(X)^\top = kI_2$$

over $\mathcal{S}$, where $k \neq 0$ is not a zero divisor in $\mathcal{S}$. If $\mathcal{S}$ is commutative, then Λ is α-transposable by Lemma 7.4.1: for if D is a PCD(Λ) then

$$\alpha(D)^\top \alpha(\alpha(D)^\top)^\top = \alpha(D)^\top D = D\alpha(D)^\top = kI_v.$$

This result establishes transposability for the orthogonality sets as in (2)–(8) of Definition 2.13.1 (which have the Gram Property over a commutative ambient ring). A simplified version of the argument in the proof of Lemma 7.4.1 was used earlier (Section 2.10) to prove that $\Lambda_{\mathrm{GH}(n;G)}$ is transposable for abelian G. We now verify the rest of the transposability assertion in Theorem 7.3.5.

7.4.2. THEOREM. *Suppose that G is abelian. Then $\Lambda_{\mathrm{GH}(n;G)}$, $\Lambda_{\mathrm{GW}(v,k;G)}$, and $\Lambda_{\mathrm{BGW}(v,k,\lambda;G)}$ are transposable.*

PROOF. Let $\alpha : \mathbb{Z}[G]/\mathbb{Z}G \to \mathbb{Z}[G]/\mathbb{Z}G$ be the map induced by the inversion map $G \to G$ extended to $\mathbb{Z}[G]$. Since G is abelian, α is a ring homomorphism. The condition for membership in the first or second orthogonality set is (7.1) viewed as an equation over $\mathbb{Z}[G]/\mathbb{Z}G$. For $\Lambda_{\mathrm{BGW}(v,k,\lambda;G)}$ we have the extra requirement that every pair of different rows of X has precisely λ columns not containing a zero. By Lemma 7.4.1, $\Lambda_{\mathrm{GH}(n;G)}$ and $\Lambda_{\mathrm{GW}(v,k;G)}$ are α-transposable. Together with transposability of $\Lambda_{\mathrm{SBIBD}(v,k,\lambda)}$, this in turn gives that $\Lambda_{\mathrm{BGW}(v,k,\lambda;G)}$ is α-transposable. As α induces a homomorphism of order 2 from any matrix ring over $\mathbb{Z}[G]/\mathbb{Z}G$ into itself, transposability in all three cases follows. $\square$

We move on finally to showing why $\Lambda_{\mathrm{GW}(v,k;G)}$ and $\Lambda_{\mathrm{BGW}(v,k,\lambda;G)}$ are conjugate transposable when G is non-abelian.

7.4.3. LEMMA. *Let $\mathbb{F}$ be a field and let $\mathcal{R}$ be an $\mathbb{F}$-algebra of dimension m. Then there is an injective ring homomorphism $\psi : \mathrm{Mat}(n, \mathcal{R}) \to \mathrm{Mat}(nm, \mathbb{F})$ that maps each scalar kI_n, $k \in \mathbb{F}$, to kI_{nm}.*

PROOF. Choose an $\mathbb{F}$-basis $x_1, \ldots, x_m$ of $\mathcal{R}$. If $r \in \mathcal{R}$ then $\psi(r) \in \mathrm{Mat}(m, \mathbb{F})$ will be the matrix with ith column $(a_1, \ldots, a_m)^\top$ where $rx_i = \sum_{j=1}^m a_j x_j$. This defines a ring homomorphism $\psi : \mathcal{R} \to \mathrm{Mat}(m, \mathbb{F})$. Since $\psi(r) = \psi(s)$ if and only if $rx_i = sx_i$ for all i, ψ is injective. Also $\psi(r) = rI_m$ if $r \in \mathbb{F}$. Then we define ψ in general by $\psi : [x_{ij}] \mapsto [\psi(x_{ij})]$. $\square$

7.4.4. COROLLARY. *With the notation of Lemma 7.4.3, suppose that $A, B \in \mathrm{Mat}(n, \mathcal{R})$ and $AB = kI_n$ for some non-zero $k \in \mathbb{F}$. Then $BA = kI_n$.*

PROOF. By Lemma 7.4.3, $\psi(A)\psi(B) = kI_{nm}$. Thus $\psi(A)$, $\psi(B)$ are invertible and $\psi(B)\psi(A) = kI_{nm}$, i.e., $\psi(AB) = \psi(BA)$. Since ψ is injective on $\mathrm{Mat}(n, \mathcal{R})$, the result follows. $\square$

Thus, disregarding commutativity or otherwise of $\mathcal{S}$, the orthogonality sets Λ defined by (7.1) and the attendant hypotheses are α-transposable. This holds in particular for the remaining orthogonality sets in Theorem 7.3.5.

7.4.5. THEOREM. *Suppose that G is non-abelian. Then $\Lambda_{\mathrm{GH}(n;G)}$, $\Lambda_{\mathrm{GW}(v,k;G)}$, and $\Lambda_{\mathrm{BGW}(v,k,\lambda;G)}$ are conjugate transposable but not transposable.*

PROOF. We mimic the proof of Theorem 7.4.2, using Corollary 7.4.4 with $\mathcal{R} = \mathbb{Q}[G]/\mathbb{Q}G$ rather than Lemma 7.4.1. Reasoning as in the discussion before Theorem 2.10.7 proves that the orthogonality sets are not transposable. $\qquad\square$

7.5. A different kind of transposable orthogonality set

In this section we give an example of a transposable orthogonality set that does not satisfy (7.1) for any ring $\mathcal{S}$.

Let $n = 2m + 1$ and $\mathcal{A} = \{a, b\}$. Denote by Λ_n the set of $2 \times n$ $\mathcal{A}$-arrays such that the number of columns with distinct entries is one less or one more than the number of columns with identical entries. Thus Λ_n consists of the arrays

$$
\begin{bmatrix}
\overbrace{a \quad a \quad \cdots \quad a}^{w} & \overbrace{a \quad a \quad \cdots \quad a}^{x} & \overbrace{b \quad b \quad \cdots \quad b}^{y} & \overbrace{b \quad b \quad \cdots \quad b}^{z} \\
a \quad a \quad \cdots \quad a & b \quad b \quad \cdots \quad b & a \quad a \quad \cdots \quad a & b \quad b \quad \cdots \quad b
\end{bmatrix}
$$

up to column permutation, where $w, x, y, z \geq 0$ and $w + z = (x + y) \pm 1$. If we take $a = 1$ and $b = -1$ in the ring $\mathbb{Z}$, then a $2 \times n$ $(1, -1)$-array X is in Λ_n if and only if

$$(7.2) \qquad XX^\top = nI_2 \pm (J_2 - I_2).$$

In other words, a $\mathrm{PCD}(\Lambda_n)$ is an $n \times n$ $(1, -1)$-array such that the inner products of its distinct rows are all ± 1.

7.5.1. EXAMPLE. The arrays

$$
\begin{bmatrix}
1 & -1 & -1 \\
-1 & 1 & -1 \\
-1 & -1 & 1
\end{bmatrix}, \qquad
\begin{bmatrix}
-1 & 1 & 1 & 1 & 1 \\
1 & -1 & 1 & 1 & 1 \\
1 & 1 & -1 & 1 & 1 \\
1 & 1 & 1 & -1 & 1 \\
1 & 1 & 1 & 1 & -1
\end{bmatrix}, \qquad
\begin{bmatrix}
1 & 1 & 1 & -1 & 1 & -1 & -1 \\
-1 & 1 & 1 & 1 & -1 & 1 & -1 \\
-1 & -1 & 1 & 1 & 1 & -1 & 1 \\
1 & -1 & -1 & 1 & 1 & 1 & -1 \\
-1 & 1 & -1 & -1 & 1 & 1 & 1 \\
1 & -1 & 1 & -1 & -1 & 1 & 1 \\
1 & 1 & -1 & 1 & -1 & -1 & 1
\end{bmatrix}
$$

are circulant $\mathrm{PCD}(\Lambda_n)$ for $n = 3, 5, 7$.

We will show that Λ_n (for $\{a, b\} = \{1, -1\}$) is transposable for all odd n. Note that if there is a Hadamard matrix of order $n + 1$ then $v_{\max}(\Lambda_n) = n + 1$. So this is an example of a transposable orthogonality set where $v_{\max}$ exceeds v.

Equation (7.2) implies that Λ_n has the Gram Property over $\mathbb{Z}$. However, since $\mathrm{Gram}_{\mathbb{Z}}(\Lambda_n)$ contains two elements, (7.2) does not have the form of (7.1). We take a moment to prove that Λ_n does not satisfy (7.1) over any commutative involutory ring $\mathcal{S}$. Note that Λ_n contains

$$
\begin{bmatrix}
\overbrace{a \quad a \quad \cdots \quad a}^{w} & \overbrace{a \quad a \quad \cdots \quad a}^{x} \\
a \quad a \quad \cdots \quad a & b \quad b \quad \cdots \quad b
\end{bmatrix}
$$

for $(w, x) = (m, m + 1)$ and $(w, x) = (m + 1, m)$. Suppose that $\{XX^* \mid X \in \Lambda_n\} = \{kI_2\}$ where $k \neq 0$ is not a zero divisor in $\mathcal{S}$. Working over $\mathcal{S}$ we then have

$$
\begin{bmatrix}
naa^* & waa^* + xab^* \\
waa^* + xba^* & waa^* + xbb^*
\end{bmatrix} =
\begin{bmatrix}
k & 0 \\
0 & k
\end{bmatrix}.
$$

Thus $k = naa^*$, and so a is not a zero divisor. Also $maa^* + (m+1)ab^* = 0 = (m+1)aa^* + mab^*$, giving $a(a^* - b^*) = 0$. Since a is not a zero divisor, we must have $a^* = b^*$, and hence $a = b$. This contradiction proves that Λ_n does not satisfy (7.1) for any suitable ring $\mathcal{S}$.

Henceforth Λ_n is the set of all $(1, -1)$-matrices X satisfying (7.2). We determine the local equivalence operations for Λ_n.

7.5.2. THEOREM. $\Pi_{\Lambda_n} = \Pi_{\Lambda_n}^{\text{row}} = \Pi_{\Lambda_n}^{\text{col}} = \langle -1 \rangle$.

PROOF. Negating a row changes the sign of its inner products with other rows. Negating a column does not change the inner products of rows. $\square$

Next, we uncover an interesting special property of $\text{PCD}(\Lambda_n)$s.

7.5.3. THEOREM. A $\text{PCD}(\Lambda_n)$, D, is Λ_n-equivalent to a $\text{PCD}(\Lambda_n)$ whose every pair of distinct rows has inner product $(-1)^{(n+3)/2}$.

PROOF. We make a new array $X \approx_{\Lambda_n} D$ from D by first negating each column starting with a -1, then negating each non-initial row whose inner product with the first row is 1. Denote the ith row of X by r_i, and the inner product of r_i and r_j by s_{ij}. If x_{ab} $(a, b \in \{0, 1\})$ is the number of times $(-1)^b$ appears in r_j under $(-1)^a$ in r_i, $1 < i < j$, then

$$x_{00} + x_{01} + x_{10} + x_{11} = n,$$
$$x_{00} + x_{01} - x_{10} - x_{11} = -1,$$
$$x_{00} - x_{01} + x_{10} - x_{11} = -1,$$
$$x_{00} - x_{01} - x_{10} + x_{11} = s_{ij}.$$

Therefore $s_{ij} = 4x_{00} + 2 - n$, and so

$$s_{ij} = \begin{cases} -1 & \text{if } n \equiv 3 \pmod 4 \\ 1 & \text{if } n \equiv 1 \pmod 4 \end{cases}$$

for $1 < i < j \le n$, i.e., $s_{ij} = (-1)^{(n+3)/2}$. If $n \equiv 3 \pmod 4$ then we are done; else we negate the first row of X. $\square$

We will now prove that Λ_n is transposable. Let $\mathcal{X}_{m,n}$ be the set of all $m \times n$ $(1, -1)$-matrices, and for $X = [x_{ij}]_{ij} \in \mathcal{X}_{m,n}$ let

$$(7.3) \qquad \pi(X)_{(i,j)} = \sum_{l=1}^{n} x_{il} x_{jl}, \ 1 \le i < j \le m, \quad \text{and} \quad \eta(X) = \pi(X) \cdot \pi(X).$$

That is, $\pi(X)$ is the vector of length $\binom{m}{2}$ whose (i, j)th component for $j > i$ is the inner product of rows i and j of X, and $\eta(X)$ is the inner product of $\pi(X)$ with itself. The following lemma is taken from de Launey and Levin's paper [52]. We delay its proof until after the theorem.

7.5.4. LEMMA. $\eta(X) = \frac{1}{2}mn(m - n) + \eta(X^\top)$.

7.5.5. THEOREM. Λ_n is transposable for all odd n.

PROOF. Observe that D is a $\text{PCD}(\Lambda_n)$ if and only if $D \in \mathcal{X}_{n,n}$ and $\eta(D) = \binom{n}{2}$. (Here we use the fact that, because n is odd, every inner product of distinct rows of $D \in \mathcal{X}_{n,n}$ has absolute value at least 1.) By Lemma 7.5.4, $\eta(D^\top) = \eta(D)$. Thus $D^\top$ is a $\text{PCD}(\Lambda_n)$ if and only if D is a $\text{PCD}(\Lambda_n)$. $\square$

PROOF OF LEMMA 7.5.4. Let $w(y)$ denote the sum of the components of a real vector y. Note that $x \cdot y = w(x \wedge y)$. If y has length m, let $I(y)$ be the vector of length $M := \binom{m}{2}$ whose (i,j)th component, $1 \leq i < j \leq m$, is

$$(7.4) \qquad\qquad I(y)_{(i,j)} = y_i y_j.$$

Equations (7.3) and (7.4) imply that $\pi(X) = \sum_{l=1}^{n} I(c_l)$, where c_l is the lth column of X. Then

$$\eta(X) = \pi(X) \cdot \pi(X) = \sum_{l=1}^{n} (I(c_l) \cdot I(c_l)) + 2 \sum_{1 \leq l < r \leq n} I(c_l) \cdot I(c_r).$$

Since $I(c_l) \wedge I(c_r) = I(c_l \wedge c_r)$, we get

$$I(c_l) \cdot I(c_l) = w(I(c_l \wedge c_l)) = M$$

and hence

$$(7.5) \qquad\qquad \eta(X) = nM + 2 \sum_{1 \leq l < r \leq n} w(I(c_l \wedge c_r)).$$

For a $(1, -1)$-vector y of length m, let s be the number of components of y that are 1, and let t be the number of components that are -1. Then

$$w(I(y)) = \binom{s}{2} + \binom{t}{2} - st = \frac{(s-t)^2}{2} - \frac{(s+t)}{2} = \tfrac{1}{2}(w(y)^2 - m).$$

Now by (7.5),

$$\eta(X) = nM + 2 \sum_{1 \leq l < r \leq n} \tfrac{1}{2}(w(c_l \wedge c_r)^2 - m)$$

$$= n\binom{m}{2} - m\binom{n}{2} + \eta(X^{\top})$$

$$= \tfrac{1}{2}mn(m-n) + \eta(X^{\top}),$$

as claimed. $\square$

CHAPTER 8

New Designs from Old

This chapter explores ways of obtaining larger designs from smaller known designs.

In the first section we discuss composition. We allow designs over different and novel orthogonality sets to be composed. A notable feature of composition is that it destroys symmetry and equivalence, resulting in a plethora of inequivalent designs with no non-trivial automorphisms.

The second section is on 'transference': our term for the general situation where designs over seemingly quite different orthogonality sets are combinatorially equivalent. Some instances of transference are not at all obvious.

8.1. Composition

Composition often involves taking a sum of Kronecker products.

8.1.1. The Kronecker product. It is very well-known that the Kronecker product of two Hadamard matrices H_1 and H_2 of orders n_1 and n_2 is a Hadamard matrix of order $n_1 n_2$:

$$
\begin{aligned}
(H_1 \otimes H_2)(H_1 \otimes H_2)^\top &= (H_1 \otimes H_2)(H_1^\top \otimes H_2^\top) \\
&= H_1 H_1^\top \otimes H_2 H_2^\top \\
&= n_1 I_{n_1} \otimes n_2 I_{n_2} \\
&= n_1 n_2 I_{n_1 n_2}.
\end{aligned}
$$

There are many variations of this method.

8.1.1. EXAMPLE. The first design below is a complex Hadamard matrix, and the second is an orthogonal design of type $(1, 1)$.

$$
\begin{bmatrix} 1 & i \\ i & 1 \end{bmatrix}, \quad
\begin{bmatrix} x_1 & x_2 \\ x_2 & -x_1 \end{bmatrix}.
$$

We have

$$
\begin{bmatrix} 1 & i \\ i & 1 \end{bmatrix} \otimes
\begin{bmatrix} x_1 & x_2 \\ x_2 & -x_1 \end{bmatrix} =
\begin{bmatrix}
x_1 & x_2 & ix_1 & ix_2 \\
x_2 & -x_1 & ix_2 & -ix_1 \\
ix_1 & ix_2 & x_1 & x_2 \\
ix_2 & -ix_1 & x_2 & -x_1
\end{bmatrix},
$$

which is a complex orthogonal design of type $(2, 2)$.

We recall from Chapter 6 the definition of Kronecker product of orthogonality sets.

8.1.2. CONSTRUCTION. For $i = 1, 2$, let Λ_i be an orthogonality set of $2 \times v_i$ arrays ($v_i \geq 2$) with alphabet $\mathcal{A}_i$ and ambient ring $\mathcal{R}_i$. Suppose that $\mathcal{R}_3$ is a ring containing $\mathcal{R}_1$ and $\mathcal{R}_2$ as subrings. Define $\mathcal{A}_3 = \{a_1 a_2 \mid a_1 \in \mathcal{A}_1, a_2 \in \mathcal{A}_2\} \subseteq \mathcal{R}_3$, and let Λ_3 be the Kronecker product $\Lambda_1 \otimes \Lambda_2$ over $\mathcal{R}_3$ as in Definition 6.3.4.

8.1.3. EXAMPLE. Let $\mathcal{R}_1 = \mathbb{Z}[\mathrm{i}]$ and $\mathcal{R}_2 = \mathbb{Z}[x_1, x_2]$; then in Example 8.1.1 we may take $\mathcal{R}_3$ to be $\mathcal{R}_1[x_1, x_2]$.

The following theorem formalizes composition of designs via the Kronecker product.

8.1.4. THEOREM. *In the notation of Construction 8.1.2, if D_i is a $\mathrm{PCD}(\Lambda_i)$ for $i = 1, 2$, and $D_3 = D_1 \otimes D_2$ has no repeated rows, then D_3 is a $\mathrm{PCD}(\Lambda_3)$.*

PROOF. Index the rows of D_3 by the ordered pairs (i, j) where $1 \leq i \leq v_1$ and $1 \leq j \leq v_2$. Let W be any $2 \times v_1 v_2$ submatrix of D_3, with rows indexed by (i_1, j_1) and (i_2, j_2). Then

  (a) $i_1 \neq i_2$ and $j_1 = j_2$, or
  (b) $i_1 = i_2$ and $j_1 \neq j_2$, or
  (c) $i_1 \neq i_2$ and $j_1 \neq j_2$.

We now define integers i_3, i_4, j_3, j_4, where $1 \leq i_3, i_4 \leq v_1$ and $1 \leq j_3, j_4 \leq v_2$. In case (a), let $i_3 = i_1$, $i_4 = i_2$, $j_3 = j_1$, and $j_4 \neq j_1$. In case (b), let $i_3 = i_1$, $i_4 \neq i_1$, $j_3 = j_1$, and $j_4 = j_2$. In case (c), let $i_3 = i_1$, $i_4 = i_2$, $j_3 = j_1$, and $j_4 = j_2$. Then in all cases W is a $2 \times v_1 v_2$ submatrix of $X \otimes Y$, where X is the $2 \times v_1$ submatrix of D_1 with rows indexed by i_3 and i_4, and Y is the $2 \times v_2$ submatrix of D_2 with rows indexed by j_3 and j_4. Since $X \in \Lambda_1$ and $Y \in \Lambda_2$, this means that $W \in \Lambda_3$. Hence, D_3 is a $\mathrm{PCD}(\Lambda_3)$. $\square$

8.1.2. Plug-in techniques. There is an old and celebrated method due to Williamson for obtaining larger orthogonal designs from smaller ones, that first appeared in [**157**]. This construction was generalized by Seberry [**156**] to prove her landmark theorem on the asymptotic existence of Hadamard matrices. The plug-in technique is also used widely in the book [**71**] on orthogonal designs by Geramita and Seberry.

Williamson's construction is stated in the following theorem.

8.1.5. THEOREM. *If W, X, Y, Z are symmetric circulant $(1, -1)$-matrices such that*

(8.1) $$W W^\top + X X^\top + Y Y^\top + Z Z^\top = 4t I_t,$$

then

$$H = \begin{bmatrix} W & X & Y & Z \\ X & -W & Z & -Y \\ Y & -Z & -W & X \\ Z & Y & -X & -W \end{bmatrix}$$

is a Hadamard matrix of order $4t$.

8.1.6. EXAMPLE. Let

$$W = \begin{bmatrix} 1 & 1 & 1 \\ 1 & 1 & 1 \\ 1 & 1 & 1 \end{bmatrix} \quad \text{and} \quad X = Y = Z = \begin{bmatrix} -1 & 1 & 1 \\ 1 & -1 & 1 \\ 1 & 1 & -1 \end{bmatrix}.$$

Then

$$\begin{bmatrix}
1 & 1 & 1 & -1 & 1 & 1 & -1 & 1 & 1 & -1 & 1 & 1 \\
1 & 1 & 1 & 1 & -1 & 1 & 1 & -1 & 1 & 1 & -1 & 1 \\
1 & 1 & 1 & 1 & 1 & -1 & 1 & 1 & -1 & 1 & 1 & -1 \\
-1 & 1 & 1 & -1 & -1 & -1 & -1 & 1 & 1 & 1 & -1 & -1 \\
1 & -1 & 1 & -1 & -1 & -1 & 1 & -1 & 1 & -1 & 1 & -1 \\
1 & 1 & -1 & -1 & -1 & -1 & 1 & 1 & -1 & -1 & -1 & 1 \\
-1 & 1 & 1 & 1 & -1 & -1 & -1 & -1 & -1 & -1 & 1 & 1 \\
1 & -1 & 1 & -1 & 1 & -1 & -1 & -1 & -1 & 1 & -1 & 1 \\
1 & 1 & -1 & -1 & -1 & 1 & -1 & -1 & -1 & 1 & 1 & -1 \\
-1 & 1 & 1 & -1 & 1 & 1 & 1 & -1 & -1 & -1 & -1 & -1 \\
1 & -1 & 1 & 1 & -1 & 1 & -1 & 1 & -1 & -1 & -1 & -1 \\
1 & 1 & -1 & 1 & 1 & -1 & -1 & -1 & 1 & -1 & -1 & -1
\end{bmatrix}$$

is a Hadamard matrix of order 12.

The following remark contains a proof of Theorem 8.1.5.

8.1.7. REMARK. We have

$$HH^\top = \begin{bmatrix}
W & X & Y & Z \\
X & -W & Z & -Y \\
Y & -Z & -W & X \\
Z & Y & -X & -W
\end{bmatrix}\begin{bmatrix}
W^\top & X^\top & Y^\top & Z^\top \\
X^\top & -W^\top & -Z^\top & Y^\top \\
Y^\top & Z^\top & -W^\top & -X^\top \\
Z^\top & -Y^\top & X^\top & -W^\top
\end{bmatrix}$$

$$= \begin{bmatrix}
S & -T & -U & -V \\
T & S & -V & U \\
U & V & S & -T \\
V & -U & T & S
\end{bmatrix},$$

where

$$S = WW^\top + XX^\top + YY^\top + ZZ^\top, \quad T = XW^\top - WX^\top + ZY^\top - YZ^\top,$$
$$U = YW^\top - ZX^\top - WY^\top + XZ^\top, \quad V = ZW^\top + YX^\top - XY^\top - WZ^\top.$$

Thus H is Hadamard if and only if W, X, Y, Z are $(1, -1)$-matrices such that $S = 4tI_t$ and $T = U = V = 0$. This is certainly all true if W, X, Y, Z are commuting symmetric $(1, -1)$-matrices satisfying (8.1). Note that (forward) circulant matrices commute.

Williamson's construction fails for some odd t: see [85]. However, to obtain $HH^\top = 4tI_{4t}$, any W, X, Y, Z such that

(8.2)
$$WW^\top + XX^\top + YY^\top + ZZ^\top = 4tI_t, \quad XW^\top - WX^\top + ZY^\top - YZ^\top = 0,$$
$$YW^\top - ZX^\top - WY^\top + XZ^\top = 0, \quad ZW^\top + YX^\top - XY^\top - WZ^\top = 0$$

will do. It is not known whether there exist $(1, -1)$-matrices satisfying (8.2) for all odd t.

A *substitution scheme* is a general form of the plug-in technique. One begins with a $v \times v$ $(0, \mathcal{A})$-array $D = [d_{ij}]$, called the *template array*, and a set of *plug-in matrices* that depend on how D is interpreted.

We interpret D as a PCD(Λ) where Λ has alphabet $\mathcal{A}$. Let $\mathcal{R}$ be an ambient ring for Λ with involution $*$, row group $R \cong \Pi_\Lambda^{\mathrm{row}}$, and column group $C \cong \Pi_\Lambda^{\mathrm{col}}$. Pick a complete and irredundant set of orbit representatives $\mathcal{T}$ for the action of Π_Λ on $\mathcal{A}$. So each non-zero entry d_{ij} of D may be written as $r_{ij}t_{ij}c_{ij}$, where $r_{ij} \in R$, $c_{ij} \in C$, and $t_{ij} \in \mathcal{T}$. Suppose that $\mathcal{X} = \{X_t \mid t \in \mathcal{T}\}$ is a set of plug-in matrices over some ring $\mathcal{S}$ with involution $*$. We require $\mathcal{S}$-matrices M_r, M_c such that the assignments

$$rtc \mapsto M_r X_t M_c, \qquad c^{-1} t^* r^{-1} \mapsto M_c^{-1} X_t^* M_r^{-1}$$

extend to a homomorphism from $\mathcal{R}$ to the matrix ring over $\mathcal{S}$ generated by M_r, M_c, X_t, X_t^* ($r \in R$, $c \in C$, $t \in \mathcal{T}$). The composite design is the block matrix

$$D(\mathcal{X}) = [M_{r_{ij}} X_{t_{ij}} M_{c_{ij}}].$$

We control the Grammian of $D(\mathcal{X})$ by placing constraints on the plug-in matrices.

Now we look at instances of this versatile method.

Williamson's array may be regarded as the following $\mathrm{OD}(4; 1^4)$:

$$D(w, x, y, z) = \begin{bmatrix} w & x & y & z \\ x & -w & z & -y \\ y & -z & -w & x \\ z & y & -x & -w \end{bmatrix}.$$

In this case, $\mathcal{A} = \{\pm w, \pm x, \pm y, \pm z\}$, $\Pi_\Lambda^{\mathrm{row}} = \Pi_\Lambda^{\mathrm{col}} = \langle -1 \rangle$, and $\mathcal{T} = \{w, x, y, z\}$. The usual ambient ring for this type of design is $\mathcal{R} = \mathbb{Z}[w, x, y, z]$, with identity involution. Over $\mathcal{R}$,

(8.3) $D(w, x, y, z) D(w, x, y, z)^* = (w^2 + x^2 + y^2 + z^2) I_4.$

Put $M_{-1} = -I$, and choose plug-in matrices W, X, Y, Z over an involutory ring $\mathcal{S}$ such that

(8.4) $W^* = W, \quad X^* = X, \quad Y^* = Y, \quad Z^* = Z,$

$$WX = XW, \quad WY = YW, \quad WZ = ZW, \quad XY = YX, \quad XZ = ZX, \quad YZ = ZY.$$

Then, over $\mathcal{S}$,

$$D(W, X, Y, Z) \, D(W, X, Y, Z)^* = I_4 \otimes (W^2 + X^2 + Y^2 + Z^2).$$

The array $D(w, x, y, z)$ together with (8.4) comprise the substitution scheme. If we require

$$D(W, X, Y, Z) \, D(W, X, Y, Z)^* = 4t I_{4t}$$

then we would add the constraint $W^2 + X^2 + Y^2 + Z^2 = 4t I_t$.

The monoid ring $\mathbb{Z}[M]$ with involution $*$, where M is the free monoid generated by $w, x, y, z, w^*, x^*, y^*, z^*$, is also an ambient ring for D. Each diagonal entry of the Grammian of D is equal to $ww^* + xx^* + yy^* + zz^*$, and the off-diagonal entries are in

$$\mathrm{df}(D) = \pm \{xw^* - wx^* + zy^* - yz^*, \; yw^* - zx^* - wy^* + xz^*, \; zw^* + yx^* - xy^* - wz^*\}.$$

We may treat D as a $\mathrm{PCD}(\Lambda)$ where rows are Λ-orthogonal if their inner product lies in $\mathrm{df}(D)$. If $\mathcal{I}$ is the ideal of $\mathbb{Z}[M]$ generated by $\mathrm{df}(D)$, then over $\mathbb{Z}[M]/\mathcal{I}$ we have (8.3) as before, with plug-in conditions (8.2).

Note that (8.4) implies the three homogeneous equations in (8.2). The perhaps rarer plug-in matrices satisfying (8.4) are more useful because they may be plugged into any $\mathrm{OD}(v; k^4)$.

There are weaker plug-in conditions that will work for any $\mathrm{OD}(v; k^4)$. Let $\mathcal{R}$ be a ring of characteristic zero with involution $*$ containing w, x, y, z, such that

$$wx^* = xw^*, \; wy^* = yw^*, \; wz^* = zw^*, \; xy^* = yx^*, \; xz^* = zx^*, \; yz^* = zy^*.$$

The plug-in conditions are now

(8.5) $WW^* + XX^* + YY^* + ZZ^* = 4tI_t;$

(8.6) $WX^* = XW^*,\ WY^* = YW^*,\ WZ^* = ZW^*,\ XY^* = YX^*,$

  $XZ^* = ZX^*,\ YZ^* = ZY^*.$

If D is an OD$(v; k^4)$, and W, X, Y, Z satisfy (8.5) and (8.6), then the Grammian of $D(W, X, Y, Z)$ is $4ktI_t$.

Next we consider the effect of changing the target orthogonality set for $D(\mathcal{X})$. Suppose that we want to make an OD$(12; 9, 2, 1)$ using Williamson's array. It suffices to work over the polynomial ring $\mathbb{Z}[a, b, c]$, for which $*$ is the identity. We must have plug-in matrices W, X, Y, Z such that

$$D(\mathcal{X})D(\mathcal{X})^* = (9a^2 + 2b^2 + c^2)I_{12}$$

and

$$WW^* + XX^* + YY^* + ZZ^* = (9a^2 + 2b^2 + c^2)I_3.$$

However, there is a choice for the other constraints on W, X, Y, Z. The weakest are (8.2), and the strongest are (8.4). Equations (8.6) offer an intermediate possibility. The following circulant matrices satisfy (8.6) (but not (8.4)):

$$W = \begin{bmatrix} a & a & a \\ a & a & a \\ a & a & a \end{bmatrix}, \qquad X = \begin{bmatrix} c & -a & a \\ a & c & -a \\ -a & a & c \end{bmatrix}, \qquad Y = Z = \begin{bmatrix} a & -a & b \\ -a & b & a \\ b & a & -a \end{bmatrix}.$$

A rather more elaborate substitution scheme arises from the Goethals-Seidel array

$$D = \begin{bmatrix} w & xr & yr & zr \\ -xr & w & rz & -ry \\ -yr & -rz & w & rx \\ -zr & ry & -rx & w \end{bmatrix}.$$

Take the alphabet $\mathcal{A}$ to be

$$\{\pm r^i w r^j, \pm r^i x r^j, \pm r^i y r^j, \pm r^i z r^j \mid i, j = 0, 1\}.$$

Choose an involutory ring containing r, w, x, y, z, such that

(8.7) $r^2 = 1,\ r^* = r,\ rwr = w^*,\ rxr = x^*,\ ryr = y^*,\ rzr = z^*.$

The off-diagonal entries of DD^* are zero if $w, x, y, z, w^*, x^*, y^*, z^*$ commute pairwise. So we regard D as a PCD(Λ) where Λ is the set of 2×4 $\mathcal{A}$-arrays X such that

$$XX^* = (ww^* + xx^* + yy^* + zz^*)I_2$$

over an involutory ring $\mathcal{R}$ containing r, w, x, y, z with relations (8.7), and relations specifying that $w, x, y, z, w^*, x^*, y^*, z^*$ commute. A complete and irredundant set of orbit representatives for the action of Π_Λ on $\mathcal{A}$ is $\{w, x, y, z\}$.

Let $\mathcal{S}$ be a commutative ring with identity involution. For $U \in \{W, X, Y, Z\}$, let U be an $\mathcal{S}$-matrix of the form $[h_U(ab^{-1})]_{a,b \in G}$ indexed by the abelian group G. Then W, X, Y, Z and their transposes commute pairwise (see Lemma 3.7.4). Furthermore, if $M_r = [\delta_b^{a^{-1}}]_{a,b \in G}$ then

$$M_r^2 = I,\quad M_r W M_r = W^\top,\quad M_r X M_r = X^\top,\quad M_r Y M_r = Y^\top,\quad M_r Z M_r = Z^\top.$$

So the Grammian of

$$D(W,X,Y,Z) = \begin{bmatrix} W & XM_r & YM_r & ZM_r \\ -XM_r & W & M_rZ & -M_rY \\ -YM_r & -M_rZ & W & M_rX \\ -ZM_r & M_rY & -M_rX & W \end{bmatrix}$$

is $I_4 \otimes (WW^\top + XX^\top + YY^\top + ZZ^\top)$.

8.1.8. EXAMPLE. Plugging the circulant matrices

$$W = \begin{bmatrix} b & b & b \\ b & b & b \\ b & b & b \end{bmatrix}, \ X = \begin{bmatrix} a & -b & b \\ b & a & -b \\ -b & b & a \end{bmatrix}, \ Y = \begin{bmatrix} c & -b & b \\ b & c & -b \\ -b & b & c \end{bmatrix}, \ Z = \begin{bmatrix} d & -b & b \\ b & d & -b \\ -b & b & d \end{bmatrix}$$

into the Goethals-Seidel array, we get an $OD(12; 9, 1^3)$.

The next example recurs in the section on transference. Let X, Y be $(\pm 1, \pm i)$-matrices of order n such that $XX^* + YY^* = 2nI_n$ and $XY^\top = YX^\top$ (X, Y is an amicable *complementary pair*—we say a lot more about these pairs in Chapter 19). Then

$$D(X,Y) = \begin{bmatrix} X & Y \\ Y^{(-1)} & -X^{(-1)} \end{bmatrix}$$

is a complex Hadamard matrix of order $2n$.

8.1.9. EXAMPLE. If X and Y are the circulants with first rows $[1, i, i]$ and $[-i, i, i]$ then

$$D(X,Y) = \begin{bmatrix} 1 & i & i & -i & i & i \\ i & 1 & i & i & -i & i \\ i & i & 1 & i & i & -i \\ i & -i & -i & -1 & i & i \\ -i & i & -i & i & -1 & i \\ -i & -i & i & i & i & -1 \end{bmatrix}$$

is a complex Hadamard matrix.

8.1.3. Orthogonal sets. *Orthogonal sets* comprise an abundant class of plug-in matrices. (This terminology is unfortunate, but we are stuck with it.) The corresponding orthogonality set is all $2 \times n$ $(a_1, \ldots, a_n)$-arrays such that every indeterminate a_i appears in each row, and each column has distinct entries. An ambient ring $\mathcal{R}$ for this orthogonality set contains $a_1, \ldots, a_n, a_1^*, \ldots, a_n^*$ subject to the relations $a_i a_j^* = 0$ for all $i \neq j$. Thus L is a Latin square of order n if and only if $LL^* = (\sum_{i=1}^n a_i a_i^*) I_n$ over $\mathcal{R}$. So if we wish to obtain a design by replacing the indeterminates a_i in L by $v \times v$ matrices A_i over some alphabet $\mathcal{A}$, then the rows of A_i must be Λ-orthogonal to the rows of A_j whenever $i \neq j$.

8.1.10. NOTATION. Let Λ be an orthogonality set. The orthogonality set $\sqcup^n \Lambda$ is the closure under column permutations of the set of all arrays $[X_1 \mid \cdots \mid X_n]$ where $X_1, \ldots, X_n \in \Lambda$.

8.1.11. DEFINITION. A set of $v \times v$ $(0, \mathcal{A})$-arrays $A_1, \ldots, A_n$ is a Λ-*orthogonal set* if

(1) for $i \neq j$ the rows of A_i are Λ-orthogonal to the rows of A_j;
(2) every two distinct rows of $[A_1 \mid \cdots \mid A_n]$ are $\sqcup^n \Lambda$-orthogonal.

When $n = 2$, A_1, A_2 is called an *orthogonal pair*.

8.1.12. EXAMPLE. We exhibit an orthogonal pair for the orthogonality set Λ of 2×4 $(\pm 1, \pm i)$-arrays X such that $XX^* = 4I_2$:

$$(8.8) \qquad U = \begin{bmatrix} i & 1 & i & 1 \\ 1 & i & 1 & i \\ i & 1 & i & 1 \\ 1 & i & 1 & i \end{bmatrix}, \qquad V = \begin{bmatrix} i & i & -i & -i \\ -i & i & i & -i \\ -i & -i & i & i \\ i & -i & -i & i \end{bmatrix}.$$

Next, let $G = \langle \omega \rangle \cong C_3$, and let Λ be the set of 2×3 G-arrays X such that $XX^* = 3I_2$ modulo the ideal $(1 + \omega + \omega^2)\mathbb{Z}[G]$ of $\mathbb{Z}[G]$. Then

$$(8.9) \qquad A = \begin{bmatrix} 1 & 1 & 1 \\ 1 & 1 & 1 \\ 1 & 1 & 1 \end{bmatrix}, \quad B = \begin{bmatrix} 1 & \omega & \omega^2 \\ \omega & \omega^2 & 1 \\ \omega^2 & 1 & \omega \end{bmatrix}, \quad C = \begin{bmatrix} 1 & \omega^2 & \omega \\ \omega^2 & \omega & 1 \\ \omega & 1 & \omega^2 \end{bmatrix}$$

is a Λ-orthogonal set.

8.1.13. PROPOSITION. *Suppose that L is a Latin square with indeterminates $a_1, \ldots, a_n$, and $\{A_1, \ldots, A_n\}$ is a Λ-orthogonal set. Then substituting A_i for a_i in L yields a $\mathrm{PCD}(\sqcup^n \Lambda)$.*

8.1.14. EXAMPLE. Substituting the matrices U and V of (8.8) into $L = \begin{bmatrix} U & V \\ V & U \end{bmatrix}$, we get the complex Hadamard matrix

$$\begin{bmatrix} i & 1 & i & 1 & i & i & -i & -i \\ 1 & i & 1 & i & -i & i & i & -i \\ i & 1 & i & 1 & -i & -i & i & i \\ 1 & i & 1 & i & i & -i & -i & i \\ i & i & -i & -i & i & 1 & i & 1 \\ -i & i & i & -i & 1 & i & 1 & i \\ -i & -i & i & i & 1 & 1 & i & i \\ i & -i & -i & i & 1 & i & 1 & i \end{bmatrix}.$$

Using the template $\begin{bmatrix} A & B & C \\ C & A & B \\ B & C & A \end{bmatrix}$ and A, B, C as in (8.9) yields

$$\begin{bmatrix} 1 & 1 & 1 & 1 & \omega & \omega^2 & 1 & \omega^2 & \omega \\ 1 & 1 & 1 & \omega & \omega^2 & 1 & \omega^2 & \omega & 1 \\ 1 & 1 & 1 & \omega^2 & 1 & \omega & \omega & 1 & \omega^2 \\ 1 & \omega^2 & \omega & 1 & 1 & 1 & 1 & \omega & \omega^2 \\ \omega^2 & \omega & 1 & 1 & 1 & 1 & \omega & \omega^2 & 1 \\ \omega & 1 & \omega^2 & 1 & 1 & 1 & \omega^2 & 1 & \omega \\ 1 & \omega^2 & \omega & 1 & \omega & \omega^2 & 1 & 1 & 1 \\ \omega^2 & \omega & 1 & \omega & \omega^2 & 1 & 1 & 1 & 1 \\ \omega & 1 & \omega^2 & \omega^2 & 1 & \omega & 1 & 1 & 1 \end{bmatrix},$$

which is a $\mathrm{GH}(9; C_3)$.

No discussion of orthogonal sets would be complete without reference to the construction of Craigen, Seberry, and Zhang [27], which gives a Hadamard matrix of order $16abcd$ from Hadamard matrices of orders $4a$, $4b$, $4c$, $4d$. This discovery was made more than 150 years after Sylvester noted that the Kronecker product of Hadamard matrices is another Hadamard matrix. Kronecker multiplication gives a matrix of order $256abcd$, so the construction of [27] drastically reduces the size of the resulting design.

Let A, B, C, D be Hadamard matrices of orders $4a, 4b, 4c, 4d$ respectively. We divide C into $4c \times c$ subarrays, and D into $d \times 4d$ subarrays, as indicated:

$$C = \begin{bmatrix} C_1 & C_2 & C_3 & C_4 \end{bmatrix}, \qquad D = \begin{bmatrix} D_1 \\ D_2 \\ D_3 \\ D_4 \end{bmatrix}.$$

Then

$$V = \tfrac{1}{4}((C_1 + C_2) \otimes D_1 + (C_1 - C_2) \otimes D_2 + (C_3 + C_4) \otimes D_3 + (C_3 - C_4) \otimes D_4)$$

and

$$W = \tfrac{1}{4}((C_1 + C_2) \otimes D_1 + (C_1 - C_2) \otimes D_2 - (C_3 + C_4) \otimes D_3 - (C_3 - C_4) \otimes D_4)$$

are weighing matrices of weight $2cd$. Moreover, V and W are disjoint, $V \pm W$ is a $(1, -1)$-matrix, and $VW^\top = WV^\top$.

Similarly, divide A into $4a \times a$ subarrays, and B into $b \times 4b$ subarrays:

$$A = \begin{bmatrix} A_1 & A_2 & A_3 & A_4 \end{bmatrix}, \qquad B = \begin{bmatrix} B_1 \\ B_2 \\ B_3 \\ B_4 \end{bmatrix}.$$

Define

$$X = \tfrac{1}{2}((A_1 + A_2) \otimes B_1 + (A_1 - A_2) \otimes B_2),$$
$$Y = \tfrac{1}{2}((A_3 + A_4) \otimes B_3 + (A_3 - A_4) \otimes B_4).$$

Then $XX^\top + YY^\top = 8abI_{4ab}$ and $XY^\top = YX^\top = 0$. So X, Y is an orthogonal pair of order $4ab$. It is straightforward to check that $(V \otimes X) + (W \otimes Y)$ is a Hadamard matrix of order $16abcd$.

8.1.4. Weaving. We come now to the 'weaving' technique of Craigen and Kharaghani [26].

Recall Construction 8.1.2. Suppose that $r \geq 1$ divides v_1 and v_2. For $i = 1, 2$, let D_i be a PCD(Λ_i), and write

$$D_1 = \begin{bmatrix} A_1 & A_2 & \cdots & A_r \end{bmatrix}, \qquad D_2 = \begin{bmatrix} B_1 \\ B_2 \\ \vdots \\ B_r \end{bmatrix},$$

where the A_i are $v_1 \times v_1/r$ matrices and the B_i are $v_2/r \times v_2$ matrices. We form the $(0, \mathcal{A}_3)$-matrices $X_i = A_i \otimes B_i$. Let $L = \sum_{i=1}^r a_i P_i$, where the P_i are $(0, 1)$-matrices, be a Latin square of order r. Then the matrix $D_3 = \sum_{i=1}^r P_i \otimes X_i$ (with no repeated rows) is a PCD(Λ_3).

8.1.15. EXAMPLE. Weaving

$$\begin{bmatrix} 1 & 1 & 1 & 1 & 1 & 1 \\ 1 & 1 & \omega & \omega^2 & \omega^2 & \omega \\ 1 & \omega & 1 & \omega & \omega^2 & \omega^2 \\ 1 & \omega^2 & \omega & 1 & \omega & \omega^2 \\ 1 & \omega^2 & \omega^2 & \omega & 1 & \omega \\ 1 & \omega & \omega^2 & \omega^2 & \omega & 1 \end{bmatrix}, \qquad \begin{bmatrix} 1 & 1 \\ 1 & -1 \end{bmatrix}$$

with $r = 2$, we form

$$
\begin{bmatrix}
1 & 1 & 1 \\
1 & 1 & \omega \\
1 & \omega & 1 \\
1 & \omega^2 & \omega \\
1 & \omega^2 & \omega^2 \\
1 & \omega & \omega^2
\end{bmatrix} \otimes \begin{bmatrix} 1 & 1 \end{bmatrix}, \qquad
\begin{bmatrix}
1 & 1 & 1 \\
\omega^2 & \omega^2 & \omega \\
\omega & \omega^2 & \omega^2 \\
1 & \omega & \omega^2 \\
\omega & 1 & \omega \\
\omega^2 & \omega & 1
\end{bmatrix} \otimes \begin{bmatrix} 1 & -1 \end{bmatrix}.
$$

If γ is any root of unity then we get the complex generalized Hadamard matrix

$$
\begin{bmatrix}
1 & 1 & 1 & 1 & 1 & 1 & 1 & 1 & 1 & -1 & -1 & -1 \\
1 & 1 & \omega & 1 & 1 & \omega & \omega^2 & \omega^2 & \omega & -\omega^2 & -\omega^2 & -\omega \\
1 & \omega & 1 & 1 & \omega & 1 & \omega & \omega^2 & \omega^2 & -\omega & -\omega^2 & -\omega^2 \\
1 & \omega^2 & \omega & 1 & \omega^2 & \omega & 1 & \omega & \omega^2 & -1 & -\omega & -\omega^2 \\
1 & \omega^2 & \omega^2 & 1 & \omega^2 & \omega^2 & \omega & 1 & \omega & -\omega & -1 & -\omega \\
1 & \omega & \omega^2 & 1 & \omega & \omega^2 & \omega^2 & \omega & 1 & -\omega^2 & -\omega & -1 \\
1 & 1 & 1 & -1 & -1 & -1 & \gamma & \gamma & \gamma & \gamma & \gamma & \gamma \\
\omega^2 & \omega^2 & \omega & -\omega^2 & -\omega^2 & -\omega & \gamma & \gamma & \gamma\omega & \gamma & \gamma & \gamma\omega \\
\omega & \omega^2 & \omega^2 & -\omega & -\omega^2 & -\omega^2 & \gamma & \gamma\omega & \gamma & \gamma & \gamma\omega & \gamma \\
1 & \omega & \omega^2 & -1 & -\omega & -\omega^2 & \gamma & \gamma\omega^2 & \gamma\omega & \gamma & \gamma\omega^2 & \gamma\omega \\
\omega & 1 & \omega & -\omega & -1 & -\omega & \gamma & \gamma\omega^2 & \gamma\omega^2 & \gamma & \gamma\omega^2 & \gamma\omega^2 \\
\omega^2 & \omega & 1 & -\omega^2 & -\omega & -1 & \gamma & \gamma\omega & \gamma\omega^2 & \gamma & \gamma\omega & \gamma\omega^2
\end{bmatrix}.
$$

8.1.16. EXAMPLE. Weaving can be used when there is no common divisor $r > 1$. Let

$$
D_1 = \begin{bmatrix} 1 & 1 \\ 1 & -1 \end{bmatrix}, \qquad
D_2 = \begin{bmatrix} i & 1 \\ 1 & i \end{bmatrix}, \qquad
D_3 = \begin{bmatrix} 1 & 1 & 1 \\ 1 & \omega & \omega^2 \\ 1 & \omega^2 & \omega \end{bmatrix}, \qquad
D_4 = \begin{bmatrix} 1 & 1 & 1 \\ \omega & 1 & \omega^2 \\ \omega^2 & 1 & \omega \end{bmatrix}.
$$

Then

$$
D_5 = \left[\begin{array}{c|c}
\begin{bmatrix} 1 \\ 1 \end{bmatrix} \otimes \begin{bmatrix} 1 & 1 & 1 \\ 1 & \omega^2 & \omega \end{bmatrix} & \begin{bmatrix} 1 \\ -1 \end{bmatrix} \otimes \begin{bmatrix} 1 & 1 & 1 \\ \omega^2 & 1 & \omega \end{bmatrix} \\
\hline
\begin{bmatrix} i \\ 1 \end{bmatrix} \otimes \begin{bmatrix} 1 & \omega & \omega^2 \end{bmatrix} & \begin{bmatrix} 1 \\ i \end{bmatrix} \otimes \begin{bmatrix} \omega & 1 & \omega^2 \end{bmatrix}
\end{array}\right]
$$

is a CGH$(6; 12)$.

8.2. Transference

This section gives examples of *transference*: converting a design with one type of orthogonality to a design with another type of orthogonality. Often the composition methods for one kind of design preserve the structure needed for transference. In that event, we obtain a new method for composing designs of the other type.

8.2.1. Regular Hadamard matrices and SBIBDs. Suppose that H is a regular Hadamard matrix of order $v = n^2$, and let $A = (H + J_v)/2$. Now

$$
4AA^\top = (H + J_v)(H^\top + J_v) = n^2 I_v + (n^2 + 2n)J_v.
$$

So A is an incidence matrix of an SBIBD (as noted in Section 2.3). Conversely, if A is an incidence matrix of an SBIBD$(n^2, n(n+1)/2, n(n+2)/4)$ then $H = 2A - J$ is Hadamard. Thus, a class of Hadamard matrices maps to a class of SBIBDs, and vice versa. This leads to a composition method for SBIBDs. If H_i is a regular Hadamard matrix of order n_i^2, $i = 1, 2$, then $H_1 \otimes H_2$ is a regular Hadamard matrix of order $(n_1 n_2)^2$. So if there are SBIBD$(n_i^2, n_i(n_i + 1)/2, n_i(n_i + 2)/4)$s, then there is an SBIBD$(n^2, n(n+1)/2, n(n+2)/4)$ where $n = n_1 n_2$.

8.2.1. EXAMPLE.

$$H_1 = \begin{bmatrix} -1 & 1 & 1 & 1 \\ 1 & -1 & 1 & 1 \\ 1 & 1 & -1 & 1 \\ 1 & 1 & 1 & -1 \end{bmatrix} \quad \text{and} \quad H_2 = \begin{bmatrix} 1 & -1 & -1 & -1 \\ -1 & 1 & -1 & -1 \\ -1 & -1 & 1 & -1 \\ -1 & -1 & -1 & 1 \end{bmatrix}$$

are regular Hadamard matrices. Replacing the -1 entries of $H_1 \otimes H_2$ with zeros results in the following incidence matrix for an SBIBD$(16, 6, 2)$:

$$\begin{bmatrix} 0 & 1 & 1 & 1 & 1 & 0 & 0 & 0 & 1 & 0 & 0 & 0 & 1 & 0 & 0 & 0 \\ 1 & 0 & 1 & 1 & 0 & 1 & 0 & 0 & 0 & 1 & 0 & 0 & 0 & 1 & 0 & 0 \\ 1 & 1 & 0 & 1 & 0 & 0 & 1 & 0 & 0 & 0 & 1 & 0 & 0 & 0 & 1 & 0 \\ 1 & 1 & 1 & 0 & 0 & 0 & 0 & 1 & 0 & 0 & 0 & 1 & 0 & 0 & 0 & 1 \\ 1 & 0 & 0 & 0 & 0 & 1 & 1 & 1 & 1 & 0 & 0 & 0 & 1 & 0 & 0 & 0 \\ 0 & 1 & 0 & 0 & 1 & 0 & 1 & 1 & 0 & 1 & 0 & 0 & 0 & 1 & 0 & 0 \\ 0 & 0 & 1 & 0 & 1 & 1 & 0 & 1 & 0 & 0 & 1 & 0 & 0 & 0 & 1 & 0 \\ 0 & 0 & 0 & 1 & 1 & 1 & 1 & 0 & 0 & 0 & 0 & 1 & 0 & 0 & 0 & 1 \\ 1 & 0 & 0 & 0 & 1 & 0 & 0 & 0 & 0 & 1 & 1 & 1 & 1 & 0 & 0 & 0 \\ 0 & 1 & 0 & 0 & 0 & 1 & 0 & 0 & 1 & 0 & 1 & 1 & 0 & 1 & 0 & 0 \\ 0 & 0 & 1 & 0 & 0 & 0 & 1 & 0 & 1 & 1 & 0 & 1 & 0 & 0 & 1 & 0 \\ 0 & 0 & 0 & 1 & 0 & 0 & 0 & 1 & 1 & 1 & 1 & 0 & 0 & 0 & 0 & 1 \\ 1 & 0 & 0 & 0 & 1 & 0 & 0 & 0 & 1 & 0 & 0 & 0 & 0 & 1 & 1 & 1 \\ 0 & 1 & 0 & 0 & 0 & 1 & 0 & 0 & 0 & 1 & 0 & 0 & 1 & 0 & 1 & 1 \\ 0 & 0 & 1 & 0 & 0 & 0 & 1 & 0 & 0 & 0 & 1 & 0 & 1 & 1 & 0 & 1 \\ 0 & 0 & 0 & 1 & 0 & 0 & 0 & 1 & 0 & 0 & 0 & 1 & 1 & 1 & 1 & 0 \end{bmatrix}.$$

8.2.2. Complex Hadamard matrices and Hadamard matrices. The proof of Theorem 2.7.7 gives another example of transference. Let A and B be $v \times v$ $(1, -1)$-matrices. Define

$$C(A, B) = \tfrac{1+i}{2}(A + iB) \quad \text{and} \quad H(A, B) = \begin{bmatrix} A & B \\ B & -A \end{bmatrix}.$$

Note that $C(A, B)$ is a $(\pm 1, \pm i)$-matrix. Since

$$4C(A, B)C(A, B)^* = 2(AA^\top + BB^\top) + 2i(BA^\top - AB^\top),$$

$H(A, B)$ is Hadamard if and only if $C(A, B)$ is a complex Hadamard matrix.

The Kronecker product for complex Hadamard matrices yields a product for Hadamard matrices of the form $H(A, B)$, which gives a matrix whose order is half that of the Kronecker product. We have

$$2C(A, B) \otimes C(A', B') = i(A \otimes A' - B \otimes B') - (B \otimes A' + A \otimes B').$$

So

$$2(1 - i)C(A, B) \otimes C(A', B') = (A - B) \otimes A' - (A + B) \otimes B' \\ + i((A + B) \otimes A' + (A - B) \otimes B').$$

Therefore, if $H(A, B)$ and $H(A', B')$ are Hadamard matrices of orders $2v$ and $2v'$, and we define

$$A'' = \tfrac{1}{2}((A - B) \otimes A' - (A + B) \otimes B'), \quad B'' = \tfrac{1}{2}((A + B) \otimes A' + (A - B) \otimes B'),$$

then $H(A'', B'')$ is a Hadamard matrix of order $2vv'$.

8.2.2. EXAMPLE. The matrices

$$A + iB = (1 - i)\begin{bmatrix} 1 & i \\ i & 1 \end{bmatrix} \quad \text{and} \quad A' + iB' = (1 - i)\begin{bmatrix} 1 & 1 \\ 1 & -1 \end{bmatrix}$$

are complex Hadamard matrices. The corresponding Hadamard matrices are

$$H(A,B) = \begin{bmatrix} 1 & 1 & -1 & 1 \\ 1 & 1 & 1 & -1 \\ -1 & 1 & -1 & -1 \\ 1 & -1 & -1 & -1 \end{bmatrix}, \qquad H(A',B') = \begin{bmatrix} 1 & 1 & -1 & -1 \\ 1 & -1 & -1 & 1 \\ -1 & -1 & -1 & -1 \\ -1 & 1 & -1 & 1 \end{bmatrix}.$$

Now

$$A'' = \begin{bmatrix} 1 & 1 \\ 1 & 1 \end{bmatrix} \otimes \begin{bmatrix} 1 & 1 \\ 1 & -1 \end{bmatrix} \qquad \text{and} \qquad B'' = \begin{bmatrix} -1 & 1 \\ 1 & -1 \end{bmatrix} \otimes \begin{bmatrix} 1 & 1 \\ 1 & -1 \end{bmatrix},$$

so

$$H(A'',B'') = \begin{bmatrix} 1 & 1 & 1 & 1 & -1 & -1 & 1 & 1 \\ 1 & -1 & 1 & -1 & -1 & 1 & 1 & -1 \\ 1 & 1 & 1 & 1 & 1 & 1 & -1 & -1 \\ 1 & -1 & 1 & -1 & 1 & -1 & -1 & 1 \\ -1 & -1 & 1 & 1 & -1 & -1 & -1 & -1 \\ -1 & 1 & 1 & -1 & -1 & 1 & -1 & 1 \\ 1 & 1 & -1 & -1 & -1 & -1 & -1 & -1 \\ 1 & -1 & -1 & 1 & -1 & 1 & -1 & 1 \end{bmatrix}$$

is a Hadamard matrix of order 8.

Next, we show that each Hadamard matrix obtained by plugging any $(1,-1)$-matrices W, X, Y, Z into the Williamson array implies the existence of a complex Hadamard matrix. Let

$$A = \tfrac{1+\mathrm{i}}{2}(W + \mathrm{i}X) \qquad \text{and} \qquad B = \tfrac{1+\mathrm{i}}{2}(Y + \mathrm{i}Z),$$

so that A and B are $(\pm 1, \pm \mathrm{i})$-matrices. We have

$$4AB^\top = (1+\mathrm{i})(W+\mathrm{i}X)(1+\mathrm{i})(Y^\top + \mathrm{i}Z^\top) = 2\mathrm{i}(WY^\top - XZ^\top + \mathrm{i}(XY^\top + WZ^\top)).$$

Then $2BA^\top - 2AB^\top$ equals

$$(XY^\top + WZ^\top - YX^\top - ZW^\top) - \mathrm{i}(WY^\top - XZ^\top - YW^\top + ZX^\top)$$

and $2AA^* + 2BB^*$ equals

$$(WW^\top + XX^\top + YY^\top + ZZ^\top) + \mathrm{i}(XW^\top - WX^\top + ZY^\top - YZ^\top).$$

Define

$$D(A,B) = \begin{bmatrix} A & B \\ B^{(-1)} & -A^{(-1)} \end{bmatrix}, \qquad H(W,X,Y,Z) = \begin{bmatrix} W & X & Y & Z \\ X & -W & Z & -Y \\ Y & -Z & -W & X \\ Z & Y & -X & -W \end{bmatrix}.$$

We see that $D(A,B)$ is a complex Hadamard matrix if and only if H is a Hadamard matrix.

8.2.3. EXAMPLE. As in Example 8.1.6, let

$$W = \begin{bmatrix} 1 & 1 & 1 \\ 1 & 1 & 1 \\ 1 & 1 & 1 \end{bmatrix} \qquad \text{and} \qquad X = Y = Z = \begin{bmatrix} -1 & 1 & 1 \\ 1 & -1 & 1 \\ 1 & 1 & -1 \end{bmatrix}.$$

Then $D(A, B)$ is a complex Hadamard matrix of order 6:

$$\begin{bmatrix} 1 & i & i & -i & i & i \\ i & 1 & i & i & -i & i \\ i & i & 1 & i & i & -i \\ i & -i & -i & -1 & i & i \\ -i & i & -i & i & -1 & i \\ -i & -i & i & i & i & -1 \end{bmatrix}.$$

8.2.3. Unreal matrices. The ideas in this subsection are due to Compton, Craigen, and de Launey [20]. Let i, j, k be the usual generators for the quaternions. If $Q = [q_{ij}]$ is a $(\pm 1 \pm i \pm j \pm k)$-matrix then we may define $Q^* := [q_{ji}^{-1}]$. Say that Q is 'orthogonal' if its Grammian has all off-diagonal entries zero, and all diagonal entries real. Let $Q = W + iX + jY + kZ$ where W, X, Y, Z are $(1, -1)$-matrices; then Q is orthogonal if and only if the Williamson matrix H of Theorem 8.1.5 is Hadamard. For many orders we have circulant plug-in matrices W, X, Y, Z, in which case Q is orthogonal and circulant.

8.2.4. EXAMPLE. When $X = Y = Z$, we get transference between Hadamard matrices and $\mathrm{CGH}(v; 6)$ with no real entries. Here are examples (γ is a primitive cube root of 1):

$$\begin{bmatrix} \gamma & \gamma^2 & \gamma^2 \\ \gamma^2 & \gamma & \gamma^2 \\ \gamma^2 & \gamma^2 & \gamma \end{bmatrix}, \qquad \begin{bmatrix} -\gamma^2 & \gamma & \gamma^2 & \gamma & \gamma^2 & \gamma^2 & \gamma \\ \gamma & -\gamma & \gamma & \gamma^2 & \gamma^2 & \gamma & \gamma \\ \gamma^2 & \gamma & -\gamma & \gamma^2 & \gamma & \gamma & \gamma \\ \gamma & \gamma^2 & \gamma^2 & -\gamma^2 & \gamma & \gamma & \gamma^2 \\ \gamma^2 & \gamma^2 & \gamma & \gamma & -\gamma^2 & \gamma^2 & \gamma \\ \gamma^2 & \gamma & \gamma & \gamma & \gamma^2 & -\gamma^2 & \gamma^2 \\ \gamma & \gamma & \gamma & \gamma^2 & \gamma & \gamma^2 & -\gamma \end{bmatrix}.$$

8.2.4. SBIBDs and generalized Hadamard matrices over symmetric groups. Finite geometries provide many examples of transference. An incidence matrix A of an $\mathrm{SBIBD}(n^2 + n + 1, n + 1, 1)$ may be put into the form

$$A = \begin{array}{|c|c|c|c|c|}
\hline
\begin{matrix}111\cdots 1\\100\cdots 0\\100\cdots 0\\ \vdots\\100\cdots 0\end{matrix} & \begin{matrix}00\cdots 0\\11\cdots 1\\00\cdots 0\\ \vdots\\00\cdots 0\end{matrix} & \begin{matrix}00\cdots 0\\00\cdots 0\\11\cdots 1\\ \vdots\\00\cdots 0\end{matrix} & \begin{matrix}\cdots\\ \cdots\\ \cdots\\ \\ \cdots\end{matrix} & \begin{matrix}00\cdots 0\\00\cdots 0\\00\cdots 0\\ \vdots\\11\cdots 1\end{matrix}\\
\hline
\begin{matrix}010\cdots 0\\010\cdots 0\\ \vdots\\010\cdots 0\end{matrix} & P_{11} & P_{12} & \cdots & P_{1n}\\
\hline
\begin{matrix}001\cdots 0\\001\cdots 0\\ \vdots\\001\cdots 0\end{matrix} & P_{21} & P_{22} & \cdots & P_{2n}\\
\hline
\vdots & \vdots & \vdots & & \vdots\\
\hline
\begin{matrix}000\cdots 1\\000\cdots 1\\ \vdots\\000\cdots 1\end{matrix} & P_{n1} & P_{n2} & \cdots & P_{nn}\\
\hline
\end{array}.$$

The P_{ij} are $n \times n$ permutation matrices such that

(8.10)
$$\sum_{k=1}^{n} P_{ik} P_{jk}^{\top} = J_n$$

for all $i \neq j$. Let $\mathcal{A}$ be the set of $n \times n$ permutation matrices, and say two rows $[P_{ik}]_k$, $[P_{jk}]_k$ of the $n \times n$ array $P = [P_{ij}]$ are Λ-orthogonal if and only if (8.10) holds. Then P is a PCD(Λ) if and only if A is an incidence matrix of a projective plane of order n.

CHAPTER 9

Automorphism Groups

In broad terms, the automorphism group of a combinatorial design is its group of symmetries. More precisely, it is the stabilizer of the design under the action of the group generated by all elementary local equivalence operations. It is believed that most large designs have no non-trivial symmetries. In part this is because of the composition techniques, described in Section 8.1, which allow one to construct many inequivalent designs without preserving symmetry. Nevertheless, most small designs and many important infinite families of designs do have interesting automorphism groups.

This chapter provides a firm algebraic foundation for studying automorphism groups of pairwise combinatorial designs. Throughout, Λ is an orthogonality set of $2 \times v$ $(0, \mathcal{A})$-arrays, and $\mathcal{R}$ is an ambient ring for Λ satisfying the requirements listed in 5.2.3. That is, $\mathcal{R}$ is a ring with involution $*$ containing $\mathcal{A}$, and its group of units contains the row group $R \cong \Pi_\Lambda^{\text{row}}$ and column group $C \cong \Pi_\Lambda^{\text{col}}$. Furthermore, $\rho : R \to \Pi_\Lambda^{\text{row}}$ and $\kappa : C \to \Pi_\Lambda^{\text{col}}$ are isomorphisms such that

$$(9.1) \quad ra = \rho_r(a), \quad ac^{-1} = \kappa_c(a), \quad r^* = r^{-1}, \quad c^* = c^{-1} \quad \forall a \in \mathcal{A}, r \in R, c \in C.$$

We write $D' \approx_\Lambda D$ if $D' = PDQ^*$ for some $P \in \text{Mon}(v, R)$ and $Q \in \text{Mon}(v, C)$; and $D' \approx D$ when $P, Q \in \text{Perm}(v)$. Note that $Q^* = Q^{-1}$. As in Subsection 3.7.3, for a finite group G and $g \in G$ we define $S_g = [\delta_{gy}^x]_{x,y \in G}$ and $T_g = [\delta_y^{xg}]_{x,y \in G}$.

9.1. Automorphism groups of pairwise combinatorial designs

As a consequence of (9.1), the direct product $\text{Mon}(v, R) \times \text{Mon}(v, C)$ acts on the (supposedly non-empty) set of all $\text{PCD}(\Lambda)$s: if D is a $\text{PCD}(\Lambda)$, $P \in \text{Mon}(v, R)$, and $Q \in \text{Mon}(v, C)$, define

$$(P, Q)D = PDQ^*.$$

The orbits of this action are the Λ-equivalence classes of $\text{PCD}(\Lambda)$s. The stabilizer of D is its automorphism group.

9.1.1. DEFINITION. An element (P, Q) of $\text{Mon}(v, R) \times \text{Mon}(v, C)$ such that $PDQ^* = D$ is an *automorphism* of D. The group of all automorphisms of D is denoted $\text{Aut}(D)$.

By Theorem 5.2.5, Definitions 9.1.1 and 5.1.1 are equivalent.

9.1.2. EXAMPLE. To calculate the automorphism group of the smallest Hadamard matrix

$$H_2 = \begin{bmatrix} 1 & 1 \\ 1 & -1 \end{bmatrix},$$

we simply check which of the eight $(1,-1)$-monomial matrices P yield a $(1,-1)$-monomial matrix $Q = (H_2^{-1}P^{-1}H_2)^\top = H_2PH_2^{-1}$. We find that all eight choices for P are valid. Indeed, if

$$\alpha = \left(\begin{bmatrix} -1 & 0 \\ 0 & 1 \end{bmatrix}, \begin{bmatrix} 0 & -1 \\ -1 & 0 \end{bmatrix} \right), \qquad \beta = \left(\begin{bmatrix} 0 & -1 \\ 1 & 0 \end{bmatrix}, \begin{bmatrix} 0 & 1 \\ -1 & 0 \end{bmatrix} \right)$$

then

$$\mathrm{Aut}(H_2) = \langle \alpha, \beta \mid \alpha^2 = \beta^4 = 1,\ \beta^\alpha = \beta^{-1} \rangle$$

where $\beta^2 = (-I_2, -I_2)$ generates the center of $\mathrm{Aut}(H_2)$. That is, $\mathrm{Aut}(H_2)$ is the dihedral group $\mathrm{D}_8 \cong \mathrm{Mon}(2, \mathrm{C}_2)$ of order 8.

The next two results prove that $\mathrm{Aut}(D)$ is essentially independent of the choice of representative for D, and the choice of ambient ring $\mathcal{R}$.

9.1.3. THEOREM. *If $D' \approx_\Lambda D$ then $\mathrm{Aut}(D')$ and $\mathrm{Aut}(D)$ are conjugate to each other in $\mathrm{Mon}(v, R) \times \mathrm{Mon}(v, C)$.*

PROOF. If $D' = ADB^*$ for $A \in \mathrm{Mon}(v, R)$ and $B \in \mathrm{Mon}(v, C)$, and $(P, Q) \in \mathrm{Aut}(D')$, then $(A, B)^{-1}(P, Q)(A, B) \in \mathrm{Aut}(D)$. □

9.1.4. THEOREM. *The isomorphism type of $\mathrm{Aut}(D)$ is independent of the choice of ambient ring.*

PROOF. Suppose that $\mathcal{R}^{(1)}$ and $\mathcal{R}^{(2)}$ are ambient rings for Λ, with respective isomorphisms $\rho^{(i)} : R \to \Pi_\Lambda^{\mathrm{row}}$ and $\kappa^{(i)} : C \to \Pi_\Lambda^{\mathrm{col}}$. Then there are automorphisms α of R and β of C such that $\rho_r^{(2)} = \rho_{\alpha(r)}^{(1)}$ and $\kappa_c^{(2)} = \kappa_{\beta(c)}^{(1)}$. Now for $D = [d_{ij}]_{ij}$ and $([\delta_j^{\pi(i)} r_i]_{ij}, [\delta_l^{\phi(k)} c_k]_{kl}) \in \mathrm{Mon}(v, R) \times \mathrm{Mon}(v, C)$, we have

$$\begin{aligned}
[\delta_j^{\pi(i)} r_i]_{ij} [d_{jk}]_{jk} [\delta_k^{\phi(l)} c_l^{-1}]_{kl} &= [\textstyle\sum_{jk} \delta_j^{\pi(i)} r_i d_{jk} \delta_l^{\phi(l)} c_l^{-1}]_{il} \\
&= [r_i d_{\pi(i), \phi(l)} c_l^{-1}]_{il} \\
&= [\rho_{r_i}^{(2)} \kappa_{c_l}^{(2)} (d_{\pi(i), \phi(l)})]_{il} \\
&= [\rho_{\alpha(r_i)}^{(1)} \kappa_{\beta(c_l)}^{(1)} (d_{\pi(i), \phi(l)})]_{il} \\
&= [\alpha(r_i) d_{\pi(i), \phi(l)} \beta(c_l)^{-1}]_{il} \\
&= [\delta_j^{\pi(i)} \alpha(r_i)]_{ij} [d_{jk}]_{jk} [\delta_k^{\phi(l)} \beta(c_l)^{-1}]_{kl}.
\end{aligned}$$

So, working over $\mathcal{R}^{(2)}$, (P, Q) is an automorphism of D if and only if, working over $\mathcal{R}^{(1)}$, $(\alpha(P), \beta(Q))$ is an automorphism of D. □

9.1.5. REMARK. Let $\mathcal{R}^{(i)}$ $(i = 1, 2)$ be ambient rings for D, and let $\mathrm{Aut}^{(i)}(D)$ be the respective automorphism groups. From the proof of Theorem 9.1.4, we see that $\mathrm{Aut}^{(2)}(D) = \{(\alpha(P), \beta(Q)) \mid (P, Q) \in \mathrm{Aut}^{(1)}(D)\}$ for some automorphisms α of R and β of C.

9.2. A class of generalized Hadamard matrices

In this section we detour from the general theory, to consider automorphism groups of designs in an important special class. This is a pleasing instance of the algebraic definition of a design leading to a direct algebraic determination of its automorphism group.

Let p be a prime, and denote the k-dimensional vector space over $\mathbb{F} = \mathrm{GF}(p^m)$ by V. Then

$$(9.2) \qquad D_{(p,m,k)} = [xy^\top]_{x,y \in V}$$

is a $\mathrm{GH}(p^{mk}; \mathrm{C}_p^m)$, written additively.

9.2.1. EXAMPLE. We display $D_{(p,m,k)}$ for $(p,m,k) = (3,1,1), (2,1,2), (2,2,1), (5,1,1)$:

$$
\begin{bmatrix}
0 & 0 & 0 \\
0 & 1 & 2 \\
0 & 2 & 1
\end{bmatrix}, \quad
\begin{bmatrix}
0 & 0 & 0 & 0 \\
0 & 1 & 0 & 1 \\
0 & 0 & 1 & 1 \\
0 & 1 & 1 & 0
\end{bmatrix}, \quad
\begin{bmatrix}
00 & 00 & 00 & 00 \\
00 & 01 & 10 & 11 \\
00 & 10 & 11 & 01 \\
00 & 11 & 01 & 10
\end{bmatrix}, \quad
\begin{bmatrix}
0 & 0 & 0 & 0 & 0 \\
0 & 1 & 2 & 3 & 4 \\
0 & 2 & 4 & 1 & 3 \\
0 & 3 & 1 & 4 & 2 \\
0 & 4 & 3 & 2 & 1
\end{bmatrix}.
$$

For each $a \in V$, the *translation* $t_a : x \mapsto x + a$ is a permutation of V. The translations form a subgroup of $\mathrm{Sym}(V)$ isomorphic to V. The invertible linear transformations $x \mapsto xA$, $A \in \mathrm{GL}(k,\mathbb{F})$, are also permutations of V. The subgroup of $\mathrm{Sym}(V)$ generated by all translations and invertible linear transformations is the *affine general linear group*, $\mathrm{AGL}(k,\mathbb{F})$. We have $\mathrm{AGL}(k,\mathbb{F}) \cong \mathrm{GL}(k,\mathbb{F}) \ltimes V$, where the action of $\mathrm{GL}(k,\mathbb{F})$ on V is the natural one:

$$(9.3) \qquad (At_a)(Bt_b) = ABt_{aB+b}.$$

If $\mathbb{F}$ has prime size p then $\mathrm{AGL}(k,\mathbb{F})$ is the holomorph of the elementary abelian p-group of rank k.

9.2.2. LEMMA. *The center of $\mathrm{AGL}(k,\mathbb{F})$ is trivial unless $k = 1$ and $|\mathbb{F}| = 2$.*

PROOF. By (9.3), each central element At_a corresponds to a pair $a \in V$, $A \in \mathrm{GL}(k,\mathbb{F})$ such that $bA + a = aB + b$ and $AB = BA$ for all $b \in V$ and $B \in \mathrm{GL}(k,\mathbb{F})$. Taking $B = I_k$ gives $A = I_k$. Hence $aB = a$ for all $B \in \mathrm{GL}(k,\mathbb{F})$. If $k > 1$ then $a = 0$, because $\mathrm{GL}(k,\mathbb{F})$ moves every non-zero element of V. If $k = 1$ and $|\mathbb{F}| > 2$, then taking $B = \lambda$ $(\lambda \neq 0,1)$, we see again that $a = 0$. $\square$

When doing matrix algebra, it is convenient to think of $D_{(p,m,k)}$ as being over the multiplicative group C_p^m rather than the additive group $(\mathbb{F},+)$. Accordingly, we use the notation $\hat{x}$ for the element in C_p^m corresponding to $x \in \mathbb{F}$. The multiplicative version of $D_{(p,m,k)}$ is denoted $\hat{D}_{(p,m,k)}$.

Let $(P,Q) \in \mathrm{Mon}(p^{mk}, \mathrm{C}_p^m)^2$, so that

$$P = [\hat{f}(x)\delta_y^{\pi(x)}]_{x,y \in V} \quad \text{and} \quad Q = [\hat{g}(x)\delta_y^{\phi(x)}]_{x,y \in V},$$

where $\pi, \phi \in \mathrm{Sym}(V)$ and $\hat{f}, \hat{g} : V \to \mathrm{C}_p^m$. Now

$$P\hat{D}_{(p,m,k)}Q^* = [\textstyle\sum_{w,z} \hat{f}(x)\delta_w^{\pi(x)} \widehat{wz^\top} \hat{g}(y)^{-1} \delta_z^{\phi(y)}]_{x,y \in V}$$
$$= [\hat{f}(x)\widehat{\pi(x)\phi(y)^\top} \hat{g}(y)^{-1}]_{x,y \in V}.$$

So if we write $f, g : V \to (\mathbb{F},+)$ for the additive versions of $\hat{f}$ and $\hat{g}$, then $(P,Q) \in \mathrm{Aut}(D_{(p,m,k)})$ if and only if

$$(9.4) \qquad f(x) - g(y) + \pi(x)\phi(y)^\top = xy^\top \qquad \forall\, x,y \in V.$$

9.2.3. LEMMA. $\pi, \phi \in \mathrm{Sym}(V)$ and $f, g : V \to (\mathbb{F}, +)$ satisfy (9.4) if and only if there exist $A \in \mathrm{GL}(k, \mathbb{F})$, $a, b \in V$, and $c \in \mathbb{F}$ such that

$$(9.5) \qquad \pi(x) = xA + a, \qquad\qquad \phi(x) = x(A^{-1})^\top + b,$$

$$(9.6) \qquad f(x) = -xAb^\top + c, \qquad\qquad g(x) = aA^{-1}x^\top + ab^\top + c$$

for all $x \in V$.

PROOF. If (9.5) and (9.6) hold for some A, a, b, and c, then (9.4) holds. We now prove the converse.

Suppose that (9.4) holds. Put $L(x) = \pi(x) - \pi(0)$, so

$$L(x)\phi(y)^\top = xy^\top - f(x) + f(0).$$

If $y_0 = \phi^{-1}(0)$ then $f(x) - f(0) = xy_0^\top$. Hence

$$L(x)\phi(y)^\top = x(y - y_0)^\top.$$

We now show that L is $\mathbb{F}$-linear. Let $x, z \in V$ and $\lambda, \mu \in \mathbb{F}$. Then

$$\begin{aligned}
L(\lambda x + \mu z)\phi(y)^\top &= (\lambda x + \mu z)(y - y_0)^\top \\
&= \lambda x(y - y_0)^\top + \mu z(y - y_0)^\top \\
&= \lambda L(x)\phi(y)^\top + \mu L(z)\phi(y)^\top.
\end{aligned}$$

Since this is true for all $y \in V$, L is $\mathbb{F}$-linear. Similarly, $\phi(x) = M(x) + \phi(0)$ where M is $\mathbb{F}$-linear. So there are $A, B \in \mathrm{GL}(k, \mathbb{F})$ and $a, b \in V$ such that

$$\pi(x) = xA + a \qquad \text{and} \qquad \phi(y) = yB + b.$$

Returning to (9.4), we have

$$f(x) - g(y) + (xA + a)(yB + b)^\top = xy^\top \qquad \forall\, x, y \in V.$$

Simplifying,

$$(9.7) \qquad x(I - AB^\top)y^\top = f(x) - g(y) + aB^\top y^\top + xAb^\top + ab^\top.$$

Taking $x = w$ in (9.7), and subtracting from (9.7), we get

$$(9.8) \qquad (x - w)(I_k - AB^\top)y^\top = f(x) - f(w) + (x - w)Ab^\top \qquad \forall\, x, w, y \in V.$$

Similarly (taking $y = z$ in (9.8) and subtracting from (9.8)),

$$(x - w)(I_k - AB^\top)(y - z)^\top = 0 \qquad \forall\, x, w, y, z \in V.$$

This implies that $B = (A^{-1})^\top$, thereby proving (9.5).

Substituting (9.5) into (9.4) yields

$$aA^{-1}y^\top + xAb^\top + ab^\top + f(x) - g(y) = 0 \qquad \forall\, x, y \in V.$$

Putting $y = 0$, $x = 0$, and $x = y = 0$:

$$f(x) = -xAb^\top - ab^\top + g(0), \quad g(y) = aA^{-1}y^\top + ab^\top + f(0), \quad g(0) - f(0) = ab^\top.$$

So, with $c = f(0)$,

$$f(x) = -xAb^\top + c \qquad \text{and} \qquad g(x) = aA^{-1}x^\top + ab^\top + c,$$

which is (9.6). $\square$

We now determine the isomorphism class of $G_0 := \mathrm{Aut}(D_{(p,m,k)})$. If $\alpha_{\pi,\phi,f,g}$ is the automorphism corresponding to a quadruple (π, ϕ, f, g) satisfying the conditions of Lemma 9.2.3, then the maps $\psi_{\mathrm{row}}, \psi_{\mathrm{col}} : G_0 \to \mathrm{Sym}(V)$ defined by

$$\psi_{\mathrm{row}}(\alpha_{\pi,\phi,f,g}) = \pi^{-1}, \quad \psi_{\mathrm{col}}(\alpha_{\pi,\phi,f,g}) = \phi^{-1}$$

are homomorphisms. Notice that

$$\psi_{\mathrm{row}}(G_0) = \{\pi \in \mathrm{Sym}(V) \mid \pi(x) = xA + a \text{ for some } A \in \mathrm{GL}(k, \mathbb{F}), a \in V\}$$
$$= \mathrm{AGL}(k, \mathbb{F}),$$

and the kernel K_1 of ψ_{row} is

$$\{\alpha_{\pi,\phi,f,g} \mid \pi = \mathrm{id}_V, \phi(x) = x + b, f(x) = -xb^\top + c, g(x) = c \text{ for some } b, c \in V\}.$$

Moreover, ψ_{row} is an isomorphism from the subgroup

$$G_1 = \{\alpha_{\pi,\phi,f,g} \mid \pi(x) = xA + a, \phi(x) = x(A^{-1})^\top, f(x) = 0, g(x) = aA^{-1}x^\top\}$$

of G_0 to $\psi_{\mathrm{row}}(G_0)$. It follows that $G_0 = G_1 \ltimes K_1$.

Now ψ_{col} maps K_1 to

$$\{\phi \in \mathrm{Sym}(V) \mid \phi(x) = b + x \text{ for some } b \in V\} \cong \mathrm{C}_p^{mk}.$$

The kernel of ψ_{col} on K_1 is

$$K_2 = \{\alpha_{\pi,\phi,f,g} \mid \pi = \phi = \mathrm{id}_V, f(x) = g(x) = c \text{ for some } c \in V\} \cong \mathrm{C}_p^m,$$

and ψ_{col} restricted to the subgroup

$$G_2 = \{\alpha_{\pi,\phi,f,g} \mid \pi(x) = \mathrm{id}_V, \phi(x) = x + b, f(x) = -xb^\top, g(x) = 0\} \cong \mathrm{C}_p^{mk}$$

of K_1 is an isomorphism. Hence

$$G_0 = G_1 \ltimes (G_2 \ltimes K_2).$$

Finally, note that (except when $k = m = 1$ and $p = 2$) the center of $G_1 G_2$ is trivial by Lemma 9.2.2. So K_2 is the center of G_0. We have proved

9.2.4. THEOREM. *Except when $k = m = 1$ and $p = 2$,*

$$\mathrm{Aut}(D_{(p,m,k)}) \cong \mathrm{AGL}(k, \mathbb{F}) \ltimes \mathrm{C}_p^{mk} \times Z(\mathrm{Aut}(D_{(p,m,k)})),$$

where the center $Z(\mathrm{Aut}(D_{(p,m,k)}))$ is isomorphic to C_p^m.

9.3. A bound on the size of the automorphism group

This section presents a bound on the order of the automorphism group of a PCD(Λ). The idea is that Λ-orthogonality guarantees the existence of a 'small' set of columns that determines the behavior of each automorphism. This idea goes back to Leon [114].

9.3.1. DEFINITION. Let Y be a $2 \times l$ $(0, \mathcal{A})$-matrix. The two (different) rows w and u of Y are *R-distinct* if there is no $r \in R$ such that $ru = w$. If Z is an $l \times 2$ $(0, \mathcal{A})$-matrix, then its columns w and u are *C-distinct* if there is no $c \in C$ such that $wc = u$.

For each l, $2 \leq l \leq v$, let

$$r_l = \min_{X \in \Lambda} \left\{ \frac{|\{Y \mid Y \text{ is a } 2 \times l \text{ submatrix of } X \text{ with } R\text{-distinct rows }\}|}{|\{Y \mid Y \text{ is a } 2 \times l \text{ submatrix of } X\}|} \right\}.$$

9.3.2. EXAMPLE. Let $\Lambda = \Lambda_{H(2n)}$, and suppose that $X \in \Lambda$. If the rows of a $2 \times l$ submatrix Y of X are R-distinct, then Y must contain a column where the first and second rows agree, and a column where the first and second rows disagree. Since X has exactly n columns where the first and second rows disagree, and n columns where the first and second rows agree,

$$r_l = \sum_{i=1}^{l-1} \binom{n}{i}\binom{n}{l-i}\binom{2n}{l}^{-1} = 1 - 2\binom{n}{l}\binom{2n}{l}^{-1} > 1 - 2^{1-l}$$

by Vandermonde's identity. Similarly, $r_l > 1 - |G|^{1-l}$ for $\Lambda = \Lambda_{GH(v;G)}$.

If it exists, denote by l_0 the least value of l such that

$$r_l > 1 - \binom{v}{2}^{-1}.$$

9.3.3. EXAMPLE. For $\Lambda = \Lambda_{GH(v;G)}$, we have $l_0 \leq l_1$, where $l_1 > 1$ is the least integer such that

$$1 - |G|^{1-l_1} > 1 - (v^2/2)^{-1}.$$

So, in this case,

(9.9) $$l_0 \leq l_1 = \lceil 1 + \log_{|G|}(v^2/2) \rceil \leq \lceil 2(1 + \log_{|G|} v) \rceil.$$

9.3.4. LEMMA. *Let D be a PCD(Λ). If $l \geq l_0$ then there is a $v \times l$ submatrix of D whose rows are pairwise R-distinct.*

PROOF. The number of $v \times l_0$ submatrices of D is $\binom{v}{l_0}$. At most

$$\binom{v}{2}(1 - r_{l_0})\binom{v}{l_0} < \binom{v}{l_0}$$

of these submatrices have at least two rows that are not R-distinct. So D contains a $v \times l_0$ submatrix whose rows are all R-distinct. Now add $l - l_0$ columns to this submatrix.   □

9.3.5. THEOREM. *Suppose that D is a PCD(Λ) whose columns are pairwise C-distinct. If D contains a $v \times l$ submatrix with pairwise R-distinct rows, then*

$$|\mathrm{Aut}(D)| < v(v-1)(v-2)\cdots(v-l+1)|C|^l.$$

In particular, if $l \leq l_0$ then

$$|\mathrm{Aut}(D)| < (v|C|)^{l_0}.$$

PROOF. Since the columns of D are C-distinct, for each $P \in \mathrm{Mon}(v, R)$ there is at most one $Q \in \mathrm{Mon}(v, C)$ such that $PDQ^* = D$. Moreover, if $(P, Q) \in \mathrm{Aut}(D)$, then, since D contains a $v \times l$ submatrix whose rows are R-distinct, the action of Q on the l distinguished columns of D determines P; which, as just noted, uniquely determines (the rest of the action of) Q. Since there are at most

$$v(v-1)(v-2)\cdots(v-l+1)|C|^l$$

possibilities for the action of Q on those l columns, the result follows.   □

9.3.6. REMARK. Notice that if a $v \times l$ submatrix of a GH($v; G$) has G-distinct rows, then $|G|^{l-1} \geq v$. So if D is a GH($v; G$), then in Theorem 9.3.5 we must have $l \geq 1 + \log_{|G|} v$. On the other hand, we may take $l \geq \lceil 2(1 + \log_{|G|} v) \rceil$ by Example 9.3.3 and Lemma 9.3.4.

We mention two noteworthy corollaries of Theorem 9.3.5, both of which rely on (9.9).

9.3.7. COROLLARY. *Let D be a Hadamard matrix of order $2n$. Then*

$$|\mathrm{Aut}(D)| < (4n)^{3+2\log_2 n}.$$

PROOF. Equation (9.9) implies that $l_0 \leq \lceil 1 + \log_2(2n^2) \rceil < 3 + 2\log_2 n$. Now use Lemma 9.3.4 and Theorem 9.3.5.   □

9.3.8. COROLLARY. *Let D be a $\mathrm{GH}(v; G)$. Then*

$$|\mathrm{Aut}(D)| < \tfrac{1}{2}|G|^2 v^{4+\log_{|G|}(v^2/2)}.$$

Consider again the generalized Hadamard matrices $D_{(p,m,k)}$ defined by (9.2) in the previous section. Let X be the $p^{mk} \times k$ submatrix of $D_{(p,m,k)}$ whose columns correspond to a basis of V. The rows of X are R-distinct. If we adjoin the column indexed by 0 to X, then we obtain a $p^{mk} \times (k+1)$ submatrix of $D_{(p,m,k)}$ whose rows are R-distinct. Putting $v = p^{mk}$ and $G = \mathrm{C}_p^m$, we see that $D_{(p,m,k)}$ contains a $v \times l$ submatrix with G-distinct rows, where $l = 1 + \log_{|G|} v$. So, in view of Remark 9.3.6, this class of generalized Hadamard matrices is extremal in the sense that each matrix contains a very small set of columns comprising a submatrix with R-distinct rows. By Theorem 9.3.5,

(9.10) $$|\mathrm{Aut}(D_{(p,m,k)}| < p^{m(k^2+2k+1)}.$$

By Theorem 9.2.4,

$$|\mathrm{Aut}(D_{(p,m,k)})| = p^{(2k+1)m}(p^{mk} - 1)(p^{mk} - p^m) \cdots (p^{mk} - p^{m(k-1)})$$
$$= p^{(k^2+2k+1)m}(1 - p^{-mk})(1 - p^{-m(k-1)}) \cdots (1 - p^{-m})$$
$$= p^{(k^2+2k+1)m}c(p, m, k),$$

where

$$c(p, m, k) \geq (1 - p^{-m})(1 - p^{-mk} - p^{-m(k-1)} - \cdots - p^{-2m})$$
$$> (1 - p^{-m})(1 - p^{-2m}(1 - p^{-m})^{-1})$$
$$\geq (1 - p^{-m})^2.$$

So the bound (9.10) is close to being exact in this case.

9.4. Permutation automorphism groups

Let X be an $v \times b$ $(0, \mathcal{A})$-array. Sometimes we restrict attention to symmetries of X that involve only permutations of its rows and columns.

9.4.1. DEFINITION. The *permutation automorphism group* $\mathrm{PermAut}(X)$ is the subgroup of $\mathrm{Perm}(v) \times \mathrm{Perm}(b)$ consisting of the pairs (P, Q) such that

$$PXQ^\top = X.$$

As usual, the multiplication is carried out over an ambient ring containing $\mathcal{A}$.

When $X = D$ is a $\mathrm{PCD}(\Lambda)$, $\mathrm{PermAut}(D) \leq \mathrm{Aut}(D)$. Remark 9.1.5 implies the following.

9.4.2. THEOREM. $\mathrm{PermAut}(D)$ *is independent of the choice of ambient ring.*

9.4.3. EXAMPLE. We calculate the permutation automorphism group of each of the following arrays (obtained by dropping the signs from the orthogonal designs in Example 4.4.14):

$$X_2 = \begin{bmatrix} A & B \\ B & A \end{bmatrix}, \quad X_4 = \begin{bmatrix} A & B & C & D \\ B & A & D & C \\ C & D & A & B \\ D & C & B & A \end{bmatrix}, \quad X_8 = \begin{bmatrix} A & B & C & D & E & F & G & H \\ B & A & D & C & F & E & H & G \\ C & D & A & B & G & H & E & F \\ D & C & B & A & H & G & F & E \\ E & F & G & H & A & B & C & D \\ F & E & H & G & B & A & D & C \\ G & H & E & F & C & D & A & B \\ H & G & F & E & D & C & B & A \end{bmatrix}.$$

Let $(P, Q) \in \mathrm{PermAut}(X_{2^i})$. The diagonal of X_{2^i} must be fixed, so $Q = P$. Also, there are maps $g_i : \mathrm{C}_2^i \to \{A, B, C, D, E, F, G, H\}$ such that $X_{2^i} = [g_i(xy)]_{x,y \in \mathrm{C}_2^i}$. By (3.5), $X_{2^i} = \sum_{x \in \mathrm{C}_2^i} g_i(x) T_x$. Since g_i is injective, $P X_{2^i} P^{-1} = X_{2^i}$ if and only if P commutes with T_x $(= S_x)$ for all $x \in \mathrm{C}_2^i$. Therefore

$$\mathrm{PermAut}(X_{2^i}) = \{(S_x, S_x) \mid x \in \mathrm{C}_2^i\} \cong \mathrm{C}_2^i$$

by Theorem 3.7.5.

9.5. Automorphism groups of orthogonal designs

We show how to use the permutation automorphism group to calculate the automorphism group of an orthogonal design.

Let G be a group. Recall that each element of $\mathrm{Mon}(v, G)$ may be expressed uniquely as the product of a diagonal matrix with a permutation matrix (in that order). Denote by ϕ the canonical projection of $\mathrm{Mon}(v, G)$ onto $\mathrm{Perm}(v)$ with kernel $\mathrm{Diag}(v, G)$. This induces a surjective homomorphism $\Phi : \mathrm{Mon}(v, G)^2 \to \mathrm{Perm}(v)^2$, where $\Phi(P, Q) = (\phi(P), \phi(Q))$.

Given an orthogonal design D of order v, we define a bipartite graph on the vertices $\{a_1, \ldots, a_v, b_1, \ldots, b_v\}$ by drawing an edge between a_i and b_j if and only if the (i, j)th entry of D is non-zero. Let $X = [x_{ij}]$ be the array obtained by dropping the signs in D.

9.5.1. THEOREM. *For D, X, and Φ as above, $\Phi(\mathrm{Aut}(D)) \leq \mathrm{PermAut}(X)$. If the number of connected components in the graph of D is k then $\ker \Phi \cap \mathrm{Aut}(D)$ is elementary abelian of order 2^k. In particular, $\ker \Phi \cap \mathrm{Aut}(D) = \langle (-I_v, -I_v) \rangle$ if this graph is connected.*

PROOF. First observe that $PDQ^* = D$ implies $\phi(P) X \phi(Q)^\top = X$. So Φ maps $\mathrm{Aut}(D)$ into $\mathrm{PermAut}(X)$.

It remains to prove the statements about $\ker \Phi$. We discuss the case where the graph of D is connected; the general result is then an easy extension. Each element in the kernel has the form $(\mathrm{diag}\,(d_1, \ldots, d_v), \mathrm{diag}\,(e_1, \ldots, e_v))$, where $d_i, e_i = \pm 1$ and

$$\mathrm{diag}\,(d_1, \ldots, d_v) X \,\mathrm{diag}\,(e_1, \ldots, e_v) = X.$$

We show that

$$d_1 = \cdots = d_v = e_1 = \cdots = e_v.$$

For all $i, j \in \{1, \ldots, v\}$, $x_{ij} d_i e_j = x_{ij}$. Thus, if $x_{ij} \neq 0$ then $e_j = d_i$. Now if a_i is joined to b_j and b_k, then x_{ij} and x_{ik} are non-zero, and $e_j = d_i = e_k$. Indeed, if there is a path from b_1 to b_j, then because the graph is bipartite, every second node in the path is in $\{a_1, \ldots, a_v\}$, and every other node is in $\{b_1, \ldots, b_v\}$. Moreover, if

b_l appears in the path, then $e_l = e_1$. Connectivity now implies that $e_1 = \cdots = e_v$. Similarly, $d_1 = \cdots = d_v$. □

An element of $\Phi^{-1}(\mathrm{PermAut}(X))$ is an automorphism of D if and only if it preserves the signs of D. Therefore

9.5.2. THEOREM. *Let D be an orthogonal design. Let X be the array obtained by dropping the signs in D, and let W be the weighing matrix obtained by setting all indeterminates in D to 1. Then*

$$(9.11) \qquad \mathrm{Aut}(D) = \mathrm{Aut}(W) \cap \Phi^{-1}(\mathrm{PermAut}(X)).$$

PROOF. We have

$$[\textstyle\sum_{k,l} \delta^i_{\pi(k)}\, s(i)\, w_{kl}\, x_{kl}\, \delta^j_{\phi(l)} t(j)]_{ij} = [w_{ij} x_{ij}]_{ij}$$

if and only if

$$[\textstyle\sum_{k,l} \delta^i_{\pi(k)}\, s(i)\, w_{kl}\, \delta^j_{\phi(l)} t(j)]_{ij} = [w_{ij}]_{ij}$$

and

$$[\textstyle\sum_{k,l} \delta^i_{\pi(k)}\, x_{kl}\, \delta^j_{\phi(l)}]_{ij} = [x_{ij}]_{ij}. \qquad \square$$

9.5.3. EXAMPLE. We determine the automorphism groups of the orthogonal designs O_2, $O_4 = O_4^{(1)}$, and O_8 of Example 4.4.14. Let X_i be the array obtained by dropping the signs in O_i, and let W_i be the weighing matrix obtained by setting the indeterminates in O_i to 1. Since the O_i contain no zeros, their graphs are connected (and complete). Therefore $\ker \Phi \cap \mathrm{Aut}(O_i) = \langle (-I_i, -I_i) \rangle$ by Theorem 9.5.1. We found $\mathrm{PermAut}(X_i)$ in Example 9.4.3. By (9.11), it remains to see which elements of $\mathrm{PermAut}(X_i)$ lift to $\mathrm{Aut}(W_i)$.

We begin with O_2. Note that $\mathrm{Aut}(O_2)$ is a subgroup of the automorphism group $\langle \alpha, \beta \rangle \cong \mathrm{D}_8$ found in Example 9.1.2. Since β is an automorphism of O_2, but $\Phi(\alpha) \notin \mathrm{PermAut}(X_2)$, it follows that $\mathrm{Aut}(O_2) = \langle \beta \rangle \cong \mathrm{C}_4$.

Next, we calculate $\mathrm{Aut}(O_4)$. Testing the template

$$\left(\begin{bmatrix} 0 & a & 0 & 0 \\ b & 0 & 0 & 0 \\ 0 & 0 & 0 & c \\ 0 & 0 & d & 0 \end{bmatrix}, \begin{bmatrix} 0 & e & 0 & 0 \\ f & 0 & 0 & 0 \\ 0 & 0 & 0 & g \\ 0 & 0 & h & 0 \end{bmatrix} \right)$$

on W_4 yields

$$ae = de = ag = dg = bf = cf = bh = ch = -1,$$
$$af = df = be = ce = bg = cg = ah = dh = 1.$$

If $a = 1$ then $d = f = h = 1$, and $b = c = e = g = -1$. This gives the automorphism

$$\alpha = \left(\begin{bmatrix} 0 & 1 & 0 & 0 \\ - & 0 & 0 & 0 \\ 0 & 0 & 0 & - \\ 0 & 0 & 1 & 0 \end{bmatrix}, \begin{bmatrix} 0 & - & 0 & 0 \\ 1 & 0 & 0 & 0 \\ 0 & 0 & 0 & - \\ 0 & 0 & 1 & 0 \end{bmatrix} \right)$$

of O_4. Then by testing the back-diagonal template, we see that

$$\beta = \left(\begin{bmatrix} 0 & 0 & 0 & 1 \\ 0 & 0 & - & 0 \\ 0 & 1 & 0 & 0 \\ - & 0 & 0 & 0 \end{bmatrix}, \begin{bmatrix} 0 & 0 & 0 & - \\ 0 & 0 & - & 0 \\ 0 & 1 & 0 & 0 \\ 1 & 0 & 0 & 0 \end{bmatrix} \right)$$

is an automorphism of O_4. Thus

$$\mathrm{Aut}(O_4) = \langle \alpha, \beta \mid \alpha^2 = \beta^2 = (-I_4, -I_4),\ \alpha\beta = -\beta\alpha \rangle \cong \mathrm{Q}_8.$$

Finally we come to O_8. Here the only automorphisms are $(\pm I_8, \pm I_8)$. The verification of this fact relies on the observation that a non-identity element (M, M) of $\operatorname{PermAut}(X_8) \cong \operatorname{C}_2^3$ can be lifted to $(P, Q) \in \operatorname{Aut}(O_8)$ if and only if the entrywise product of MW_8M and W_8 has rank 1.

9.6. Expanded designs

In this section we discuss expanded designs of a pairwise combinatorial design D. Our aim is to set up an embedding of $\operatorname{Aut}(D)$ into the permutation automorphism group of an expanded design of D. We also derive sufficient conditions for this embedding to be an isomorphism.

9.6.1. Definitions, examples, and basic properties. Let D be a PCD(Λ). The (*full*) *expanded design of* D is the design $\mathcal{E}(D)$ with incidence matrix

$$(9.12) \qquad \mathcal{E}(D) \approx [rDc]_{r \in R, c \in C}.$$

Notice that $\mathcal{E}(D)$ need not be square. An incidence matrix for the *lesser expanded design* $\mathcal{E}_{K,L}(D)$, where $K \leq R$ and $L \leq C$, is defined by

$$(9.13) \qquad \mathcal{E}_{K,L}(D) \approx [kD\ell]_{k \in K, \ell \in L}.$$

In this chapter we work almost exclusively with the full expanded design $\mathcal{E}(D)$ (Chapter 14 deals with a certain lesser expanded design). Many of the results for $\mathcal{E}(D)$ generalize to results for lesser expanded designs $\mathcal{E}_{K,L}(D)$, and we will state some of these generalizations, without proof. Henceforth we refer to $\mathcal{E}(D)$ as the expanded design of D.

In matrix calculations, $\mathcal{E}(D)$ denotes a particular incidence matrix over an ambient ring for Λ. We point out that the definition of $\mathcal{E}(D)$ does not depend on the choice of incidence matrix for D (this is a consequence of Lemma 9.6.8 below). The right-hand sides of (9.12) and (9.13) are *block matrix forms* for the respective expanded designs.

9.6.1. EXAMPLE. The arrays

$$\begin{bmatrix} 1 & 1 & -1 & -1 \\ 1 & -1 & -1 & 1 \\ -1 & -1 & 1 & 1 \\ -1 & 1 & 1 & -1 \end{bmatrix}, \qquad \begin{bmatrix} A & B & C & D & -A & -B & -C & -D \\ B & -A & D & -C & -B & A & -D & C \\ C & -D & -A & B & -C & D & A & -B \\ D & C & -B & -A & -D & -C & B & A \\ -A & -B & -C & -D & A & B & C & D \\ -B & A & -D & C & B & -A & D & -C \\ -C & D & A & -B & C & -D & -A & B \\ -D & -C & B & A & D & C & -B & -A \end{bmatrix}$$

are block matrix forms for $\mathcal{E}(H_2)$ and $\mathcal{E}(O_4)$.

There is a nice relationship between the Grammians of D and of $\mathcal{E}(D)$:

$$\mathcal{E}(D)\mathcal{E}(D)^* = [rDc]_{r \in R, c \in C}[c^{-1}D^*s^{-1}]_{c \in C, s \in R}$$
$$= [\textstyle\sum_{c \in C} rDcc^{-1}D^*s^{-1}]_{r,s \in R}$$
$$= |C|[rDD^*s^{-1}]_{r,s \in R}.$$

Similarly, $\mathcal{E}(D)^*\mathcal{E}(D) = |R|[c^{-1}D^*Dd]_{c,d \in C}$.

Although we usually employ the block matrix form for $\mathcal{E}(D)$, sometimes it is more convenient to use other row and column orderings. Moreover, different incidence matrices for the expanded design have already appeared in the literature. We discuss some of these cases now.

Suppose that $\mathcal{A} = R = C = G$, say. We get an incidence matrix for $\mathcal{E}(D)$ by replacing each 0 in D by the $|G| \times |G|$ matrix of zeros, and replacing each entry g in D by

$$[xgy]_{x,y \in G} = [ab]_{a,b \in G}[\delta_{gy}^x]_{x,y \in G} = [ab]_{a,b \in G}S_g.$$

If we let S_0 denote the $|G| \times |G|$ matrix of zeros, then

$$\mathcal{E}(D) \approx [[ab]_{a,b \in G}S_{d_{ij}}]_{1 \leq i,j \leq v}.$$

We call this the *plug-in form for the expanded design.* If we write $[D]_I$ for the block matrix $[S_{d_{ij}}]_{ij}$ then

(9.14) $$\mathcal{E}(D) \approx (I_v \otimes [ab]_{a,b \in G})[D]_I.$$

9.6.2. EXAMPLE. We show $\mathcal{E}(H_2)$ in plug-in form, and $[H_2]_{I_2}$:

$$\mathcal{E}(H_2) \approx \begin{bmatrix} 1 & -1 & 1 & -1 \\ -1 & 1 & -1 & 1 \\ 1 & -1 & -1 & 1 \\ -1 & 1 & 1 & -1 \end{bmatrix}, \qquad [H_2]_{I_2} = \begin{bmatrix} 1 & 0 & 1 & 0 \\ 0 & 1 & 0 & 1 \\ 1 & 0 & 0 & 1 \\ 0 & 1 & 1 & 0 \end{bmatrix}.$$

Notice that

$$\begin{bmatrix} 1 & -1 & 1 & -1 \\ -1 & 1 & -1 & 1 \\ 1 & -1 & -1 & 1 \\ -1 & 1 & 1 & -1 \end{bmatrix} = \begin{bmatrix} 1 & -1 & 0 & 0 \\ -1 & 1 & 0 & 0 \\ 0 & 0 & 1 & -1 \\ 0 & 0 & -1 & 1 \end{bmatrix} [H_2]_{I_2},$$

as expected by (9.14).

When G is abelian we have $[xgy]_{x,y \in G} = g[xy]_{x,y \in G}$, and

$$\mathcal{E}(D) \approx D \otimes [ab]_{a,b \in G}.$$

We call this the *abelian plug-in form for the expanded design.* The same formula holds for orthogonal designs if we choose the usual ambient ring $\mathbb{Z}[x_1, \ldots, x_r]$, where $x_1, \ldots, x_r$ are the indeterminates appearing in the design. There is an analogous notion of $[D]_{I_2}$ for orthogonal designs such that (9.14) holds. We will examine this issue more fully when we come to the *associated design.*

9.6.3. EXAMPLE. We show $[O_4]_{I_2}$ and the plug-in form for $\mathcal{E}(O_4)$:

$$\begin{bmatrix} A & 0 & B & 0 & C & 0 & D & 0 \\ 0 & A & 0 & B & 0 & C & 0 & D \\ B & 0 & 0 & A & D & 0 & 0 & C \\ 0 & B & A & 0 & 0 & D & C & 0 \\ C & 0 & 0 & D & 0 & A & B & 0 \\ 0 & C & D & 0 & A & 0 & 0 & B \\ D & 0 & C & 0 & 0 & B & 0 & A \\ 0 & D & 0 & C & B & 0 & A & 0 \end{bmatrix}, \qquad \begin{bmatrix} A & -A & B & -B & C & -C & D & -D \\ -A & A & -B & B & -C & C & -D & D \\ B & -B & -A & A & D & -D & -C & C \\ -B & B & A & -A & -D & D & C & -C \\ C & -C & -D & D & -A & A & B & -B \\ -C & C & D & -D & A & -A & -B & B \\ D & -D & C & -C & -B & B & -A & A \\ -D & D & -C & C & B & -B & A & -A \end{bmatrix}.$$

9.6.2. The homomorphisms $\theta_G^{(1)}$, $\theta_G^{(2)}$ and $\Theta_{R,C}$. Let G be a finite group. For each $X \in \mathrm{Mon}(v, G)$, there are unique disjoint $(0,1)$-matrices X_g, $g \in G$, such that $X = \sum_{g \in G} gX_g$. Define

$$\theta_G^{(1)}(X) = \sum_{g \in G} T_g \otimes X_g \qquad \text{and} \qquad \theta_G^{(2)}(X) = \sum_{g \in G} S_g \otimes X_g.$$

Modulo row and column permutations that change the order in Kronecker products, $\theta_G^{(1)}$ (respectively, $\theta_G^{(2)}$) blows X up to a $v|G| \times v|G|$ matrix, replacing each 0 in X by the $|G| \times |G|$ zero matrix, and replacing a non-zero entry g by T_g (respectively, S_g). Hence $\theta_G^{(1)}(X)$ and $\theta_G^{(2)}(X)$ are permutation matrices.

9.6.4. LEMMA. $\theta_G^{(1)}, \theta_G^{(2)}$ are embeddings of $\mathrm{Mon}(v, G)$ into $\mathrm{Perm}(v|G|)$.

PROOF. We prove the result for $\theta_G^{(1)}$ only. If $X, Y \in \mathrm{Mon}(v, G)$ then

$$\theta_G^{(1)}(XY) = \theta_G^{(1)}\left(\sum_{g,h\in G} gh X_g Y_h\right)$$
$$= \theta_G^{(1)}\left(\sum_{a\in G} a \sum_{g\in G} X_g Y_{g^{-1}a}\right)$$
$$= \sum_{a\in G} T_a \otimes \left(\sum_{g\in G} X_g Y_{g^{-1}a}\right)$$
$$= \sum_{g\in G} \sum_{h\in G} (T_{gh} \otimes X_g Y_h)$$
$$= \left(\sum_{g\in G} T_g \otimes X_g\right)\left(\sum_{h\in G} T_h \otimes Y_h\right)$$
$$= \theta_G^{(1)}(X)\, \theta_G^{(1)}(Y).$$

Also, $\theta_G^{(1)}\left(\sum_g g X_g\right) = I_{v|G|} \Leftrightarrow X_1 = I_v$ and X_g is the zero matrix for $g \neq 1$.   □

9.6.5. REMARK. Both $\theta_G^{(1)}$ and $\theta_G^{(2)}$ may be extended to ring homomorphisms from $\mathrm{Mat}(v, \mathbb{Q}[G])$ to $\mathrm{Mat}(v|G|, \mathbb{Q})$.

We introduce notation for two subgroups of $\mathrm{Perm}(v|G|)$:

$$Z_G^{(1)} = \{S_g \otimes I_v \mid g \in G\} \qquad \text{and} \qquad Z_G^{(2)} = \{T_g \otimes I_v \mid g \in G\}.$$

That is, $Z_G^{(1)}$ (respectively, $Z_G^{(2)}$) is the image under $\theta_G^{(2)}$ (respectively, $\theta_G^{(1)}$) of the scalar subgroup $\{g I_v \mid g \in G\}$ of $\mathrm{Mon}(v, G)$.

9.6.6. LEMMA. $\theta_G^{(i)}(\mathrm{Mon}(v, G))$ is the centralizer of $Z_G^{(i)}$ in $\mathrm{Perm}(v|G|)$.

PROOF. We prove the result for $i = 1$ only. By Theorem 3.7.7, a $v|G| \times v|G|$ integer matrix A commutes with $S_a \otimes I_v$ for all $a \in G$ if and only if it has the form $\sum_{a\in G} T_a \otimes g(a)$. Moreover, A is a permutation matrix if and only if $\sum_{a\in G} g(a)a$ is a monomial matrix.   □

Lemma 9.6.4 allows us to define the injective homomorphism

$$\Theta_{R,C} : \mathrm{Mon}(v, R) \times \mathrm{Mon}(v, C) \to \mathrm{Perm}(v|R|) \times \mathrm{Perm}(v|C|)$$

by

$$\Theta_{R,C} : (P, Q) \mapsto (\theta_R^{(1)}(P), \theta_C^{(2)}(Q)).$$

Set $Z_{R,C} = \{(S_r \otimes I_v, T_c \otimes I_v) \mid r \in R, c \in C\} = Z_R^{(1)} \times Z_C^{(2)}$. Lemma 9.6.6 implies the following.

9.6.7. THEOREM. $\Theta_{R,C}(\mathrm{Mon}(v, R) \times \mathrm{Mon}(v, C))$ is the centralizer of $Z_{R,C}$ in $\mathrm{Perm}(v|R|) \times \mathrm{Perm}(v|C|)$.

9.6.3. An embedding of the automorphism group. In this subsection we embed the automorphism group of a design in the permutation automorphism group of its expanded design.

9.6.8. LEMMA. If $(P, Q) \in \mathrm{Mon}(v, R) \times \mathrm{Mon}(v, C)$ then

$$\theta_R^{(1)}(P)[rDc]_{r\in R, c\in C} = [rPDc]_{r\in R, c\in C}$$
$$[rDc]_{r\in R, c\in C}\, \theta_C^{(2)}(Q)^\top = [rDQ^*c]_{r\in R, c\in C}.$$

Therefore $(P, Q) \in \mathrm{Aut}(D)$ if and only if $\Theta_{R,C}((P, Q)) \in \mathrm{PermAut}(\mathcal{E}(D))$.

PROOF. We prove the first equation. Write $P = \sum_{g \in R} g P_g$ for $(0,1)$-matrices P_g. Then

$$\theta_R^{(1)}(P)[rDc]_{r \in R, c \in C} = (\sum_{g \in R} T_g \otimes P_g)[rDc]_{r \in R, c \in C}$$
$$= \sum_{g \in R} [\sum_{s \in R} \delta_s^{rg} P_g s Dc]_{r \in R, c \in C}$$
$$= \sum_{g \in R} [P_g rg Dc]_{r \in R, c \in C}$$
$$= [r \sum_{g \in R} g P_g Dc]_{r \in R, c \in C}$$
$$= [rPDc]_{r \in R, c \in C}. \qquad \square$$

9.6.9. REMARK. Lemma 9.6.8 generalizes. That is,

$$\theta_K^{(1)}(P) \mathcal{E}_{K,L}(D) \theta_L^{(2)}(Q)^\top = \mathcal{E}_{K,L}(PDQ^*)$$

for any $P \in \mathrm{Mon}(v, K)$ and $Q \in \mathrm{Mon}(v, L)$.

Lemma 9.6.8 (or Remark 9.6.9) now gives a result cited (but not proved) earlier; namely, that the choice of incidence matrix for a design D is irrelevant in defining an expanded design of D.

9.6.10. COROLLARY. *If D and D' are $\mathrm{PCD}(\Lambda)$s such that $D' = MDN$ for some $M \in \mathrm{Mon}(v, K)$ and $N \in \mathrm{Mon}(v, L)$, then $\mathcal{E}_{K,L}(D') \approx \mathcal{E}_{K,L}(D)$.*

The following theorem is the main result of this subsection.

9.6.11. THEOREM. *Let D be a $\mathrm{PCD}(\Lambda)$. Then $\Theta_{R,C}$ is an isomorphism from $\mathrm{Aut}(D)$ onto the centralizer of $Z_{R,C}$ in $\mathrm{PermAut}(\mathcal{E}(D))$.*

PROOF. By Lemma 9.6.8,

$$\Theta_{R,C}(\mathrm{Aut}(D)) = \mathrm{PermAut}(\mathcal{E}(D)) \cap \Theta_{R,C}(\mathrm{Mon}(v, R) \times \mathrm{Mon}(v, C)).$$

Also, by Theorem 9.6.7,

$$\Theta_{R,C}(\mathrm{Mon}(v, R) \times \mathrm{Mon}(v, C)) = C_{\mathrm{Perm}(v|R|) \times \mathrm{Perm}(v|C|)}(Z_{R,C}).$$

So $\Theta_{R,C}(\mathrm{Aut}(D))$ is the set of elements of $\mathrm{PermAut}(\mathcal{E}(D))$ that centralize $Z_{R,C}$, as required. $\qquad \square$

In many cases, $\mathrm{Aut}(D)$ is isomorphic to $\mathrm{PermAut}(\mathcal{E}(D))$. Suppose that X is a $v \times v$ $(0, \mathcal{A})$-array. Recall from Definition 6.2.1 that $\mathrm{wt}(X)$ (respectively, $\mathrm{df}(X)$) is the set of diagonal (respectively, off-diagonal) elements of the Grammian of X over $\mathcal{R}$, and that X is *non-degenerate* if the sets $r_1 \mathrm{df}(X) r_2$ and $r_3 \mathrm{wt}(X) r_4$ for all r_1, r_2, r_3, r_4 in the row group R are mutually disjoint.

9.6.12. THEOREM. *Let D be a $\mathrm{PCD}(\Lambda)$. If D and D^* are non-degenerate, then $\Theta_{R,C}$ maps $\mathrm{Aut}(D)$ onto $\mathrm{PermAut}(\mathcal{E}(D))$.*

PROOF. We work with the block matrix form of $\mathcal{E}(D)$. It suffices to show that any element (P, Q) of $\mathrm{PermAut}(\mathcal{E}(D))$ centralizes $Z_{R,C}$. We have

$$\mathcal{E}(D) \mathcal{E}(D)^* = |C|[r_1 DD^* r_2^{-1}]_{r_1, r_2 \in R},$$

so

$$P[r_1 DD^* r_2^{-1}]_{r_1, r_2 \in R} P^{-1} = [r_1 DD^* r_2^{-1}]_{r_1, r_2 \in R}.$$

Then because D is non-degenerate, and P is a permutation matrix,

$$P[r_1 I_v r_2^{-1}]_{r_1, r_2 \in R} P^{-1} = [r_1 I_v r_2^{-1}]_{r_1, r_2 \in R}.$$

That is,

$$P(\textstyle\sum_{r \in R} r[\delta^x_{ry}]_{x,y \in R} \otimes I_v)P^{-1} = \textstyle\sum_{r \in R} r[\delta^x_{ry}]_{x,y \in R} \otimes I_v.$$

So P centralizes $S_r \otimes I_v$ for all $r \in R$. Similarly,

$$Q[c_1^{-1}D^*Dc_2]_{c_1,c_2 \in C}Q^{-1} = [c_1^{-1}D^*Dc_2]_{c_1,c_2 \in C},$$

and non-degeneracy of D^* implies that Q commutes with $T_c \otimes I_v$ for all $c \in C$. $\quad\square$

9.6.13. EXAMPLE. Here is an example of the argument that was used in the proof of Theorem 9.6.12. Let $(P,Q) \in \mathrm{PermAut}(\mathcal{E}(H_2))$. We use the plug-in form of Example 9.6.2 for $\mathcal{E}(H_2)$. Since

$$\mathcal{E}(H_2)\mathcal{E}(H_2)^\top = P\mathcal{E}(H_2)Q^\top Q\mathcal{E}(H_2)^\top P^\top = P\mathcal{E}(H_2)\mathcal{E}(H_2)^\top P^\top,$$

P is a 4×4 permutation matrix that commutes with

$$\mathcal{E}(H_2)\mathcal{E}(H_2)^\top = 4\begin{bmatrix} 1 & -1 & 0 & 0 \\ -1 & 1 & 0 & 0 \\ 0 & 0 & 1 & -1 \\ 0 & 0 & -1 & 1 \end{bmatrix} = 4I_2 \otimes I_2 - 4I_2 \otimes M,$$

where M is the non-trivial element of $\mathrm{Perm}(2)$. Thus $P(I_2 \otimes M) = (I_2 \otimes M)P$. If we write

$$P = \begin{bmatrix} A & B \\ C & D \end{bmatrix},$$

where A, B, C, D are 2×2 $(0,1)$-matrices, then

$$\begin{bmatrix} AM & BM \\ CM & DM \end{bmatrix} = \begin{bmatrix} MA & MB \\ MC & MD \end{bmatrix},$$

so that A, B, C, D all commute with M. Since P is a permutation matrix, this means that either $A, D \in \{I, M\}$ and $B = C = 0$, or $B, C \in \{I, M\}$ and $A = D = 0$. Each of these eight possibilities is the image under $\theta_R^{(1)}$ of a 2×2 monomial matrix over C_2. This is in accord with Example 9.1.2, where we showed that $\mathrm{Aut}(H_2) \cong \mathrm{Mon}(2, C_2)$.

By Theorems 6.2.2 and 9.6.12, we have the following fundamental result.

9.6.14. THEOREM. *Let Λ be any one of the non-trivial orthogonality sets listed in Definition 2.13.1, and let D be a $\mathrm{PCD}(\Lambda)$. Then $\Theta_{R,C}$ is an isomorphism from $\mathrm{Aut}(D)$ onto $\mathrm{PermAut}(\mathcal{E}(D))$.*

9.7. Computing automorphism groups

Algorithms for computing the automorphism group of a $(0,1)$-array, due to Leon [114] and McKay [120], have been implemented in MAGMA [14]. This chapter has shown the need for methods to calculate the automorphism group of an array over alphabets that contain more than one non-zero element. We outline below a naive backtrack algorithm for this task.

It is reassuring that such simple ideas lead to an algorithm that is often quite effective. However, computing these more general automorphism groups offers an interesting avenue of research.

4. RESEARCH PROBLEM. *Develop efficient algorithms for computing the automorphism groups of $(0, \mathcal{A})$-arrays.*

Let $D = [d_{ij}]_{ij}$ be a $v \times v$ $(0, \mathcal{A})$-array. We give an algorithm to compute PermAut(D). Similar ideas can be used to compute the full automorphism group of D directly when D is a PCD(Λ). Alternatively, one could compute PermAut$(\mathcal{E}(D))$, and then invoke Theorem 9.6.11 or Theorem 9.6.14.

We list all pairs of permutations $(\pi, \phi) \in \mathrm{Sym}(v)^2$ such that $d_{\pi(i), \phi(j)} = d_{ij}$ for all $i, j \in \{1, 2, \ldots, v\}$. Let $\mathcal{G}$ be the group of such pairs. Sometimes we only require partial information about π and ϕ to be sure that (π, ϕ) is not in $\mathcal{G}$. Therefore, we will use a standard data structure, called a *tree*, to take advantage of this feature of the problem.

The details of the tree T depend on the order in which we unveil information defining π and ϕ. As an example, suppose that we have found

$$(9.15) \qquad \pi(1), \phi(1), \pi(2), \phi(2), \ldots, \pi(v), \phi(v)$$

in that order (of course, once we know $\pi(1), \ldots, \pi(v-1)$ and $\phi(1), \ldots, \phi(v-1)$, then $\pi(v)$ and $\phi(v)$ are determined). The corresponding (inverted) tree T begins with the *root* node $N_{()}$. This root node is connected by edges to v *daughter nodes* $N_{(a_1)}$. The daughter node $N_{(a_1)}$ corresponds to assuming that $\pi(1) = a_1$. These nodes are said to be at level one. Each of them is connected to v daughters $N_{(a_1, b_1)}$ which correspond to the assignments $\pi(1) = a_1$ and $\phi(1) = b_1$. These nodes are at level two. So there are v nodes at level one and v^2 nodes at level two. Since $\pi(2) \neq \pi(1)$, each of the nodes at level two has just $v - 1$ daughter nodes $N_{(a_1, b_1, a_2)}$ where $a_2 \neq a_1$. These nodes are at level three, and each corresponds to a possible assignment $\pi(1) = a_1$, $\phi(1) = b_1$, $\pi(2) = a_2$. Continuing down the tree, we reach nodes $N_{(a_1, b_1, a_2, \ldots, b_v)}$ at level $2v$. Each of these corresponds to a fully-determined pair of permutations (π, ϕ), where all the images in (9.15) have been assigned values. These *leaf nodes* have no daughters.

There is a standard way to traverse T. The generic step is as follows.

process(node x)

(1) *Test node.* Run the routine **test**(x), which tests the partial information assumed to reach node x.

(2) *Output.* If x is a leaf node, and it passes the test, then output some information.

(3) *Branch.* If x has daughter nodes, and it passes the test, then, in some prescribed order that may depend on x, run the routine **process**(y) for each daughter y of x.

Note that we only visit the nodes in a subtree S of T, determined by the test function **test**(x). In our case, the node $N_{(a_1, b_1, \ldots, a_k)}$ passes the test if

$$d_{a_k, b_j} = d_{kj} \qquad \text{for} \qquad j = 1, \ldots, k-1,$$

and the node $N_{(a_1, b_1, \ldots, a_k, b_k)}$ passes the test if

$$d_{a_i, b_k} = d_{ik} \qquad \text{for} \qquad i = 1, \ldots, k.$$

If a leaf node is reached, and it passes the test, then a particular element of PermAut(D) has been found, specified by the output $a_1, b_1, \ldots, a_v, b_v$.

The computing resources consumed by the algorithm depend on how many nodes are visited at each level. We can estimate how expensive the algorithm will be for a given array D. For example, using a random number generator, we can quickly estimate the number of nodes in the subtree S by modifying the function **test**(x) so that nodes fail their test.

The following heuristic gives some understanding of how much work is done when the automorphism group is trivial. For $N_{(a_1,b_1,...,a_{k+1})}$ to be reached, we must have $d_{a_i,b_j} = d_{ij}$ for all i,j. If $|\mathcal{A}| = m$, then, as a rough guide, we might suppose that about m^{-k^2} of the nodes at level $2k + 1$ are reached. Since there are fewer than v^{2k+1} nodes at level $2k + 1$, we expect that fewer than $m^{-k^2}v^{2k+1}$ nodes at level $2k + 1$ are visited. This number is maximized when k is about $\log_m v$. So, assuming that there are no automorphisms, we estimate the searched subtree to have around $w = v^{\log_m v}$ nodes. For example, if D is a Hadamard matrix of order 128 then $\mathcal{E}(D)$ has order $v = 256$, and $w = 2^{64}$; but if D is a GH$(256; C_{16})$ then $\mathcal{E}(D)$ has order $v = 4096$, and $w = 2^{36}$, which is manageable.

We now comment on approaches to Research Problem 4. Firstly, one should see whether the methods of Leon and McKay carry over to the general problem. Secondly, to handle smaller alphabets, it would probably be better to organize the search for automorphisms using a tree that reflects knowledge of π (or ϕ) only. Thirdly, of course one should try to find a generating set for the automorphism group rather than to list elements. Once a subgroup H has been found, the search for the next automorphism α might avail of what is known about the coset αH.

9.8. The associated design

The associated design provides one more way to compute the automorphism group of a small design.

Select a complete irredundant set of orbit representatives $\mathcal{T}$ for the action of $R \cong \Pi_\Lambda^{\text{row}}$ on the alphabet $\mathcal{A}$. Assume that R acts semiregularly on $\mathcal{A}$. Then for any PCD(Λ), D, we may define the $(0, \mathcal{T})$-matrices D_r $(r \in R)$ by

$$\mathcal{E}(D) = \sum_{r \in R} r D_r$$

where $\mathcal{E}(D)$ is in block matrix form (say). Then

$$A_D = D_1$$

is the *associated design of D* (determined by $\mathcal{T}$).

The size of the alphabet for A_D is significantly smaller than that of $\mathcal{E}(D)$. If the action of R on $\mathcal{A}$ is regular then we may even take $\mathcal{T} = \{1\}$, and A_D is a $(0, 1)$-array.

9.8.1. EXAMPLE. In Example 9.6.1 we gave expanded designs of the Hadamard matrix H_2 and orthogonal design O_4. Take $\mathcal{T} = \{1\}$ in the former case and $\mathcal{T} = \{A, B, C, D\}$ in the latter. We have

$$A_{H_2} = \begin{bmatrix} 1 & 1 & 0 & 0 \\ 1 & 0 & 0 & 1 \\ 0 & 0 & 1 & 1 \\ 0 & 1 & 1 & 0 \end{bmatrix}, \qquad A_{O_4} = \begin{bmatrix} A & B & C & D & 0 & 0 & 0 & 0 \\ B & 0 & D & 0 & 0 & A & 0 & C \\ C & 0 & 0 & B & 0 & D & A & 0 \\ D & C & 0 & 0 & 0 & 0 & B & A \\ 0 & 0 & 0 & 0 & A & B & C & D \\ 0 & A & 0 & C & B & 0 & D & 0 \\ 0 & D & A & 0 & C & 0 & 0 & B \\ 0 & 0 & B & A & D & C & 0 & 0 \end{bmatrix}.$$

The associated designs were obtained from the expanded designs by replacing all entries not in $\mathcal{T}$ with zeros.

The associated design A_D is independent of the choice of incidence matrix for D. We have a stronger result.

9.8.2. THEOREM. *If $D' \approx_\Lambda D$ then $A_{D'} \approx A_D$.*

PROOF. By Corollary 9.6.10, there exist permutation matrices U and V such that
$$\sum_{r \in R} r D'_r = \mathcal{E}(D') = U\mathcal{E}(D)V^\top = \sum_{r \in R} rU D_r V^\top.$$
So $A_{D'} = D'_1 = U D_1 V^\top = U A_D V^\top$. $\qquad\square$

However, in general, the equivalence class of A_D does depend on the choice of the set $\mathcal{T}$ of R-orbit representatives. We identify one important exception.

9.8.3. LEMMA. *For all $r \in R$, $D_r = (S_r \otimes I_v)D_1$.*

PROOF. We have
$$\begin{aligned}
\sum_{s \in R} s D_{rs} &= \sum_{s \in R} r^{-1}s D_s \\
&= r^{-1}[xDc]_{x \in R, c \in C} \\
&= [r^{-1}xDc]_{x \in R, c \in C} \\
&= [\delta_y^{r^{-1}x} I_v]_{x,y \in R}[yDc]_{y \in R, c \in C} \\
&= ([\delta_{ry}^x]_{x,y \in R} \otimes I_v)\sum_{s \in R} s D_s.
\end{aligned}$$
Since R acts semiregularly on $\mathcal{A}$, the result follows. $\qquad\square$

9.8.4. THEOREM. *If R acts regularly on $\mathcal{A}$ then, up to permutation equivalence, A_D is independent of the choice of $\mathcal{T}$.*

PROOF. Replacing $\mathcal{T} = \{ra\}$ with $\{a\}$ replaces D_1 with D_r, so Lemma 9.8.3 implies the result. $\qquad\square$

Below we prove

9.8.5. THEOREM.
$$\Theta_{R,C}(\mathrm{Aut}(D)) = \mathrm{PermAut}(A_D) \cap C_{\mathrm{Perm}(v|R|) \times \mathrm{Perm}(v|C|)}(Z_{R,C}).$$

Thus, although a PCD(Λ) may have several inequivalent associated designs, their permutation automorphism groups all intersect in a large common subgroup.

9.8.6. LEMMA. $\Theta_{R,C}(\mathrm{Aut}(D)) \le \mathrm{PermAut}(\mathcal{E}(D)) \le \mathrm{PermAut}(A_D)$.

PROOF. The first inclusion follows from Lemma 9.6.8. For the other, let P and Q be permutation matrices such that $P\mathcal{E}(D)Q^\top = \mathcal{E}(D)$. Then the $(0, \mathcal{T})$-entries of $P\mathcal{E}(D)Q^\top$ and $\mathcal{E}(D)$ must agree: thus $PA_D Q^\top = A_D$. $\qquad\square$

By Lemma 9.8.3,
$$\mathcal{E}(D) = \sum_{r \in R} r D_r = \sum_{r \in R} r([\delta_{ry}^x]_{x,y \in R} \otimes I_v)D_1 = ([xy^{-1}]_{x,y \in R} \otimes I_v)D_1,$$
so
$$(9.16) \qquad\qquad \mathcal{E}(D) = ([xy^{-1}]_{x,y \in R} \otimes I_v)A_D.$$

A companion identity for the designs in plug-in form is obtained by switching the order in the Kronecker product. Next we prove

$$(9.17) \qquad \theta_R^{(1)}(P)([xy^{-1}]_{x,y \in R} \otimes I_v) = ([xy^{-1}]_{x,y \in R} \otimes I_v)\theta_R^{(1)}(P)$$

for all $P \in \mathrm{Mon}(v, R)$. We have $[xy^{-1}]_{x,y \in R} = \sum_{r \in R} rS_r$, and

$$\theta_R^{(1)}(P) = \sum_{r \in R} T_r \otimes P_r$$

where the P_r are $(0,1)$-matrices such that $P = \sum_{r \in R} rP_r$. Therefore, since T_a and S_b commute,

$$\theta_R^{(1)}(P)([xy^{-1}]_{x,y \in R} \otimes I_v) = \sum_{a,b \in R} (T_a \otimes P_a)(bS_b \otimes I_v)$$

$$= \sum_{a,b \in R} (bS_b \otimes I_v)(T_a \otimes P_a) = ([xy^{-1}]_{x,y \in R} \otimes I_v)\theta_R^{(1)}(P).$$

We are ready to prove Theorem 9.8.5.

PROOF OF THEOREM 9.8.5. By Theorem 9.6.11 and Lemma 9.8.6, it suffices to show that

$$\mathrm{PermAut}(\mathrm{A}_D) \cap C_{\mathrm{Perm}(v|R|) \times \mathrm{Perm}(v|C|)}(Z_{R,C})$$

$$\leq \mathrm{PermAut}(\mathcal{E}(D)) \cap C_{\mathrm{Perm}(v|R|) \times \mathrm{Perm}(v|C|)}(Z_{R,C}).$$

This is a consequence of Theorem 9.6.7 and the following implications:

$$\theta_R^{(1)}(P)\mathrm{A}_D\theta_C^{(2)}(Q)^\top = \mathrm{A}_D$$

$$\Rightarrow \quad ([xy^{-1}]_{x,y \in R} \otimes I_v)\theta_R^{(1)}(P)\mathrm{A}_D\theta_C^{(2)}(Q)^\top = ([xy^{-1}]_{x,y \in R} \otimes I_v)\mathrm{A}_D$$

$$\Rightarrow \quad \theta_R^{(1)}(P)([xy^{-1}]_{x,y \in R} \otimes I_v)\mathrm{A}_D\theta_C^{(2)}(Q)^\top = ([xy^{-1}]_{x,y \in R} \otimes I_v)\mathrm{A}_D \quad \text{by (9.17)}$$

$$\Rightarrow \quad \theta_R^{(1)}(P)\mathcal{E}(D)\theta_C^{(2)}(Q)^\top = \mathcal{E}(D) \quad \text{by (9.16).}$$

$\square$

So when R acts regularly on $\mathcal{A}$, one gains access to $\mathrm{Aut}(D)$ by examining the automorphism group of the $(0,1)$-array A_D. This is a considerable advantage, since (as we remarked earlier) there are publicly available procedures (for example, as part of MAGMA) for computing the automorphism groups of $(0,1)$-arrays. However, the trade-off is blowing up the size of D to that of A_D.

9.9. Associated designs and group divisible designs

Suppose that $\mathcal{A} = R = C = G$; so we may choose $\mathcal{T} = \{1\}$. Since $[xy]_{x,y \in G} \approx [xy^{-1}]_{x,y \in G}$, we have from (9.16) and (9.14) that

$$([xy]_{x,y \in G} \otimes I_v)\mathrm{A}_D \approx ([xy]_{x,y \in G} \otimes I_v)[S_{d_{ij}}].$$

Thus we may take $\mathrm{A}_D = [S_{d_{ij}}] = \theta_G^{(2)}(D)$.

We show that the associated design of a balanced generalized weighing matrix is a square group divisible design.

9.9.1. DEFINITION. A *group divisible design* $\mathrm{GDD}(vm, k, \lambda; m)$ consists of

(1) a vm-set V of *points*,
(2) a vm-set of k-subsets of V called *lines*, and
(3) a v-set of m-subsets of V called *point classes*,

where

(4) each point appears in k lines,
(5) no line meets a point class in more than one point,
(6) any two points from different point classes lie together in exactly λ lines.

9.9.2. REMARK. Other authors call the 'point classes' in Definition 9.9.1 'groups'.

By grouping together rows of an incidence matrix corresponding to all points in a point class, we see that existence of a GDD$(vm, k, \lambda; m)$ is equivalent to existence of a $vm \times vm$ $(0, 1)$-array A such that

$$J_{vm}A = kJ_{vm} \qquad \text{and} \qquad AA^\top = kI_{vm} + \lambda(J_v - I_v) \otimes J_m.$$

9.9.3. THEOREM. *The associated design* A_D *of a balanced generalized weighing matrix* BGW$(v, k, \mu|G|; G)$, D, *is a group divisible design* GDD$(v|G|, k, \mu; |G|)$.

PROOF. Let $m = |G|$ and $A = A_D = \theta_G^{(2)}(D)$. By Remark 9.6.5,

$$AA^\top = \theta_G^{(2)}(DD^*) = \theta_G^{(2)}(kI_v + \mu G(J_v - I_v)) = kI_{vm} + (J_v - I_v) \otimes \mu J_m.$$

Also, $J_{vm} = \theta_G^{(2)}(GJ_v)$. Thus

$$J_{vm}A = \theta_G^{(2)}(GJ_vD) = \theta_G^{(2)}(kGJ_v) = kJ_{vm}.$$

So A_D is an incidence matrix of a GDD$(vm, k, \mu; m)$. $\square$

9.9.4. EXAMPLE. The BGW$(5, 4, 3; C_3)$

$$W = \begin{bmatrix} 0 & 1 & 1 & 1 & 1 \\ 1 & 0 & 1 & \omega & \omega^2 \\ 1 & 1 & 0 & \omega^2 & \omega \\ 1 & \omega & \omega^2 & 0 & 1 \\ 1 & \omega^2 & \omega & 1 & 0 \end{bmatrix}$$

has associated design

$$A_W = \left[\begin{array}{ccc|ccc|ccc|ccc|ccc} 0 & 0 & 0 & 1 & 0 & 0 & 1 & 0 & 0 & 1 & 0 & 0 & 1 & 0 & 0 \\ 0 & 0 & 0 & 0 & 1 & 0 & 0 & 1 & 0 & 0 & 1 & 0 & 0 & 1 & 0 \\ 0 & 0 & 0 & 0 & 0 & 1 & 0 & 0 & 1 & 0 & 0 & 1 & 0 & 0 & 1 \\ \hline 1 & 0 & 0 & 0 & 0 & 0 & 1 & 0 & 0 & 0 & 1 & 0 & 0 & 0 & 1 \\ 0 & 1 & 0 & 0 & 0 & 0 & 0 & 1 & 0 & 0 & 0 & 1 & 1 & 0 & 0 \\ 0 & 0 & 1 & 0 & 0 & 0 & 0 & 0 & 1 & 1 & 0 & 0 & 0 & 1 & 0 \\ \hline 1 & 0 & 0 & 1 & 0 & 0 & 0 & 0 & 0 & 0 & 0 & 1 & 0 & 1 & 0 \\ 0 & 1 & 0 & 0 & 1 & 0 & 0 & 0 & 0 & 1 & 0 & 0 & 0 & 0 & 1 \\ 0 & 0 & 1 & 0 & 0 & 1 & 0 & 0 & 0 & 0 & 1 & 0 & 1 & 0 & 0 \\ \hline 1 & 0 & 0 & 0 & 1 & 0 & 0 & 0 & 1 & 0 & 0 & 0 & 1 & 0 & 0 \\ 0 & 1 & 0 & 0 & 0 & 1 & 1 & 0 & 0 & 0 & 0 & 0 & 0 & 1 & 0 \\ 0 & 0 & 1 & 1 & 0 & 0 & 0 & 1 & 0 & 0 & 0 & 0 & 0 & 0 & 1 \\ \hline 1 & 0 & 0 & 0 & 0 & 1 & 0 & 1 & 0 & 1 & 0 & 0 & 0 & 0 & 0 \\ 0 & 1 & 0 & 1 & 0 & 0 & 0 & 0 & 1 & 0 & 1 & 0 & 0 & 0 & 0 \\ 0 & 0 & 1 & 0 & 1 & 0 & 1 & 0 & 0 & 0 & 0 & 1 & 0 & 0 & 0 \end{array} \right].$$

This is an incidence matrix of a GDD$(15, 4, 1; 3)$.

9.10. An isomorphism for weighing matrices

By Theorem 9.8.5, $\Theta_{R,C}$ maps Aut(D) into PermAut(A_D). We show that, for many weighing matrices, this mapping is surjective.

Let $W = [w_{ij}]$ be a $v \times v$ weighing matrix. Let $\mathcal{G}_W$ be the graph on v vertices such that the ith vertex is joined to the jth vertex whenever there is a column k of W such that $w_{ik}w_{jk} \neq 0$. So, if W has weight exceeding $v/2$, then $\mathcal{G}_W$ is complete.

9.10.1. THEOREM. *If the graph $\mathcal{G}_W$ of a weighing matrix W is complete, then $\Theta := \Theta_{C_2, C_2}$ is an isomorphism from $\mathrm{Aut}(W)$ onto $\mathrm{Aut}(A_W) = \mathrm{PermAut}(A_W) = \mathrm{PermAut}(\mathcal{E}(W))$.*

PROOF. By Lemma 9.8.6, we only need to show that

$$\mathrm{PermAut}(A_W) \leq \Theta(\mathrm{Aut}(W)).$$

Suppose that A_W is in plug-in form. Hence row $2i - 1$ and row $2i$ have zero inner product. Since $\mathcal{G}_W$ is complete, these rows have non-zero inner product with all other rows of A_W. Now suppose that there are permutation matrices U and V such that

$$U A_W = A_W V.$$

Then rows $2i - 1$ and $2i$ of $A_W V$ have zero inner product, and their inner product with all other rows of $A_W V$ is non-zero. As a consequence U permutes the rows of A_W so that the partition $1, 2 : 3, 4 : \cdots : 2v - 1, 2v$ is preserved. Therefore, because U is a permutation matrix, there exists $P \in \mathrm{Mon}(v, C_2)$ such that $U = \theta_{C_2}^{(1)}(P)$. A similar argument applies to the columns of A_W, so we conclude that there exists $Q \in \mathrm{Mon}(v, C_2)$ such that $V = \theta_{C_2}^{(2)}(Q)$. Thus $(U, V) = \Theta(P, Q)$. By Theorem 9.8.5, we are done. $\qquad\square$

If W is a $\mathrm{BW}(v, k, \lambda)$ with $\lambda > 0$, then $\mathcal{G}_W$ is complete. Hence

9.10.2. COROLLARY. *Let W be any balanced weighing matrix $\mathrm{BW}(v, k, \lambda)$ with $\lambda > 0$. Then $\Theta(\mathrm{Aut}(W)) = \mathrm{PermAut}(A_W) = \mathrm{PermAut}(\mathcal{E}(W))$.*

The reasoning in the proof of Theorem 9.10.1 also applies to orthogonal designs.

9.10.3. THEOREM. *If D is an orthogonal design with complete graph $\mathcal{G}_D$, then $\Theta(\mathrm{Aut}(D)) = \mathrm{PermAut}(A_D) = \mathrm{PermAut}(\mathcal{E}(D))$.*

CHAPTER 10

Group Development and Regular Actions on Arrays

We now commence our study of regular group actions on square arrays. Some of the theory in this chapter is carried over to Chapter 14, where we consider centrally regular actions on expanded designs, and, in particular, elicit the connection with cocyclic development.

It is a slight abuse of terminology to say that a group G 'acts' or has an 'action' on a square array A. Here we mean that G is isomorphic to a group of permutation equivalence operations that stabilize A. We also say that the group *fixes* A.

10.1. Matrix preliminaries

Let G be a finite group, and let $g_1 < \cdots < g_v$ be a (fixed but arbitrary) ordering of its elements. Our notation $M = [m_{x,y}]_{x,y \in G}$ signifies that $M = [m_{g_i,g_j}]_{1 \leq i,j \leq v}$. As noted in Subsection 3.7.2, when we multiply matrices all indexed by G under the same ordering, we can regard the indices as dummy variables without recourse to that ordering.

For each $\pi \in \mathrm{Sym}(G)$, define the permutation matrix

$$(10.1) \qquad P_\pi = [\delta^x_{\pi(y)}]_{x,y \in G}.$$

Remember that the map $\pi \mapsto P_\pi$ is a faithful representation $\mathrm{Sym}(G) \to \mathrm{Perm}(v)$. Premultiplying a matrix by P_π shifts row r to row $\pi(r)$; postmultiplying by $P_\pi^{-1} = P_{\pi^{-1}} = P_\pi^\top$ shifts column s to column $\pi(s)$. Special cases of (10.1) are the left and right regular permutation matrices S_x and T_x, $x \in G$. We have $S_x = P_{\sigma_x}$ and $T_x = P_{\tau_x}$, where $\sigma_x(a) = xa$ and $\tau_x(a) = ax^{-1}$ for all $a \in G$:

$$S_x = [\delta^a_{xb}]_{a,b \in G}, \qquad T_x = [\delta^{ax}_b]_{a,b \in G}.$$

10.2. Group-developed arrays

Consider the following incidence matrix A of an SBIBD$(7, 3, 1)$, with the rows and columns indexed by the elements of $\{1, 2, 3, 4, 5, 6, 7\}$ as shown.

	1	2	3	4	5	6	7
1	0	1	0	0	0	1	1
2	1	1	0	1	0	0	0
3	0	0	0	1	1	0	1
4	0	1	1	0	1	0	0
5	0	0	1	1	0	1	0
6	1	0	0	0	1	1	0
7	1	0	1	0	0	0	1

We may reorder the rows and columns of A to get back-circulant arrays:

	5	1	7	6	3	2	4
4	1	0	0	0	1	1	0
5	0	0	0	1	1	0	1
1	0	0	1	1	0	1	0
7	0	1	1	0	1	0	0
6	1	1	0	1	0	0	0
3	1	0	1	0	0	0	1
2	0	1	0	0	0	1	1

	1	4	3	7	5	2	6
1	0	0	0	1	0	1	1
4	0	0	1	0	1	1	0
3	0	1	0	1	1	0	0
7	1	0	1	1	0	0	0
5	0	1	1	0	0	0	1
2	1	1	0	0	0	1	0
6	1	0	0	0	1	0	1

Now the rows and columns of any $v \times v$ back-circulant array may be labeled with the residues $0, 1, \ldots, v-1$ modulo v so that the (i,j)th entry depends only on $i+j$ modulo v. Therefore we may label the rows and columns of A with the elements of the additive group $\mathbb{Z}_7$ in at least two ways, shown below, so that under those labelings the (x,y)th entry depends only on the sum $x + y$ in $\mathbb{Z}_7$.

	1	5	4	6	0	3	2
2	0	1	0	0	0	1	1
6	1	1	0	1	0	0	0
5	0	0	0	1	1	0	1
0	0	1	1	0	1	0	0
1	0	0	1	1	0	1	0
4	1	0	0	0	1	1	0
3	1	0	1	0	0	0	1

	0	5	2	1	4	6	3
0	0	1	0	0	0	1	1
5	1	1	0	1	0	0	0
2	0	0	0	1	1	0	1
1	0	1	1	0	1	0	0
4	0	0	1	1	0	1	0
6	1	0	0	0	1	1	0
3	1	0	1	0	0	0	1

In fact, if we put

$$x_1 = 2, x_2 = 6, x_3 = 5, x_4 = 0, x_5 = 1, x_6 = 4, x_7 = 3,$$

$$y_1 = 1, y_2 = 5, y_3 = 4, y_4 = 6, y_5 = 0, y_6 = 3, y_7 = 2,$$

$$z_1 = 0, z_2 = 5, z_3 = 2, z_4 = 1, z_5 = 4, z_6 = 6, z_7 = 3,$$

and let $h_1, h_2 : \mathbb{Z}_7 \to \{0,1\}$ be the maps with supports $D_1 = \{0,4,5\}$ and $D_2 = \{3,5,6\}$ respectively, then

$$A = [h_1(x_i + y_j)]_{1 \le i,j \le 7} = [h_2(z_i + z_j)]_{1 \le i,j \le 7}.$$

We seek to determine all the ways in which the rows and columns of A can be permuted to put A into back-circulant form. At first sight this seems to be a purely combinatorial problem. But as we will see, algebra provides an effective way to solve this kind of problem for $v \times v$ arrays even when v is quite large. The key idea is that each solution of our problem is paired with a regular group action on A. This action is apparent in a back-circulant array: cycling the rows down and cycling the columns to the left (the same number of positions) fixes the array.

10.2.1. DEFINITION. Let G be a finite group of order v. A $v \times v$ $(0, \mathcal{A})$-array A is *group-developed modulo G* if there are total orderings $x_1 < \cdots < x_v$ and $y_1 < \cdots < y_v$ of the elements of G such that $A = [h(x_i y_j)]_{1 \le i,j \le v}$ for some map $h : G \to \{0\} \cup \mathcal{A}$.

We say 'group-developed modulo G' because every entry is determined, via group arithmetic on row and column labels, by the entries in the first row. In the next chapter, we look at more general ways of developing an array from its initial row.

One goal of this chapter is to describe a method for determining all the ways that a given array may be written in group-developed form. The following reformulation of Definition 10.2.1 allows us to state this task more precisely.

10.2.2. LEMMA. *The $(0, \mathcal{A})$-array A is group-developed modulo G if and only if there are permutations $\pi, \phi \in \mathrm{Sym}(G)$ and a map $h : G \to \{0\} \cup \mathcal{A}$ such that*

$$(10.2) \qquad A = P_\pi^{-1} [h(xy)]_{x,y \in G} P_\phi.$$

PROOF. Using the chosen total order $g_1 < \cdots < g_v$ of G, we have

$$P_\pi^{-1} [h(xy)]_{x,y \in G} P_\phi = [\textstyle\sum_{j,k} \delta^{g_j}_{\pi(g_i)} h(g_j g_k) \delta^{g_k}_{\phi(g_l)}]_{il} = [h(\pi(g_i) \phi(g_l))]_{il}.$$

So (10.2) holds if and only if the (i, j)th entry of A is $h(\pi(g_i) \phi(g_j))$. □

Therefore, the problem of finding all indexings of A that put A into group-developed form is equivalent to the following.

10.2.3. PROBLEM. *For a given $(0, \mathcal{A})$-array A, classify all the ways of writing A in the form (10.2).*

Notice that, in the context of Problem 10.2.3, the choice of ordering of G used to index matrices on the right-hand side of (10.2) is irrelevant.

The maps h can be valuable combinatorial objects. For example, because the example array A above is an incidence matrix of an $\mathrm{SBIBD}(7, 3, 1)$, the support sets D_1 and D_2 both have the property that, for each non-zero residue d modulo 7, there is exactly one pair of elements $a, b \in D_i$ such that $a - b \equiv d \pmod 7$. This reflects the arithmetic used in developing the design from its initial row. The supports D_1 and D_2 are *difference sets*. In this case, we may view the differences as being calculated in a cyclic group of order 7. We say more about difference sets in Section 10.4.

Lemma 10.2.2 shows that an array A is group-developed modulo G if and only if $A \approx [h(xy)]_{x,y \in G}$ for some map h on G. Thus, it is natural to pose the following sharper version of Problem 10.2.3.

10.2.4. PROBLEM. *Given a $(0, \mathcal{A})$-array A, determine all groups G and maps $h : G \to \{0\} \cup \mathcal{A}$ such that $A \approx [h(xy)]_{x,y \in G}$.*

A solution of Problem 10.2.4 (or Problem 10.2.3) should not only be complete, but also irredundant, i.e., not contain any duplicates, according to some appropriate classification criteria.

10.3. Regular embeddings

Let $(P_\pi, P_\phi) \in \mathrm{PermAut}(A)$. As we noted in Section 10.1, premultiplying by P_π moves the g_ith row of A to the $\pi(g_i)$th row of $P_\pi A$, and postmultiplying by P_ϕ^{-1} moves column g_i of A to column $\phi(g_i)$ of $A P_\phi^{-1}$. So $\mathrm{PermAut}(A)$ has two induced actions: one on the set of row labels of A, and the other on the set of column labels of A. We study subgroups of $\mathrm{PermAut}(A)$ whose induced row and column actions are both regular.

10.3.1. DEFINITION. Let A be a square array. A subgroup G of $\mathrm{PermAut}(A)$ *acts regularly on A* if its induced actions on the set of row labels of A, and on the set of column labels of A, are regular. In this case, G is called a *regular subgroup* of $\mathrm{PermAut}(A)$.

Note that a subgroup of $\mathrm{PermAut}(A)$ can be regular on rows without being regular on columns, and vice versa.

Shortly we will prove that solutions of (10.2) correspond to regular subgroups of $\mathrm{PermAut}(A)$. In particular, this characterizes group development of an array in terms of actions on the array. We first reduce the problem to finding solutions where π and ϕ are normalized; that is, fix 1.

10.3.2. DEFINITION. The triple $(S_a^{-1}P, x \mapsto h(axb), T_bQ)$ is called a *translation* of the triple (P, h, Q).

10.3.3. LEMMA. *Every solution of* (10.2) *has a normalized translation satisfying* (10.2).

PROOF. By (3.6) and (3.7),

$$
\begin{aligned}
{[h(xy)]_{x,y \in G}} &= W_1 \sum_{x \in G} h(x)T_x \\
&= W_1 \sum_{x \in G} h(xb)T_{xb} \\
&= W_1 \sum_{x \in G} h(xb)T_x T_b \\
&= \sum_{x \in G} h(xb)S_x W_1 T_b \\
&= \sum_{x \in G} h(axb)S_{ax} W_1 T_b \\
&= S_a \Big(\sum_{x \in G} h(axb)S_x W_1 \Big) T_b \\
&= S_a [h(axyb)]_{x,y \in G} T_b.
\end{aligned}
$$

So (P_π, h, P_ϕ) satisfies (10.2) if and only if $(S_a^{-1}P_\pi, x \mapsto h(axb), T_b P_\phi)$ does. Since $S_a = P_{\sigma_a}$ and $T_b = P_{\tau_b}$, if we put $a = \pi(1)$ and $b = \phi(1)$ then $\alpha = \sigma_{a^{-1}}\pi$ and $\beta = \tau_b \phi$ are normalized; moreover $S_{a^{-1}}P_\pi = P_\alpha$ and $T_b P_\phi = P_\beta$. □

Lemma 10.3.3 implies

10.3.4. THEOREM. *Fix G, and suppose that the set of solutions of* (10.2) *is non-empty. Translation defines an equivalence relation on this set, hence partitions it into equivalence classes each containing $|G|^2$ elements precisely one of which is normalized.*

The translation equivalence classes for G correspond to isomorphisms of G into $\mathrm{PermAut}(A)$.

10.3.5. DEFINITION. An isomorphism ϑ from the abstract group G onto a regular subgroup of $\mathrm{PermAut}(A)$ is called a *regular embedding of G in* $\mathrm{PermAut}(A)$, and G is said to *act regularly on A* (via the embedding ϑ).

The next theorem points to a method for solving our main problem.

10.3.6. THEOREM. *Let A be a $v \times v$ array, and let G be a group of order v.*

(1) *Each regular embedding of G in $\mathrm{PermAut}(A)$ may be expressed uniquely as*

$$
(10.3) \qquad x \mapsto (P_\pi, P_\phi)^{-1}(T_x, S_x)(P_\pi, P_\phi),
$$

where (P_π, h, P_ϕ) is a normalized solution of (10.2). *Conversely,* (10.3) *is a regular embedding of G in $\mathrm{PermAut}(A)$ for each normalized solution (P_π, h, P_ϕ) of* (10.2). *The set of regular embeddings is therefore in 1-1 correspondence with the set of normalized solutions of* (10.2).

(2) *Suppose that ϑ_1 and ϑ_2 are regular embeddings of G into $\mathrm{PermAut}(A)$.
Then $\vartheta_2(G) = \vartheta_1(G)$ if and only if there is an automorphism α of G such
that $\vartheta_2 = \vartheta_1 \circ \alpha$. Thus each regular subgroup of $\mathrm{PermAut}(A)$ isomorphic
to G corresponds to $|\mathrm{Aut}(G)|$ distinct regular embeddings of G. Each of
these in turn gives rise to $|G|^2$ distinct solutions of (10.2).*

PROOF.

1. Let $\vartheta : x \mapsto (P_x, Q_x)$ be a regular embedding, so the maps defined by $x \mapsto P_x$ and $x \mapsto Q_x$ are faithful regular representations of G in $\mathrm{Perm}(v) \cong \mathrm{Sym}(G)$. By Lemma 3.3.7, $P_x = P_\pi^{-1} T_x P_\pi$ and $Q_x = P_\phi^{-1} S_x P_\phi$ for some normalized $\pi, \phi \in \mathrm{Sym}(G)$. By Lemma 3.3.1, the only normalized element of $\mathrm{Sym}(G)$ that centralizes the (left or right) regular permutation representation of G is the identity. Hence ϑ can be written uniquely as

$$\epsilon_{\pi,\phi} : x \mapsto (P_\pi, P_\phi)^{-1} (T_x, S_x)(P_\pi, P_\phi)$$

for normalized $\pi, \phi \in \mathrm{Sym}(G)$. So the regular embeddings of G in $\mathrm{PermAut}(A)$ are the maps $\epsilon_{\pi,\phi}$ where π and ϕ are normalized and $\epsilon_{\pi,\phi}(G) \subseteq \mathrm{PermAut}(A)$. Now $\epsilon_{\pi,\phi}(G)$ is in $\mathrm{PermAut}(A)$ if and only if, for all $x \in G$,

$$T_x P_\pi A P_\phi^{-1} S_x^{-1} = P_\pi A P_\phi^{-1}.$$

By Theorem 3.7.8, this is equivalent to there being a map h on G such that

$$P_\pi A P_\phi^{-1} = [h(rs)]_{r,s \in G}.$$

This completes the proof of part (1).

2. The first two statements in this part are clear. The third follows from (1) and Theorem 10.3.4. $\square$

Theorem 10.3.6 shows how knowledge of the regular subgroups of $\mathrm{PermAut}(A)$ that are isomorphic to G enables us to solve Problem 10.2.3. In Section 10.6, we will see that in fact only a complete set of representatives for the conjugacy classes of regular subgroups of $\mathrm{PermAut}(A)$ is required.

We round out the section with an abbreviated version of Theorem 10.3.6 (see Theorem 10.3.8 below), which is frequently all that is needed in practice. We first make a definition.

10.3.7. DEFINITION. Let $x_1, \ldots, x_v$ and $y_1, \ldots, y_v$ be row and column labelings that place the array A in G-developed form. The *canonical regular action of G on A* (for the specified row and column labelings) is the action such that $g \in G$ moves row x_i to row $x_i g^{-1}$ and column y_j to column gy_j.

In other words, the canonical action of G premultiplies $A = [h(x_i y_j)]_{1 \leq i,j \leq v}$ by T_g and postmultiplies A by $S_g^\top$ for each $g \in G$, where T_g has indexing $x_1, \ldots, x_v$, and S_g has indexing $y_1, \ldots, y_v$.

10.3.8. THEOREM. *The group G acts regularly on A if and only if*

$$A = P_\pi^{-1} [g(xy)]_{x,y \in G} P_\phi$$

for some map g and $\pi, \phi \in \mathrm{Sym}(G)$. The corresponding regular embedding of G in $\mathrm{PermAut}(A)$ is

$$x \mapsto (T_x, S_x)^{(P_\pi, P_\phi)}.$$

Equivalently, each regular action of G on A is the canonical regular action for a row labeling $x_1, \ldots, x_v$ and a column labeling $y_1, \ldots, y_v$ such that

$$A = [g(x_i y_j)]_{1 \le i, j \le v}.$$

10.4. Difference sets and relative difference sets

We now discuss SBIBD(v, k, λ)s and GDD$(vm, k, \lambda; m)$s with group-developed incidence matrices. The support of the development function of such an array is either a difference set or relative difference set (cf. Section 10.2). There is a large body of work on difference sets and relative difference sets (see, e.g., [**8, 11, 130, 141**]), and recent decades have witnessed major advances in the field. In this book we encounter many examples of difference sets and relative difference sets. These are used to illustrate more general ideas, but we also prove some new results.

10.4.1. Incidence matrices from $(0, 1)$-maps. Let $h : G \to \{0, 1\}$ be any map, and denote the support of h by D. So the (x, y)th entry of $A = [h(xy)]_{x, y \in G}$ is $\mathbf{1}\{xy \in D\}$, where the logical operator $\mathbf{1}$ evaluates an expression to 1 if it is true, and 0 if it is false. In what follows we will take A to be an incidence matrix of a design. Now

$$
\begin{aligned}
AA^\top &= [\textstyle\sum_y h(xy) h(zy)]_{x, z \in G} \\
&= [\textstyle\sum_y \mathbf{1}\{xy \in D\} \mathbf{1}\{zy \in D\}]_{x, z \in G} \\
&= [\textstyle\sum_{a, b \in D} \mathbf{1}\{ab^{-1} = xz^{-1}\}]_{x, z \in G}.
\end{aligned}
$$
(10.4)

Thus, each map $h : G \to \{0, 1\}$ gives us a matrix that is group-developed modulo G, whose Grammian depends only on the behavior of the quotients ab^{-1} of the elements of D.

10.4.2. Difference sets.

10.4.1. DEFINITION. Let G be a finite group of order v. A k-subset D of G is a *difference set* of cardinality k and index λ if every non-identity element is the 'difference' ab^{-1} of exactly λ pairs $a, b \in D$. We use the notation (v, k, λ)-difference set.

We recast this definition: D is a (v, k, λ)-difference set in G if and only if

$$\sum_{a, b \in D} \mathbf{1}\{ab^{-1} = x\} = \begin{cases} k & \text{if } x = 1 \\ \lambda & \text{if } x \in G \setminus \{1\}. \end{cases}$$
(10.5)

Equations (10.4) and (10.5) then imply

10.4.2. LEMMA. *Let G be a group of order v and let $h : G \to \{0, 1\}$ be a map. Then the support of h is a (v, k, λ)-difference set if and only if $[h(xy)]_{x, y \in G}$ is an incidence matrix of an* SBIBD(v, k, λ).

10.4.3. THEOREM. *There exists a (v, k, λ)-difference set in the group G if and only if G acts regularly on an incidence matrix A of some* SBIBD(v, k, λ).

PROOF. By Theorem 10.3.8, if G acts regularly on A then $A = [h(x_i y_j)]_{ij}$ for some map $h : G \to \{0, 1\}$ and labeling of rows and columns. The support of h is a (v, k, λ)-difference set by Lemma 10.4.2. The other implication is Lemma 10.4.2 again. $\square$

In summary, each (v, k, λ)-difference set D in a group G corresponds to an SBIBD(v, k, λ) with incidence matrix $[\mathbf{1}\{xy \in D\}]_{x,y \in G}$, that is group-developed modulo G. Thus $D_1 \sim D_2$ if and only if $[\mathbf{1}\{xy \in D_1\}]_{x,y \in G_1} \approx [\mathbf{1}\{xy \in D_2\}]_{x,y \in G_2}$ defines an equivalence relation on the set of all (v, k, λ)-difference sets (for fixed v, k, λ and varying G). A single SBIBD(v, k, λ) may correspond to difference sets in several non-isomorphic groups.

10.4.4. EXAMPLE. Consider the SBIBD$(16, 6, 2)$ with incidence matrix

$$A = \begin{bmatrix}
1 & 1 & 1 & 1 & 1 & 1 & 0 & 0 & 0 & 0 & 0 & 0 & 0 & 0 & 0 & 0 \\
1 & 1 & 0 & 0 & 0 & 0 & 1 & 1 & 1 & 1 & 0 & 0 & 0 & 0 & 0 & 0 \\
1 & 0 & 1 & 0 & 0 & 0 & 1 & 0 & 0 & 0 & 1 & 1 & 1 & 0 & 0 & 0 \\
1 & 0 & 0 & 1 & 0 & 0 & 0 & 1 & 0 & 0 & 1 & 0 & 0 & 1 & 1 & 0 \\
1 & 0 & 0 & 0 & 1 & 0 & 0 & 0 & 1 & 0 & 0 & 1 & 0 & 1 & 0 & 1 \\
1 & 0 & 0 & 0 & 0 & 1 & 0 & 0 & 0 & 1 & 0 & 0 & 1 & 0 & 1 & 1 \\
0 & 1 & 1 & 0 & 0 & 0 & 1 & 0 & 0 & 0 & 0 & 0 & 0 & 1 & 1 & 1 \\
0 & 1 & 0 & 1 & 0 & 0 & 0 & 1 & 0 & 0 & 1 & 1 & 0 & 0 & 0 & 1 \\
0 & 1 & 0 & 0 & 1 & 0 & 0 & 0 & 1 & 0 & 1 & 0 & 1 & 0 & 1 & 0 \\
0 & 1 & 0 & 0 & 0 & 1 & 0 & 0 & 0 & 1 & 1 & 1 & 0 & 1 & 0 & 0 \\
0 & 0 & 1 & 1 & 0 & 0 & 0 & 0 & 1 & 1 & 1 & 0 & 0 & 0 & 0 & 1 \\
0 & 0 & 1 & 0 & 1 & 0 & 0 & 1 & 0 & 1 & 0 & 1 & 0 & 0 & 1 & 0 \\
0 & 0 & 1 & 0 & 0 & 1 & 0 & 1 & 1 & 0 & 0 & 0 & 1 & 1 & 0 & 0 \\
0 & 0 & 0 & 1 & 1 & 0 & 1 & 0 & 0 & 1 & 0 & 0 & 1 & 1 & 0 & 0 \\
0 & 0 & 0 & 1 & 0 & 1 & 1 & 0 & 1 & 0 & 0 & 1 & 0 & 0 & 1 & 0 \\
0 & 0 & 0 & 0 & 1 & 1 & 1 & 1 & 0 & 0 & 1 & 0 & 0 & 0 & 0 & 1
\end{bmatrix}.$$

The matrix $2A - J$ is equivalent to the Sylvester Hadamard matrix of order 16. The automorphism group PermAut(A) has 24 conjugacy classes of regular subgroups, which fall into 12 isomorphism types. That is, all but two of the 14 groups of order 16 act regularly on A. (The missing groups are C_{16} and D_{16}. In fact these two groups cannot act regularly on any SBIBD$(16, 6, 2)$.) This is indicative of the highly symmetrical nature of the Sylvester matrix. All groups of order 16 that contain a $(16, 6, 2)$-difference set contain $(16, 6, 2)$-difference sets obtained from A. Up to equivalence there are two other SBIBD$(16, 6, 2)$s. One of these has a single conjugacy class of regular subgroups, isomorphic to $C_8 \times C_2$; whereas the other has two classes of regular subgroups, both non-abelian. So there are $(16, 6, 2)$-difference sets that do not correspond to the design with incidence matrix A.

10.4.3. Relative difference sets.

10.4.5. DEFINITION. Let G be a group of order v, with subgroup H of order m. A k-subset D of G is a *relative difference set* with index λ and *forbidden subgroup* H if

- every element of $G \setminus H$ is the quotient ab^{-1} of exactly λ pairs $a, b \in D$,
- no non-identity element of H is equal to ab^{-1} for any $a, b \in D$.

We say that D is a (v, k, λ, m)-relative difference set.

A difference set is a relative difference with trivial forbidden subgroup. Note that $D \subseteq G$ is a (v, k, λ, m)-relative difference set with forbidden subgroup H if and only if

$$(10.6) \qquad \sum_{a,b \in D} \mathbf{1}\{ab^{-1} = x\} = \begin{cases} k & \text{if } x = 1 \\ 0 & \text{if } x \in H \setminus \{1\} \\ \lambda & \text{if } x \in G \setminus H. \end{cases}$$

10.4.6. LEMMA. *Let G be a group of order v and let $h : G \to \{0, 1\}$ be a map. The support D of h is a (v, k, λ, m)-relative difference set with forbidden subgroup H if and only if $A = [h(xy)]_{x,y \in G}$ is an incidence matrix of a $\mathrm{GDD}(v, k, \lambda; m)$ with point classes $\{Hy \mid y \in G\}$.*

PROOF. By (10.4), $[AA^\top]_{x,y} = \sum_{a,b \in D} \mathbf{1}\{ab^{-1} = xy^{-1}\}$. So, by (10.6), if D is a (v, k, λ, m)-relative difference set with forbidden subgroup H, then

$$[AA^\top]_{x,y} = \begin{cases} k & \text{if } x = y \\ 0 & \text{if } x \in Hy \setminus \{y\} \\ \lambda & \text{if } x \in G \setminus Hy. \end{cases}$$

Thus A is an incidence matrix of a $\mathrm{GDD}(v, k, \lambda; m)$ with point classes $\{Hy \mid y \in G\}$ (and lines the 'translates' Dy of D, $y \in G$).

Conversely, if $A = [h(xy)]_{x,y \in G}$ is an incidence matrix of a $\mathrm{GDD}(v, k, \lambda; m)$, then G acts (via the canonical regular action) on A. As this action respects inner products, the point classes form a system of imprimitivity for G. If H denotes the stabilizer of the first point class, then the point classes are $\{Hx \mid x \in G\}$. Moreover,

$$\sum_{a,b \in D} \mathbf{1}\{ab^{-1} = xy^{-1}\} = [AA^\top]_{x,y} = \begin{cases} k & \text{if } x = y \\ 0 & \text{if } x \in Hy \setminus \{y\} \\ \lambda & \text{if } x \in G \setminus Hy. \end{cases}$$

Therefore D is a (v, k, λ, m)-relative difference set with forbidden subgroup H. $\square$

A relative difference set in G is said to be *normal* if its forbidden subgroup is normal in G. Drawing on the connection between regular actions on group divisible designs and relative difference sets evinced by Lemma 10.4.6, the next theorem gives a necessary and sufficient condition for the forbidden subgroup to be normal. In Chapter 15, we will consider the special case where the forbidden subgroup is central.

10.4.7. THEOREM. *Let H be a subgroup of order m of a finite group G. There exists a (v, k, λ, m)-relative difference set in the group G with normal forbidden subgroup H if and only if there is a regular action of G on an incidence matrix A of some $\mathrm{GDD}(v, k, \lambda; m)$ such that H fixes every point class.*

PROOF. Suppose that there is a (v, k, λ, m)-relative difference set D in G with forbidden subgroup H. By Lemma 10.4.6, if $h : G \to \{0, 1\}$ is the map with support D then $A = [h(xy)]_{x,y \in G}$ is an incidence matrix of a $\mathrm{GDD}(v, k, \lambda; m)$. Furthermore, the point classes are the right cosets of H. Under the canonical regular action of G on A, $h \in H$ sends row x to row xh^{-1}; so H fixes every point class if and only if $Hxh = Hx$ for all $x \in G$ and $h \in H$, i.e., H is normal.

Next, let A be an incidence matrix of a $\mathrm{GDD}(v, k, \lambda; m)$ and suppose that G acts regularly on A with H fixing every point class. By Theorem 10.3.8 we may assume that $A = [h(x_i y_j)]_{1 \le i, j \le v}$ for some map $h : G \to \{0, 1\}$ and row and column labelings $x_1, \ldots, x_v$ and $y_1, \ldots, y_v$ of A, so that G acts via the canonical regular action. By Lemma 10.4.6, the support of h is a (v, k, λ, m)-relative difference set with forbidden subgroup $\tilde{H}$ say. As we saw in the proof of Lemma 10.4.6, the point classes are the right cosets of $\tilde{H}$. Since H fixes $\tilde{H}$ and $|H| = |\tilde{H}|$, we must have $H = \tilde{H}$. Indeed, $\tilde{H}xh = \tilde{H}x$ for all $h \in H$ and $x \in G$ implies that H is normal, as required. $\square$

10.4.4. Group ring equations for (relative) difference sets. A difference set or relative difference set is equivalent to a solution of a certain group ring equation.

Let D be a subset of the finite group G of order v. By (10.5), D is a (v, k, λ)-difference set if and only if, in the group ring $\mathbb{Z}[G]$,

$$\left(\sum_{x \in D} x\right)\left(\sum_{x \in D} x^{-1}\right) = k - \lambda + \lambda \sum_{x \in G} x.$$

The following notation is customary. Let $A \subseteq G$, and denote the element $\sum_{x \in A} x$ of $\mathbb{Z}[G]$ also by A. We write $A^{(-1)}$ for $\sum_{x \in A} x^{-1}$. The above equation becomes

$$(10.7) \qquad\qquad DD^{(-1)} = k - \lambda + \lambda G.$$

Similarly, by (10.6), D is a (v, k, λ, m)-relative difference set in G with forbidden subgroup H if and only if

$$(10.8) \qquad\qquad DD^{(-1)} = k + \lambda(G - H).$$

10.5. Group ring equations and associates

We have seen in the previous section how a group-developed $\mathrm{SBIBD}(v, k, \lambda)$ or $\mathrm{GDD}(v, k, \lambda; m)$ corresponds to a solution of a group ring equation. In this section, we show that for any orthogonality set Λ there is a correspondence between group-developed $\mathrm{PCD}(\Lambda)$s and solutions, with restricted coefficients, of a group ring equation.

10.5.1. Associates of group-developed arrays.

10.5.1. DEFINITION. Let A be a $v \times v$ $(0, \mathcal{A})$-array where $\mathcal{A}$ is contained in a ring $\mathcal{R}$, and let G be a group of order v. If $A \approx [h(xy)]_{x,y \in G}$ then the element $\sum_{x \in G} h(x)x$ of the group ring $\mathcal{R}[G]$ is a *G-associate* of A.

Associates of Menon difference sets were introduced by Dillon [**59**] to study group-developed Hadamard matrices. In this section, we extend some of Dillon's ideas. Later (in Section 10.7), we show how associates may be used to construct and classify all solutions of (10.2).

10.5.2. NOTATION. Let S be a subset of a ring $\mathcal{R}$ such that $0 \notin S$. We say that $\alpha = \sum_{x \in G} a_x x \in \mathcal{R}[G]$ has *(S)-coefficients* if $a_x \in S$ for all x; α has *(0, S)-coefficients* if every a_x is in $\{0\} \cup S$.

A group-developed array can have associates for several non-isomorphic groups.

10.5.3. EXAMPLE. Let $\mathrm{C}_4 = \langle a \rangle$ and $\mathrm{C}_2^2 = \langle a, b \rangle$. The back-circulant matrix with initial row $[1, 1, 1, -1]$ is Hadamard. It has C_4-associate $1 + a + a^2 - a^3$ and C_2^2-associate $1 + a + b - ab$.

10.5.2. Derivation of the group ring equation. Let $\mathcal{R}$ be an ambient involutory ring for the orthogonality set Λ with alphabet $\mathcal{A}$, and suppose that Λ has the Gram Property over $\mathcal{R}$. So there is a subset $\mathrm{Gram}_{\mathcal{R}}(\Lambda)$ of $\mathrm{Mat}(v, \mathcal{R})$ such that a $v \times v$ $(0, \mathcal{A})$-array D with distinct rows is a $\mathrm{PCD}(\Lambda)$ if and only if

$$(10.9) \qquad\qquad DD^* = B$$

for some $B \in \mathrm{Gram}_{\mathcal{R}}(\Lambda)$. We show that there is a group ring counterpart of (10.9) whose solutions with $(0, \mathcal{A})$-coefficients correspond to group-developed $\mathrm{PCD}(\Lambda)$s.

10.5.4. NOTATION. If $\alpha = \sum_{x \in G} h(x)x \in \mathcal{R}[G]$ then $\alpha^{(*)}$ denotes the element $\sum_{x \in G} h(x)^* x^{-1}$ of $\mathcal{R}[G]$.

10.5.5. THEOREM. *Let D be the array $[h(xy)]_{x,y \in G}$ with distinct rows, where $h : G \to \{0\} \cup \mathcal{A}$ is a map. Then D is a $\mathrm{PCD}(\Lambda)$ if and only if $\alpha = \sum_{u \in G} h(u)u$ satisfies a group ring equation*

$$(10.10) \qquad \alpha\alpha^{(*)} = \sum_{u \in G} b(u)u$$

such that

$$\sum_{u \in G} b(u)S_u \in \mathrm{Gram}_{\mathcal{R}}(\Lambda).$$

PROOF. For $u \in G$, let $b(u) = \sum_{x \in G} h(ux)h(x)^*$. By (3.7), if $D = [h(xy)]_{x,y \in G}$ then $D = \sum_{u \in G} h(u)S_u W$ where $W = [\delta^x_{y^{-1}}]_{x,y \in G}$, and we have

$$DD^* = \left(\sum_{u \in G} h(u)S_u\right)\left(\sum_{u \in G} h(u)S_u\right)^* = \sum_{u \in G} b(u)S_u.$$

Also, if $\alpha = \sum_{u \in G} h(u)u$ then

$$\alpha\alpha^{(*)} = \sum_{u \in G} b(u)u.$$

Thus D is a $\mathrm{PCD}(\Lambda)$ if and only if $\alpha = \sum_{u \in G} h(u)u$ is a solution of (10.10) with $(0, \mathcal{A})$-coefficients such that $\sum_{u \in G} b(u)S_u \in \mathrm{Gram}_{\mathcal{R}}(\Lambda)$. $\square$

10.5.3. Group ring equations and associates for the familiar designs. Theorem 10.5.5 and Example 6.1.3 yield the following result, which lists group ring equations for the familiar kinds of group-developed designs, except orthogonal designs (which we omit by Theorem 2.6.8). Notice that the condition for a group-developed $\mathrm{SBIBD}(v, k, \lambda)$ is the same as (10.7).

10.5.6. THEOREM. *For each $\mathrm{PCD}(\Lambda)$ below, the group ring element α is a G-associate of that design if and only if α has $(0, \mathcal{A})$-coefficients (or $(\mathcal{A})$-coefficients if the arrays in Λ contain no zeros) and satisfies the indicated group ring equation.*

(1) $\mathrm{SBIBD}(v, k, \lambda)$: $\alpha\alpha^{(-1)} = k - \lambda + \lambda G$ *in* $\mathbb{Z}[G]$.
(2) $\mathrm{H}(n)$: $\alpha\alpha^{(-1)} = n$ *in* $\mathbb{Z}[G]$.
(3) $\mathrm{W}(v, k)$: $\alpha\alpha^{(-1)} = k$ *in* $\mathbb{Z}[G]$.
(4) $\mathrm{BW}(v, k, \lambda)$: $\alpha\alpha^{(-1)} = k + \frac{\lambda}{2}C_2(G - 1)$ *in* $\mathbb{Z}[C_2 \times G]$.
(5) $\mathrm{CH}(n)$: $\alpha\alpha^{(*)} = n$ *in* $\mathbb{Z}[\mathrm{i}][G]$.
(6) $\mathrm{CGH}(n; m)$: $\alpha\alpha^{(*)} = n$ *in* $\mathbb{Z}[\zeta_m][G]$.
(7) $\mathrm{CGW}(v, k; m)$: $\alpha\alpha^{(*)} = k$ *in* $\mathbb{Z}[\zeta_m][G]$.
(8) $\mathrm{GH}(n; H)$: $\alpha\alpha^{(*)} = n$ *in* $(\mathbb{Z}[H]/\mathbb{Z}H)[G]$.
(9) $\mathrm{GW}(v, k; H)$: $\alpha\alpha^{(*)} = k$ *in* $(\mathbb{Z}[H]/\mathbb{Z}H)[G]$.
(10) $\mathrm{BGW}(v, k, \lambda; H)$: $\alpha\alpha^{(*)} = k + \frac{\lambda}{|H|}H(G - 1)$ *in* $\mathbb{Z}[H \times G]$.

We know that the support of any G-associate of an $\mathrm{SBIBD}(v, k, \lambda)$ is a (v, k, λ)-difference set in G. Next, we note two other connections between associates and relative difference sets (the first result extends Theorem 9.9.3).

10.5.7. THEOREM. *Let G and H be groups of orders v and m respectively. A G-developed $\mathrm{BGW}(v, k, \lambda; H)$ is equivalent to a $(vm, k, \lambda/m, m)$-relative difference set in $G \times H$ with forbidden subgroup H.*

PROOF. The group ring equation for a BGW$(v, k, \lambda; H)$ is the same as (10.8) with $\lambda = \lambda/|H|$ and $G = G \times H$. $\qquad \square$

10.5.8. THEOREM. *Let G be a group of order v, and let $H = G \times \langle g \mid g^4 = 1 \rangle$. Suppose that $\alpha \in \mathbb{C}[G]$ has $(\pm 1, \pm i)$-coefficients, and define $a, b \in \mathbb{C}[G]$ with (± 1)-coefficients via the equation $a + ib = (1 + i)\alpha$. Also define*

$$D_\alpha = \tfrac{1}{2}((1 - g^2)(a + gb) + \langle g \rangle G).$$

Then D_α is a $(4v, 2v, v, 2)$-relative difference set in H with forbidden subgroup $\langle g^2 \rangle$ if and only if α is a G-associate of a complex Hadamard matrix.

PROOF. We have

$$aa^{(-1)} + bb^{(-1)} + i(ba^{(-1)} - ab^{(-1)}) = (a + ib)(a^{(-1)} - ib^{(-1)})$$
$$= (1 + i)(1 - i)\alpha\alpha^{(*)}$$
$$= 2\alpha\alpha^{(*)}.$$

Hence α is an associate of a G-developed complex Hadamard matrix if and only if

$$aa^{(-1)} + bb^{(-1)} = 2v \qquad \text{and} \qquad ab^{(-1)} = ba^{(-1)}.$$

This holds if and only if

$$D_\alpha D_\alpha^{(-1)} = v\langle g \rangle G + (1 - g^2)(a + gb)(a^{(-1)} + g^3 b^{(-1)})/2$$
$$= v\langle g \rangle G + (1 - g^2)(aa^{(-1)} + bb^{(-1)} + g(ba^{(-1)} - ab^{(-1)}))/2$$
$$= v\langle g \rangle G + v(1 - g^2)$$
$$= 2v + v(\langle g \rangle G - (1 + g^2)).$$

Therefore D_α is a $(4v, 2v, v, 2)$-relative difference set in $G \times \langle g \rangle$ with forbidden subgroup $\langle g^2 \rangle$ if and only if α is a G-associate of a complex Hadamard matrix. $\quad \square$

10.6. Finding all associates of an array

Associates have interesting algebraic and combinatorial properties: they are solutions of group ring equations, and sometimes they are equivalent to (relative) difference sets. In this section we show how the associates of a group-developed array A are related to the regular subgroups of PermAut(A). This will allow us to find all the associates.

Let G be a group of order v, and let A be a $v \times v$ $(0, \mathcal{A})$-array with ambient ring $\mathcal{R}$. We define an action of the holomorph Hol(G) on the set of G-associates of A, and then we prove that the orbits are in 1-1 correspondence with the conjugacy classes of regular subgroups of PermAut(A) isomorphic to G.

For $\pi \in \text{Sym}(G)$ and $\sum_{x \in G} a_x x \in \mathcal{R}[G]$, define

$$(10.11) \qquad m_\pi\left(\sum_{x \in G} a_x x\right) = \sum_{x \in G} a_x \pi(x).$$

10.6.1. LEMMA. *Equation (10.11) defines an action of $\text{Sym}(G)$ on the set of elements of $\mathcal{R}[G]$ with $(0, \mathcal{A})$-coefficients.*

PROOF. First note that m_π merely permutes the coefficients of an element of $\mathcal{R}[G]$. Also, it is readily checked that $m_\pi \circ m_\phi = m_{\pi\phi}$. Hence the map $\pi \mapsto m_\pi$ is a homomorphism from $\text{Sym}(G)$ into the symmetric group on the set of group ring elements with $(0, \mathcal{A})$-coefficients. $\qquad \square$

Recall that the holomorph of G can be embedded in $\mathrm{Sym}(G)$:

$$\mathrm{Hol}(G) = \langle \sigma_a, \alpha \mid a \in G,\ \alpha \in \mathrm{Aut}(G) \rangle.$$

So $\mathrm{Hol}(G)$ acts on the set of elements of $\mathcal{R}[G]$ with $(0, \mathcal{A})$-coefficients. Moreover, we have

10.6.2. LEMMA. *Let A be a $(0, \mathcal{A})$-array. Equation (10.11) defines an action of $\mathrm{Hol}(G)$ on the set of G-associates of A.*

PROOF. Since

$$(10.12) \qquad P_\alpha^{-1} S_r^{-1} [h(xy)]_{x,y \in G} P_\alpha = [\textstyle\sum_{b,c,d} \delta^b_{\alpha(x)} \delta^c_{rb} h(cd) \delta^d_{\alpha(y)}]_{x,y \in G}$$
$$= [h(r\alpha(x)\alpha(y))]_{x,y \in G}$$
$$= [h(r\alpha(xy))]_{x,y \in G},$$

$m_{\sigma_r \alpha}(\sum_{x \in G} h(x)x)$ is an associate of A if and only if $\sum_{x \in G} h(x)x$ is.   □

The action of $\mathrm{Hol}(G)$ in Lemma 10.6.2 defines an equivalence relation on the set of associates of a G-developed array.

10.6.3. DEFINITION. *The associates $\sum_{x \in G} h_i(x)x$, $i = 1, 2$, are equivalent if there are $a \in G$ and $\alpha \in \mathrm{Aut}(G)$ such that $h_2(x) = h_1(a\alpha(x))$ for all $x \in G$.*

10.6.4. REMARK. When A is an $\mathrm{SBIBD}(v, k, \lambda)$, Definition 10.6.3 is in accord with the standard equivalence relation for difference sets in the same group.

The next two lemmas pave the way for Theorem 10.6.7 below, which is the main result of this section.

10.6.5. LEMMA. *If the set $\{(P, Q) \in \mathrm{Perm}(v)^2 \mid A = P^{-1}[h(xy)]_{x,y \in G} Q\}$ is non-empty, then it is a left coset of $\mathrm{PermAut}(A)$ in $\mathrm{Perm}(v)^2$.*

PROOF. Let $\sum_{x \in G} h(x)x$ be an associate of A. Then there exist $\pi, \phi \in \mathrm{Sym}(G)$ such that $P_\pi^{-1}[h(xy)]_{x,y \in G} P_\phi = A$. So if $P^{-1}[h(xy)]_{x,y \in G} Q = A$ then $P_\pi A P_\phi^{-1} = PAQ^{-1}$. Thus $(P, Q) \in (P_\pi, P_\phi)\mathrm{PermAut}(A)$.   □

As in Subsection 3.7.4, let $K = \{(T_x, S_x) \mid x \in G\} \cong G$. For each G-associate $\sum_{x \in G} h(x)x$ of A, define

$$\mathcal{K}_h = \{K^{(P,Q)} \mid A = P^{-1}[h(xy)]_{x,y \in G} Q\}.$$

10.6.6. LEMMA. *$\mathcal{K}_h$ is a conjugacy class in $\mathrm{PermAut}(A)$, consisting of regular subgroups of $\mathrm{PermAut}(A)$ isomorphic to G.*

PROOF. This follows from Lemma 10.6.5.   □

10.6.7. THEOREM. *Let A be a $(0, \mathcal{A})$-array that is group-developed modulo G. Two associates $\sum_{x \in G} h_1(x)x$ and $\sum_{x \in G} h_2(x)x$ of A are equivalent if and only if $\mathcal{K}_{h_1} = \mathcal{K}_{h_2}$. Hence there is a 1-1 correspondence between the equivalence classes of G-associates of A, and the conjugacy classes of regular subgroups of $\mathrm{PermAut}(A)$ isomorphic to G.*

PROOF. Suppose that

$$(10.13) \qquad P_{\pi_1}^{-1}[h_1(xy)]_{x,y \in G} P_{\phi_1} = A \qquad \text{and} \qquad P_{\pi_2}^{-1}[h_2(xy)]_{x,y \in G} P_{\phi_2} = A$$

for some $\pi_i, \phi_i \in \mathrm{Sym}(G)$. Then

$$\mathcal{K}_{h_1} = \mathcal{K}_{h_2}$$

$\Leftrightarrow$ $\exists\,(P,Q) \in \mathrm{PermAut}(A)$ s.t.

$\quad K^{(P_{\pi_1}, P_{\phi_1})(P,Q)} = K^{(P_{\pi_2}, P_{\phi_2})}$ (Lemma 10.6.6)

$\Leftrightarrow$ $\exists\,(P,Q) \in \mathrm{PermAut}(A), a,b \in G, \alpha \in \mathrm{Aut}(G)$ s.t.

$\quad (P_{\pi_1}, P_{\phi_1})(P,Q) = (S_a, T_b)(P_\alpha, P_\alpha)(P_{\pi_2}, P_{\phi_2})$ (Lemma 3.7.10)

$\Leftrightarrow$ $\exists\, a,b \in G, \alpha \in \mathrm{Aut}(G)$ s.t.

$\quad P_{\pi_1} A P_{\phi_1}^{-1} = S_a P_\alpha P_{\pi_2} A P_{\phi_2}^{-1} P_\alpha^{-1} T_b^{-1}$ (since (P,Q) fixes A)

$\Leftrightarrow$ $\exists\, a,b \in G, \alpha \in \mathrm{Aut}(G)$ s.t.

$\quad [h_1(xy)]_{x,y \in G} = S_a P_\alpha [h_2(xy)]_{x,y \in G} P_\alpha^{-1} T_b^{-1}$ (by (10.13))

$\Leftrightarrow$ $\exists\, a,b \in G, \alpha \in \mathrm{Aut}(G)$ s.t.

$\quad [h_1(xy)]_{x,y \in G} = [h_2(a\alpha(xy)b)]_{x,y \in G}$ (direct calculation)

$\Leftrightarrow$ $\exists\, a \in G, \alpha \in \mathrm{Aut}(G)$ s.t.

$\quad h_1(x) = h_2(a\alpha(x)) \quad \forall\, x \in G$ ($\sigma_{b^{-1}} \tau_{b^{-1}} \in \mathrm{Aut}(G)$)

$\Leftrightarrow$ $\sum_{x \in G} h_1(x)x$ is equivalent to $\sum_{x \in G} h_2(x)x$ (Definition 10.6.3). $\square$

10.7. An algorithm for solving Problems 10.2.3 and 10.2.4

Let A be a G-developed $v \times v$ $(0, \mathcal{A})$-array. The following sequence of five steps constitutes a procedure whose output is

- a complete and irredundant set of representatives $\sum_{x \in G} h_i(x)x$ of the equivalence classes of G-associates; and
- for each h_i, a pair $P_i, Q_i \in \mathrm{Perm}(v)$ such that $A = P_i^{-1}[h_i(xy)]_{x,y \in G} Q_i$.

Step 1. Determine $\mathrm{PermAut}(A)$ as $\{(\pi, \phi) \mid P_\pi A P_\phi^{-1} = A\} \le \mathrm{Sym}(v)^2$.

Step 2. Select a representative from each conjugacy class of regular subgroups of $\mathrm{PermAut}(A)$ isomorphic to G.

Step 3. For each conjugacy class representative H found in Step 2, let $\gamma : G \to H$ be any isomorphism (regular embedding). Define orderings $x_1, \ldots, x_v$ and $y_1, \ldots, y_v$ of G as follows:

$\quad\quad \gamma(x_i)$ is the element of H that moves row i of A to row 1;

$\quad\quad \gamma(y_j)$ is the element of H that moves column 1 of A to column j.

Step 4. Define a map $h : G \to \{0\} \cup \mathcal{A}$ by setting $h(x_i)$ to be the ith entry in the first column of A. Then

$$A = [h(x_i y_j)]_{1 \le i, j \le v}.$$

Step 5. Fix the ordering $x_1, \ldots, x_v$ of G to label rows and columns of matrices in this step. Let $Q = P_\phi = [\delta_{\phi(y)}^x]_{x,y \in G}$ where $\phi \in \mathrm{Sym}(G)$ sends x_i to y_i, $1 \le i \le n$. Then

$$A = [h(xy)]_{x,y \in G} Q.$$

10.7.1. REMARK. With regard to Steps 1–3, we note again that MAGMA has facilities for computing $\mathrm{PermAut}(A)$ and its regular subgroups for any $(0,1)$-array A. Other alphabets must be handled by other means.

10.7.2. REMARK. We justify the claim made in Step 4. By Step 3, $\gamma(x_s^{-1}x_r) = \gamma(x_s)^{-1}\gamma(x_r)$ moves row r to row s. Thus, since γ is a regular embedding, $\gamma(g)$ moves row r of A to row s of A if and only if $g = x_s^{-1}x_r$. Now the automorphism $\gamma(y_j^{-1})$ of A moves column j to column 1, so moves row i to row t where $x_t = x_iy_j$. Thus $\gamma(y_j^{-1})$ moves the (i,j)th entry a_{ij} of A to row t of column 1, which (by the definition of h) contains the entry $h(x_t) = h(x_iy_j)$. So $a_{ij} = h(x_iy_j)$.

We solve Problem 10.2.4, up to associate-equivalence, by repeating the above procedure for each isomorphism type G of regular subgroup of $\mathrm{PermAut}(A)$. Then we can infer a complete solution of Problem 10.2.4 via (10.12). For each G we obtain a list $\mathcal{H}_G$ of triples (P_h, h, Q_h) such that $A = P_h^{-1}[h(xy)]_{x,y\in G}Q_h$. If $A \approx [h'(xy)]_{x,y\in G}$ then $(P_{h'}, h', Q_{h'}) \in \mathcal{H}_G$ (notice that $\mathcal{H}_G$ absorbs the variability of γ in Step 3). Then to solve Problem 10.2.3 fully, we enlarge $\mathcal{H}_G$ by replacing each of its elements (P_h, h, Q_h) by $\{(P', h, Q') \mid (P', Q') \in (P_h, Q_h)\mathrm{PermAut}(A)\}$.

10.7.3. EXAMPLE. Let A be the $(0,1)$-matrix at the beginning of Section 10.2. Since 7 is prime, only a cyclic group can act regularly on A. Indeed, C_7 does act regularly on A, because A is equivalent to a circulant. (Of course, an arbitrary circulant array can have regular actions by non-cyclic groups.) We compute that $\mathrm{PermAut}(A) \cong \mathrm{PSL}(2,7)$ has order 168. Sylow's theorem then implies that there is a single conjugacy class of regular subgroups of $\mathrm{PermAut}(A)$. Thus, up to associate-equivalence, there is just one function $h : \mathrm{C}_7 \to \{0,1\}$ such that $A \approx [h(xy)]_{x,y\in\mathrm{C}_7}$.

We find that $x = (P_{(1,7,6,5,4,3,2)}, P_{(1,6,3,5,4,2,7)}) \in \mathrm{PermAut}(A)$. So we let $G = \langle x \rangle \cong \mathrm{C}_7$ and label rows and columns $1, x, \ldots, x^6$. Let ϕ be the permutation of the elements of G defined by $x_i \mapsto y_i$, where $x_i = x^{i-1}$ and y_i is the element of G that moves the first column to column i. Then $AP_{(2,6)(4,5)}$ is back-circulant, hence of the form $[h(rs)]_{r,s\in G}$ as we expect. Reading off the first column, we get the (unique up to equivalence) associate $\sum_{i=0}^{6} h(x^i)x^i = x + x^5 + x^6$. This associate is already visible in the original array, whose first row and column are not changed by the column permutation $(2,6)(4,5)$.

10.8. Composition via associates

In this final section we demonstrate the enormous utility that associates have in the composition of group-developed designs. In particular, we prove that each finite group of odd square-free exponent acts regularly on a complex generalized Hadamard matrix.

10.8.1. General composition results. Assume the notation set down in Construction 8.1.2. Also assume that no array $X \otimes Y$ where $X \in \Lambda_1$ and $Y \in \Lambda_2$ has a repeated row.

10.8.1. THEOREM. *Suppose that $\alpha_i = \sum_{x_i\in G_i} h_i(x_i)x_i$ is a G_i-associate of a* $\mathrm{PCD}(\Lambda_i)$ *for $i = 1, 2$. Then $\alpha_3 = \alpha_1\alpha_2$ is a $(G_1 \times G_2)$-associate of a $\mathrm{PCD}(\Lambda_3)$.*

PROOF. Put $G_3 = G_1 \times G_2$, and define $h_3 : G_3 \to \{0\} \cup \mathcal{A}_3$ by $h_3(g_1g_2) = h_1(g_1)h_2(g_2)$ for $g_1 \in G_1$ and $g_2 \in G_2$. Then

$$\alpha_3 = \sum_{x_1\in G_1} h_1(x_1)x_1 \sum_{x_2\in G_2} h_2(x_2)x_2$$
$$= \sum_{x_1\in G_1} \sum_{x_2\in G_2} h_1(x_1)h_2(x_2)x_1x_2$$
$$= \sum_{x_1\in G_1} \sum_{x_2\in G_2} h_3(x_1x_2)x_1x_2$$

is an element of $\mathcal{R}_3[G_3]$ with $(0, \mathcal{A}_3)$-coefficients. Further,

$$
\begin{aligned}
[h_3(x_3 y_3)]_{x_3, y_3 \in G_3} &= [h_3(x_1 x_2 y_1 y_2)]_{x_1, y_1 \in G_1, x_2, y_2 \in G_2} \\
&= [h_3(x_1 y_1 x_2 y_2)]_{x_1, y_1 \in G_1, x_2, y_2 \in G_2} \\
&= [h_1(x_1 y_1) h_2(x_2 y_2)]_{x_1, y_1 \in G_1, x_2, y_2 \in G_2} \\
&\approx [h_1(x_1 y_1)]_{x_1, y_1 \in G_1} \otimes [h_2(x_2 y_2)]_{x_2, y_2 \in G_2}
\end{aligned}
$$

is a PCD(Λ_3), by Theorem 8.1.4. $\qquad\square$

So we obtain a PCD(Λ) on which there is a regular action by $G_1 \times G_2$ as the Kronecker product of a G_1-developed design and a G_2-developed design.

We now work toward a version of Theorem 10.8.1 for the more general group product $G = G_1 \cdot G_2$ (i.e., G *factors*: $G = G_1 G_2$ and $G_1 \cap G_2 = 1$). This entails replacing $\Lambda_1 \otimes \Lambda_2$ by its Gram completion.

Recall Definition 6.3.5. Suppose that $\mathcal{R}_4$ is a ring with involution $*$ containing $\mathcal{R}_1$ and $\mathcal{R}_2$ as involutory subrings (the restriction of $*$ to $\mathcal{R}_i$ is the involution of that ring). Suppose also that $\mathcal{R}_1$ and $\mathcal{R}_2$ centralize each other. Denote $\Lambda_1 \curlywedge \Lambda_2$ by Λ_4. Thus: if $\mathrm{wt}_i = \mathrm{wt}(\Lambda_i)$ and $\mathrm{df}_i = \mathrm{df}(\Lambda_i)$ for $i = 1, 2$, then Λ_4 has alphabet $\mathcal{A}_4 = \{a_1 a_2 \mid a_1 \in \mathcal{A}_1, a_2 \in \mathcal{A}_2\} \setminus \{0\}$, and consists of all $2 \times v_1 v_2$ $(0, \mathcal{A}_4)$-arrays X with distinct rows such that

$$
\mathrm{wt}(X) \subseteq \mathrm{wt}(\Lambda_4) = \mathrm{wt}_1 \mathrm{wt}_2 \quad \text{and} \quad \mathrm{df}(X) \subseteq \mathrm{df}(\Lambda_4) = \mathrm{wt}_1 \mathrm{df}_2 \cup \mathrm{df}_1 \mathrm{wt}_2 \cup \mathrm{df}_1 \mathrm{df}_2.
$$

10.8.2. THEOREM. *Suppose that*

(i) $\mathrm{wt}(\Lambda_4)$ *and* $\mathrm{df}(\Lambda_4)$ *are disjoint,*

(ii) $\alpha_i = \sum_{x_i \in G_i} h_i(x_i) x_i$ *is a* G_i-*associate of a* PCD(Λ_i) *for* $i = 1, 2$, *and*

(iii) α_1 *commutes with* $\alpha_2 \alpha_2^{(*)}$.

Then $\alpha_4 = \alpha_1 \alpha_2 \in \mathcal{R}_4[G_1 \cdot G_2]$ *is a* $(G_1 \cdot G_2)$-*associate of a* PCD(Λ_4).

PROOF. Let

$$
\beta_i = \alpha_i \alpha_i^{(*)} = \sum_{x_i \in G_i} b_i(x_i) x_i.
$$

By hypothesis (iii), $\alpha_1 \beta_2 = \beta_2 \alpha_1$, and, since $\beta_2^{(*)} = \beta_2$, we have $\alpha_1^{(*)} \beta_2 = \beta_2 \alpha_1^{(*)}$. Now $[h_i(x_i y_i)]_{x_i, y_i \in G_i}$ is a PCD(Λ_i), so

$$
D_i = \sum_{x_i \in G_i} h_i(x_i) S_{x_i}
$$

is a PCD(Λ_i) by (3.4). Thus

$$
D_i D_i^* = \sum_{x_i \in G_i} b_i(x_i) S_{x_i}
$$

where

$$
(10.14) \qquad b_i(x_i) \in \begin{cases} \mathrm{wt}_i & \text{if } x_i = 1 \\ \mathrm{df}_i & \text{if } x_i \neq 1. \end{cases}
$$

Next we define a $v_1 v_2 \times v_1 v_2$ $(0, \mathcal{A}_4)$-array D_4 indexed by $G = G_1 \cdot G_2$:

$$
D_4 = \left(\sum_{x_1 \in G_1} h_1(x_1) S_{x_1 \cdot 1} \right) \left(\sum_{x_2 \in G_2} h_2(x_2) S_{1 \cdot x_2} \right).
$$

Then

$$
\begin{aligned}
D_4 D_4^* &= \left(\sum_{x_1 \in G_1} h_1(x_1) S_{x_1 \cdot 1} \right) \left(\sum_{x_2 \in G_2} b_2(x_2) S_{1 \cdot x_2} \right) \left(\sum_{x_1 \in G_1} h_1(x_1) S_{x_1 \cdot 1} \right)^* \\
&= \left(\sum_{x_1 \in G_1} b_1(x_1) S_{x_1 \cdot 1} \right) \left(\sum_{x_2 \in G_2} b_2(x_2) S_{1 \cdot x_2} \right) \\
&= \sum_{x_1 x_2 \in G_1 \cdot G_2} b_1(x_1) b_2(x_2) S_{x_1 x_2}.
\end{aligned}
$$

Hence D_4 is a PCD(Λ_4) by (i) and (10.14). $\qquad\square$

10.8.3. REMARK. The proofs of Theorems 10.8.1 and 10.8.2 tell us explicitly how to construct a larger group-developed design from the pair of smaller group-developed designs.

10.8.2. Composition for some of the familiar designs. In this subsection we give a few of the many applications of Theorem 10.8.2.

10.8.4. THEOREM. *Let $G = G_1 \cdot G_2$. For each triple D_1, D_2, D_3 listed below, if there are a G_1-developed design D_1 and a G_2-developed design D_2, then there is a G-developed design D_3.*

(1) $\mathrm{H}(n_1), \mathrm{H}(n_2), \mathrm{H}(n_1 n_2)$.

(2) $\mathrm{W}(n_1, k_1), \mathrm{W}(n_2, k_2), \mathrm{W}(n_1 n_2, k_1 k_2)$.

(3) $\mathrm{CH}(n_1), \mathrm{CH}(n_2), \mathrm{CH}(n_1 n_2)$.

(4) $\mathrm{CGH}(n_1; m_1), \mathrm{CGH}(n_2; m_2), \mathrm{CGH}(n_1 n_2; \mathrm{lcm}(m_1, m_2))$.

(5) $\mathrm{CGW}(n_1, k_1; m_1), \mathrm{CGW}(n_2, k_2; m_2), \mathrm{CGW}(n_1 n_2, k_1 k_2; \mathrm{lcm}(m_1, m_2))$.

(6) $\mathrm{GH}(n_1; H), \mathrm{GH}(n_2; H), \mathrm{GH}(n_1 n_2; H)$.

(7) $\mathrm{GW}(n_1, k_1; H), \mathrm{GW}(n_2, k_2; H), \mathrm{GW}(n_1 n_2, k_1 k_2; H)$.

PROOF. Everything follows smoothly from Theorems 10.5.6 and 10.8.2. We prove part (4) only. In this case, $\mathcal{R}_i = \mathbb{Z}[\zeta_{m_i}]$ for $i = 1, 2$, $\mathcal{R}_4 = \mathbb{Z}[\zeta_m]$, and $\Lambda_4 = \Lambda_{\mathrm{CGH}(n_1 n_2; m)}$, where $m = \mathrm{lcm}(m_1, m_2)$. Let α_i be a G_i-associate: then $\alpha_i \alpha_i^{(*)} = n_i$. So α_1 commutes with $\alpha_2 \alpha_2^{(*)}$ in $\mathcal{R}_4[G_1 \cdot G_2]$, and Theorem 10.8.2 implies that $\alpha_1 \alpha_2$ is a $(G_1 \cdot G_2)$-associate of a $\mathrm{CGH}(n_1 n_2; m)$. $\square$

As promised, we now prove

10.8.5. THEOREM. *Each group of odd order $n > 1$ whose exponent is square-free acts regularly on a $\mathrm{CGH}(n; n)$.*

PROOF. By Theorem 3.1.11, such a group factors into cyclic subgroups of odd prime order. So by Theorem 10.8.4 (3), it suffices to prove that there is a circulant $\mathrm{CGH}(p; p)$ for all odd primes p.

Let $\alpha = \sum_{k=0}^{p-1} \zeta^{k^2} x^k$, where ζ is a primitive pth root of unity and $\langle x \rangle$ is a cyclic group of order p. Then

$$\begin{aligned}
\alpha \alpha^{(*)} &= \sum_{j,k=0}^{p-1} \zeta^{k^2} x^k \zeta^{-j^2} x^{-j} \\
&= \sum_{j,k=0}^{p-1} \zeta^{(k+j)(k-j)} x^{k-j} \\
&= \sum_{l=0}^{p-1} \sum_{j=0}^{p-1} \zeta^{(l+2j)l} x^l \\
&= p,
\end{aligned}$$

because $\sum_{j=0}^{p-1} (\zeta^{2l})^j = 0$ if $1 \le l \le p - 1$. By Theorem 10.5.6, we are done. $\square$

CHAPTER 11

Origins of Cocyclic Development

For some orthogonality sets Λ, there do not exist any group-developed PCD(Λ)s. As just one example, group-developed Hadamard matrices can exist only at square orders. We would like to have a generalization of group development that retains the richness of the theory of group-developed designs, but places less severe restrictions on Λ. In this chapter, we give such a generalization.

Let $P_{\mathcal{A}}$ be the group of permutations of $\{0\} \cup \mathcal{A}$ that fix 0. We modify group development by a *development function* $f : G \times G \to P_{\mathcal{A}}$, so that designs have the form

$$[f(x, y)(g(xy))]_{x, y \in G}.$$

Clearly any $(0, \mathcal{A})$-array may be so written. Design-theoretic considerations lead to the following condition on f:

(11.1) $$f(x, y)f(xy, z) = f(y, z)f(x, yz) \qquad \forall\, x, y, z \in G.$$

This *cocycle identity* is the well-known defining equation for 2-cocycles.

We show how cocycles arise in two separate ways in combinatorial design theory. Chapter 12 contains background on cocycles, while Chapters 14 and 15 put forth a general theory of cocyclic designs that parallels and subsumes the theory of group-developed designs in Chapter 10.

11.1. First derivation

In this section we consider higher-dimensional pairwise combinatorial designs. Higher-dimensional Hadamard matrices were introduced by Shlichta [**145**]; the reader who is interested in learning more about them could consult [**165**].

Before we define an n-dimensional pairwise combinatorial design, we need to recite some terminology. An *n-dimensional $(0, \mathcal{A})$-array A of order v* is a $v \times \cdots \times v$ array

$$A = [a(i_1, \ldots, i_n)]_{1 \le i_1, \ldots, i_n \le v}.$$

A 2-dimensional *section* of A is any $v \times v$ subarray $B = [b(i, j)]_{1 \le i, j \le v}$ where

$$b(i, j) = a(u_1, u_2, \ldots, u_{k-1}, i, u_{k+1}, \ldots, u_{l-1}, j, u_{l+1}, \ldots, u_n).$$

This section is obtained by fixing all indices except i_k and i_l. The array A has a total of $\binom{n}{2} v^{n-2}$ 2-dimensional sections.

11.1.1. DEFINITION. Let Λ be a transposable orthogonality set with alphabet $\mathcal{A}$ such that $\Pi_{\Lambda}^{\mathrm{row}} = \Pi_{\Lambda}^{\mathrm{col}}$. An n-dimensional $(0, \mathcal{A})$-array is a *proper n-dimensional pairwise combinatorial design*, denoted $\mathrm{PCD}^n(\Lambda)$, if every 2-dimensional section is a PCD(Λ).

We show how the cocycle identity arises naturally when constructing higher-dimensional designs from 2-dimensional designs. Many of the ideas in this section were first discussed by de Launey in [**40**].

Seberry and Hammer [**81**] noted that if G is an additive abelian group and there is a function $f : G \to \{1, -1\}$ such that $H = [f(x + y)]_{x,y \in G}$ is a Hadamard matrix, then $[f(x_1 + \cdots + x_n)]_{x_i \in G}$ is a proper n-dimensional Hadamard matrix. To see why this is true, fix the values of all but two co-ordinates; the resulting subarray has the form $[f(x + y + a)]_{x,y \in G}$, which is permutation equivalent to H.

11.1.2. EXAMPLE. We display a $\mathrm{CH}^3(2)$, a $\mathrm{GH}^3(3; C_3)$, and a $\mathrm{H}^3(4)$.

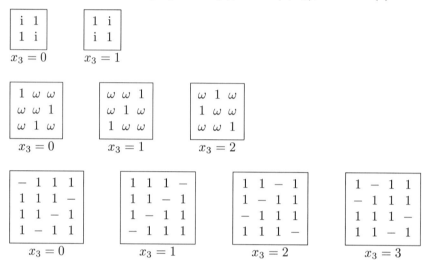

These were all found by the Seberry and Hammer argument.

The drawback of this method is that the starting arrays H are too rare. Our goal now is to identify a larger class of $\mathrm{PCD}(\Lambda)$s that can be extended to give, for each $n > 2$, a $\mathrm{PCD}^n(\Lambda)$ all of whose 2-dimensional sections are Λ-equivalent to the originating design.

11.1.3. DEFINITION. Let G be a finite group, and let $\mathcal{G}$ denote the set of finite tuples of elements of G. A function $f : \mathcal{G} \to \{0\} \cup \mathcal{A}$ is *collapsible* if

(1) for each integer $n \geq 2$, there exists a function $p_n : G \times G \to P_{\mathcal{A}}$ such that
$$f(h_1, h_2, \ldots, h_n) = p_n(h_1, h_2)(f(h_1 h_2, h_3, \ldots, h_n));$$
(2) for each integer $n \geq 2$, there exists a function $q_n : G \times G \to P_{\mathcal{A}}$ such that
$$f(h_1, h_2, \ldots, h_n) = q_n(h_{n-1}, h_n)(f(h_1, h_2, \ldots, h_{n-2}, h_{n-1} h_n))$$

for all $h_1, \ldots, h_n \in G$.

So the value of a collapsible function f is determined by one of the functions p_n or q_n, and the values of f on $(n-1)$-tuples. Therefore, it is easy to obtain a function f that satisfies either (1) or (2), by applying the condition iteratively. It is not so obvious that (1) and (2) can both be satisfied at the same time. However, this will happen if p_n and q_n have special properties, which are implied by the cocycle identity. For the moment, we suppose that it is possible to satisfy (1) and (2) simultaneously.

11.1.4. THEOREM. *Suppose that $[f(h_1, h_2)]_{h_1, h_2 \in G}$ is a $\mathrm{PCD}(\Lambda)$, where f is a collapsible function with $p_i(G \times G)$, $q_i(G \times G) \subseteq \Pi_\Lambda$. Then $[f(h_1, \ldots, h_n)]_{h_i \in G}$ is a $\mathrm{PCD}^n(\Lambda)$.*

PROOF. We prove that every 2-dimensional subarray Y of the form

$$[y(h_i, h_j)]_{h_i, h_j \in G} = [f(a_1, \ldots, a_{i-1}, h_i, a_{i+1}, \ldots, a_{j-1}, h_j, a_{j+1}, \ldots, a_n)]_{h_i, h_j \in G}$$

is a $\mathrm{PCD}(\Lambda)$. Define maps $Q : G \to P_{\mathcal{A}}$ and $P : G \to P_{\mathcal{A}}$ by

$$Q(h_j) = q_{n-j+2}(a_{n-1}, a_n) \circ q_{n-j+1}(a_{n-2}, a_{n-1}a_n) \circ \cdots \circ q_3(h_j, a_{j+1} \cdots a_n),$$

and

$$P(h_i) = p_n(a_1, a_2) \circ p_{n-1}(a_1 a_2, a_3) \circ \cdots \circ p_{n-j+3}(a_1 \cdots a_{i-1} h_i a_{i+1} \cdots a_{j-2}, a_{j-1}).$$

Then

$$y(h_i, h_j) = P(h_i) \circ Q(h_j)(f(a_1 \cdots a_{i-1} h_i a_{i+1} \cdots a_{j-1}, h_j a_{j+1} \cdots a_n)).$$

So $Y \approx_\Lambda [f(h_1, h_2)]_{h_1, h_2 \in G}$, and Y is a $\mathrm{PCD}(\Lambda)$. $\qquad\square$

We will construct a collapsible function by starting with the map $f : G \to \{0\} \cup \mathcal{A}$ and extending it incrementally to all finite tuples of elements of G using p_i and q_i so that (1) and (2) of Definition 11.1.3 hold. To simplify matters, we will confine our attention to *uniform collapsible functions*, where there is a single function $p : G \times G \to P_{\mathcal{A}}$ such that for all $h_1, h_2 \in G$, and $n \geq 2$,

$$q_n(h_1, h_2) = p_n(h_1, h_2) = p(h_1, h_2).$$

Now given any maps $f : G \to \{0\} \cup \mathcal{A}$ and $p : G \times G \to P_{\mathcal{A}}$, we may define two functions $C_{p,f}$ and $D_{p,f}$ which satisfy at least one of (1) or (2) in Definition 11.1.3: put

$$C_{p,f}(h_1) = D_{p,f}(h_1) = f(h_1),$$

and

$$C_{p,f}(h_1, h_2, \ldots, h_n) = p(h_1, h_2)(C_{p,f}(h_1 h_2, h_3, h_4, \ldots, h_n)),$$
$$D_{p,f}(h_1, h_2, \ldots, h_n) = p(h_{n-1}, h_n)(D_{p,f}(h_1, h_2, \ldots, h_{n-2}, h_{n-1} h_n)).$$

If we could be sure that $C_{p,f} = D_{p,f}$, then we would have our collapsible function. Before determining circumstances under which this equality occurs, we clarify our aims with a definition.

11.1.5. DEFINITION. The map $p : G \times G \to P_{\mathcal{A}}$ is an *extension function* if $C_{p,f}$ is collapsible for all $f : G \to \{0\} \cup \mathcal{A}$.

Suppose that p is an extension function. By Definition 11.1.3,

$$p(h_1, h_2) \circ p(h_1 h_2, h_3)(f(h_1 h_2 h_3)) = p(h_2, h_3) \circ p(h_1, h_2 h_3)(f(h_1 h_2 h_3)).$$

Since f is arbitrary,

(11.2) $p(h_1, h_2) \circ p(h_1 h_2, h_3) = p(h_2, h_3) \circ p(h_1, h_2 h_3) \quad \forall\, h_1, h_2, h_3 \in G,$

which we recognize as the cocycle identity. We will have $C_{p,f} = D_{p,f}$ if p satisfies a succession of increasingly complicated identities involving more and more variables. For example, after (11.2) the next identity is

$$p(h_1, h_2) \circ p(h_1 h_2, h_3) \circ p(h_1 h_2 h_3, h_4) = p(h_3, h_4) \circ p(h_2, h_3 h_4) \circ p(h_1, h_2 h_3 h_4).$$

However, if p is *abelian* in the following sense, then (as shown below) only (11.2) is needed.

11.1.6. DEFINITION. A map $p : G \times G \to P_{\mathcal{A}}$ is *abelian* if

$$p(h_1, h_2) \circ p(h_3, h_4) = p(h_3, h_4) \circ p(h_1, h_2)$$

for all $h_1, h_2, h_3, h_4 \in G$.

11.1.7. THEOREM. *An abelian map $p : G \times G \to P_{\mathcal{A}}$ is an extension function if and only if* (11.2) *holds.*

PROOF. Assuming (11.2), it suffices to prove that $C_{p,f} = D_{p,f}$, i.e.,

(11.3) $p(h_1, h_2) \circ p(h_1 h_2, h_3) \circ \cdots \circ p(h_1 h_2 \cdots h_{n-1}, h_n)$

$$= p(h_{n-1}, h_n) \circ p(h_{n-2}, h_{n-1} h_n) \circ \cdots \circ p(h_1, h_2 h_3 \cdots h_n).$$

By (11.2), for $i = 1, \ldots, n - 2$ we have

$$p(h_1 h_2 \cdots h_i, h_{i+1}) \circ p(h_1 h_2 \cdots h_{i+1}, h_{i+2} h_{i+3} \cdots h_n)$$

$$= p(h_{i+1}, h_{i+2} h_{i+3} \cdots h_n) \circ p(h_1 h_2 \cdots h_i, h_{i+1} h_{i+2} \cdots h_n).$$

Combining these equations, and using that p is abelian, gives (11.3).   □

Theorems 11.1.4 and 11.1.7 now imply

11.1.8. THEOREM. *If $p : G \times G \to \Pi_\Lambda$ is an abelian map satisfying* (11.2) *and*

$$[p(h_1, h_2)(f(h_1 h_2))]_{h_1, h_2 \in G}$$

is a PCD(Λ) *for some map $f : G \to \{0\} \cup \mathcal{A}$, then*

$$[C_{p,f}(h_1, \ldots, h_n)]_{h_1, \ldots, h_n \in G}$$

is a PCD^n(Λ).

11.1.9. EXAMPLE. Let $G \cong C_2 \times C_2$ be the group $(\{00, 01, 10, 11\}, +)$ under addition modulo 2. Let $\mathcal{A} = \{\pm a, \pm b, \pm c, \pm d\}$. Define the map $g : G \to \mathcal{A}$ by

$$g(00) = a, \quad g(01) = b, \quad g(10) = c, \quad g(11) = d.$$

Let $p : G \times G \to \text{Sym}(\{\pm 1\})$ be the map such that $p(x, y)$ is multiplication by -1 or 1 according to the table

	00	01	10	11
00	1	1	1	1
01	1	–	1	–
10	1	–	–	1
11	1	1	–	–

Then p satisfies the cocycle identity, and

$$[p(x, y)g(x + y)]_{x,y \in G} = \begin{bmatrix} a & b & c & d \\ b & -a & d & -c \\ c & -d & -a & b \\ d & c & -b & -a \end{bmatrix}$$

is an OD($4; 1^4$). So by Theorem 11.1.8, the 3-dimensional array

$$[p(x, y)p(x + y, z)g(x + y + z)]_{x,y,z \in G}$$

is an $\mathrm{OD}^3(4; 1^4)$. We show the sections of this design below.

$$
\begin{array}{|cccc|}
a & b & c & d \\
b & -a & d & -c \\
c & -d & -a & b \\
d & c & -b & -a \\
\end{array}
\quad
\begin{array}{|cccc|}
b & -a & -d & c \\
-a & -b & c & d \\
-d & -c & -b & -a \\
c & -d & a & -b \\
\end{array}
\quad
\begin{array}{|cccc|}
c & d & -a & -b \\
d & -c & -b & a \\
-a & b & -c & d \\
-b & -a & -d & -c \\
\end{array}
\quad
\begin{array}{|cccc|}
d & -c & b & -a \\
-c & -d & -a & -b \\
b & a & -d & -c \\
-a & b & c & -d \\
\end{array}
$$

$$z = 00 \qquad\qquad z = 01 \qquad\qquad z = 10 \qquad\qquad z = 11$$

$$
\begin{array}{|cccc|}
a & b & c & d \\
b & -a & d & -c \\
c & -d & -a & b \\
d & c & -b & -a \\
\end{array}
\quad
\begin{array}{|cccc|}
b & -a & d & -c \\
-a & -b & -c & -d \\
-d & -c & b & a \\
c & -d & -a & b \\
\end{array}
\quad
\begin{array}{|cccc|}
c & -d & -a & b \\
d & c & -b & -a \\
-a & -b & -c & -d \\
-b & a & -d & c \\
\end{array}
\quad
\begin{array}{|cccc|}
d & c & -b & -a \\
-c & d & a & -b \\
b & -a & d & -c \\
-a & -b & -c & -d \\
\end{array}
$$

$$y = 00 \qquad\qquad y = 01 \qquad\qquad y = 10 \qquad\qquad y = 11$$

$$
\begin{array}{|cccc|}
a & b & c & d \\
b & -a & d & -c \\
c & -d & -a & b \\
d & c & -b & -a \\
\end{array}
\quad
\begin{array}{|cccc|}
b & -a & d & -c \\
-a & -b & -c & -d \\
d & c & -b & -a \\
-c & d & a & -b \\
\end{array}
\quad
\begin{array}{|cccc|}
c & -d & -a & b \\
-d & -c & b & a \\
-a & -b & -c & -d \\
b & -a & d & -c \\
\end{array}
\quad
\begin{array}{|cccc|}
d & c & -b & -a \\
c & -d & -a & b \\
-b & a & -d & c \\
-a & -b & -c & -d \\
\end{array}
$$

$$x = 00 \qquad\qquad x = 01 \qquad\qquad x = 10 \qquad\qquad x = 11$$

The Seberry and Hammer construction cannot be applied to orthogonal designs (with more than one indeterminate), by Theorem 2.6.8.

As the latter chapters of this book demonstrate, there is a lot of evidence suggesting that for any order divisible by 4 (i.e., not necessarily square), there exists a Hadamard matrix that can be employed in Theorem 11.1.8.

Functions $p : G \times G \to \Pi_\Lambda$ satisfying (11.2) are at the heart of our method for constructing higher-dimensional designs. The connection between (11.2) and the second cohomology of G was first noted by Horadam; cf. [**87**, p. 107]. Subsequently de Launey and Horadam [**50, 88**] made considerable use of Horadam's realization that (11.2) defines a cocycle.

11.1.10. DEFINITION. Let G be a finite group and let C be an abelian group. A map $f : G \times G \to C$ satisfying the equation (11.1) is a *cocycle*, or more formally a *2-cocycle*.

Theorem 11.1.7 shows that abelian extension functions are 2-cocycles. We have said very little about non-uniform collapsible functions and non-abelian extension functions—but at some point these ought to be examined.

5. RESEARCH PROBLEM. *Investigate non-uniform collapsible functions or non-abelian extension functions.*

Finally, we remark that cocycles may be defined in dimensions greater than 2 (for more on this, see Section 20.4 of Chapter 20). Having seen how 2-dimensional cocycles enter into the construction of proper higher-dimensional designs, we therefore ask

6. RESEARCH PROBLEM. *Do n-dimensional cocycles for $n > 2$ also play a role in the construction of higher-dimensional designs?*

11.2. Second derivation

Once the group is specified, all the information in a group-developed array is contained in its first row (or, indeed, any row). We say that the array is *developed from its first row*. Note that the search space for such a design is smaller than the search space for a general design. Another useful property of a group-developed design D is that the array resulting from development of any row of D is equivalent to D; actually, it may be obtained from D by row permutations only.

In this section we look at developing a design from one of its rows in such a way that the choice of row to develop does not affect the Λ-equivalence class of the design. Some of these ideas appeared previously in [**88**].

11.2.1. DEFINITION. Let G be a group of order v, with a fixed total ordering $x_1 < \cdots < x_v$ of its elements. Let $f : G \times G \to P_\mathcal{A}$ and $g : G \to \{0\} \cup \mathcal{A}$ be maps. The $(0, \mathcal{A})$-array

$$[f(x, y)(g(xy))]_{x,y \in G} = [f(x_i, x_j)(g(x_i x_j))]_{1 \leq i,j \leq v}$$

is *f-developed modulo G from the row* $[g(x)]_{x \in G}$. The map f is called a *development function*; it is *normalized* if $f(1, x) = f(x, 1)$ is the identity permutation for all $x \in G$.

11.2.2. EXAMPLE. Let f_{id} be the identity development function, i.e., $f_{\mathrm{id}}(x, y)$ is the identity permutation for all $x, y \in G$. Then group development modulo G is f_{id}-development modulo G.

Any $\mathcal{A}$-array is f-developed for some f. So if f is unrestricted, then the fact that an array is f-developed tells us nothing.

Let Λ be an orthogonality set of $2 \times v$ $(0, \mathcal{A})$-arrays, and let G be a group of order v. We are interested in development functions such as those in the following definition.

11.2.3. DEFINITION. A normalized development function $f : G \times G \to P_\mathcal{A}$ is *Λ-row-invariant* if for any map $g : G \to \{0\} \cup \mathcal{A}$ and $a \in G$ there is a sequence of elementary Λ-equivalence row operations independent of g that transforms

$$X = [f(x, y)(g(xy))]_{x,y \in G}$$

into

$$Y = [f(x, y) \circ f(a, xy)(g(axy))]_{x,y \in G}.$$

11.2.4. REMARK. The arrays X and Y in Definition 11.2.3 are both f-developed modulo G (in the latter case, from the ath row of X). We rephrase this definition in the language of Chapter 5: the development function f is Λ-row-invariant if and only if, over an ambient ring $\mathcal{R}$ with row group R, for all maps $g : G \to \{0\} \cup \mathcal{A}$ and $a \in G$ there is $M_a \in \mathrm{Mon}(v, R)$ such that $Y = M_a X$.

11.2.5. EXAMPLE. Group development is Λ-row-invariant for any Λ:

$$[\delta_{ax}^y]_{x,y \in G}[g(xy)]_{x,y \in G} = [\textstyle\sum_w \delta_{ax}^w g(wy)]_{x,y \in G} = [g(axy)]_{x,y \in G}.$$

11.2.6. LEMMA. *Let $f : G \times G \to P_\mathcal{A}$ be a normalized development function. Then f is Λ-row-invariant if and only if for each $a \in G$, there exist a map $\rho_a : G \to \Pi_\Lambda^{\mathrm{row}}$ and permutation $\pi_a \in \mathrm{Sym}(G)$ such that for any map $g : G \to \{0\} \cup \mathcal{A}$,*

$$(11.4) \quad f(x, y) \circ f(a, xy)(g(axy)) = \rho_a(x) \circ f(\pi_a(x), y)(g(\pi_a(x)y)) \quad \forall x, y \in G.$$

PROOF. The operations allowed in Definition 11.2.3 amount to a permutation π_a of the row labels, followed by the application, for each $x \in G$, of some $\rho_a(x) \in \Pi_\Lambda^{\text{row}}$ to the entries in row x of

$$[f(x, y)(g(xy))]_{x,y \in G}.$$

This sequence of operations yields

$$[\rho_a(x) \circ f(\pi_a(x), y)(g(\pi_a(x)y))]_{x,y \in G}$$

which must equal

$$[f(x, y) \circ f(a, xy)(g(axy))]_{x,y \in G}$$

by Definition 11.2.3. $\square$

11.2.7. DEFINITION. A development function $f : G \times G \to P_{\mathcal{A}}$ is Λ-*suitable* if $f(G \times G) \subseteq \Pi_\Lambda^{\text{row}}$.

11.2.8. THEOREM. *Let f be a Λ-suitable normalized development function. Then f is Λ-row-invariant if and only if f satisfies (11.1).*

PROOF. If f satisfies (11.1) then

$$[f(x, y) \circ f(a, xy)(g(axy))]_{x,y \in G} = [f(a, x) \circ f(ax, y)(g(axy))]_{x,y \in G}$$
$$\approx_\Lambda [f(x, y)(g(xy))]_{x,y \in G}.$$

Next suppose that f is Λ-row-invariant. By Lemma 11.2.6, for all maps $g : G \to \{0\} \cup \mathcal{A}$ and each $a \in G$ there exist $\pi_a \in \text{Sym}(G)$ and a map $\rho_a : G \to \Pi_\Lambda^{\text{row}}$ such that (11.4) holds. Letting g run over all constant maps, we see that

$$(\dagger) \qquad\qquad f(x, y)f(a, xy) = \rho_a(x)f(\pi_a(x), y) = \phi_{a,x,y},$$

say. So

$$(\ddagger) \qquad\qquad \phi_{a,x,y}(g(axy)) = \phi_{a,x,y}(g(\pi_a(x)y)) \quad \forall\, a, x, y, g.$$

If $\pi_a(x) \neq ax$ then we may choose g such that $g(\pi_a(x)y) \neq g(axy)$. Since $\phi_{a,x,y}$ is injective, this contradicts $(\ddagger)$. Thus $\pi_a(x) = ax$. Equation $(\dagger)$ with $y = 1$ gives that $\rho_a(x) = f(a, x)$, because f is normalized. Then $(\dagger)$ becomes

$$f(x, y)f(a, xy) = f(a, x)f(ax, y),$$

which is the cocycle identity. $\square$

11.2.9. DEFINITION. A development function f is *abelian* if $f(a, b)f(c, d) = f(c, d)f(a, b)$ for all $a, b, c, d \in G$.

Thus, an abelian Λ-suitable normalized development function is Λ-row-invariant if and only if it is a cocycle. We point out that any f-developed array, for a Λ-suitable cocycle f, is Λ-equivalent to an f'-developed array where f' is a normalized cocycle. Indeed, we can define f' by $f'(x, y) = f(1, 1)^{-1} \circ f(x, y)$; this works because (11.1) implies that $f(x, 1) = f(1, y) = f(1, 1)$ for all $x, y \in G$.

11.2.10. EXAMPLE. Bilinear maps are cocycles. Let V be a vector space over a finite field $\mathbb{F}$, and regard $G = V$ and $C = \mathbb{F}$ as additive groups. Define $f : G \times G \to C$ by $f(x, y) = xy^\top$. Then

$$f(x, y) + f(x + y, z) = xy^\top + (x + y)z^\top = xy^\top + xz^\top + yz^\top$$
$$= yz^\top + x(y + z)^\top = f(y, z) + f(x, y + z).$$

11.2.11. EXAMPLE. Let V be the k-dimensional vector space over $\mathrm{GF}(p^m)$. Recall from Section 9.2 that $D = [xy^\top]_{x,y \in V}$ is a $\mathrm{GH}(p^{mk}; C_p^m)$. For $x, y \in V$, define $f(x,y) \in \mathrm{Sym}(\mathrm{GF}(p^m))$ by $f(x,y) : w \mapsto w + xy^\top$. By Example 11.2.10, f is a cocycle. We have $D = [f(x,y)(g(xy))]_{x,y \in V}$ where g is the zero map.

Assuming that $[\Pi_\Lambda^{\mathrm{row}}, \Pi_\Lambda^{\mathrm{col}}] = 1$, $\Pi_\Lambda^{\mathrm{row}} \cap \Pi_\Lambda^{\mathrm{col}}$ is a central subgroup of $\Pi_\Lambda^{\mathrm{row}}$ and so is abelian. The next theorem indicates that, for a given cocycle $f : G \times G \to \Pi_\Lambda^{\mathrm{row}} \cap \Pi_\Lambda^{\mathrm{col}}$, it is easier to test whether an f-developed array is a $\mathrm{PCD}(\Lambda)$ than it is to test whether an arbitrary array is a $\mathrm{PCD}(\Lambda)$.

11.2.12. THEOREM. *Let $f : G \times G \to \Pi_\Lambda^{\mathrm{row}} \cap \Pi_\Lambda^{\mathrm{col}}$ be a cocycle, and let $g : G \to \{0\} \cup \mathcal{A}$ be a map. The f-developed array $D = [f(x,y)(g(xy))]_{x,y \in G}$ is a $\mathrm{PCD}(\Lambda)$ if and only if its first row is Λ-orthogonal to every other row.*

PROOF. Let a_0 and $a_1 \neq a_0$ be elements of G. By (11.1),

$$f(a_0, a_0^{-1}c) = f(a_0, a_0^{-1})f(1,c)f(a_0^{-1}, c)^{-1}$$

and

$$f(a_1, a_0^{-1}c) = f(a_1, a_0^{-1})f(a_1 a_0^{-1}, c)f(a_0^{-1}, c)^{-1}$$

for all $c \in G$. Thus

     row a_0 is Λ-orthogonal to row a_1

$\iff$ $[f(a_0,b)(g(a_0 b))]_{b \in G}$ is Λ-orthogonal to $[f(a_1,b)(g(a_1 b))]_{b \in G}$

$\iff$ $[f(a_0, a_0^{-1}c)(g(c))]_{c \in G}$ is Λ-orthogonal to $[f(a_1, a_0^{-1}c)(g(a_1 a_0^{-1}c))]_{c \in G}$

$\iff$ $[f(a_0, a_0^{-1})f(1,c)f(a_0^{-1}, c)^{-1}(g(c))]_{c \in G}$ is Λ-orthogonal to $[f(a_1, a_0^{-1})f(a_1 a_0^{-1}, c)f(a_0^{-1}, c)^{-1}(g(a_1 a_0^{-1}c))]_{c \in G}$

$\iff$ $[f(1,c)(g(c))]_{c \in G}$ is Λ-orthogonal to $[f(a_1 a_0^{-1}, c)(g(a_1 a_0^{-1}c))]_{c \in G}$

$\iff$ row 1 is Λ-orthogonal to row $a_1 a_0^{-1}$.   □

11.3. Cocycles for cyclic groups

In the next chapter we will see that there is a solid algebraic theory of cocycles. We conclude the present chapter with a direct calculation of all cocycles of a finite group G, in the special but fundamental case that G is cyclic.

11.3.1. EXAMPLE. Let C be an abelian group. If $s, t \in \{0, 1, \ldots, n-1\}$, write $s \oplus t$ for the residue in $\{0, 1, \ldots, n-1\}$ of $s + t$ modulo n. Fix an element ω of C, and define $f : \mathbb{Z}_n \times \mathbb{Z}_n \to C$ by $f(s,t) = \omega^{\lfloor (s+t)/n \rfloor}$. Then f is a cocycle:

$$f(s,t)f(s \oplus t, u) = \omega^{\lfloor (s+t)/n \rfloor} \omega^{\lfloor (u+(s \oplus t))/n \rfloor} = \omega^{\lfloor (u+s+t)/n \rfloor}$$

$$= \omega^{\lfloor (t+u)/n \rfloor} \omega^{\lfloor (s+(t \oplus u))/n \rfloor} = f(t,u)f(s, t \oplus u).$$

11.3.2. THEOREM. *Let G be a finite cyclic group, written as the integers modulo $|G|$, and let C be an abelian group, also written additively. Then $f : G \times G \to C$ is a cocycle if and only if, for all $s, t \in G$,*

$$(11.5) \qquad f(s,t) = f(0,0) + \sum_{i=0}^{t-1} [f(s+i, 1) - f(i, 1)].$$

PROOF. Suppose that

$$(11.6) \qquad f(a,b) + f(a+b,c) = f(b,c) + f(a,b+c)$$

for all $a, b, c \in G$. Equation (11.6) implies that $f(s,0) = f(0,0)$, which is (11.5) for $t = 0$. Now assume that (11.5) is true for $0 \le t \le n$. By (11.6) again,

$$(11.7) \qquad f(s, n+1) = f(s,n) + f(s+n, 1) - f(n,1).$$

Substituting $f(0,0) + \sum_{i=0}^{n-1}[f(s+i,1) - f(i,1)]$ for $f(s,n)$ in (11.7) results in (11.5) for $t = n + 1$. This proves (11.5) by induction.

We now prove the other direction. Suppose that (11.5) holds for all s and t. Then

$$\begin{aligned}
f(s,t) + f(s+t, u) &= f(0,0) + \sum_{i=0}^{t-1}[f(s+i,1) - f(i,1)] \\
&\quad + f(0,0) + \sum_{i=0}^{u-1}[f(s+t+i,1) - f(i,1)] \\
&= f(0,0) + \sum_{i=0}^{t-1}[f(s+i,1) - f(i,1)] \\
&\quad + f(0,0) + \sum_{i=0}^{u-1}[f(t+i,1) - f(i,1)] \\
&\quad + \sum_{i=0}^{u-1}[f(s+t+i,1) - f(t+i,1)] \\
&= f(0,0) + \sum_{i=0}^{u-1}[f(t+i,1) - f(i,1)] \\
&\quad + f(0,0) + \sum_{i=0}^{u+t-1}[f(s+i,1) - f(i,1)] \\
&= f(t,u) + f(s, t+u),
\end{aligned}$$

as required. $\qquad\qquad\qquad\qquad\qquad\qquad\qquad\qquad\qquad\qquad\qquad\square$

Thus, if G is cyclic and we know

$$(11.8) \qquad f(0,0) \quad \text{and} \quad f(i,1), \quad 1 \le i \le |G| - 1,$$

then we know the cocycle $f : G \times G \to C$ completely. All possibilities for f may be displayed in a $|G| \times |G|$ 'development table', whose (s,t)th entry is $f(s,t)$ in terms of the quantities (11.8).

11.3.3. EXAMPLE. By Theorem 11.3.2 we get the following development tables for $\mathbb{Z}_2$, $\mathbb{Z}_3$, and $\mathbb{Z}_4$:

$$\begin{bmatrix} Z & Z \\ Z & W \end{bmatrix}, \quad \begin{bmatrix} Z & Z & Z \\ Z & W & X \\ Z & X & W^{-1}XZ \end{bmatrix}, \quad \begin{bmatrix} Z & Z & Z & Z \\ Z & W & X & Y \\ Z & X & W^{-1}XY & W^{-1}YZ \\ Z & Y & W^{-1}YZ & X^{-1}YZ \end{bmatrix}.$$

Setting $Z = 1$ gives normalized cocycles.

A development table displays all cocyclic development functions for a given finite group G. In Chapter 20 we show how to calculate these tables when G is solvable.

Group Extensions and Cocycles

In the study of cocyclic development of combinatorial designs, it is imperative to have a good working acquaintance with cocycles. This chapter provides some of the (mostly well-known) theory of cocycles required in the book.

12.1. Central extensions

We begin by discussing what it means for a group to be an extension.

12.1.1. DEFINITION. Let E and G be groups, and let C be a subgroup of E. Then E is a *central extension of C by G* if $C \leq Z(E)$ and $E/C \cong G$.

Since it is central, the subgroup C of E in Definition 12.1.1 is abelian. Any group E with normal subgroup N (not necessarily central, nor abelian) such that $E/N \cong G$ is said to be an extension of N by G. The theory of central extensions is a special case, i.e., when the conjugation action of E on N is trivial.

Often we will want to treat E as a central extension of an abstract group C, meaning that E has a central subgroup isomorphic to C. For example, we can say that a group is a central extension of C_p if its center has order divisible by the prime p. So we have the following weaker version of Definition 12.1.1.

12.1.2. DEFINITION. A group E is a *central extension of C by G* if E contains an isomorphic copy Z of C as a central subgroup, and $E/Z \cong G$.

From now on G is a finite group and C is an abelian group.

12.1.1. Canonical central extension for a cocycle. Let $f : G \times G \to C$ be a cocycle; so

$$(12.1) \qquad f(x,y)f(xy,w) = f(x,yw)f(y,w) \quad \forall\, x,y,w \in G.$$

The cocycle f' defined by $f'(x,y) = f(1,1)^{-1}f(x,y)$ is *normalized*, i.e., $f'(1,1) = 1$. Unless stated otherwise, our cocycles are normalized. The group G is the *ground* (or *indexing*) *group* for f, and C is the *coefficient group*.

The set of ordered pairs $\{(x,a) \mid x \in G, a \in C\}$ becomes a group $E(f)$ under the operation

$$(x,a)(y,b) = (xy, f(x,y)ab).$$

Associativity of this operation is equivalent to f satisfying (12.1). The identity of $E(f)$ is $(1,1)$, and $(x,a)^{-1} = (x^{-1}, a^{-1}f(x,x^{-1})^{-1})$. If f is the *trivial cocycle*—i.e., $f(x,y) = 1$ for all x,y—then $E(f) = G \times C$.

We call $E(f)$ the *canonical (central) extension of C by G for the cocycle f.* The subgroup $Z_0 = \{(1,c) \mid c \in C\}$ of $E(f)$ is central and isomorphic to C. We show that $E(f)/Z_0 \cong G$ in the next subsection.

12.1.2. Exact sequences. A sequence

$$\cdots \longrightarrow H_i \xrightarrow{\phi_i} H_{i+1} \xrightarrow{\phi_{i+1}} H_{i+2} \longrightarrow \cdots$$

of group homomorphisms is *exact* if $\ker \phi_{i+1} = \operatorname{im} \phi_i$ for all i. An exact sequence of the form

$$(12.2) \qquad\qquad 1 \longrightarrow N \xrightarrow{\iota} E \xrightarrow{\pi} G \longrightarrow 1$$

is a *short exact sequence*. The first arrow embeds the trivial group in N, ι embeds N as a subgroup of the group E, π has kernel $\iota(N)$, and the final arrow denotes the trivial homomorphism. Hence π is surjective, and $E/\iota(N) \cong G$. We have an exact sequence (12.2) precisely when E contains a normal subgroup $\bar{N}$ isomorphic to N, such that $E/\bar{N} \cong G$: take ι to be inclusion of $\bar{N}$ in E composed with an isomorphism $N \to \bar{N}$, and π to be the composite of an isomorphism $E/\bar{N} \to G$ with the natural surjection of E onto $E/\bar{N}$. When $\iota(N) \leq Z(E)$ we say that (12.2) is a *central short exact sequence*. Thus, existence of a central short exact sequence (12.2) is equivalent to E being a central extension of N by G. We will sometimes also call (12.2) a 'central extension'.

We now describe a central short exact sequence corresponding to the canonical central extension group $E(f)$ for a cocycle $f : G \times G \to C$. Define the embedding $\iota_0 : C \to Z(E(f))$ by $\iota_0(a) = (1, a)$. The map $\pi_0 : E(f) \to G$ such that $\pi_0((x, a)) = x$ is a homomorphism from $E(f)$ onto G whose kernel is $\iota_0(C)$. It follows that

$$(12.3) \qquad\qquad 1 \longrightarrow C \xrightarrow{\iota_0} E(f) \xrightarrow{\pi_0} G \longrightarrow 1$$

is exact. This is the *canonical central short exact sequence for the cocycle f*.

12.1.3. Extracting a cocycle from a central extension. We have seen how a cocycle determines a central short exact sequence. We show next how this process reverses. (Eventually we will see that the two processes are inverses of each other, modulo notions of equivalence for cocycles and for central extensions.) Let

$$(12.4) \qquad\qquad 1 \longrightarrow C \xrightarrow{\iota} E \xrightarrow{\pi} G \longrightarrow 1$$

be a central short exact sequence. Let $\tau : G \to E$ be a map satisfying

$$\pi\tau = \operatorname{id}_G.$$

The set map τ is injective (if it is a homomorphism then $E \cong C \times G$ splits over $\iota(C)$). The image of τ is a transversal of $\iota(C)$ in E. For this reason we say that τ is a *transversal map*. We assume that τ is normalized, i.e., $\tau(1_G) = 1_E$; this is our default assumption for all transversal maps.

Since $\tau(x)\tau(y)\tau(xy)^{-1} \in \ker \pi = \iota(C)$, we can define $f_{\iota,\tau} : G \times G \to C$ by

$$(12.5) \qquad\qquad f_{\iota,\tau}(x, y) = \iota^{-1}(\tau(x)\tau(y)\tau(xy)^{-1}).$$

The following calculation proves that $f_{\iota,\tau}$ is a cocycle:

$$\iota(f_{\iota,\tau}(x,y)f_{\iota,\tau}(xy,z)) = \tau(x)\tau(y)\tau(xy)^{-1}\tau(xy)\tau(z)\tau(xyz)^{-1}$$
$$= \tau(x)\tau(y)\tau(z)\tau(xyz)^{-1}$$
$$= \tau(x)\tau(y)\tau(z)\tau(yz)^{-1}\tau(yz)\tau(xyz)^{-1}$$
$$= \tau(x)\tau(yz)\tau(xyz)^{-1}\tau(y)\tau(z)\tau(yz)^{-1}$$
$$= \iota(f_{\iota,\tau}(x,yz)f_{\iota,\tau}(y,z)),$$

where we used $\tau(y)\tau(z)\tau(yz)^{-1} \in \iota(C) \leq Z(E)$ in moving from the third to fourth line. Now although there may be many choices for τ, each giving rise to a different cocycle as above for a fixed central extension (12.4) of C by G, all such cocycles are essentially equivalent. We deal with this variability and equivalence in a later subsection.

12.1.3. REMARK. Since

$$\tau(x)\tau(y)\tau(xy)^{-1}\tau(y)^{-1}\tau(x)^{-1}\tau(xy) = \tau(y)^{-1}\tau(x)^{-1}\tau(x)\tau(y)\tau(xy)^{-1}\tau(xy) = 1,$$

we have that $\tau(x)\tau(y)\tau(xy)^{-1} = \tau(xy)^{-1}\tau(x)\tau(y)$.

Let $f : G \times G \to C$ be any (normalized) cocycle. We apply the above process to extract a cocycle from $E(f)$. A transversal map $\tau_0 : G \to E(f)$ for the canonical central short exact sequence (12.3) is given by $\tau_0(x) = (x, 1)$. Then

$$\tau_0(x)\tau_0(y) = (1, f(x, y))\tau_0(xy),$$

and so

(12.6) $$f_{\iota_0, \tau_0}(x, y) = \iota_0^{-1}(\tau_0(x)\tau_0(y)\tau_0(xy)^{-1}) = f(x, y).$$

Hence $f = f_{\iota_0, \tau_0}$ for this choice τ_0 of transversal map.

In the next few subsections we examine the cocycles that arise when we vary the choices of embedding ι, surjection π, and transversal map τ. We write $\mathcal{C}_{E,C}(\iota, \pi, \tau)$ for the cocycle $f_{\iota, \tau}$ determined by (12.4) and a particular transversal map τ. Similar notation will denote sets of cocycles obtained by specifying only part of the triple (ι, π, τ). For example, $\mathcal{C}_{E,C}(\iota, \pi)$ is the set of cocycles $f_{\iota, \tau}$ obtained from a fixed central extension (12.4) for all choices of τ. We denote by $\mathcal{C}_{E,C}$ the set of cocycles obtained when ι, π, and τ are all free to be varied.

12.1.4. Varying the transversal map. If τ is a transversal map for (12.4), and $\rho : G \to C$ is a normalized map, then $\sigma : G \to E$ defined by $\sigma(x) = \tau(x)\iota(\rho(x))$ is a transversal map for (12.4). Conversely, let τ and σ be any two transversal maps for (12.4). Since $\tau(x)^{-1}\sigma(x) = \sigma(x)\tau(x)^{-1} \in \iota(C)$, we get a normalized map $\rho : G \to C$ defined by $\rho(x) = \iota^{-1}(\tau(x)^{-1}\sigma(x))$. Then

$$\iota(f_{\iota,\tau}^{-1}(x, y)f_{\iota,\sigma}(x, y)) = \tau(xy)\tau(y)^{-1}\tau(x)^{-1}\sigma(x)\sigma(y)\sigma(xy)^{-1}$$
$$= \tau(xy)(\tau(x)^{-1}\sigma(x))\tau(y)^{-1}\sigma(y)\sigma(xy)^{-1}$$
$$= (\tau(x)^{-1}\sigma(x))(\tau(y)^{-1}\sigma(y))\tau(xy)\sigma(xy)^{-1}$$
$$= \iota(\rho(x)\rho(y)\rho(xy)^{-1}).$$

Thus $f_{\iota,\sigma}(x, y) = f_{\iota,\tau}(x, y)\partial\rho(x, y)$, where $\partial\rho(x, y) = \rho(x)\rho(y)\rho(xy)^{-1}$. Since $f_{\iota,\tau}$ and $f_{\iota,\sigma}$ satisfy (12.1), so does $\partial\rho$. The cocycles $\partial\rho$ arising from arbitrary set maps $\rho : G \to C$ are known as *coboundaries*. Two cocycles f and f' are *cohomologous* if there is a coboundary $\partial\rho$ such that

$$f(x, y) = f'(x, y)\partial\rho(x, y) \quad \forall x, y \in G.$$

This defines an equivalence relation on the set of cocycles $G \times G \to C$. Let $[f]$ denote the *cohomology class* of f (the equivalence class of f under this relation). We deduce that $\mathcal{C}_{E,C}(\iota, \pi)$ is a cohomology class:

$$f \in \mathcal{C}_{E,C}(\iota, \pi) \iff \mathcal{C}_{E,C}(\iota, \pi) = [f].$$

In particular, $[f] = \mathcal{C}_{E(f),C}(\iota_0, \pi_0)$. Also $\mathcal{C}_{E,C}(\iota, \pi) = [f_{\iota,\tau}]$.

12.1.5. Equivalence of extensions. Consider two central extensions

$$\text{(12.7)} \qquad 1 \longrightarrow C \xrightarrow{\iota_1} E_1 \xrightarrow{\pi_1} G \longrightarrow 1, \qquad 1 \longrightarrow C \xrightarrow{\iota_2} E_2 \xrightarrow{\pi_2} G \longrightarrow 1$$

of C by G. Each of these extensions gives rise to a cocycle $f_i : G \times G \to C$, where $f_i = f_{\iota_i, \tau_i}$ for transversal maps $\tau_i : G \to E_i$, $i = 1, 2$. We have seen that varying τ_i does not affect the cohomology class $[f_i]$. When do $[f_1]$ and $[f_2]$ coincide? In fact $[f_1] = [f_2]$ if and only if there is a homomorphism $\theta : E_1 \to E_2$ such that $\theta\iota_1 = \iota_2$ and $\pi_2\theta = \pi_1$ (the proof of this claim is part of the proof of Theorem 12.1.5 below). Another way to say this is that the diagram

$$\text{(12.8)} \qquad \begin{array}{ccccccccc} 1 & \longrightarrow & C & \xrightarrow{\iota_1} & E_1 & \xrightarrow{\pi_1} & G & \longrightarrow & 1 \\ & & \| & & \downarrow{\scriptstyle\theta} & & \| & & \\ 1 & \longrightarrow & C & \xrightarrow{\iota_2} & E_2 & \xrightarrow{\pi_2} & G & \longrightarrow & 1 \end{array}$$

commutes: any two paths in the diagram with the same origin and destination represent equal composite maps. If (12.8) commutes then the extensions (12.7) are said to be *equivalent*. In this case, θ is an isomorphism. However, if E_1 and E_2 are isomorphic then the extensions (12.7) may not be equivalent. It is easily checked that equivalence of extensions is an equivalence relation on the set of all central extensions of C by G.

12.1.4. LEMMA. *Let τ be a transversal map for the central extension (12.4), so that each element of E may be expressed uniquely as $\tau(g)\iota(c)$ for some $g \in G$ and $c \in C$. Then $\theta : \tau(g)\iota(c) \mapsto (g, c) \in E(f_{\iota,\tau})$ defines an equivalence between (12.4) and the canonical central extension (12.3) for $f_{\iota,\tau}$.*

PROOF. We have $\theta\iota(c) = (1, c) = \iota_0(c)$ and $\pi_0\theta(\tau(g)\iota(c)) = \pi_0((g, c)) = g = \pi(\tau(g)\iota(c))$. That θ is a homomorphism follows from the definitions of $f_{\iota,\tau}$ and multiplication in $E(f_{\iota,\tau})$. $\qquad\square$

12.1.5. THEOREM. *The set of cohomology classes of cocycles $G \times G \to C$ is bijective with the set of equivalence classes of central extensions of C by G.*

PROOF. We map the cohomology class $[f]$ of a cocycle $f : G \times G \to C$ to the extension-equivalence class of (12.3). This map is well-defined. For suppose that $[f] = [f']$, say $f = f'\partial\rho$ where $\rho : G \to C$ is normalized. Define an isomorphism $\theta : E(f) \to E(f')$ by $\theta : (x, c) \mapsto (x, \rho(x)c)$. Then the canonical central extensions for f and f' are equivalent via θ.

Consider the map that assigns the equivalence class of a central extension (12.4) to $[f_{\iota,\tau}]$, where τ is a transversal map for the extension. To see that this map is well-defined, assume that (12.8) commutes, and let τ_1 be a transversal map for the top extension in the diagram. Then $\tau_2 = \theta\tau_1$ is a transversal map for the bottom extension. Furthermore, $f_{\iota_1,\tau_1} = f_{\iota_2,\tau_2}$, so certainly $[f_{\iota_1,\tau_1}] = [f_{\iota_2,\tau_2}]$.

We have defined two maps between the set of cohomology classes of cocycles $G \times G \to C$ and the set of equivalence classes of central extensions of C by G. These maps are inverses of each other by (12.6) and Lemma 12.1.4. $\qquad\square$

The 'central extension problem' is to find all possible (isomorphism types of) central extensions of C by G, given the kernel C and quotient G. Theorem 12.1.5 shows that a solution of this problem could begin by calculating cohomology classes of cocycles $G \times G \to C$. For methods to calculate cocycles, see Chapter 20 and Section 12.4 below.

The following result characterizes the cocycles $f : G \times G \to C$ that stem from the same embedding of C into $Z(E)$.

12.1.6. PROPOSITION. *Fix an embedding $\iota : C \to Z(E)$. Let $f : G \times G \to C$ be a cocycle. Then $f = f_{\iota,\tau}$ for some transversal map $\tau : G \to E$ if and only if there is an isomorphism $\phi : E(f) \to E$ such that $\phi((1, c)) = \iota(c) \; \forall c \in C$.*

PROOF. If $f = f_{\iota,\tau}$ then the required isomorphism $E(f) \to E$ is provided by Lemma 12.1.4. Conversely, an isomorphism ϕ as stated is an equivalence between the canonical extension (12.3) and the central extension (12.4) with $\pi = \pi_0 \phi^{-1}$. Then $\tau = \phi \tau_0$ is a transversal map for (12.4), and $f_{\iota,\tau} = f$ by (12.6) just as in the proof of Theorem 12.1.5. $\qquad\square$

12.1.7. COROLLARY. *Let $f : G \times G \to C$ be a cocycle, and let Z be a central copy of C in the group E. Then $f = f_{\iota,\tau}$ for some isomorphism $\iota : C \to Z$ and transversal map $\tau : G \to E$ if and only if there is an isomorphism $\phi : E(f) \to E$ mapping $Z_0 = \{(1, c) \mid c \in C\}$ onto Z.*

PROOF. The forward implication is immediate from Proposition 12.1.6. Conversely, if ϕ exists then we may take $\iota = \phi \iota_0$ in Proposition 12.1.6. $\qquad\square$

We use the $\mathcal{C}$ notation to restate the above.

12.1.8. PROPOSITION.
   (1) *Fix an embedding $\iota : C \to Z(E)$. There is an isomorphism $\phi : E(f) \to E$ such that $\phi((1, c)) = \iota(c) \; \forall c \in C$ if and only if $f \in \mathcal{C}_{E,C}(\iota)$.*
   (2) *Fix a central subgroup Z of E isomorphic to C. There is an isomorphism $\phi : E(f) \to E$ that maps Z_0 to Z if and only if $f \in \mathcal{C}_{E,C}$.*

12.1.6. Varying the inclusion. Let ι and ι' be embeddings of C into $Z(E)$ such that $\iota(C) = \iota'(C)$. Then $\beta = \iota^{-1}\iota' \in \operatorname{Aut}(C)$, and $f_{\iota,\tau} = \beta f_{\iota',\tau}$ by (12.5). Indeed,

$$f \in \mathcal{C}_{E,C}(\pi, \tau) \iff \mathcal{C}_{E,C}(\pi, \tau) = \{\beta f \mid \beta \in \operatorname{Aut}(C)\}.$$

Ranging over all transversal maps τ, we get

$$(12.9) \qquad f \in \mathcal{C}_{E,C}(\pi) \iff \mathcal{C}_{E,C}(\pi) = \bigcup_{\beta \in \operatorname{Aut}(C)} [\beta f].$$

12.1.7. Varying the projection. Let π and ψ be surjections $E \to G$ with kernel $Z = \iota(C)$. Let τ be a transversal map for π, and define $\alpha = \psi \tau$. Since $\tau(x)\tau(y)\tau(xy)^{-1} \in Z = \ker \psi$, α is a homomorphism. Moreover $\alpha \in \operatorname{Aut}(G)$: if $\alpha(x) = 1$ then $\tau(x) \in Z$, and because τ is normalized, $x = 1$. Now $\psi\sigma = \operatorname{id}_G$ where $\sigma = \tau\alpha^{-1}$, so that σ is a transversal map for ψ. Then

$$\iota f_{\iota,\tau}(x, y) = \sigma\alpha(x)\sigma\alpha(y)\sigma\alpha(xy)^{-1}$$
$$= \sigma\alpha(x)\sigma\alpha(y)\sigma(\alpha(x)\alpha(y))^{-1}$$
$$= \iota f_{\iota,\sigma}(\alpha(x), \alpha(y)).$$

For a cocycle $f : G \times G \to C$ and $\gamma \in \operatorname{Aut}(G)$, define the cocycle $f_\gamma : G \times G \to C$ by $(x, y) \mapsto f(\gamma(x), \gamma(y))$. So $f_{\iota,\tau} = (f_{\iota,\sigma})_\alpha$, and

$$(12.10) \qquad f \in \mathcal{C}_{E,C}(\iota) \iff \mathcal{C}_{E,C}(\iota) = \bigcup_{\alpha \in \operatorname{Aut}(G)} [f_\alpha].$$

12.1.8. Central isomorphism. By (12.9) and (12.10), $\mathcal{C}_{E,C}(\pi)$ and $\mathcal{C}_{E,C}(\iota)$ are unions of cohomology classes. The same is therefore true of $\mathcal{C}_{E,C}$:

$$f \in \mathcal{C}_{E,C} \quad \Longleftrightarrow \quad \mathcal{C}_{E,C} = \bigcup_{\substack{\alpha \in \mathrm{Aut}(G) \\ \beta \in \mathrm{Aut}(C)}} [\beta f_\alpha].$$

Let us define a *central isomorphism* between two central extensions (12.7) to be an isomorphism $\mu : E_1 \to E_2$ such that $\mu(\iota_1(C)) = \iota_2(C)$. Extension equivalence is a special case, but more generally a central isomorphism is merely an isomorphism that respects the distinguished central copies of C in the extensions. The set $\mathcal{C}_{E,C}$ consists of the cocycles $G \times G \to C$ whose corresponding central extensions have extension group E, and are pairwise isomorphic by a central isomorphism. Thus $\mathcal{C}_{E,C}$ captures some, but not necessarily all, central extensions of C by G isomorphic to E.

12.2. Cocycles for product groups

Let $G = G_1 \times G_2$ where G_1 and G_2 are finite groups; so each element of G may be written uniquely in the form $x_1 x_2$ where $x_1 \in G_1$ and $x_2 \in G_2$. Given cocycles $f_i : G_i \times G_i \to C$, $i = 1, 2$, define $f : G \times G \to C$ by

$$f(x_1 x_2, y_1 y_2) = f_1(x_1, y_1) f_2(x_2, y_2).$$

The function f inherits the cocycle property (12.1) from f_1 and f_2. We write $f = f_1 \times f_2$, and call f a *product cocycle*. Note that

$$[f(x,y)]_{x,y \in G} = [f_1(x_1, y_1)]_{x_1, y_1 \in G_1} \otimes [f_2(x_2, y_2)]_{x_2, y_2 \in G_2}.$$

12.2.1. LEMMA. $E(f_1 \times f_2) \cong E(f_1) \curlyvee_{Z_0} E(f_2)$ where Z_0 denotes the central subgroup $\{(1, a) \mid a \in C\}$ of $E(f_1 \times f_2)$.

PROOF. Let $E_i = \{(x_i, a) \mid x_i \in G_i,\ a \in C\} \subseteq E(f_1 \times f_2)$, $i = 1, 2$. We have $E_i \cong E(f_i)$. Since $(x_1 y_2, a) = (x_1, a)(y_2, 1)$ and $(x_1, a)(y_2, b) = (y_2, b)(x_1, a)$, it follows that $E(f) = E_1 E_2$ and $[E_1, E_2] = 1$. Finally,

$$E_1 \cap E_2 = \{(x, a) \mid x \in G_1 \cap G_2,\ a \in C\} = Z_0. \qquad \square$$

In effect we put together two canonical central short exact sequences

$$1 \longrightarrow C \xrightarrow{\iota_0} Z_0 \le E(f_i) \xrightarrow{\pi_0} G_i \longrightarrow 1$$

to get a canonical central short exact sequence

$$1 \longrightarrow C \xrightarrow{\iota_0} Z_0 \le E(f) \xrightarrow{\pi_0} G_1 \times G_2 \longrightarrow 1$$

which contains each of the original sequences by restriction (of π_0 on $E(f)$ to E_1 and E_2). In the opposite direction, a cocycle f of any group G gives rise to a cocycle f_i by restriction to $G_i \times G_i$ for $G_i \le G$; and with a more general product group G in place, a central short exact sequence corresponding to f restricts to central short exact sequences for the cocycles f_i.

12.2.2. LEMMA. *Suppose that $G = G_1 \cdot G_2$, where G_1 and G_2 are finite groups, and let $f = f_{\iota, \tau}$ be a cocycle arising from the central extension*

$$1 \longrightarrow Z \xrightarrow{\iota} E \xrightarrow{\pi} G \longrightarrow 1$$

for some transversal map τ. Suppose also that Z_1 and Z_2 are subgroups of Z such that $Z = Z_1 Z_2$ and $f(G_i \times G_i) \subseteq Z_i$ for $i = 1, 2$. Then the following hold.

(1) *The restrictions $f_i : G_i \times G_i \to Z_i$ for $i = 1, 2$ are cocycles.*
(2) *The cocycle f_i arises from the central extension*

$$1 \longrightarrow Z_i \overset{\iota_i}{\longrightarrow} E_i \overset{\pi_i}{\longrightarrow} G_i \longrightarrow 1,$$

where $\iota_i = \iota_{|Z_i}$, E_i is the subgroup $\langle \tau(G_i), \iota(Z_i) \rangle$ of E, and $\pi_i = \pi_{|E_i}$.
(3) $E = E_1 \diamond_{Z_1 \cap Z_2} E_2$.
(4) $E = E_1 \curlyvee_{Z_1 \cap Z_2} E_2$ *if and only if $f \in [f_1 \times f_2]$ and $G = G_1 \times G_2$.*

PROOF. Parts (1) and (2) are routine. We now show that $E = E_1 \diamond_{Z_1 \cap Z_2} E_2$, i.e., $E = E_1 E_2$ and $E_1 \cap E_2 = \iota(Z_1 \cap Z_2)$. Since

$$\pi(E_1 \cap E_2) \leq \pi(E_1) \cap \pi(E_2) = G_1 \cap G_2 = 1,$$

we have $E_1 \cap E_2 \leq \ker \pi = \iota(Z)$. Thus $\iota(Z) \cap E_i = \iota(Z_i)$ implies that $E_1 \cap E_2 = \iota(Z_1) \cap \iota(Z_2) = \iota(Z_1 \cap Z_2)$. Since $|Z| = |Z_1||Z_2|/|Z_1 \cap Z_2|$, we then have

$$|E_1 E_2| = |E_1||E_2|/|E_1 \cap E_2| = |E_1||E_2|/|Z_1 \cap Z_2| = |Z|(|E_1|/|Z_1|)(|E_2|/|Z_2|)$$
$$= |G_1||G_2||Z| = |G||Z| = |E|.$$

Therefore $E = E_1 E_2$.

Next we prove part (4). First suppose that $E = E_1 \curlyvee E_2$. Then $[G_1, G_2] = [\pi(E_1), \pi(E_2)] = \pi([E_1, E_2]) = 1$, so $G = G_1 \times G_2$. Let $x_i, y_i \in G_i$, $i = 1, 2$; then

$$\tau(x_1 y_1 x_2 y_2)\, \iota f(x_1 x_2, y_1 y_2)$$
$$= \tau(x_1 x_2)\tau(y_1 y_2)$$
$$= \tau(x_1)\tau(y_1)\tau(x_2)\tau(y_2)\iota f(x_1, x_2)^{-1} \iota f(y_1, y_2)^{-1}$$
$$= \tau(x_1 y_1)\iota f_1(x_1, y_1)\tau(x_2 y_2)\iota f_2(x_2, y_2)\iota f(x_1, x_2)^{-1}\iota f(y_1, y_2)^{-1}$$
$$= \tau(x_1 y_1 x_2 y_2)\iota f(x_1 y_1, x_2 y_2)\iota f_1(x_1, y_1)\iota f_2(x_2, y_2)\iota f(x_1, x_2)^{-1}\iota f(y_1, y_2)^{-1},$$

where we are using that $[\tau(G_1), \tau(G_2)] \leq [E_1, E_2] = 1$. Therefore

$$f(x_1 x_2, y_1 y_2) = \partial \rho(x_1 x_2, y_1 y_2)f_1(x_1, y_1)f_2(x_2, y_2)$$
$$= \partial \rho(x_1 x_2, y_1 y_2)(f_1 \times f_2)(x_1 x_2, y_1 y_2)$$

where $\rho(x_1 x_2) = f(x_1, x_2)^{-1}$. This proves one implication. For the other, suppose that $[G_1, G_2] = 1$; then $\partial \rho(x_1, x_2) = \partial \rho(x_2, x_1)$ for all coboundaries $\partial \rho$. Hence, as $(f_1 \times f_2)(x_1, x_2) = (f_1 \times f_2)(x_2, x_1) = 1$, any cocycle f cohomologous to $f_1 \times f_2$ satisfies $f(x_1, x_2) = f(x_2, x_1)$. So for $z_1 \in Z_1$ and $z_2 \in Z_2$ we have

$$[\tau(x_1)z_1, \tau(x_2)z_2] = [\tau(x_1), \tau(x_2)] = \iota f(x_2, x_1)^{-1}\tau(x_2 x_1)^{-1}\iota f(x_1, x_2)\tau(x_1 x_2) = 1.$$

Thus $[E_1, E_2] = 1$, as required. $\square$

12.3. Polycyclic presentations

Algorithms in computational group theory depend on how the input group is given: whether by a finite presentation, or by a generating set of permutations or matrices. In this section we look at a special type of presentation, called a *polycyclic presentation*, which is ideal for computing.

References for computing with polycyclic groups are [83, pp. 273–284] and [147, Sections 9.3, 9.4]. Many algorithms for polycyclic groups have been implemented in the computer algebra systems GAP [79] and MAGMA [14].

This section is a brisk introduction to polycyclic presentations. Section 12.4 shows how to compute cocycles $f : G \times G \to C$ if we have a consistent polycyclic presentation for G. Also, Chapter 20 outlines a method for computing the Schur multiplier of a finite solvable group given by a consistent polycyclic presentation.

A group G is *polycyclic* if it has a chain of subgroups

$$(12.11) \qquad\qquad G = G_1 \geq G_2 \geq \cdots \geq G_{n+1} = 1$$

where G_{i+1} is normal in G_i, and G_i/G_{i+1} is cyclic for $1 \leq i \leq n$. Subgroups and homomorphic images of polycyclic groups are polycyclic. The chain (12.11) is a *polycyclic series* of G. A finitely generated abelian group, as the direct product of a finite number of cyclic groups, is polycyclic. If G is a finite solvable group then it is polycyclic (decompose each abelian factor $G^{(i)}/G^{(i+1)}$ as a direct product of cyclic groups), and a polycyclic group is solvable.

Given a polycyclic series (12.11) of G, select $g_i \in G_i \setminus G_{i+1}$ for each i. Clearly $G = \langle g_1, \ldots, g_n \rangle$ (so a polycyclic group is finitely generated). The facts that G_{i+1} is normal in G_i, and G_i/G_{i+1} is cyclic, imply relations among the chosen generators. That is, if G_i/G_{i+1} is finite then there exists a positive integer m_i such that $g_i^{m_i}$ is a word in the $g_{i+1}, \ldots, g_n$; and for any $i > j$, $g_i^{g_j}$ is a word in the $g_{j+1}, \ldots, g_n$. Indeed, it can be proved that every polycyclic group has a *polycyclic presentation*: a presentation $\langle x_1, \ldots, x_n \mid Y \rangle$ such that each element of Y is either a 'power relation'

$$(12.12) \qquad\qquad x_i^{r_i} = x_{i+1}^{l_{i,i+1}} \cdots x_n^{l_{i,n}}, \quad r_i > 0$$

or a 'conjugate relation'

$$(12.13) \qquad\qquad x_j^{-1} x_i x_j = x_{j+1}^{r_{i,j,j+1}} \cdots x_n^{r_{i,j,n}}, \quad 1 \leq j < i \leq n$$

$$(12.14) \qquad\qquad x_j x_i x_j^{-1} = x_{j+1}^{s_{i,j,j+1}} \cdots x_n^{s_{i,j,n}}, \quad 1 \leq j < i \leq n$$

where all exponents are integers. Moreover, Y must contain the relations (12.13) and (12.14) for all i, j, $1 \leq j < i \leq n$. A polycyclic presentation with generating set $\{x_1, \ldots, x_n\}$ defines a polycyclic series (12.11): take $G_i = \langle x_i, \ldots, x_n \rangle$, and then $|G_k/G_{k+1}| \leq r_k$ if there is a power relation (12.12) in the presentation for $i = k$, whereas G_k/G_{k+1} may be finite or infinite if a power relation (12.12) does not appear for $i = k$.

12.3.1. EXAMPLE. Let G be the group with polycyclic presentation

$$\langle a, b \mid a^{-1} b a = b^{-1}, \, a b a^{-1} = b^2 \rangle.$$

Since there are no power relations, a and b might have infinite order. However

$$b = a(a^{-1} b a) a^{-1} = a b^{-1} a^{-1} = (a b a^{-1})^{-1} = b^{-2}$$

so that $b^3 = 1$.

The exponent on the left-hand side of a power relation is only an upper bound on the order of the corresponding cyclic factor. This wrinkle can be removed by a slight adjustment of the presentation.

Let $\langle x_1, \ldots, x_n \mid Y \rangle$ be a polycyclic presentation, and let I be the set of subscripts appearing on the left-hand sides of the power relations (12.12) in Y. The presentation is *consistent* if for all i the cyclic quotient $\langle x_i, \ldots, x_n \rangle / \langle x_{i+1}, \ldots, x_n \rangle$ has order r_i when $i \in I$, and is infinite when $i \notin I$. Note that there exist methods to obtain a consistent polycyclic presentation from any polycyclic presentation; see [**83**, Section 12.4].

12.3.2. EXAMPLE. If we suppress conjugate relations implying that generators commute, then
$$\langle a, b, x_1, x_2, x_3 \mid a^2 = x_1,\, b^2 = x_2,\, a^{-1}ba = bx_3^{-1},\, aba^{-1} = bx_3 \rangle.$$
is a polycyclic presentation. However, it is not consistent, because
$$1 = [a, x_2] = [a, b^2] = (a^{-1}b^{-1}a)^2 b^2 = (x_3 b^{-1})^2 b^2 = x_3^2.$$
Adding the relation $x_3^2 = 1$ gives a consistent polycyclic presentation.

Computing with a consistent polycyclic presentation for a polycyclic group G has several advantages. For example, if G is finite then $|G|$ can be read off the presentation. We are also able to solve the *word problem* in G: we can decide in a finite number of steps whether any given pair of words in the generators of G represent the same group element. There are known finitely presented groups in which the word problem cannot be solved.

12.3.3. LEMMA. *Let G be given by a consistent polycyclic presentation, with power relations* (12.12). *Then every element g of G has a* unique *collected form:*
$$g = x_1^{e_1} \cdots x_n^{e_n}$$
where the e_i are integers and $0 \le e_i < r_i$ if x_i appears in the left-hand side of a power relation (12.12).

PROOF. See [**83**, Lemma 8.3, p. 276]. □

12.3.4. THEOREM. *Let G be given by a consistent polycyclic presentation. There exists an algorithm, called the* collection algorithm, *which accepts any word w in the generators of G as input, and outputs the collected form of w.*

PROOF. See [**83**, pp. 280–283]. The basic operation of the algorithm is to use the relations to eliminate uncollected subwords $x_i x_j$ and $x_i x_j^{-1}$, $i > j$, in w. □

12.3.5. COROLLARY. *By means of the collection algorithm, the word problem can be solved in any polycyclic group given by a consistent polycyclic presentation.*

12.4. Cocycles from collection in polycyclic groups

The collection process in a polycyclic group G yields a way to calculate cocycles for central extensions of a finite abelian group C by G. To describe the ideas, we will assume for the sake of simplicity that G is abelian. Cf. also [**83**, pp. 307–310].

Suppose that $G \cong C_{m_1} \times \cdots \times C_{m_k}$ with all m_is greater than 1, and

(12.15) $$1 \longrightarrow C \xrightarrow{\iota} E \xrightarrow{\pi} G \longrightarrow 1$$

is a central short exact sequence. So G has presentation

(12.16) $$\langle a_1, \ldots, a_k \mid a_i^{m_i} = 1\,(1 \le i \le k),\, [a_j, a_i] = 1\,(1 \le i < j \le k) \rangle.$$

(We get a polycyclic presentation for G after replacing each commutator relation $[a_j, a_i] = 1$ in (12.16) by the conjugate relations $a_i^{-1} a_j a_i = a_j$ and $a_i a_j a_i^{-1} = a_j$.) Let b_i be an element of E such that $\pi(b_i) = a_i$. Then
$$\pi(b_i^{m_i}) = a_i^{m_i} = 1 \qquad \text{and} \qquad \pi([b_j, b_i]) = [a_j, a_i] = 1.$$
Therefore
$$z_{ii} = b_i^{m_i} \qquad \text{and} \qquad z_{ij} = [b_j, b_i]$$

are elements of $Z = \iota(C)$. Suppressing relations that define Z as a central subgroup of E isomorphic to C, it is then not hard to see that E has presentation

$$(12.17) \qquad \langle b_1, \ldots, b_k, z \mid b_i^{m_i} = z_{ii}\ (1 \le i \le k),\ [b_j, b_i] = z_{ij}\ (1 \le i < j \le k)\rangle,$$

where z ranges over a generating set for Z.

Every element of G is uniquely expressible as a word $a_1^{r_1} \cdots a_k^{r_k}$ where $0 \le r_i < m_i$. We denote this product by a^r, where a stands for the 'generator vector' $(a_1, \ldots, a_k)$ and r stands for the 'exponent vector' $(r_1, \ldots, r_k)$. A transversal map $\tau : G \to E$ for the central extension (12.15) is defined by

$$\tau : a^r = a_1^{r_1} \cdots a_k^{r_k} \mapsto b^r = b_1^{r_1} \cdots b_k^{r_k}.$$

Then

$$(12.18) \qquad\qquad f_{\iota,\tau}(a^r, a^s) = \iota^{-1}(b^r \cdot b^s \cdot b^{-\overline{r+s}})$$

where overlining denotes reduction modulo m_i in the ith component of the exponent vector, $1 \le i \le k$. Now the relations in (12.17) may be used to convert any product of elements in E to its collected form $b_1^{r_1} \cdots b_k^{r_k} w$ where $0 \le r_i < m_i$ and $w \in Z$. That is, moving b_j to the right of b_i for $j > i$ introduces the central element z_{ij} (by the commutator relation $[b_j, b_i] = z_{ij}$), and reducing the exponent in b_i^r modulo m_i introduces z_{ii} (by the power relation $b_i^{m_i} = z_{ii}$). This is essentially the collection process in E; see Theorem 12.3.4. The element $b^{\overline{r+s}}$ is already in collected form. After applying collection to $b^r \cdot b^s$ we may then complete the evaluation of $f_{\iota,\tau}(a^r, a^s)$ in C, since all occurrences of the b_is will cancel each other out.

We call $f_{\iota,\tau}$ as in (12.18) a *collection cocycle* for the central extension (12.15). We can be more explicit. Suppose that $j \ge i$, $0 \le r_i < m_i$, and $0 \le s_j < m_j$. Then

$$(12.19) \qquad\qquad b_j^{r_j} b_i^{s_i} = b_i^{s_i} b_j^{r_j} z_{ij}^{s_i r_j}$$

if $j > i$, using $b_j b_i = b_i b_j z_{ij}$ repeatedly, and the fact that z_{ij} is central in E. If $i = j$ then $b_i^{m_i} = z_{ii}$ implies that

$$(12.20) \qquad\qquad b_i^{r_i} b_i^{s_i} = b_i^{\overline{r_i+s_i}} z_{ii}^{\left\lfloor \frac{r_i+s_i}{m_i} \right\rfloor}.$$

Therefore, by (12.18), (12.19), and (12.20),

$$(12.21) \qquad f_{\iota,\tau}(a^r, a^s) = \iota^{-1}\left(\prod_{i=1}^{k}\left(z_{ii}^{\left\lfloor \frac{r_i+s_i}{m_i} \right\rfloor} \prod_{j>i} z_{ij}^{s_i r_j}\right)\right).$$

The set of collection cocycles for a (fixed presentation of a) polycyclic group G contains at least one representative from each cohomology class of cocycles $f : G \times G \to C$. The next proposition sums up the situation when G is abelian.

12.4.1. PROPOSITION. *Let G be abelian, with presentation* (12.16). *Select an element c_{ii} of C for each power relation $a_i^{m_i} = 1$, and an element c_{ij} of C for each commutator relation $[a_j, a_i] = 1$. Then $f_c : G \times G \to C$ defined by*

$$f_c(a_1^{r_1} \cdots a_k^{r_k}, a_1^{s_1} \cdots a_k^{s_k}) = \prod_{i=1}^{k}\left(c_{ii}^{\left\lfloor \frac{r_i+s_i}{m_i} \right\rfloor} \prod_{j>i} c_{ij}^{s_i r_j}\right)$$

is a collection cocycle. Moreover, any cocycle $f : G \times G \to C$ is cohomologous to some f_c.

PROOF. Every choice of c_{ii}, c_{ij} is valid, because (12.17) with $z_{ii} = \iota(c_{ii})$ and $z_{ij} = \iota(c_{ij})$ for any isomorphism $\iota : C \to Z$ defines a central extension of C by G. Let $f : G \times G \to C$ be a cocycle, and take (12.15) to be the canonical central extension for f. Then (12.17), with $b_i = (a_i, 1)$ and z_{ii}, z_{ij} determined by f, is a presentation for $E(f)$. Since τ and τ_0 are both transversal maps for this extension, $f = f_{\iota_0, \tau_0}$ and $f_{\iota_0, \tau}$ are cohomologous. By (12.21), f is then cohomologous to the collection cocycle $f_c = f_{\iota_0, \tau}$. □

We have a closed formula (12.21) for a collection cocycle when G is abelian. If G is polycyclic but non-abelian, the same method of calculating collection cocycles goes through, but it is not so straightforward to obtain such a nice closed formula.

The presentation (12.16) of the abelian group G has $n = k + \binom{k}{2}$ relations. By Proposition 12.4.1, $|C|^n$ is an upper bound on the number of cohomology classes of cocycles $f : G \times G \to C$. Sometimes the number of collection cocycles matches the number of cohomology classes, but in general there could be repetition of classes. So it is natural to ask for a characterization of collection coboundaries, at least in the case that G is abelian. Note that if G is abelian and f is a coboundary then f is a *symmetric cocycle*: $f(x, y) = f(y, x)$ for all $x, y \in G$.

12.4.2. LEMMA. *A collection cocycle $f_c : G \times G \to C$ as per Proposition 12.4.1 is a coboundary if and only if $c_{ij} = 1$ and c_{ii} is an m_ith power of an element of C for all $j > i$ and $1 \leq i, j \leq k$.*

PROOF. Suppose that f_c is a coboundary, so $f_c(x, y) = \phi(x)\phi(y)\phi(xy)^{-1}$ for some normalized map $\phi : G \to C$. If $j > i$ then $f_c(a_i, a_j) = 1$, whereas $f_c(a_j, a_i) = c_{ij}$, so that $c_{ij} = 1$ since f_c is symmetric. Next, $f_c(a_i^l, a_i) = \phi(a_i^l)\phi(a_i)\phi(a_i^{l+1})^{-1}$ implies that $\prod_{l=1}^{m_i-1} f_c(a_i^l, a_i) = \phi(a_i)^{m_i}$. Since $f_c(a_i^l, a_i)$ is equal to c_{ii} if $l + 1 = m_i$ and is 1 otherwise, $c_{ii} = \phi(a_i)^{m_i}$.

Now suppose that all c_{ij}s are 1, and that for each i there exists $d_i \in C$ such that $c_{ii} = d_i^{m_i}$. Define $\varphi : G \to C$ by $\varphi(a_1^{r_1} \cdots a_k^{r_k}) = \prod_{i=1}^k d_i^{r_i}$. Then $f_c(a^r, a^s) = \varphi(a^r)\varphi(a^s)\varphi(a^{\overline{r+s}})^{-1}$. □

12.4.3. COROLLARY. *If G and C are elementary abelian p-groups, then the full set of collection cocycles for the presentation (12.16) of G contains one and only one representative from each cohomology class of cocycles $G \times G \to C$.*

PROOF. By Lemma 12.4.2, the only collection cocycle that is a coboundary in this case is the trivial one. Also, the (pointwise) product of two collection cocycles is a collection cocycle. Hence the set of collection cocycles contains no duplication of cohomology class. □

Let G be an elementary abelian p-group of rank k, written additively, i.e., G is a vector space of dimension k over $\mathrm{GF}(p)$. Choose a basis of G. If $M \in \mathrm{Mat}(k, p)$ and $\omega \in \mathrm{GF}(p) \setminus \{0\}$ then the bilinear form f on G defined by $f(r, s) = \omega^{rMs^\top}$ is a cocycle. All collection cocycles are bilinear when $p = 2$.

12.4.4. PROPOSITION. *Suppose that $G \cong \mathrm{C}_2^k$ and $C = \langle -1 \rangle$. For $i \leq j$, let the c_{ij} be chosen as in Proposition 12.4.1, and define $q_{ij} \in \{0, 1\}$ by $c_{ij} = (-1)^{q_{ij}}$. Then*

(12.22)
$$f_c(a^r, a^s) = (-1)^{sQr^\top},$$

where Q is the $k \times k$ upper triangular $(0, 1)$-matrix with (i, j)th entry q_{ij} for $i \leq j$.

PROOF. By Proposition 12.4.1, $f_c(a^r, a^s) = (-1)^e$ where

$$e = \sum_{i=1}^{k} \left(q_{ii} \lfloor (r_i + s_i)/2 \rfloor + \sum_{j>i} q_{ij} s_i r_j \right).$$

Since $r_i, s_i \in \{0, 1\}$ we have $\lfloor (r_i + s_i)/2 \rfloor = r_i s_i$, so $e = \sum_{1 \leq i \leq j} q_{ij} s_i r_j = sQr^{\top}$, and the proof is complete. $\square$

12.4.5. REMARK. Collection cocycles for an elementary abelian p-group G are defined according to a chosen basis of G. In the context of Proposition 12.4.4, a change of basis is given by $a^r \mapsto a^{rB}$ for some $B \in \mathrm{GL}(k, 2)$. The set of collection cocycles with respect to any basis consists of all cocycles (12.22) as Q ranges over the upper triangular elements of $\mathrm{Mat}(k, 2)$. However, the sets of collection cocycles will be different for different bases. Since $\mathrm{Aut}(G) \cong \mathrm{GL}(k, 2)$, changing basis has the effect of $\mathrm{Aut}(G)$-action on cocycles.

The matrix Q in Proposition 12.4.4 may be thought of as a 'relation matrix' for $E(f_c)$. The diagonal entries of Q give the power relations in a presentation for $E(f_c)$, while the non-zero upper triangular entries give the commutator relations.

12.4.6. LEMMA. *If f_c is the collection cocycle defined as in (12.22) then $E(f_c)$ has presentation*

(*)
$$\langle b_1, \ldots, b_k, z \mid b_i^2 = z^{q_{ii}}, z^2 = 1, [b_j, b_i] = z^{q_{ij}} \ (i < j), [z, b_i] = 1, 1 \leq i, j \leq k \rangle.$$

Conversely, suppose that (12.15) is a central extension of $C = \langle -1 \rangle$ by G, and E has presentation (*). Set $a_i = \pi(b_i)$. Then the (collection) cocycle $f : G \times G \to \langle -1 \rangle$ defined by $f(a^r, a^s) = (-1)^{sQr^{\top}}$ for $Q = [q_{ij}]$ is a cocycle of (12.15).

As we observed in Corollary 12.4.3, if G and C are elementary abelian p-groups, then the set of collection cocycles accounts irredundantly for all cohomology classes. Together with Proposition 12.4.4, the following lemma also provides a verification of this fact (at least when $p = 2$).

12.4.7. LEMMA. *Let G be an elementary abelian 2-group of rank k. Define $f : G \times G \to \langle -1 \rangle$ by $f(a^r, a^s) = (-1)^{sMr^{\top}}$ where M is a $k \times k$ $(0, 1)$-matrix. Then f is a coboundary if and only if M is symmetric and has zero main diagonal.*

PROOF. Suppose that $f(a^r, a^s) = \phi(a^r)\phi(a^s)\phi(a^{r+s})^{-1}$ for some map $\phi : G \to \langle -1 \rangle$. Since then $f(a^r, a^r) = 1$, we have $rMr^{\top} \equiv 0 \pmod 2$ for all $\mathrm{GF}(2)$-vectors r. Taking r to be each of the vectors with 1 in a single position, this implies that M has zero diagonal. Similarly, if we take r to be the vector with 1 in positions i and j and zeros elsewhere then we see that M is symmetric.

Now suppose that M is symmetric and has zero diagonal, so $M = S + S^{\top}$ for some $(0, 1)$-matrix S. Define the map $\phi : G \to \langle -1 \rangle$ by $\phi(a^r) = (-1)^{rSr^{\top}}$. Then f is a coboundary:

$$
\begin{aligned}
f(a^r, a^s) &= (-1)^{sSr^{\top} + sS^{\top}r^{\top}} \\
&= (-1)^{sSr^{\top} + rSs^{\top} + sSs^{\top} + rSr^{\top}} (-1)^{sSs^{\top}} (-1)^{rSr^{\top}} \\
&= (-1)^{(r+s)S(r+s)^{\top}} (-1)^{sSs^{\top}} (-1)^{rSr^{\top}} \\
&= \phi(a^r)\phi(a^s)\phi(a^{r+s}).
\end{aligned}
$$
$\square$

Our earlier observation is an easy corollary of this lemma, because the only symmetric upper triangular matrix with zero diagonal is the zero matrix. So for an elementary abelian 2-group G, any cocycle $f : G \times G \to \langle -1 \rangle$ is cohomologous to a bilinear cocycle (but f need not be bilinear itself).

12.5. Monomial representations and cocycles

A *monomial representation* of a finite group G over a ring $\mathcal{R}$ is a representation of G as a group of monomial matrices in some $\mathrm{GL}(n, \mathcal{R})$. As the only operation involved in multiplication of monomial matrices is multiplication of its non-zero entries, we can speak of monomial representations of G over a group (e.g., the group of units of $\mathcal{R}$).

Let $f : G \times G \to C$ be a cocycle, where G has order v. We define a faithful monomial representation of $E(f)$ of degree v over C. For $(x, c) \in E(f)$, let

$$(12.23) \qquad P^f_{(x,c)} = c^{-1}[\delta^{rx}_s f(r, x)^{-1}]_{r,s \in G}.$$

Clearly $P^f_{(x,c)} \in \mathrm{Mon}(v, C)$. We drop the superscript from the notation $P^f_{(x,c)}$ when the cocycle is clear, or is unimportant. Also, $P_x := P_{(x,1)}$. Note that the Hadamard product $P_x \wedge P_y$ is 0 if $x \neq y$.

12.5.1. LEMMA. *The following hold.*

(1) $P^{-1}_{(x,c)} = P^*_{(x,c)}$.

(2) *The map given by* $(x, c) \mapsto P^f_{(x,c)}$ *is a faithful monomial representation of* $E(f)$.

(3) *For any* $c \in C$, *the map* $x \mapsto P_{(x,c)} \wedge \cdots \wedge P_{(x,c)}$, *with* $|C|$ *terms in the Hadamard product, is a regular representation of* G.

(4) *The map* $x \mapsto P^f_x = P^f_{(x,1)}$ *is a monomial representation of* G *if and only if* f *is the trivial cocycle (so* $E(f) \cong G \times C$).

PROOF.

1. See Lemma 3.5.3 (1).

2. The stated map is a homomorphism: $(x, c)(y, d)$ is sent to

$$(cdf(x, y))^{-1}[\delta^{rxy}_s f(r, xy)^{-1}]_{r,s \in G} = c^{-1}d^{-1}[\delta^{rxy}_s f(r, x)^{-1} f(rx, y)^{-1}]_{r,s \in G}$$

$$= c^{-1}d^{-1}[\textstyle\sum_t \delta^{rx}_t \delta^{ty}_s f(r, x)^{-1} f(t, y)^{-1}]_{r,s \in G}$$

$$= c^{-1}[\delta^{rx}_t f(r, x)^{-1}]_{r,t \in G} \, d^{-1}[\delta^{ty}_s f(t, y)^{-1}]_{t,s \in G}$$

$$= P_{(x,c)} P_{(y,d)}.$$

Also, if $P^f_{(x,c)} = I_v$ then, by (12.23), $\delta^{rx}_r = 1$ forces $x = 1$ and $c = 1$.

3. $P_{(x,c)} \wedge \cdots \wedge P_{(x,c)} = c^{-|C|}[\delta^{rx}_s f(r, x)^{-|C|}]_{r,s \in G} = [\delta^{rx}_s]_{r,s \in G} = T_x$ where T_x is as in Subsection 3.7.3.

4. By (2), $P_{(x,1)} P_{(y,1)} = P_{(xy, f(x,y))}$. Thus $x \mapsto P_{(x,1)}$ is a homomorphism if and only if $P_{(xy,1)} = P_{(xy, f(x,y))}$ for all $x, y \in G$. By faithfulness, this latter premise is equivalent to f being trivial. $\qquad \square$

The matrix representation defined in Lemma 12.5.1 (2) is simultaneously a faithful regular permutation representation, and as such is unique up to similarity.

12.5.2. LEMMA. *Denote by* Ω *the set of* $(0, C)$-*column vectors of length* v *with a single non-zero entry. The following hold.*

(i) *The monomial group* $\langle P^f_{(x,c)} : (x,c) \in E(f) \rangle$ *acts regularly by ordinary matrix multiplication on* Ω.

(ii) *Let* μ *be any isomorphism of* $E(f)$ *onto a subgroup of* $\mathrm{Mon}(v,C)$ *whose image acts regularly on* Ω. *Then, regarding* $\mathrm{Mon}(v,C)$ *as a subgroup of* $\mathrm{Sym}(\Omega)$ *in the obvious way, there exists* $\sigma \in \mathrm{Sym}(\Omega)$ *such that* $\mu((x,c)) = \sigma^{-1} P^f_{(x,c)} \sigma$ *for all* $(x,c) \in E(f)$.

PROOF. The column of $P_{(x^{-1},c^{-1}f(x,x^{-1})^{-1})}$ labelled by 1_G has c in position x and zeros everywhere else. This shows that the monomial group acts transitively. Therefore, as the group has order $|E(f)| = v|C| = |\Omega|$, it acts regularly. For part (ii), see Lemma 3.3.7. $\qquad\square$

As one might expect, there is a relationship between cocyclic development and monomial representations of a corresponding extension group. Lemma 12.5.5 gives the details.

12.5.3. DEFINITION. For a cocycle $f : G \times G \to C$ and element (x,c) of $E(f)$, denote the monomial matrix $c^{-1}[\delta^{rs}_x f(r,s)]_{r,s\in G}$ by $U^f_{(x,c)}$. Set $U^f_{(1,1)} = U$.

12.5.4. LEMMA. *If* C *has exponent dividing* 2 *then* U *is an involution.*

PROOF. Since $f(a,a^{-1}) = f(a^{-1},a)$,
$$U^2 = [\delta^a_b f(a,a^{-1}) f(b^{-1},b)]_{a,b\in G} = [\delta^a_b f(a,a^{-1})^2]_{a,b\in G} = I_v. \qquad\square$$

12.5.5. LEMMA. *Let* $f : G \times G \to C$ *be a cocycle and let* $g : G \to \mathcal{R}$ *be a map, where* $\mathcal{R}$ *is a ring containing* C *in its group of units. Then* $[g(ab)f(a,b)]_{a,b\in G}$ *may be decomposed into a disjoint sum of monomial* $\mathcal{R}$-matrices:
$$[g(ab)f(a,b)]_{a,b\in G} = U\sum_{x\in G} g(x)P^f_x.$$

PROOF. We have
$$UP^f_x = [\delta^{ac}_1 f(a,a^{-1})]_{a,c\in G} [\delta^{cx}_b f(c,x)^{-1}]_{c,b\in G}$$
$$= [\delta^{ab}_x f(a^{-1},a) f(a^{-1},ab)^{-1}]_{a,b\in G}$$
$$= [\delta^{ab}_x f(a,b)]_{a,b\in G}.$$

Thus
$$U\sum_{x\in G} g(x)P^f_x = [\sum_{x\in G} g(x)\delta^{ab}_x f(a,b)]_{a,b\in G} = [g(ab)f(a,b)]_{a,b\in G},$$
as claimed. $\qquad\square$

Finally, we say a bit about monomial representations of central products.

12.5.6. LEMMA. *Let* $f : G \times G \to C$ *and* $h : H \times H \to C$ *be cocycles. Then, for all* $(x,c) \in E(f)$ *and* $(y,d) \in E(h)$,

(1) $P^{f\times h}_{(x\cdot y,cd)} = P^f_{(x,c)} \otimes P^h_{(y,d)}$,

(2) $U^{f\times h}_{(x\cdot y,cd)} = U^f_{(x,c)} \otimes U^h_{(y,d)}$.

PROOF. We have
$$P^f_{(x,c)} \otimes P^h_{(y,d)} = [c^{-1} \delta^{ax}_b f(a,x)^{-1}]_{a,b\in G} \otimes [d^{-1} \delta^{ry}_s h(r,y)^{-1}]_{r,s\in H}$$
$$= [c^{-1}d^{-1} \delta^{ax\cdot ry}_{b\cdot s} f(a,x)^{-1}h(r,y)^{-1}]_{a\cdot r, b\cdot s\in G\times H}$$

$$= [(cd)^{-1} \, \delta_{b \cdot s}^{(a \cdot r)(x \cdot y)}(f \times h)(a \cdot r, x \cdot y)^{-1}]_{a \cdot r, b \cdot s \in G \times H}$$
$$= P_{(x \cdot y, cd)}^{f \times h},$$

proving (1). The proof of (2) is similar. $\square$

Lemma 12.5.6 indicates how monomial representations of $E(f \times h) \cong E(f) \curlyvee E(h)$ arise from monomial representations of $E(f)$ and $E(h)$. Forming a direct product of the indexing groups amounts to Kronecker multiplication of the matrices.

CHAPTER 13

Cocyclic Pairwise Combinatorial Designs

This chapter sets out preliminary concepts for cocyclic pairwise combinatorial designs—those designs that are f-developed for some cocycle f.

13.1. The main definitions

Let $\mathcal{A}$ be an alphabet.

13.1.1. DEFINITION. A $v \times v$ $(0, \mathcal{A})$-array X is in *cocyclic form* if

$$X = [f(x,y)g(xy)]_{x,y \in G}$$

for some group G of order v, (normalized) cocycle $f : G \times G \to Z$ where Z is an abelian subgroup of $\mathrm{Sym}(\mathcal{A} \cup \{0\})$ fixing 0, and map $g : G \to \{0\} \cup \mathcal{A}$.

13.1.2. EXAMPLE. We display the 3×3 (± 1)-arrays that are in cocyclic form:

$$\begin{bmatrix} C & D & E \\ D & AE & BC \\ E & BC & ABD \end{bmatrix} = \begin{bmatrix} 1 & 1 & 1 \\ 1 & A & B \\ 1 & B & AB \end{bmatrix} \wedge \begin{bmatrix} C & D & E \\ D & E & C \\ E & C & D \end{bmatrix}.$$

So 32 of the 512 such arrays are in cocyclic form. (The first array on the right-hand side is a development table for cocycles with ground group C_3; see Example 11.3.3.) For instance, $X = \begin{bmatrix} 1 & 1 & 1 \\ - & 1 & - \\ 1 & 1 & - \end{bmatrix}$ is not in cocyclic form, whereas $Y = \begin{bmatrix} 1 & 1 & 1 \\ 1 & 1 & - \\ 1 & - & - \end{bmatrix} \approx X$ is in cocyclic form.

The definition of cocyclic design is more subtle. Let Λ be an orthogonality set of $2 \times v$ $(0, \mathcal{A})$-arrays. Recall that $\Pi_\Lambda^{\mathrm{row}}$ can be assumed to commute elementwise with $\Pi_\Lambda^{\mathrm{col}}$, so that $\Pi_\Lambda^{\mathrm{row}} \cap \Pi_\Lambda^{\mathrm{col}}$ is a central (hence abelian) subgroup of $\Pi_\Lambda = \Pi_\Lambda^{\mathrm{row}} \Pi_\Lambda^{\mathrm{col}}$. A *cocyclic* $\mathrm{PCD}(\Lambda)$ is one that is Λ-equivalent to an array in cocyclic form.

13.1.3. DEFINITION. Let $D = [d_{x,y}]_{x,y \in G}$ be a $\mathrm{PCD}(\Lambda)$ indexed by the group G, and let $f : G \times G \to \Pi_\Lambda^{\mathrm{row}} \cap \Pi_\Lambda^{\mathrm{col}}$ be a cocycle. Then D is *cocyclic*, with cocycle f, if there exist

(1) a map $g : G \to \{0\} \cup \mathcal{A}$,
(2) $\pi, \phi \in \mathrm{Sym}(G)$,
(3) elements μ_x of $\Pi_\Lambda^{\mathrm{row}}$ and ν_y of $\Pi_\Lambda^{\mathrm{col}}$,

such that

$$\mu_x \nu_y (d_{\pi(x),\phi(y)}) = f(x,y)(g(xy)) \qquad \forall \, x, y \in G.$$

We call f a Λ-*cocycle*, and G a Λ-*indexing group*.

Definition 13.1.3 may appear to be needlessly complicated. However, since a $\mathrm{PCD}(\Lambda)$ is a Λ-equivalence class of arrays, our definition must take Λ-equivalence into account.

13.1.4. EXAMPLE. As designs, the arrays X and Y of Example 13.1.2 are both cocyclic, even though X is not in cocyclic form. Here Λ could be all 2×3 $(1, -1)$-arrays such that the inner product of the first row and the second row is ± 1.

13.2. Ambient rings with a central group

To define cocyclic PCD(Λ)s in the more natural setting of matrices over an ambient ring $\mathcal{R}$, we need $\mathcal{R}$ to have not only a row group and a column group, but also a *central group* $Z \cong \Pi_\Lambda^{\text{row}} \cap \Pi_\Lambda^{\text{col}}$.

13.2.1. REQUIREMENTS FOR AN AMBIENT RING WITH A CENTRAL GROUP Z

(1) $\mathcal{R}$ is an ambient ring for Λ such that the conditions of 5.2.3 all hold; in particular $\mathcal{R}$ has row group R and column group C equipped with isomorphisms $\rho : R \to \Pi_\Lambda^{\text{row}}$ and $\kappa : C \to \Pi_\Lambda^{\text{col}}$.
(2) $Z = \rho^{-1}(\Pi_\Lambda^{\text{row}} \cap \Pi_\Lambda^{\text{col}}) = \kappa^{-1}(\Pi_\Lambda^{\text{row}} \cap \Pi_\Lambda^{\text{col}})$.
(3) For all $a \in \mathcal{A}$ and $z \in Z$, $za = az$.

Such an ambient ring $\mathcal{R}$ supports a matrix model for Λ-orthogonality. Note that $Z \leq R \cap C$ will be determined by Λ and ρ (or κ).

To show that $\mathcal{R}$ as in 13.2.1 always exists, we use a variant of the ambient ring construction in Chapter 5.

13.2.2. LEMMA. *There exist a group $E = R \curlyvee C$ and isomorphisms $\rho : R \to \Pi_\Lambda^{\text{row}}$, $\kappa : C \to \Pi_\Lambda^{\text{col}}$ such that $\rho(z) = \kappa(z^{-1})$ for all $z \in R \cap C$.*

PROOF. Let τ be a normalized transversal map for the central extension

$$1 \longrightarrow K \overset{\iota}{\longrightarrow} \Pi_\Lambda^{\text{col}} \overset{\pi}{\longrightarrow} \Pi_\Lambda^{\text{col}}/K \longrightarrow 1$$

where $K = \Pi_\Lambda^{\text{row}} \cap \Pi_\Lambda^{\text{col}}$, ι is inclusion, and π is natural surjection. Denote the inversion automorphism of K by α. We define E to be $(\Pi_\Lambda^{\text{row}} \times E(f))/N$ where $f = \alpha \circ f_{\iota,\tau}$ and N is the central subgroup $\{(u^{-1}, (1, u)) \mid u \in K\}$.

Next, note that

$$R = \{(r, (1, 1))N \mid r \in \Pi_\Lambda^{\text{row}}\} \quad \text{and} \quad C = \{(1, (x, u))N \mid (x, u) \in E(f)\}$$

are subgroups of E such that $E = R \curlyvee C$. Furthermore, the maps $\rho : R \to \Pi_\Lambda^{\text{row}}$ and $\kappa : C \to \Pi_\Lambda^{\text{col}}$ defined by

$$\rho : (r, (1, 1))N \mapsto r \quad \text{and} \quad \kappa : (1, (x, u))N \mapsto \tau(x)u^{-1}$$

are isomorphisms as required. $\square$

Thus, we may choose a monoid presentation $\langle X_R, X_C \mid W \rangle$ where X_R and X_C are generating sets for the subgroups R and C respectively, and isomorphisms ρ, κ such that $\rho_z = \kappa_{z^{-1}}$ for all $z \in R \cap C$. Let $M(\Lambda)$ be the monoid defined as in Construction 5.2.7 for these choices. Put $\mathcal{R} = \mathbb{Z}[M(\Lambda)]$ and $Z = R \cap C$. Then (1) and (2) of 13.2.1 clearly hold. Since

$$za = \rho_z(a) = \kappa_{z^{-1}}(a) = az \qquad \forall a \in \mathcal{A}, \, z \in Z,$$

(3) holds too.

We make a few remarks about the above construction. First, the central group $Z = R \cap C$ is central in the whole ring. Also, if D is a PCD(Λ) then the scalar pairs (zI_v, zI_v) for $z \in Z$ are automorphisms of D:

$$zI_v D(zI_v)^* = zI_v D z^{-1} I_v = zz^{-1} I_v D = D.$$

This is one possible construction of ambient ring with a central group, that is valid for any orthogonality set Λ. For specific Λ, we almost always prefer to work with more appropriate and obvious rings.

We next state the central groups and their corresponding rings for each of the familiar designs of Chapter 2.

13.2.3. PROPOSITION. *Let Λ one of the orthogonality sets of Definition 2.13.1. The ambient rings $\mathcal{R}$ in Theorem 5.3.1 (3) satisfy the requirements listed in 13.2.1, where $Z = R = C$; except possibly when Λ is one of $\Lambda_{\mathrm{GH}(n;G)}$, $\Lambda_{\mathrm{BGW}(v,k,\lambda;G)}$, or $\Lambda_{\mathrm{GW}(v,k;G)}$, in which cases $R = C = G$ and $Z = Z(G)$.*

Now we return to the definition of cocyclic PCD(Λ). The following theorem achieves the main aim of this section.

13.2.4. THEOREM. *Let $f : G \times G \to \Pi_\Lambda^{\mathrm{row}} \cap \Pi_\Lambda^{\mathrm{col}}$ be a cocycle, and let $\mathcal{R}$ be a ring satisfying 13.2.1. Write f' for the cocycle $\rho^{-1} \circ f$. A PCD(Λ), D, is cocyclic with cocycle f if and only if there are $P \in \mathrm{Mon}(v, R)$, $Q \in \mathrm{Mon}(v, C)$, and a map $g : G \to \{0\} \cup \mathcal{A}$, such that*

$$D = P[f'(x,y)g(xy)]_{x,y \in G} Q$$

over $\mathcal{R}$.

PROOF. By Definition 13.1.3, $D = [d_{x,y}]_{x,y \in G}$ is cocyclic with cocycle f if and only if

$$d_{\pi(x),\phi(y)} = \mu_x \nu_y f(x,y)(g(xy))$$

for all $x, y \in G$, a map $g : G \to \{0\} \cup \mathcal{A}$, permutations $\pi, \phi \in \mathrm{Sym}(G)$, $\mu_x \in \Pi_\Lambda^{\mathrm{row}}$, and $\nu_y \in \Pi_\Lambda^{\mathrm{col}}$. That is,

$$d_{\pi(x),\phi(y)} = r_x f'(x,y)g(xy)c_y$$

in $\mathcal{R}$, where $\rho(r_x) = \mu_x$ and $\kappa(c_y) = \nu_y^{-1}$. This is equivalent to

$$[d_{x,y}]_{x,y \in G} = [\delta_w^{\pi^{-1}(x)} r_w]_{x,w \in G} [f'(w,z)g(wz)]_{w,z \in G} [\delta_{\phi^{-1}(y)}^z c_z]_{z,y \in G}. \qquad \square$$

By Theorem 13.2.4, henceforth we assume the following definition of cocyclic design (note: this differs from Horadam's definition of cocyclic matrix [87, p. 116]).

13.2.5. DEFINITION. Let D be a PCD(Λ), and let $\mathcal{R}$ be a ring satisfying 13.2.1. Then D is cocyclic, with cocycle $f : G \times G \to Z$, if

$$D = P[f(x,y)g(xy)]_{x,y \in G} Q$$

for some $P \in \mathrm{Mon}(v, R)$, $Q \in \mathrm{Mon}(v, C)$, and $g : G \to \{0\} \cup \mathcal{A}$. As before, we call f a Λ-*cocycle* and G a Λ-*indexing group* (or just *indexing group*, if Λ is implicit). We may also say that D *is cocyclic over G.*

Occasionally the form of a cocyclic PCD(Λ) may be simplified.

13.2.6. DEFINITION. The cocycle f is a *pure Λ-cocycle* if $[f(x,y)a]_{x,y \in G}$ is a PCD(Λ) for some $a \in \mathcal{A}$.

13.2.7. THEOREM. *Suppose that $\Pi_\Lambda^{\mathrm{row}} \cap \Pi_\Lambda^{\mathrm{col}}$ acts regularly on $\mathcal{A}$, and that no array in Λ contains a zero. Then every Λ-cocycle is cohomologous to a pure Λ-cocycle.*

PROOF. Let $f : G \times G \to Z$ be a Λ-cocycle, and let $[f(x,y)g(xy)]_{x,y \in G}$ be a PCD(Λ). Since $\Pi_\Lambda^{\text{row}} \cap \Pi_\Lambda^{\text{col}}$ acts regularly on $\mathcal{A}$, for each $x \in G$ there is a unique element $\gamma(x)^{-1}$ of Z such that $g(x) = \gamma(x)^{-1}a$, where $a = g(1)$. Then

$$[f(x,y)g(xy)]_{x,y \in G} = [f(x,y)\gamma(xy)^{-1}a]_{x,y \in G}$$
$$\approx_\Lambda [f(x,y)\gamma(x)\gamma(y)\gamma(xy)^{-1}a]_{x,y \in G}$$
$$= [f(x,y)\partial\gamma(x,y)a]_{x,y \in G}.$$

Therefore f is cohomologous to the pure Λ-cocycle $f\partial\gamma$. $\qquad\square$

13.3. Some big problems

Fix an ambient ring with central group Z for the orthogonality set Λ. Let D be a cocyclic PCD(Λ). We denote by $\mathcal{C}(D)$ the set of cocycles of D with coefficient group Z, and by $\mathcal{C}(\Lambda)$ the set of cocycles of all PCD(Λ)s. So, allowing D to range over the Λ-inequivalent PCD(Λ)s,

$$\mathcal{C}(\Lambda) = \bigcup_D \mathcal{C}(D).$$

We state two major problems concerning cocyclic designs.

13.3.1. PROBLEM. *Find $\mathcal{C}(D)$ for a given* PCD(Λ), D.

13.3.2. PROBLEM. *Given a cocycle $f : G \times G \to Z$, decide whether $f \in \mathcal{C}(\Lambda)$.*

As will be shown the next chapter, if $E(D)$ is a particular lesser expanded design of D, and PermAut($E(D)$) is not too large, then we can solve Problem 13.3.1 with the aid of available computational algebra procedures. There does not seem to be such a practical or straightforward means of handling the second problem. In important cases, Problem 13.3.2 leads to a 'group ring equation' which has a solution if and only if the requisite design exists (see Section 10.5 and Chapter 15).

13.4. Central extensions of a design

Let D be a PCD(Λ).

13.4.1. THEOREM. *Suppose that $f : G \times G \to Z$ is a cocycle of D. Let f' be a cocycle cohomologous to f. Then*

(i) *f' is a cocycle of D,*
(ii) *D is group-developed modulo G if f is a coboundary.*

PROOF. We have $f'(x,y) = \beta(x)\beta(y)\beta(xy)^{-1}f(x,y)$ for some $\beta : G \to Z$, and then (cf. the proof of Theorem 13.2.7)

$$[f'(x,y)g'(xy)]_{x,y \in G} \approx_\Lambda [f(x,y)g(xy)]_{x,y \in G},$$

where $g(x) = \beta(x)^{-1}g'(x)$. The second claim follows by taking f' to be trivial. $\quad\square$

Thus, if $f \in \mathcal{C}(D)$ then the entire cohomology class $[f]$ of f is contained in $\mathcal{C}(D)$. We may also state this result in the terms of Section 12.1. Suppose that $f = f_{\iota,\tau}$ is a cocycle of the central short exact sequence

(13.1) $1 \longrightarrow Z \xrightarrow{\iota} E \xrightarrow{\pi} G \longrightarrow 1$

If $f \in \mathcal{C}(D)$ then we say that (13.1) is a *central short exact sequence of D*.

13.4.2. THEOREM. *A cocyclic design D has the central short exact sequence (13.1) if and only if every cocycle arising from (13.1) is a cocycle of D. Therefore, if (13.1) is a central short exact sequence of a $\mathrm{PCD}(\Lambda)$, then so too is any equivalent sequence.*

We strengthen Theorem 13.4.2. Equation (12.10) asserts that

$$f \in \mathcal{C}_{E,Z}(\iota) \quad \Longleftrightarrow \quad \mathcal{C}_{E,Z}(\iota) = \bigcup_{\alpha \in \mathrm{Aut}(G)} [f_\alpha]$$

where f_α is the cocycle defined by $f_\alpha(x,y) = f(\alpha(x), \alpha(y))$.

13.4.3. THEOREM. *If $\mathcal{C}_{E,Z}(\iota) \cap \mathcal{C}(D) \neq \emptyset$ then $\mathcal{C}_{E,Z}(\iota) \subseteq \mathcal{C}(D)$.*

PROOF. For any cocycle $f : G \times G \to Z$ and $\alpha \in \mathrm{Aut}(G)$,

$$[f(x,y)g(xy)]_{x,y \in G} \approx [f_\alpha(x,y)\, g \circ \alpha(xy)]_{x,y \in G}.$$

So f is a cocycle of D if and only if f_α is a cocycle of D. $\qquad \square$

Consequently, $\mathcal{C}(\Lambda) \cap \mathcal{C}_{E,Z}(\iota) \neq \emptyset$ implies that $\mathcal{C}(\Lambda) \supseteq \mathcal{C}_{E,Z}(\iota)$. For many of the orthogonality sets in Chapter 2, we can say more.

13.4.4. THEOREM. *Let Λ be one of the orthogonality sets in Definition 2.13.1 with $\Pi_\Lambda^{\mathrm{row}} = \Pi_\Lambda^{\mathrm{col}}$. If $\mathcal{C}_{E,Z} \cap \mathcal{C}(\Lambda) \neq \emptyset$ then $\mathcal{C}_{E,Z} \subseteq \mathcal{C}(\Lambda)$.*

PROOF. For any $\beta \in \mathrm{Aut}(Z)$, if $[f(x,y)g(xy)]_{x,y \in G}$ is a $\mathrm{PCD}(\Lambda)$, then so too is $[\beta \circ f(x,y)g'(xy)]_{x,y \in G}$ for some g'. $\qquad \square$

Let $f \in \mathcal{C}(D)$. Any group E isomorphic to $E(f)$ is an *extension group of D*. Note that if E is an extension group of D, with cocycle $f : G \times G \to Z$, then there is an embedding $\iota : Z \to E$ and projection π such that f arises from the central short exact sequence (13.1). Indeed, even if we stipulate the central subgroup of E, there is latitude in the choice of ι, and it is possible that (13.1) determines a cocycle of D for some but not all choices of ι. But in the important cases covered by Theorem 13.4.4, this cannot happen.

13.5. Approaches to cocyclic designs

We recount below four approaches employed in the study of cocyclic pairwise combinatorial designs.

Via extension group. We may investigate the extension groups of a $\mathrm{PCD}(\Lambda)$. This approach has been adopted by Noboru Ito for Hadamard matrices [**98, 99, 100, 101**]. We touch on this work in Chapters 15 and 19.

Via indexing group. Let G be a finite group. Which cocycles of G are cocycles of a $\mathrm{PCD}(\Lambda)$? Quite a few authors have written about this more general version of Problem 13.3.2: Alvarez, Armario, Frau, and Real [**1**], Baliga and Horadam [**7**], de Launey and Smith [**54**], Flannery [**64**], Horadam [**86**], Horadam and de Launey [**88**], and de Launey, Flannery, and Horadam [**47**]. Chapters 19, 21, and 22 deal with this problem.

Known designs. We may ask for all cocycles of a known design or family of designs: Problem 13.3.1. In Chapter 14 we explain how to solve this problem by determining the 'centrally' regular actions on $\mathcal{E}(D)$ for a given design D.

Kantor [**107**] considered a certain 'symplectic' design equivalent to the Sylvester Hadamard matrix of order 2^{2m}. Assmus and Salwach [**5**] proved that 12 of the groups of order 16 act regularly on this design at order 16; while Schibell [**137**] classified the 171 groups acting regularly on this design at order 64.

Ó Catháin and Röder [**126**] recently classified all cocyclic Hadamard matrices of orders less than 40. Ito [**99**] found several interesting sporadic regular actions for the Paley type I Hadamard matrix. Some of de Launey and Stafford's work [**55, 56, 57**] in this vein for the Paley matrices is discussed in Chapter 17. Also, drawing on the paper [**54**] by de Launey and Smith, Chapter 21 determines the cocycles with elementary abelian ground group that are cocycles of the Sylvester Hadamard matrices.

Composition. In Chapters 18 and 23, we use methods from Chapter 15 for composing cocyclic designs to obtain many cocyclic Hadamard matrices.

CHAPTER 14

Centrally Regular Actions

In Chapter 10, we studied regular actions on square arrays. This chapter is concerned with *centrally regular actions* on a certain lesser expanded design of a pairwise combinatorial design D. Such actions characterize cocyclic development of D. They may be found by applying techniques of Chapter 10 to the expanded design. For many designs D of moderate size, this provides an efficient solution of Problem 13.3.1.

14.1. Cocyclic forms

Let Λ be an orthogonality set with alphabet $\mathcal{A}$. Chapter 10 describes how to determine all the ways that a $(0, \mathcal{A})$-array A may be expressed in group-developed form

$$(14.1) \qquad A = P^{-1}[h(xy)]_{x,y \in H} Q$$

where P, Q are permutation matrices. Our method is predicated on being able to compute the regular subgroups of $\mathrm{PermAut}(A)$. We will give an analogous method for determining all the ways that a $\mathrm{PCD}(\Lambda)$, D, is cocyclic.

Let $f : G \times G \to Z$ be a cocycle, where G has order v. Over an ambient ring $\mathcal{R}$ with row group R, column group C, and central group Z (see Section 13.2), D is cocyclic with cocycle f if

$$(14.2) \qquad D = P^*[f(x,y)g(xy)]_{x,y \in G} Q$$

for some map $g : G \to \{0\} \cup \mathcal{A}$, $P \in \mathrm{Mon}(v, R)$, and $Q \in \mathrm{Mon}(v, C)$. We say that $[f(x,y)g(xy)]_{x,y \in G}$ is a *cocyclic form of* D, and (P, f, g, Q) is a *solution* of (14.2). Now if

$$(14.3) \qquad \Pi_\Lambda^{\mathrm{row}} = \Pi_\Lambda^{\mathrm{col}}$$

then $R = C = Z$ by 13.2.1, and the problem of finding the cocyclic forms of D may be stated as follows.

14.1.1. PROBLEM. *Find all cocycles f and maps g such that (P, f, g, Q) is a solution of* (14.2) *for some $P, Q \in \mathrm{Mon}(v, Z)$.*

In the rest of this chapter we assume that (14.3) holds; this is certainly the case for nearly all the orthogonality sets in Definition 2.13.1.

14.2. A lesser expanded design

Let D be a $\mathrm{PCD}(\Lambda)$. In Section 9.6, we defined the (full) expanded design

$$\mathcal{E}(D) = [rDc]_{r \in R, c \in C}$$

and the (lesser) expanded design

$$\mathcal{E}_{K,L}(D) = [aDb]_{a\in K, b\in L}$$

for subgroups K of R and L of C. Since Z is central in $\mathcal{R}$,

$$\mathcal{E}_{Z,Z}(D) = [ab]_{a,b\in Z} \otimes D.$$

We denote this expanded design $\mathrm{E}(D)$. Assuming (14.3), $\mathrm{E}(D) = \mathcal{E}(D)$.

It turns out that we are able to solve Problem 14.1.1 once we know the regular actions on $\mathrm{E}(D)$. In the many cases where Λ satisfies (14.3), this yields all the cocyclic forms of D, thereby solving Problem 13.3.1.

Our initial aim is to prove that each solution of (14.2) lifts to a solution of (14.1), where $A = \mathrm{E}(D)$. Before proceeding to do that, it is necessary to specify indexings for our matrices, and establish some properties of homomorphisms that were defined back in Chapter 9.

14.3. A pair of lifting homomorphisms

Let $f : G \times G \to Z$ be a cocycle, with $|G| = v$ and $|Z| := m > 1$. Recall that the canonical central extension $E(f)$ of Z by G has elements (x, a) where $x \in G$ and $a \in Z$, and multiplication given by

$$(14.4) \qquad\qquad (x, a)(y, b) = (xy, f(x, y)ab).$$

Fix an ordering $x_1, x_2, \ldots, x_v$ of the elements of G, and use this to index the rows and columns of $v \times v$ matrices. Next fix an ordering $z_1, z_2, \ldots, z_m$ of Z, and use this to index $m \times m$ matrices. Then the elements of $E(f)$ under the ordering

$$(x_1, z_1), (x_2, z_1), \ldots, (x_v, z_1), (x_1, z_2), (x_2, z_2), \ldots, (x_v, z_2), \ldots$$
$$\ldots, (x_1, z_m), (x_2, z_m), \ldots, (x_v, z_m)$$

will index $vm \times vm$ matrices.

In particular, if D is a PCD(Λ) indexed by G, then we can use the above ordering of the elements of $E(f)$ to index the rows and columns of the block matrix form $[cd]_{c,d\in Z} \otimes D$ for $\mathrm{E}(D)$. The permutation matrix

$$P_\pi = [\delta^{(x,c)}_{\pi((y,d))}]_{(x,c),(y,d)\in E(f)}$$

can also be indexed by the same ordering.

In Section 9.6 of Chapter 9 we defined injective homomorphisms $\theta_Z^{(1)}$ and $\theta_Z^{(2)}$ from $\mathrm{Mon}(v, Z)$ into $\mathrm{Perm}(vm)$: if $X = \sum_{z\in Z} zX_z \in \mathrm{Mon}(v, Z)$ where the X_z are disjoint $v \times v$ $(0,1)$-matrices, then

$$\theta_Z^{(1)}(X) = \sum_{z\in Z} [\delta_b^{az}]_{a,b\in Z} \otimes X_z = \sum_{z\in Z} T_z \otimes X_z,$$
$$\theta_Z^{(2)}(X) = \sum_{z\in Z} [\delta_{zb}^a]_{a,b\in Z} \otimes X_z = \sum_{z\in Z} S_z \otimes X_z.$$

By Remark 9.6.9,

$$(14.5) \qquad\qquad \theta_Z^{(1)}(P)\,\mathrm{E}(D)\,\theta_Z^{(2)}(Q)^\top = \mathrm{E}(PDQ^*)$$

for any $P, Q \in \mathrm{Mon}(v, Z)$.

Each matrix in $\mathrm{Mon}(v, Z)$ may be written uniquely as

$$M_{\zeta,\mu} = [\mu(y)\delta^x_{\zeta(y)}]_{x,y\in G}$$

for some $\zeta \in \mathrm{Sym}(G)$ and map $\mu : G \to Z$. Then

$$(14.6) \qquad \theta_Z^{(1)}(M_{\zeta,\mu}) = [\delta_{(\zeta(y),\mu(y)^{-1}d)}^{(x,c)}]_{(x,c),(y,d)\in E(f)}$$

and

$$(14.7) \qquad \theta_Z^{(2)}(M_{\zeta,\mu}) = [\delta_{(\zeta(y),\mu(y)d)}^{(x,c)}]_{(x,c),(y,d)\in E(f)}.$$

That is

$$\theta_Z^{(1)}(M_{\zeta,\mu}) = P_\pi \qquad \text{and} \qquad \theta_Z^{(2)}(M_{\zeta,\mu}) = P_\phi$$

where $\pi, \phi \in \mathrm{Sym}(E(f))$ are defined by

$$\pi : (y,d) \mapsto (\zeta(y), \mu(y)^{-1}d), \qquad \phi : (y,d) \mapsto (\zeta(y), \mu(y)d).$$

The following lemma connects $\theta_Z^{(1)}$ and $\theta_Z^{(2)}$ with the regular representations of $E(f)$ and Z.

14.3.1. LEMMA. *With the above labelings of rows and columns,*

$$T_{(x,c)} = \theta_Z^{(1)}([f(r,x)\delta_s^{rx}]_{r,s\in G}) (T_c \otimes I_v),$$

$$S_{(x,c)} = \theta_Z^{(2)}([f(x,s)\delta_{xs}^{r}]_{r,s\in G}) (S_c \otimes I_v).$$

PROOF. By (14.4) and (14.6),

$$T_{(x,c)} = [\delta_{(s,b)}^{(r,a)(x,c)}]_{(r,a),(s,b)\in E(f)}$$

$$= [\delta_{(s,b)}^{(rx,f(r,x)ac)}]_{(r,a),(s,b)\in E(f)}$$

$$= \theta_Z^{(1)}([cf(r,x)\delta_s^{rx}]_{r,s\in G})$$

$$= \theta_Z^{(1)}([f(r,x)\delta_s^{rx}]_{r,s\in G})\theta_Z^{(1)}(cI_v)$$

$$= \theta_Z^{(1)}([f(r,x)\delta_s^{rx}]_{r,s\in G})(T_c \otimes I_v).$$

The other identity is proved similarly, using (14.7). □

14.4. The lift

The purpose of this section is to show that if D is cocyclic then its expanded design $\mathrm{E}(D)$ is group-developed. We do this by lifting (14.2) to a group-developed form of $\mathrm{E}(D)$.

First, some more technicalities.

14.4.1. DEFINITION. A monomial matrix $M_{\zeta,\mu}$ indexed by a group is *normalized* if $\zeta(1) = 1$ and $\mu(1) = 1$, i.e., its $(1,1)$th entry is 1.

14.4.2. LEMMA. *For $i = 1,2$, $\theta_Z^{(i)}(M_{\zeta,\mu})$ is normalized if and only if $M_{\zeta,\mu}$ is too.*

PROOF. This is immediate from (14.6) and (14.7). □

14.4.3. DEFINITION. A solution (P,f,g,Q) of (14.2) is *normalized* if both P and Q are normalized.

Given a map $g : G \to \{0\} \cup \mathcal{A}$ and cocycle $f : G \times G \to Z$, define $h_g^f : E(f) \to \{0\} \cup \mathcal{A}$ by $h_g^f : (x,c) \mapsto cg(x)$.

14.4.4. LEMMA. *Let D be a $\mathrm{PCD}(\Lambda)$, and fix a cocycle $f : G \times G \to Z$.*

(1) *For each solution* $(P_\pi = \theta_Z^{(1)}(M_{\zeta,\mu}), h, P_\phi = \theta_Z^{(2)}(M_{\xi,\nu}))$ *of*

(14.8) $\qquad$ $\mathrm{E}(D) = P_\pi^{-1}[h((x,c)(y,d))]_{(x,c),(y,d)\in E(f)}P_\phi,$

there is a map $g : G \to \{0\} \cup \mathcal{A}$ *such that* $h = h_g^f$.

(2) *The solutions* $(P = M_{\zeta,\mu}, f, g, Q = M_{\xi,\nu})$ *of*

(14.9) $\qquad\qquad$ $D = P^*[f(x,y)g(xy)]_{x,y\in G}Q$

and the solutions $(P_\pi = \theta_Z^{(1)}(M_{\zeta,\mu}), h, P_\phi = \theta_Z^{(2)}(M_{\xi,\nu}))$ *of (14.8) are in 1-1 correspondence.*

(3) *The monomial matrices* $M_{\zeta,\mu}$ *and* $M_{\xi,\nu}$ *are normalized if and only if the permutation matrices* $\theta_Z^{(1)}(M_{\zeta,\mu})$ *and* $\theta_Z^{(2)}(M_{\xi,\nu})$ *are normalized.*

PROOF. Suppose that $(P = M_{\zeta,\mu}, f, g, Q = M_{\xi,\nu})$ is a solution of (14.9). By (14.5),

$$\mathrm{E}(D) = \theta_Z^{(1)}(M_{\zeta,\mu})^{-1}\mathrm{E}(M_{\zeta,\mu}DM_{\xi,\nu}^*)\theta_Z^{(2)}(M_{\xi,\nu})$$
$$= P_\pi^{-1}\mathrm{E}([f(x,y)g(xy)]_{x,y\in G})P_\phi$$
$$= P_\pi^{-1}[f(x,y)g(xy)cd]_{(x,c),(y,d)\in E(f)}P_\phi$$
$$= P_\pi^{-1}[h_g^f((x,c)(y,d))]_{(x,c),(y,d)\in E(f)}P_\phi,$$

where $\pi, \phi \in \mathrm{Sym}(E(f))$ are uniquely determined. This shows that there is an injective map from the set of solutions $(P = M_{\zeta,\mu}, f, g, Q = M_{\xi,\nu})$ of (14.9) to the set of solutions $(P_\pi = \theta_Z^{(1)}(M_{\zeta,\mu}), h, P_\phi = \theta_Z^{(2)}(M_{\xi,\nu}))$ of (14.8). Notice that $\mathrm{E}(D) = \mathrm{E}(D')$ implies $D = D'$. Hence, to prove that this injection is surjective, and so complete the proof of part (2), we must show that $h((x,c)) = ch((x,1))$ for each solution $(P_\pi = \theta_Z^{(1)}(M_{\zeta,\mu}), h, P_\phi = \theta_Z^{(2)}(M_{\xi,\nu}))$ of (14.8). That is, we must prove part (1). By (14.5),

$$[h((x,c)(y,d))]_{(x,c),(y,d)\in E(f)} = [cd]_{c,d\in Z} \otimes M_{\zeta,\mu}DM_{\xi,\nu}^*.$$

Comparing the $((x,c),(1,1))$th and $((x,1),(1,1))$th entries of these two identically indexed matrices gives $h((x,c)) = ch((x,1))$. This proves (1) and (2). Part (3) is a restatement of Lemma 14.4.2. $\qquad\square$

14.5. Translation

A solution $\alpha = (P, f, g, Q)$ of (14.2) for $P, Q \in \mathrm{Mon}(v, Z)$ lifts to the solution $\bar{\alpha} = (P_\pi = \theta_Z^{(1)}(P), h_g^f, P_\phi = \theta_Z^{(2)}(Q))$ of (14.8). Also, by Lemma 14.3.1, $T_{(x,c)}$ and $S_{(x,c)}$ lie in the images of $\theta_Z^{(1)}$ and $\theta_Z^{(2)}$ respectively. The solutions of (14.2) therefore inherit the concept of translation as in Definition 10.3.2. The translations of $\bar{\alpha}$ have the form

$$(S_{(u,a)}^{-1}P_\pi, h', T_{(w,b)}P_\phi),$$

where $(u,a), (w,b) \in E(f)$ and $h'((x,c)) = h_g^f((u,a)(x,c)(w,b))$ for $(x,c) \in E(f)$. Now

$$h'((x,c)) = h_g^f((uxw, f(u,x)f(ux,w)abc))$$
$$= g(uxw, 1)f(u,x)f(ux,w)abc$$
$$= cg_{(u,a),(w,b)}(x)$$

where $g_{(u,a),(w,b)}(x) = g(uxw)f(u,x)f(ux,w)ab$. So

$$(S_{(u,a)}^{-1}\theta_Z^{(1)}(P), h_{g_{(u,a),(w,b)}}^f, T_{(w,b)}\theta_Z^{(2)}(Q))$$

is a translation of $\bar{\alpha}$. This corresponds to the solution

$$(\theta_Z^{(1)-1}(S_{(u,a)}^{-1})P, f, g_{(u,a),(w,b)}, \theta_Z^{(2)-1}(T_{(w,b)})Q)$$

of (14.2), which we call a *translation* of α. By Theorem 10.3.4 and Lemma 14.4.4 we then have

14.5.1. THEOREM. *Translation is an equivalence relation that partitions the set of solutions of* (14.2) *for fixed* f *(if non-empty) into equivalence classes of size* $|E(f)|^2$, *each containing exactly one normalized solution.*

Hence, it is enough to enumerate the normalized solutions of (14.2).

14.6. Centrally regular embeddings

Let $f : G \times G \to Z$ be a cocycle. We have seen in Section 14.4 that if D is a cocyclic PCD(Λ) with cocycle f, then the expanded design E(D) is group-developed modulo $E(f)$; thus $E(f)$ acts regularly on E(D). More specifically, it follows from Lemma 14.4.4 and Theorem 10.3.6 that if (P, f, g, Q) is a normalized solution of (14.2) with $P, Q \in \text{Mon}(v, Z)$, then

$$(14.10) \qquad \eta_{P,Q} : (x,c) \mapsto (\theta_Z^{(1)}(P), \theta_Z^{(2)}(Q))^{-1}(T_{(x,c)}, S_{(x,c)})(\theta_Z^{(1)}(P), \theta_Z^{(2)}(Q))$$

is a regular embedding of $E(f)$ in PermAut(E(D)). By Lemma 14.3.1, $\eta_{P,Q}$ maps $(1,c)$ to $(T_c \otimes I_v, S_c \otimes I_v)$. As a consequence, we are only interested in regular embeddings of $E(f)$ that map the central subgroup $\{(1,c) \mid c \in Z\} \cong Z$ of $E(f)$ to the subgroup

$$\Theta_Z = \{(T_c \otimes I_v, S_c \otimes I_v) \mid c \in Z\}$$

of PermAut(E(D)) via the assignment $(1,c) \mapsto (T_c \otimes I_v, S_c \otimes I_v)$.

14.6.1. DEFINITION. We call a regular subgroup of PermAut(E(D)) containing Θ_Z in its center *centrally regular*.

We note the following refinement. The proof of this theorem uses the same idea as in the proof of Theorem 9.6.12 to show that if $P^{-1}\text{E}(D)Q = \text{E}(D)$, then (P, Q) centralizes Θ_Z.

14.6.2. THEOREM. *Suppose that* D *is a* PCD(Λ) *such that* D *and* D^* *are non-degenerate over* $\mathcal{R}$. *Then any regular subgroup of* PermAut(E(D)) *containing* Θ_Z *is centrally regular.*

A centrally regular subgroup E of PermAut(E(D)) is a central extension of $\Theta_Z \cong Z$ by $G = E/\Theta_Z$. So we have the short exact sequence

$$(14.11) \qquad\qquad 1 \longrightarrow Z \overset{\iota}{\longrightarrow} E \overset{\pi}{\longrightarrow} E/\Theta_Z \longrightarrow 1$$

where ι maps $c \in Z$ to $(T_c \otimes I_v, S_c \otimes I_v)$ and π is natural surjection $E \to E/\Theta_Z$. If $f : G \times G \to Z$ is a cocycle of (14.11) then (14.11) is equivalent to the canonical central short exact sequence

$$1 \longrightarrow Z \longrightarrow E(f) \longrightarrow G \longrightarrow 1,$$

and so there is an isomorphism $E(f) \to E$ mapping $(1,c)$ to $(T_c \otimes I_v, S_c \otimes I_v)$.

14.6.3. DEFINITION. A *centrally regular embedding of $E(f)$ in* $\mathrm{PermAut}(\mathrm{E}(D))$ is an injective homomorphism $\eta : E(f) \to \mathrm{PermAut}(\mathrm{E}(D))$ such that

- $\eta(E(f))$ is a centrally regular subgroup of $\mathrm{PermAut}(\mathrm{E}(D))$, and
- $\eta((1,c)) = (T_c \otimes I_v, S_c \otimes I_v)$ for all $c \in Z$.

The next result spells out relationships between the centrally regular subgroups of $\mathrm{PermAut}(\mathrm{E}(D))$, the centrally regular embeddings in $\mathrm{PermAut}(\mathrm{E}(D))$, and the normalized solutions of (14.2).

14.6.4. THEOREM. *Let D be a* $\mathrm{PCD}(\Lambda)$, *and let $f : G \times G \to Z$ be a cocycle.*

(1) *Any centrally regular embedding of $E(f)$ in* $\mathrm{PermAut}(\mathrm{E}(D))$ *has the form $\eta_{P,Q}$ as in (14.10) for some normalized $P, Q \in \mathrm{Mon}(v, Z)$.*

(2) *Let $P, Q \in \mathrm{Mon}(v, Z)$ be normalized. Then $\eta_{P,Q}$ is a centrally regular embedding of $E(f)$ in* $\mathrm{PermAut}(\mathrm{E}(D))$ *if and only if there is a map $g : G \to \{0\} \cup \mathcal{A}$ such that*

$$(14.12) \qquad D = P^*[f(x,y)g(xy)]_{x,y \in G} Q.$$

(3) *Let E be a centrally regular subgroup of* $\mathrm{PermAut}(\mathrm{E}(D))$. *Then the set of solution cocycles f for (14.12) corresponding to the centrally regular embeddings of $E(f)$ with image E is the set $\mathcal{C}_{E,Z}(\iota)$ of cocycles determined by the central short exact sequences*

$$1 \longrightarrow Z \overset{\iota}{\longrightarrow} E \longrightarrow G \longrightarrow 1$$

where $\iota : c \mapsto (T_c \otimes I_v, S_c \otimes I_v)$ for all $c \in Z$.

PROOF. By part (1) of Theorem 10.3.6, any regular embedding η of $E(f)$ into $\mathrm{PermAut}(\mathrm{E}(D))$ has the form

$$\eta : (x,c) \mapsto (P_\pi, P_\phi)^{-1}(T_{(x,c)}, S_{(x,c)})(P_\pi, P_\phi),$$

where $P_\pi, P_\phi \in \mathrm{Perm}(vm)$ are normalized. If η is centrally regular then

$$(P_\pi, P_\phi)^{-1}(T_c \otimes I_v, S_v \otimes I_v)(P_\pi, P_\phi)$$
$$= (P_\pi, P_\phi)^{-1}(T_{(1,c)}, S_{(1,c)})(P_\pi, P_\phi) = \eta((1,c)) = (T_c \otimes I_v, S_v \otimes I_v)$$

for all $c \in Z$. So (P_π, P_ϕ) centralizes Θ_Z. Hence by Lemma 9.6.6 and Lemma 14.4.2, there are normalized $P, Q \in \mathrm{Mon}(v, Z)$ such that $P_\pi = \theta_Z^{(1)}(P)$ and $P_\phi = \theta_Z^{(2)}(Q)$. This completes the proof of part (1).

Next, we prove part (2). By Lemma 14.4.4, (14.12) holds if and only if

$$\mathrm{E}(D) = P_\pi^{-1}[h_g^f((x,a)(y,b))]_{(x,a),(y,b) \in E(f)} P_\phi$$

where $P_\pi = \theta_Z^{(1)}(P)$ and $P_\phi = \theta_Z^{(2)}(Q)$. By part (1) of Theorem 10.3.6, the latter is true if and only if $\eta_{P,Q}$ is a regular embedding of $E(f)$ in $\mathrm{PermAut}(\mathrm{E}(D))$. Since $\eta_{P,Q}((1,c)) = (T_c \otimes I_v, S_c \otimes I_v)$, $\eta_{P,Q}$ is a regular embedding if and only if it is centrally regular.

We now prove part (3). When we say that "f is a solution cocycle for (14.12) corresponding to a centrally regular embedding with image E", we mean that there is a solution (P, f, g, Q) of (14.12) with corresponding centrally regular embedding $\eta_{P,Q}$ such that $\eta_{P,Q}(E(f)) = E$. Let $\mathcal{C}$ denote the set of such cocycles. By parts (1) and (2), and Definition 14.6.3, $f \in \mathcal{C}$ if and only if there is an isomorphism $\eta : E(f) \to E$ such that $\eta((1,c)) = (T_c \otimes I_v, S_c \otimes I_v) = \iota(c)$. But by part (1) of

Proposition 12.1.8, $f \in \mathcal{C}_{E,Z}(\iota)$ if and only if there is an isomorphism $\eta : E(f) \to E$ such that $\eta((1,c)) = \iota(c)$. Thus $\mathcal{C} = \mathcal{C}_{E,Z}(\iota)$. $\qquad\square$

14.7. Finding cocyclic forms

In this section we outline a strategy for solving Problem 14.1.1.

14.7.1. THEOREM. *Let D be a $\mathrm{PCD}(\Lambda)$, and let τ be a transversal map for the central short exact sequence*

$$(14.13) \qquad\qquad 1 \longrightarrow Z \xrightarrow{\ \iota\ } E \longrightarrow G \longrightarrow 1.$$

The following are equivalent.

(1) *There are normalized monomials $M, N \in \mathrm{Mon}(v, Z)$ such that, for some map $g : G \to \{0\} \cup \mathcal{A}$,*

$$D = M^*[f_{\iota,\tau}(x,y)g(xy)]_{x,y \in G}N.$$

(2) *There is an isomorphism ρ of E onto a centrally regular subgroup of $\mathrm{PermAut}(\mathrm{E}(D))$, where $\rho(\iota(c)) = (T_c \otimes I_v, S_c \otimes I_v)$.*

(3) *The rows and columns of $\mathrm{E}(D)$ can be labeled (perhaps differently) by the elements of E so that*

$$\mathrm{E}(D) = [h(rs)]_{r,s \in E}$$

where the elements of E labeling the rows or columns of each $v \times v$ block in $\mathrm{E}(D)$ run over a transversal of $\iota(C)$ in E, and $h : E \to \{0\} \cup \mathcal{A}$ is a map satisfying

$$h(\iota(c)\tau(x)) = ch(\tau(x)) \quad \forall\, c \in Z, x \in G.$$

PROOF. Suppose that (1) holds. Then by Theorem 14.6.4, there is a centrally regular embedding ψ of $E(f_{\iota,\tau})$ in $\mathrm{PermAut}(\mathrm{E}(D))$. By equivalence of the central extensions, there is an isomorphism $\alpha : E \to E(f_{\iota,\tau})$ such that $\alpha\iota(c) = (1,c)$ for all $c \in Z$. Thus $\rho = \psi\alpha$ is as stated in (2).

Now suppose that (2) holds. We can index the rows and columns of $\mathrm{E}(D)$ (possibly differently) so that $\mathrm{E}(D) = [h(rs)]_{r,s \in E}$. Furthermore, each element $\iota(c)$ of E acts on $\mathrm{E}(D)$ under this indexing in the same way that $(T_{(1,c)}, S_{(1,c)})$ acts on $\mathrm{E}(D)$ under the standard indexing by $E(f_{\iota,\tau})$. That is, for each $x \in G$ and $c, d \in Z$, row/column $\iota(cd)\tau(x)$ of $\mathrm{E}(D)$ is d times row/column $\iota(c)\tau(x)$ of $\mathrm{E}(D)$; also, if $d \neq 1$ then these two rows (or columns) are not in the same $v \times v$ block. Hence the entry in row $\iota(d)\tau(x)$, column 1_E is $h(\iota(d)\tau(x)) = dh(\tau(x))$, and h has the desired property. This gives us (3).

Finally, suppose that (3) holds. Say row i of the block D in $\mathrm{E}(D)$ is labeled $\iota(z_i)\tau(x_i)$, and column j is labeled $\iota(w_j)\tau(y_j)$, where $x_1, \ldots, x_v$ and $y_1, \ldots, y_v$ are total orderings of G. Let $M = [\delta_j^i z_i^{-1}]_{ij}$ and $N = [\delta_j^i w_i^{-1}]_{ij}$. The (i,j)th entry of MDN is

$$h(\iota(z_i)\tau(x_i)\iota(w_j)\tau(y_j))z_i^{-1}w_j^{-1} = h(\iota(z_iw_j)\tau(x_i)\tau(y_j))z_i^{-1}w_j^{-1}$$
$$= h(\iota(z_iw_jf_{\iota,\tau}(x_i,y_j))\tau(x_iy_j))z_i^{-1}w_j^{-1}$$
$$= f_{\iota,\tau}(x_i,y_j)h(\tau(x_iy_j))$$
$$= f_{\iota,\tau}(x_i,y_j)g(x_iy_j),$$

where $g : G \to \{0\} \cup \mathcal{A}$ is defined by $g(x) = h(\tau(x))$. Thus (3) implies (1). $\qquad\square$

14.7.2. REMARK. The equivalence (1) $\Leftrightarrow$ (2) of Theorem 14.7.1 completely characterizes cocyclic development, as a special kind of group development. That is, D is a cocyclic PCD(Λ) if and only if the permutation automorphism group of its expanded design E(D) contains a *centrally* regular subgroup.

To find all cocyclic forms of D we follow the prescription of Theorem 14.7.1, after computing all conjugacy classes and presentations for the isomorphism types of centrally regular subgroups of PermAut(E(D)). For each such conjugacy class of groups isomorphic to the central extension E, say, we invoke the algorithm of Section 10.7 to find h as in Theorem 14.7.1 (3). Let $\iota : Z \to E$ be the embedding determined by the chosen presentation and centrally regular embedding of E in PermAut(E(D)). Let τ be any transversal map for (14.13) with $G = E/\iota(Z)$ (we get a presentation for G from the presentation for E by setting every element of $\iota(Z)$ that appears to 1). Then $D \approx_\Lambda [f_{\iota,\tau}(x,y)g(xy)]_{x,y \in G}$ where $g = h \circ \tau$.

The above procedure finds one cocycle f of D for each conjugacy class of centrally regular subgroups of PermAut(E(D)). This may be expanded to a full list of cocycles of D if required. All cocycles cohomologous to f are cocycles of D; this includes cocycles obtained by varying the transversal map for a fixed projection in (14.13). Varying the projection, with all other items fixed, also yields cocycles of D. We have a precise specification of those cocycles—see Subsection 12.1.7 and the next section—and their extension groups are (centrally) isomorphic to each other.

14.7.3. EXAMPLE. Denote the Hadamard matrix of order 12 by H_{12}. Then $\text{Aut}(H_{12}) \cong \text{PermAut}(\text{E}(H_{12})) = \text{Aut}(A_{H_{12}})$ is an extension of C_2 by the Mathieu simple group M_{12}. By Theorem 14.6.2, a regular subgroup of PermAut(E(H_{12})) is centrally regular if and only it contains $\Theta_{\langle -1 \rangle}$. Using the facilities in MAGMA for computing the automorphism group of a $(0,1)$-design, we find that there are three isomorphism types of centrally regular subgroups: $Q_8 \rtimes C_3 \cong Q_{24}$; $Q_8 \times C_3$; and $E \cong \text{SL}(2,3)$, which is an extension of C_2 by Alt(4). These subgroups are distributed into three conjugacy classes. In Chapter 19 we will show that the first two classes of groups arise via constructions due to Ito [**100**] and Williamson [**157**]. Here we discuss the third class.

The indexing groups $G \cong \text{Alt}(4)$ of H_{12} and E of E(H_{12}) have presentations

$$G = \langle w, x, y \mid w^2 = x^2 = y^3 = 1, \ w^y = x, \ x^y = wx \rangle,$$
$$E = \langle a, b, c, z \mid a^2 = b^2 = z, \ c^3 = 1, \ a^b = az, \ a^c = b, \ b^c = ab \rangle,$$

where z corresponds to the unique central involution, which swaps each row and column of E(H_{12}) with its negation. From the presentations, we see that an element of E (respectively, G) is uniquely expressible as $z^i a^j b^k c^l$ (respectively, $w^j x^k y^l$) for some $0 \leq i, j, k \leq 1$ and $0 \leq l \leq 2$. We have the short exact sequence

$$1 \longrightarrow \langle -1 \rangle \overset{\iota}{\longrightarrow} E \overset{\pi}{\longrightarrow} G \longrightarrow 1$$

where $\iota(-1) = z$ and π maps $z^i a^j b^k c^l$ to $w^j x^k y^l$. A transversal map for this central extension is defined by $\tau : w^j x^k y^l \mapsto a^j b^k c^l$.

The algorithm of Section 10.7 yields that $\text{E}(H_{12}) = [h(rs)]_{r,s \in E}$ where $h(r) = 1$ if and only if $r \in D = \{1, a, b, ab, c, zac, zbc, zabc, c^2, ac^2, zbc^2, zabc^2\}$. (Note that D is a relative difference set in E with forbidden subgroup $\langle z \rangle$. Chapter 15 explains more about the passage between relative difference sets and centrally regular actions on the expanded design.) In this group-developed form, the first twelve rows and twelve columns of $E(H_{12})$ are indexed $1, a, b, ab, c, ac, bc, abc, c^2, ac^2, bc^2, abc^2$. The relevant portion of the multiplication table for E is

	1	a	b	ab	c	ac	bc	abc	c^2	ac^2	bc^2	abc^2
1	1	a	b	ab	c	ac	bc	abc	c^2	ac^2	bc^2	abc^2
a	a	z	ab	zb	ac	zc	abc	zbc	ac^2	zc^2	abc^2	zbc^2
b	b	zab	z	a	bc	$zabc$	zc	ac	bc^2	$zabc^2$	zc^2	ac^2
ab	ab	b	az	z	abc	bc	zac	zc	abc^2	bc^2	zac^2	zc^2
c	c	abc	ac	bc	c^2	abc^2	ac^2	bc^2	1	ab	a	b
ac	ac	zbc	zc	abc	ac^2	zbc^2	zc^2	abc^2	a	zb	z	ab
bc	bc	ac	$zabc$	zc	bc^2	ac^2	$zabc^2$	zc^2	b	a	zab	z
abc	abc	zc	bc	zac	abc^2	zc^2	bc^2	zac^2	ab	z	b	az
c^2	c^2	bc^2	abc^2	ac^2	1	b	ab	a	c	bc	abc	ac
ac^2	ac^2	abc^2	zbc^2	zc^2	a	ab	zb	z	ac	abc	zbc	zc
bc^2	bc^2	zc^2	ac^2	$zabc^2$	b	z	a	zab	bc	zc	ac	$zabc$
abc^2	abc^2	zac^2	zc^2	bc^2	ab	az	z	b	abc	zac	zc	bc

Applying h to each entry produces the Hadamard matrix

	1	w	x	wx	y	wy	xy	wxy	y^2	wy^2	xy^2	wxy^2
1	1	1	1	1	1	$-$	$-$	$-$	1	1	$-$	$-$
w	1	$-$	1	$-$	$-$	$-$	$-$	1	1	$-$	$-$	1
x	1	$-$	$-$	1	$-$	1	$-$	$-$	$-$	1	$-$	1
wx	1	1	$-$	$-$	$-$	$-$	1	$-$	$-$	$-$	$-$	$-$
y	1	$-$	$-$	$-$	1	$-$	1	$-$	1	1	1	1
wy	$-$	1	$-$	$-$	1	1	$-$	$-$	1	$-$	$-$	1
xy	$-$	$-$	1	$-$	$-$	1	1	$-$	1	1	$-$	$-$
wxy	$-$	$-$	$-$	1	$-$	$-$	$-$	$-$	1	$-$	1	$-$
y^2	1	$-$	$-$	1	1	1	1	1	1	$-$	$-$	$-$
wy^2	1	$-$	1	$-$	1	1	$-$	$-$	$-$	$-$	1	$-$
xy^2	$-$	$-$	1	1	1	$-$	1	$-$	$-$	$-$	$-$	1
wxy^2	$-$	$-$	$-$	$-$	1	$-$	$-$	1	$-$	1	$-$	$-$

This is in cocyclic form $[f(r,s)g(rs)]_{r,s \in G}$, where the look-up table

	1	w	x	wx	y	wy	xy	wxy	y^2	wy^2	xy^2	wxy^2
1	1	1	1	1	1	1	1	1	1	1	1	1
w	1	$-$	1	$-$	1	$-$	1	$-$	1	$-$	1	$-$
x	1	$-$	$-$	1	1	$-$	$-$	1	1	$-$	$-$	1
wx	1	1	$-$	$-$	1	1	$-$	$-$	1	1	$-$	$-$
y	1	1	1	1	1	1	1	1	1	1	1	1
wy	1	$-$	$-$	1	1	$-$	$-$	1	1	$-$	$-$	1
xy	1	1	$-$	$-$	1	1	$-$	$-$	1	1	$-$	$-$
wxy	1	$-$	1	$-$	1	$-$	1	$-$	1	$-$	1	$-$
y^2	1	1	1	1	1	1	1	1	1	1	1	1
wy^2	1	1	$-$	$-$	1	1	$-$	$-$	1	1	$-$	$-$
xy^2	1	$-$	1	$-$	1	$-$	1	$-$	1	$-$	1	$-$
wxy^2	1	$-$	$-$	1	1	$-$	$-$	1	1	$-$	$-$	1

for the cocycle $f = f_{\iota, \tau}$ is obtained by replacing each entry $z^i a^j b^k c^l$ of the partial multiplication table above by $(-1)^i$, and $g = h \circ \tau$ may be read off the first row of the Hadamard matrix. That is, $g(r) = 1$ if and only if $r \in \{1, w, x, wx, y, y^2, wy^2\}$.

14.8. All the cocycles of a design

Let D be a PCD(Λ), where Λ satisfies (14.3). To decide whether D is cocyclic, we search for regular subgroups of PermAut(E(D)) containing Θ_Z as a central subgroup. In the event that D is cocyclic, we might then want to know all of its cocycles (Problem 13.3.1). We could find these by using the method sketched after Theorem 14.7.1. However, Problem 13.3.1 asks for less information, which may be acquired more directly.

14.8.1. THEOREM. *Define* $\iota : Z \to \Theta_Z$ *by* $\iota(c) = (T_c \otimes I_v, S_c \otimes I_v)$. *Then*

$$\mathcal{C}(D) = \bigcup_E \mathcal{C}_{E,Z}(\iota),$$

where E *ranges over the centrally regular subgroups of* PermAut(E(D)).

PROOF. A cocycle f is in $\mathcal{C}(D)$ if and only if there exists a normalized solution (P, f, g, Q) of (14.2). Then Theorem 14.6.4 implies that $f \in \mathcal{C}(D)$ if and only if $f \in \mathcal{C}_{E,Z}(\iota)$ for some centrally regular subgroup E of PermAut(E(D)). □

Let $f : G \times G \to Z$ be a cocycle in $\mathcal{C}_{E,Z}(\iota)$. Note that G is determined up to isomorphism by E. By (12.10), we obtain $\mathcal{C}(D)$ by selecting a single cocycle f of a central extension

$$1 \longrightarrow Z \overset{\iota}{\longrightarrow} \Theta_Z \leq E \longrightarrow G \longrightarrow 1$$

for each regular subgroup E of PermAut(E(D)) with $\Theta_Z \leq Z(E)$, and then taking all cocycles cohomologous to f_α as α ranges over Aut(G).

CHAPTER 15

Cocyclic Associates

Group development is a special case of cocyclic development. So the notion of G-associate will generalize to cocyclic pairwise combinatorial designs. In this chapter we consider the use of cocyclic associates and their group ring equations to answer existence questions for cocyclic designs.

15.1. Definition of cocyclic associates

In this section and the next, Λ will denote an orthogonality set of $2 \times v$ $(0, \mathcal{A})$-arrays, and $\mathcal{R}$ is an ambient involutory ring with row group R, column group C, and central group Z. Assume that $\Pi_\Lambda^{\mathrm{row}} = \Pi_\Lambda^{\mathrm{col}} = \Pi_\Lambda$, so $R = C = Z$. Also assume that Λ has the Gram Property over $\mathcal{R}$. The familiar orthogonality sets of Chapter 2 certainly have such ambient rings.

15.1.1. DEFINITION. Let f be the cocycle $f_{\iota,\tau}$ where τ is a transversal map for the central short exact sequence

$$(15.1) \qquad\qquad 1 \longrightarrow Z \overset{\iota}{\longrightarrow} E \overset{\pi}{\longrightarrow} G \longrightarrow 1.$$

Suppose that D is a cocyclic PCD(Λ) with cocycle f, i.e.,

$$MDN^* = [f(x,y)g(xy)]_{x,y \in G}$$

for some map $g : G \to \{0\} \cup \mathcal{A}$, and $M, N \in \mathrm{Mon}(v, Z)$. Then we call the element

$$\alpha = \alpha_{\iota,\tau,g} = \sum_{c \in Z} c\iota(c) \sum_{x \in G} g(x)\tau(x)$$

of the group ring $\mathcal{R}[E]$ an f-associate of D or cocyclic associate of D.

Write $\alpha = \gamma\beta$, where

$$\gamma = \sum_{c \in Z} c\iota(c) \qquad \text{and} \qquad \beta = \sum_{x \in G} g(x)\tau(x).$$

Notice that γ commutes with every element of $\mathcal{R}[E]$, and

$$(15.2) \qquad z\gamma = \sum_{c \in Z} zc\iota(c) = \sum_{d \in Z} d\iota(z^{-1}d) = \iota(z)^{-1}\gamma \quad \forall\, z \in Z.$$

15.1.2. EXAMPLE. Recall the OD($2; 1^2$) in Example 8.1.1:

$$O_2 = \begin{bmatrix} x_1 & x_2 \\ x_2 & -x_1 \end{bmatrix}.$$

The ambient ring here is $\mathbb{Z}[x_1, x_2]$, and $Z = \langle -1 \rangle$. Let $G = \langle a \rangle \cong C_2$. We have $O_2 = [f(x,y)g(xy)]_{x,y \in G}$ where $f(1,1) = f(1,a) = f(a,1) = 1$, $f(a,a) = -1$, and

$g(1) = x_1$, $g(a) = x_2$. Next we determine a central short exact sequence (15.1) and transversal map τ such that $f = f_{\iota,\tau}$. Put $b = \tau(a)$. Then

$$\iota(f(a,a)) = \tau(a)\tau(a)\tau(1)^{-1} = b^2.$$

So $\iota(-1) = b^2$, $|b| = 4$, and since E must have order 4, $E = \langle b \rangle$. Thus

$$\gamma = 1 - b^2, \quad \beta = x_1 + x_2 b,$$

and $\alpha = x_1 + x_2 b - x_1 b^2 - x_2 b^3$ is a cocyclic associate of O_2.

Cocyclic associates comprise a subclass of the associates of an expanded design.

15.1.3. THEOREM. *Let D be a cocyclic $\mathrm{PCD}(\Lambda)$ with cocycle f and extension group E. Then an f-associate of D is an E-associate of the expanded design $\mathrm{E}(D)$.*

PROOF. Suppose that $D \approx_\Lambda [f(x,y)g(xy)]_{x,y\in G}$, and (15.1) is a central short exact sequence of D with transversal map τ. Then by Theorem 14.7.1, there are permutation matrices P, Q such that

$$P\mathrm{E}(D)Q^{-1} = [h(rs)]_{r,s\in E},$$

where $h(\iota(c)\tau(x)) = cg(x)$ for $x \in G$, $c \in Z$. Thus $\alpha = \sum_{e\in E} h(e)e$ is an E-associate of $\mathrm{E}(D)$. Also

$$\alpha = \sum_{c\in Z}\sum_{x\in G} cg(x)\iota(c)\tau(x) = \sum_{c\in Z} c\iota(c)\sum_{x\in G} g(x)\tau(x)$$

is the f-associate $\alpha_{\iota,\tau,g}$ of D. $\square$

15.2. The group ring equation for cocyclic associates

In Section 10.5, we saw that each G-associate of a $\mathrm{PCD}(\Lambda)$ corresponds to a solution of an equation over the group ring $\mathcal{R}[G]$. The correspondence between cocyclic associates and solutions of a group ring equation is stated in the following theorem.

15.2.1. THEOREM. *Let f be the cocycle $f_{\iota,\tau}$ of the central short exact sequence (15.1) with transversal map τ, and let $g : G \to \{0\} \cup \mathcal{A}$ be any map. Define the elements*

$$\gamma = \sum_{c\in Z} c\iota(c) \quad and \quad \beta = \sum_{x\in G} g(x)\tau(x)$$

of $\mathcal{R}[E]$, and let $h : G \to \mathcal{R}$ be the map such that

$$\gamma\beta\beta^{(*)} = \gamma \sum_{x\in G} h(x)\tau(x).$$

Then the array $D = [f(x,y)g(xy)]_{x,y\in G}$ (assuming that it has distinct rows) is a $\mathrm{PCD}(\Lambda)$ if and only if

(15.3) $[f(uw^{-1},w)^{-1}h(uw^{-1})]_{u,w\in G} \in \mathrm{Gram}_\mathcal{R}(\Lambda).$

Before tackling the theorem, we prove a subsidiary lemma.

15.2.2. LEMMA.

$$\sum_{c\in Z} cS_{\iota(c)} \sum_{x\in G} h(x)S_{\tau(x)} = [ab^{-1}]_{a,b\in Z} \otimes [f(uw^{-1},w)^{-1}h(uw^{-1})]_{u,w\in G}.$$

PROOF. We have

$$S_{\iota(c)} = [\delta^{\iota(a)\tau(u)}_{\iota(c)\iota(b)\tau(w)}]_{a,b \in Z, u,w \in G} = [S_c \delta^u_w]_{u,w \in G}$$

and

$$S_{\tau(x)} = [\delta^{\iota(a)\tau(w)}_{\tau(x)\iota(b)\tau(y)}]_{a,b \in Z, w,y \in G}$$
$$= [\delta^{\iota(a)\tau(w)}_{\iota(f(x,y)b)\tau(xy)}]_{a,b \in Z, w,y \in G}$$
$$= [S_{f(x,y)} \delta^w_{xy}]_{w,y \in G}.$$

So

$$S_{\iota(c)} S_{\tau(x)} = [\sum_{w \in G} S_c \delta^u_w S_{f(x,y)} \delta^w_{xy}]_{u,y \in G} = [S_c f(x,y) \delta^u_{xy}]_{u,y \in G}$$

and thus

$$\sum_{c \in Z} c S_{\iota(c)} S_{\tau(x)} = [\sum_{c \in Z} c S_c f(x,y)^{-1} \delta^u_{xy}]_{u,y \in G}$$
$$= \sum_{c \in Z} c S_c \otimes [f(x,y)^{-1} \delta^u_{xy}]_{u,y \in G}.$$

Since

$$\sum_{c \in Z} c S_c = [ab^{-1}]_{a,b \in Z}$$

(see Lemma 3.7.4), we then get

$$\sum_{c \in Z} c S_{\iota(c)} \sum_{x \in G} h(x) S_{\tau(x)} = \sum_{c \in Z} c S_c \otimes \sum_{x \in G} h(x) [f(x,y)^{-1} \delta^u_{xy}]_{u,y \in G}$$
$$= [ab^{-1}]_{a,b \in Z} \otimes [f(uy^{-1},y)^{-1} h(uy^{-1})]_{u,y \in G},$$

as required. $\qquad \square$

PROOF OF THEOREM 15.2.1. Using (15.2) we find that

(15.4) $$\gamma \beta \beta^{(*)} = \gamma \sum_{x \in G} h(x) \tau(x)$$

where

(15.5) $$h(x) = \sum_{y \in G} g(xy) g(y)^* f(x,y).$$

Next, if $\alpha \in \mathcal{R}[E]$ is an E-associate of $E(D)$ then

$$\alpha \alpha^{(*)} = \sum_{e \in E} b(e) e \qquad \text{where} \qquad \sum_{e \in E} b(e) S_e = E(D) E(D)^*,$$

as per the proof of Theorem 10.5.5.

Suppose that D is a PCD(Λ). By Theorem 15.1.3, (15.4), and the fact that $\gamma^2 = |Z| \gamma$,

$$|Z| \sum_{c \in Z} c S_{\iota(c)} \sum_{x \in G} h(x) S_{\tau(x)} = E(D) E(D)^*.$$

Now as we saw in Subsection 9.6.1,

$$E(D) E(D)^* = |Z| [rc^{-1}]_{r,c \in Z} \otimes DD^*.$$

Thus, (15.3) holds by Lemma 15.2.2.

For the converse, we calculate that DD^* has (u,w)th entry

$$\sum_{y \in G} f(u,y) g(uy) g(wy)^* f(w,y)^{-1} = \sum_y g(uw^{-1}y) g(y)^* f(u, w^{-1}y) f(w, w^{-1}y)^{-1}$$
$$= f(uw^{-1}, w)^{-1} \sum_y g(uw^{-1}y) g(y)^* f(uw^{-1}, y)$$
$$= f(uw^{-1}, w)^{-1} h(uw^{-1}),$$

by the cocycle identity and (15.5). So D is a PCD(Λ) if (15.3) holds. $\qquad \square$

15.2.3. EXAMPLE. In Example 15.1.2, by (15.3) we must have $h(1) = x_1^2 + x_2^2$ and $h(a) = 0$. Thus $\gamma\beta\beta^{(*)} = \gamma(x_1^2 + x_2^2)$. Also $[f(xy^{-1}, y)^{-1}h(xy^{-1})]_{x,y\in\langle a\rangle} \in \mathrm{Gram}_{\mathcal{R}}(\Lambda) = \{(x_1^2 + x_2^2)I_2\}$, as Theorem 15.2.1 predicts.

15.3. The familiar designs

We apply Theorem 15.2.1 to the PCD(Λ)s of Chapter 2 (with $\mathrm{Gram}_{\mathcal{R}}(\Lambda)$ non-empty and $|Z| > 1$).

Assume the notation of Theorem 15.2.1; so

$$\gamma\beta\beta^{(*)} = \gamma h$$

where

$$h = \sum_{x\in G} h(x)\tau(x)$$

is one of finitely many possibilities determined by Λ, i.e.,

$$[f(xy^{-1}, y)^{-1}h(xy^{-1})]_{x,y\in G} \in \mathrm{Gram}_{\mathcal{R}}(\Lambda).$$

For all Λ and $\mathcal{R}$ as in Example 6.1.3, $\mathrm{Gram}_{\mathcal{R}}(\Lambda)$ is a singleton. When this element is aI_v for some $a \in \mathcal{R}$,

$$h(x) = \begin{cases} a & \text{if } x = 1 \\ 0 & \text{if } x \neq 1. \end{cases}$$

Balanced generalized weighing matrices are a little more interesting. Here

$$\mathrm{Gram}_{\mathcal{R}}(\Lambda) = kI_v + \frac{\lambda}{|H|}H(J_v - I_v)$$

and we take $H = Z$ to be abelian. So

$$h(x) = \begin{cases} k & \text{if } x = 1 \\ \frac{\lambda}{|H|}H & \text{if } x \neq 1, \end{cases}$$

and thus

$$\beta\beta^{(*)}\gamma = \gamma k + \frac{\lambda}{|H|}\gamma H \sum_{x\in G\setminus\{1\}} \tau(x).$$

The right-hand side of this equation simplifies further:

$$\gamma H{\textstyle\sum_{x\in G\setminus\{1\}}} \tau(x) = H{\textstyle\sum_{c\in H}} \iota(c) \cdot {\textstyle\sum_{x\in G\setminus\{1\}}} \tau(x) = H(E \setminus \iota(H)).$$

The above discussion yields

15.3.1. THEOREM. *Let* (15.1) *be a central short exact sequence with transversal map* $\tau : G \to E$. *In each case below, there exists a cocyclic PCD(Λ) with cocycle* $f_{\iota,\tau}$ *if and only if there is an element* $\beta = \sum_{x\in G} g(x)\tau(x)$ *of* $\mathcal{R}[E]$ *with* $(0, \mathcal{A})$-*coefficients (or* $(\mathcal{A})$-*coefficients if the arrays in* Λ *do not contain zeros) satisfying the stated group ring equation (H is abelian in* (8)–(10)).

  (1) H(n): $\beta\beta^{(-1)}\gamma = n\gamma$ *over* $\mathbb{Z}[E]$, *where* $\gamma = 1 - \iota(-1)$.
  (2) W(v, k): $\beta\beta^{(-1)}\gamma = k\gamma$ *over* $\mathbb{Z}[E]$, *where* $\gamma = 1 - \iota(-1)$.
  (3) BW(v, k, λ): $\beta\beta^{(-1)}\gamma = k\gamma + \frac{\lambda}{2}H(E\setminus\iota(H))$ *over* $\mathbb{Z}[H][E]$, *where* $H = \langle z\rangle$, $z^2 = 1$, *and* $\gamma = 1 + z\iota(z)$.
  (4) OD(n; $a_1, \ldots, a_r$): $\beta\beta^{(*)}\gamma = \gamma\sum_{i=1}^r a_i x_i^2$ *over* $\mathbb{Z}[x_1, \ldots, x_r][E]$, *where* $\gamma = 1 - \iota(-1)$.
  (5) CH(n): $\beta\beta^{(*)}\gamma = n\gamma$ *over* $\mathbb{Z}[\mathrm{i}][E]$, *where* $\gamma = (1 - \iota(-1))(1 + \mathrm{i}\iota(\mathrm{i}))$.
  (6) CGH(n; m): $\beta\beta^{(*)}\gamma = n\gamma$ *over* $\mathbb{Z}[\zeta_m][E]$, *where* $\gamma = \sum_{j=0}^{m-1} \zeta_m^j \iota(\zeta_m)^j$.

(7) CGW$(v, k; m)$: $\beta\beta^{(*)}\gamma = k\gamma$ over $\mathbb{Z}[\zeta_m][E]$, where $\gamma = \sum_{j=0}^{m-1} \zeta_m^j \iota(\zeta_m)^j$.

(8) GH$(n; H)$: $\gamma\beta\beta^{(*)} = n\gamma$ over $(\mathbb{Z}[H]/\mathbb{Z}H)[E]$, where $\gamma = \sum_{h \in H} h\iota(h)$.

(9) GW$(v, k; H)$: $\beta\beta^{(*)}\gamma = k\gamma$ over $(\mathbb{Z}[H]/\mathbb{Z}H)[E]$, where $\gamma = \sum_{h \in H} h\iota(h)$.

(10) BGW$(v, k, \lambda; H)$: $\beta\beta^{(*)}\gamma = k\gamma + \frac{\lambda}{|H|}H(E \setminus \iota(H))$ over $\mathbb{Z}[H][E]$, where $\gamma = \sum_{h \in H} h\iota(h)$.

15.4. Cocyclic designs and relative difference sets

A relative difference set in a group E is *central* if its forbidden subgroup lies in $Z(E)$. (So central and normal relative difference sets are the same thing when the forbidden subgroup has order 2.) This section shows that certain cocyclic designs are equivalent to central relative difference sets. Results obtained later in this book thereby imply the existence of large families of relative difference sets.

Let H be a finite abelian group.

15.4.1. THEOREM (Cf. [**67**], Theorem 5.1). *A cocyclic* BGW$(v, k, \mu|H|; H)$ *with cocycle* $f : G \times G \to H$ *is equivalent to a central relative difference set of size k in $E = E(f)$ with forbidden subgroup $\{(1, h) \mid h \in H\} \cong H$ and index μ.*

PROOF. Let $\iota = \iota_0$ and $\tau = \tau_0$ be the canonical injection and transversal map. By Theorem 15.3.1 (10), first suppose that $\gamma = \sum_{h \in H} h\iota(h)$ and $\beta = \sum_{x \in \bar{G}} a_x\tau(x)$, where $\bar{G} \subseteq G$ and $a_x \in H$, satisfy

$$(15.6) \qquad \beta\beta^{(*)}\gamma = \gamma k + \mu H(E \setminus \iota(H)).$$

The homomorphism $E \times H \to E$ defined by $\tau(x)\iota(h_1) \cdot h_2 \mapsto \tau(x)\iota(h_1 h_2^{-1})$ for h_1, $h_2 \in H$ and $x \in G$ extends to a ring homomorphism ρ from $\mathbb{Z}[H][E] = \mathbb{Z}[H \times E]$ onto $\mathbb{Z}[E]$. Applying ρ to (15.6), we get

$$\rho(\beta)\rho(\beta)^{(-1)} = k + \mu(E - \iota(H))$$

where

$$\rho(\beta) = \sum_{x \in \bar{G}} \iota(a_x)^{-1}\tau(x).$$

This is the group ring equation for the relative difference set $\{\iota(a_x)^{-1}\tau(x) \mid x \in \bar{G}\}$ in E with central forbidden subgroup $\iota(H)$ and index μ (see (10.8)).

Conversely, a central relative difference set as stated is a subset of a transversal of $\iota(H)$ in E. Let $\beta \in \mathbb{Z}[E]$ denote the sum of the elements of the difference set, so that $\beta\beta^{(-1)} = k + \mu(E - \iota(H))$ holds in $\mathbb{Z}[E]$. Multiplying both sides of this equation by γ gives (15.6) in $\mathbb{Z}[H][E]$. $\square$

15.4.2. COROLLARY (Cf. [**129**], Theorem 4.1). *There is a cocyclic* GH$(n; H)$ *with cocycle f if and only if there is a central relative difference set in $E(f)$ of size n with forbidden subgroup $\{(1, h) \mid h \in H\}$ and index $n/|H|$.*

PROOF. Each GH$(n; H)$ is a BGW$(n, n, n; H)$. $\square$

15.4.3. COROLLARY. *A cocyclic Hadamard matrix of order n with cocycle f is equivalent to a central relative difference set of size n in $E(f)$ with forbidden subgroup $\langle(1, -1)\rangle$ and index $n/2$.*

PROOF. This is simply Corollary 15.4.2 with $H = \langle-1\rangle$. $\square$

15.4.4. COROLLARY. *Let $n = a_1 + \cdots + a_l$ where $a_i \geq 1$ for all i. If there is a cocyclic $OD(n; a_1, \ldots, a_l)$ with cocycle f then there are $2l$ mutually disjoint subsets*

$$X_1^{(0)}, X_1^{(1)}, X_2^{(0)}, X_2^{(1)}, \ldots, X_l^{(0)}, X_l^{(1)}$$

of $E(f)$ where $|X_i^{(0)}| = |X_i^{(1)}| = a_i$, such that for all $(c_1, \ldots, c_l) \in \{0, 1\}^l$,

$$X(c_1, \ldots, c_l) = \bigcup_{i=1}^{l} X_i^{(c_i)}$$

is a relative difference set with central forbidden subgroup $\langle(1, -1)\rangle$ and index $n/2$.

PROOF. Let D be a cocyclic $OD(n; a_1, \ldots, a_l)$ with indeterminates $x_1, \ldots, x_l$. For each choice of $(c_1, \ldots, c_l)$, we obtain a cocyclic Hadamard matrix from D by setting $x_i = (-1)^{c_i}$. The result then follows from Corollary 15.4.3.   □

15.5. Normal p-complements

Let G be a finite group. If $G = M \rtimes P$ for some $M \trianglelefteq G$ and Sylow p-subgroup P, then M is said to be a *normal p-complement*. This (entirely group-theoretic) section presents assorted results that tell us when G has a normal p-complement. The focus is on $p = 2$, in preparation for Section 15.6. However, knowledge of p-complements for arbitrary p is likely to be useful in the analysis of the group ring equations identified in this chapter and Chapter 10.

15.5.1. LEMMA. *A finite group G has at most one normal p-complement.*

PROOF. If M and N are normal p-complements then $N/(N \cap M) \cong MN/M$ is a p-group. But $N/(N \cap M)$, as a quotient of N, has order coprime to p. Thus $N = N \cap M = M$.   □

The proof of Lemma 15.5.1 shows that a normal p-complement is the unique subgroup (not just normal subgroup) of its order.

15.5.2. LEMMA. *If $N_G(P)/C_G(P)$ is a p-group for every p-subgroup P of G, then G has a normal p-complement.*

PROOF. See [**77**, Theorem 4.5, p. 253].   □

Let $K \leq G$. The map $N_G(K) \to \text{Aut}(K)$ defined by $x \mapsto \theta_x$, where $\theta_x(k) = xkx^{-1}$, is a homomorphism with kernel $C_G(K)$. Thus $N_G(K)/C_G(K)$ is isomorphic to a subgroup of $\text{Aut}(K)$. So by Lemma 15.5.2 we have

15.5.3. LEMMA. *If $\text{Aut}(P)$ is a p-group for every p-subgroup P of G, then G has a normal p-complement.*

15.5.4. COROLLARY. *If G has a cyclic Sylow 2-subgroup, then it has a normal 2-complement.*

PROOF. Since the automorphism group of a cyclic p-group of order p^l has order $p^{l-1}(p-1)$, the claim follows from Lemma 15.5.3.   □

The next lemma is another simple criterion for $p = 2$.

15.5.5. LEMMA. *Suppose that G is a finite group with dihedral Sylow 2-subgroup S such that $S \cap Z(G) \neq 1$. Then G has a normal 2-complement.*

PROOF. A subgroup T of S is either cyclic, dihedral of order greater than 4, or elementary abelian of order 4. In all cases but the last, the automorphism group of T is a 2-group. Let T be elementary abelian of order 4. Then T must contain a central involution of G, and $x \in N_G(T)$ can only permute the other two involutions in T. Hence $N_G(T)/C_G(T)$ divides 2. By Lemma 15.5.2, we are done. $\square$

The rest of this section is geared toward Theorem 15.5.8 below.

15.5.6. LEMMA. *An automorphism of a finite p-group P that has order coprime to p and induces the identity automorphism on $P/\mathrm{Frat}(P)$ is the identity.*

PROOF. See [**77**, Theorem 1.4, p. 174]. $\square$

15.5.7. LEMMA. *Let P be a finite p-group, and let*

$$1 = N_0 \leq N_1 \leq \cdots \leq N_k = P$$

be a series of normal subgroups of P. If ϕ is an automorphism of P such that, for all i, $\phi(N_i) = N_i$ and ϕ induces the identity on N_i/N_{i-1}, then $|\phi|$ is a power of p.

PROOF. See [**77**, Corollary 3.3, p. 179]. $\square$

The following theorem generalizes Corollary 15.5.4, but is not quite as strong as Lemma 15.5.5 in the special case of that lemma.

15.5.8. THEOREM. *Let G be a finite group with center Z and Sylow 2-subgroup S. Suppose that $|G : S|$ is coprime to $|\mathrm{GL}(k, 2)|$, where k is the length of a polycyclic series of $S/S \cap Z$. Then G contains a normal 2-complement.*

PROOF. We will show that $N_G(S_0)/C_G(S_0)$ is a 2-group for every subgroup S_0 of S.

Suppose that $xC_G(S_0)$ is a non-trivial element of $N_G(S_0)/C_G(S_0)$ of odd order. Let θ be the automorphism of S_0 that is conjugation by x, i.e., $\theta(s) = x^{-1}sx$. So the order $m > 1$ of θ in $\mathrm{Aut}(S_0)$ divides $|N_G(S_0) : C_G(S_0)|$, and thus divides the odd part $|G : S|$ of $|G|$.

Each term of the series $1 \leq S_0 \cap Z \leq S_0$ is a θ-invariant normal subgroup of S_0. Since m is odd, and θ acts trivially on $S_0 \cap Z$, by Lemma 15.5.7 the induced action of θ on $\bar{S}_0 := S_0/S_0 \cap Z$ is not trivial. It follows from Lemma 15.5.6 that the order m' of the automorphism induced by θ on $\bar{S}_0/\mathrm{Frat}(\bar{S}_0)$ is a non-trivial divisor of $|\mathrm{Aut}(\bar{S}_0/\mathrm{Frat}(\bar{S}_0))|$.

Now $\bar{S}_0 \cong S_0 Z/Z$ is isomorphic to a subgroup of $S/S \cap Z \cong SZ/Z$. Hence $\bar{S}_0$ has a polycyclic series of length $\leq k$, and so $\bar{S}_0/\mathrm{Frat}(\bar{S}_0)$ is an elementary abelian 2-group of rank $\leq k$. Its automorphism group is therefore isomorphic to a subgroup of $\mathrm{GL}(k, 2)$. Then, since the divisor m' of m divides $|\mathrm{GL}(k, 2)|$, we get the contradiction that $|G : S|$ is not coprime to $|\mathrm{GL}(k, 2)|$. $\square$

15.6. Existence conditions for cocyclic Hadamard matrices

In this section we give conditions on the structure of an extension group E of a cocyclic Hadamard matrix. These results, which exclude two possibilities for the isomorphism type of a Sylow 2-subgroup of E, come from Ito's pioneering paper [**100**]. The proof of each result is a very good example of how arguments within the relevant group ring can yield necessary conditions for the existence of a cocyclic design.

15.6.1. Ito's first non-existence result.

15.6.1. THEOREM (Cf. [**100**], Proposition 7). *Suppose that a cocyclic Hadamard matrix with cocycle f has extension group $E(f) = N \rtimes \langle s \mid s^{2^{k+1}} = 1 \rangle$, $k > 1$, where $s^{2^k} = (1, -1)$. Then $|N|$ is even.*

PROOF. Put $s^{2^k} = z$, and denote the indexing group of the matrix by G. By part (1) of Theorem 15.3.1, there is a group ring element

$$\beta = \sum_{b \in N} \left(\sum_{i=0}^{2^k-1} \lambda_{b,i} s^i \right) b$$

with (± 1)-coefficients satisfying

$$(1 - z)\beta\beta^{(-1)} = (1 - z)|G|.$$

Thus

$$(15.7) \qquad (1 - z) \sum_{b \in N} \left(\sum_{i,j=0}^{2^k-1} \lambda_{b,i}\lambda_{b,j} s^{i-j} \right) = (1 - z)|G|.$$

The coefficient of s on the left-hand side of (15.7) is

$$\sum_{b \in N} \left(-\lambda_{b,0}\lambda_{b,2^k-1} + \sum_{i=1}^{2^k-1} \lambda_{b,i}\lambda_{b,i-1} \right) = 0.$$

Now the number of disparities $\lambda_{b,i} \neq \lambda_{b,i-1}$ for $1 \leq i \leq 2^k - 1$ must be even or odd, depending on whether $\lambda_{b,0} = \lambda_{b,2^k-1}$ or $\lambda_{b,0} \neq \lambda_{b,2^k-1}$, respectively. The number m_b of terms in

$$c_b = -\lambda_{b,0}\lambda_{b,2^k-1} + \sum_{i=1}^{2^k-1} \lambda_{b,i}\lambda_{b,i-1}$$

equal to -1 is therefore odd. So $c_b = 2^k - 2m_b \equiv 2 \pmod 4$ and $2|N| \equiv \sum_{b \in N} c_b \equiv 0 \pmod 4$. Thus $|N|$ is even, as claimed. □

15.6.2. COROLLARY. *Suppose that H is a cocyclic Hadamard matrix of order greater than 2. If the indexing group G of H has a cyclic Sylow 2-subgroup, then H is group-developed modulo G. Equivalently, the extension group of H cannot have cyclic Sylow 2-subgroups.*

PROOF. Let S be a Sylow 2-subgroup of G. By Corollary 15.5.4, $G = S \ltimes N$ for some 2-complement N. There are only two possibilities for a Sylow 2-subgroup of an extension E of $\langle z \rangle \cong C_2$ by G: either it is cyclic, or it splits (as $S \times \langle z \rangle$). The first possibility is ruled out by Theorem 15.6.1, because $|N|$ is odd. The second possibility implies that the cocycle is a coboundary, and then the matrix is group-developed modulo G. □

15.6.3. REMARK. In particular, a cocyclic Hadamard matrix indexed by a cyclic group of order greater than 2 must be group-developed, hence circulant. There is an old conjecture that circulant Hadamard matrices do not exist beyond order 4. This has been verified for all orders less than or equal to 10^{11}, with three possible exceptions (see the paper [**140**] by Schmidt).

15.6.2. Ito's second non-existence result.

15.6.4. THEOREM (Cf. [**100**], Proposition 6). *Suppose that a cocyclic Hadamard matrix has extension group $E = E(f) = N \rtimes S$, where S is a dihedral 2-group of order greater than 4 containing the central involution $z = (1, -1)$ of E. Then $|N|$ is even.*

PROOF. We have $S = \langle a, b \mid a^{2^k} = b^2 = 1, a^b = a^{-1} \rangle$ for some $k \geq 2$. Then $z = a^{2^{k-1}}$, and the element

$$\beta = \sum_{i=0}^{2^{k-1}-1} a^i \sum_{x \in N} \lambda_{i,x} x + \sum_{i=0}^{2^{k-1}-1} a^i b \sum_{x \in N} \mu_{i,x} x$$

of $\mathbb{Z}[E]$ with (± 1)-coefficients satisfies

(15.8) $$(1 - z)\beta\beta^{(-1)} = 2^k |N|(1 - z).$$

Let $\zeta \in \mathbb{C}$ be a primitive 2^kth root of unity, and define the representation $\phi : E \to \mathrm{GL}(2, \mathbb{C})$ by

$$\phi(x) = I_2 \ \forall\, x \in N, \qquad \phi(a) = \begin{bmatrix} \zeta & 0 \\ 0 & \zeta^{-1} \end{bmatrix}, \qquad \phi(b) = \begin{bmatrix} 0 & 1 \\ 1 & 0 \end{bmatrix}.$$

Note that $\phi(z) = -I_2$. We extend ϕ to a ring homomorphism $\mathbb{Z}[E] \to \mathrm{Mat}(2, \mathbb{C})$. Applying ϕ to (15.8) gives

$$2^k |N| I_2 = \begin{bmatrix} X & Y \\ Y^{(-1)} & X \end{bmatrix},$$

where

$$X = \left(\sum_{i=0}^{2^{k-1}-1} \zeta^i \sum_{x \in N} \lambda_{x,i}\right)\left(\sum_{i=0}^{2^{k-1}-1} \zeta^{-i} \sum_{x \in N} \lambda_{x,i}\right)$$
$$+ \left(\sum_{i=0}^{2^{k-1}-1} \zeta^i \sum_{x \in N} \mu_{x,i}\right)\left(\sum_{i=0}^{2^{k-1}-1} \zeta^{-i} \sum_{x \in N} \mu_{x,i}\right)$$

and

$$Y = 2\left(\sum_{i=0}^{2^{k-1}-1} \zeta^i \sum_{x \in N} \lambda_{x,i}\right)\left(\sum_{i=0}^{2^{k-1}-1} \zeta^{-i} \sum_{x \in N} \mu_{x,i}\right).$$

But the minimal polynomial of ζ has degree 2^{k-1}; so

$$\sum_{x \in N} \lambda_{x,i} = 0 \ \forall\, i \qquad \text{or} \qquad \sum_{x \in N} \mu_{x,i} = 0 \ \forall\, i.$$

Since $\lambda_{x,i}, \mu_{x,i} \in \{\pm 1\}$, in either case $|N|$ is even. $\square$

By Lemma 15.5.5, we get

15.6.5. COROLLARY. *No cocyclic Hadamard matrix has an extension group whose Sylow 2-subgroups are dihedral.*

15.7. Cyclotomic rings and circulant complex Hadamard matrices

The nth cyclotomic field $\mathbb{Q}(\zeta_n)$ is the subfield of $\mathbb{C}$ generated by $\mathbb{Q}$ and the primitive nth root of unity $\zeta_n = e^{2\pi i/n}$. Algebraic arguments in cyclotomic fields (see, e.g., [**8**]) led to the earliest existence conditions for difference sets in abelian groups. Relying on the paper [**3**] by Arasu, de Launey, and Ma, we describe some of these ideas, and use them to obtain necessary conditions for the existence of circulant complex Hadamard matrices.

A reference for the standard algebraic number theory in this section is [**112**].

15.7.1. Preamble. Circulant complex Hadamard matrices of order n are known to exist only for $n = 2, 4, 8, 16$.

15.7.1. EXAMPLE. The circulants with first rows

$$[1, i],$$
$$[1, -i, 1, i],$$
$$[1, 1, i, 1, 1, -1, i, -1],$$
$$[1, 1, i, -i, i, 1, 1, i, -1, 1, -i, -i, -i, 1, -1, i]$$

are complex Hadamard matrices.

Let $C = \langle c \mid c^4 = 1 \rangle$ and let n be an even positive integer. Theorem 10.5.8 and (10.8) imply that a circulant complex Hadamard matrix of order n exists only if there is a solution D with $(0, 1)$-coefficients of the equation

$$(15.9) \qquad\qquad DD^{(-1)} = n(1 - c^2) + n(C \times C_n)$$

in $\mathbb{C}[C \times C_n]$. The following consequence of Theorem 2.7.8 further restricts n.

15.7.2. THEOREM. *If there is a circulant complex Hadamard matrix of order n, then n is a sum of two squares.*

Arasu, de Launey, and Ma [3] proved

15.7.3. THEOREM. *For all $n \leq 1000$ and $n \notin \{2, 4, 8, 16\}$ there does not exist a circulant complex Hadamard matrix of order n, except possibly for $n \in \{260, 340, 442, 468, 520, 580, 680, 754, 820, 884, 890\}$.*

The next few subsections reproduce some results from [3] that validate part of Theorem 15.7.3. In light of this theorem, and the conjecture that there are no circulant real Hadamard matrices of order greater than 4, it is tempting to posit

5. CONJECTURE. *There is a circulant complex Hadamard matrix of order n if and only if $n \in \{2, 4, 8, 16\}$.*

15.7.2. Characters and Fourier inversion. Let G be a finite abelian group of exponent m. Recall that a *character* of G is a homomorphism from G into $\langle \zeta_m \rangle$. The characters of G form a group isomorphic to G, whose elements we denote χ_g, for g ranging over G. The identity element is the *principal character*, which maps all elements of G to 1. Each character of G extends to a homomorphism from $\mathbb{C}[G]$ into $\mathbb{C}$.

For $g \in G$, define

$$\hat{g} = \frac{1}{|G|} \sum_{x \in G} \chi_g(x^{-1}) x.$$

Then, for any $a = \sum_{x \in G} a_x x \in \mathbb{C}[G]$,

$$\sum_{g \in G} \chi_g(a) \hat{g} = \frac{1}{|G|} \sum_{g \in G} \sum_{x \in G} a_x \chi_g(x) \sum_{y \in G} \chi_g(y^{-1}) y$$
$$= \frac{1}{|G|} \left(\sum_{g \in G} \sum_{x \neq y} \chi_g(xy^{-1}) a_x y + \sum_{g \in G} \sum_{x \in G} a_x x \right)$$
$$= a,$$

because $\sum_{g \in G} \chi_g(x) = 0$ if $x \in G \setminus \{1\}$. The identity

$$a = \sum_{g \in G} \chi_g(a)\hat{g}$$

is the *Fourier inversion formula*.

15.7.3. Analysis of the group ring equation.

15.7.4. LEMMA ([**3**], Lemma 2.1). *Let G be an abelian group of order v, m be a positive integer, and $\{p_1, \ldots, p_r\}$ be the set of prime divisors of $\gcd(m, v)$. Suppose that each p_i-subgroup of G is cyclic, $i = 1, \ldots, r$, and denote the subgroup of G of order p_i by Q_i. If y is an element of $\mathbb{Z}[G]$ such that $\chi(y) \equiv 0 \pmod{m}$ for all characters χ of G that are non-principal on $Q_1, \ldots, Q_r$, then*

(15.10) $$y = mx_0 + \sum_{i=1}^{r} Q_i x_i$$

for some $x_0, x_1, \ldots, x_r \in \mathbb{Z}[G]$.

PROOF. Let s_i be a generator of the Sylow p_i-subgroup of G, say $|s_i| = p_i^{e_i}$, and denote the subgroup of G of order $w := v/(p_1^{e_1} \cdots p_r^{e_r})$ by H. Let $\psi : \mathbb{Z}[G] \to \mathbb{Z}[\zeta_{v/w}][H]$ be the ring homomorphism such that $\psi(s_i) = \zeta_{p_i^{e_i}}$ and $\psi(h) = h$ for $h \in H$. If χ is a character of H then $\chi \circ \psi$ is a character of G that is non-principal on $Q_1, \ldots, Q_r$. By the Fourier inversion formula, and hypothesis,

$$w\psi(y) = \sum_{h,k \in H} \chi_h(\psi(y))\chi_h(k^{-1})k \equiv 0 \pmod{m}.$$

Let a, b be integers such that $aw + bm = 1$; then $\psi(y) = aw\psi(y) + bm\psi(y) \equiv 0 \pmod{m}$. Now (15.10) follows, because ψ has kernel $\{\sum_{i=1}^{r} Q_i x_i \mid x_i \in \mathbb{Z}[G]\}$. $\square$

The ring of integers $\mathbb{Z}[\zeta_m]$ of $\mathbb{Q}(\zeta_m)$ is a Dedekind domain. Thus, each non-zero ideal of $\mathbb{Z}[\zeta_m]$ factors, essentially uniquely, into a product of prime ideals. Recall that in a Dedekind domain, for one ideal A to contain another ideal B is the same as A dividing B, i.e., $B = AC$ for some ideal C. Also, if P is a prime ideal dividing a product AB of ideals then P divides A or P divides B.

An ideal of $\mathbb{Z}[\zeta_m]$ that is invariant under complex conjugation is said to be *self-conjugate*. With respect to our current problem, it is useful to know when complex conjugation in $\mathbb{Z}[\zeta_w]$ fixes all the prime ideal divisors of the ideal $\langle m \rangle$.

15.7.5. DEFINITION. Let p, w be positive integers, where p is prime. Write $w = p^s u$ where $s \geq 0$ and u is coprime to p. Then p is *self-conjugate modulo w* if there is an integer r such that $p^r \equiv -1 \pmod{u}$. An arbitrary integer m is self-conjugate modulo w if all of its prime divisors are self-conjugate modulo w.

15.7.6. THEOREM. *Let m, w be positive integers. If m is self-conjugate modulo w then every prime ideal of $\mathbb{Z}[\zeta_w]$ dividing $\langle m \rangle$ is self-conjugate.*

PROOF. This is well-known; cf. [**141**], pp. 20–21]. $\square$

15.7.7. THEOREM ([**3**], Theorem 2.2). *Suppose that there is a circulant complex Hadamard matrix of order $n = wu$, and let m^2 be a divisor of $2n$ such that*

    (i) *if m is odd then m is self-conjugate modulo $4w/\gcd(4, w)$;*

    (ii) *if m is even then m is self-conjugate modulo $4w$;*

    (iii) *if m and w are even then $u \equiv 0 \pmod{4}$.*

Then $m \leq 2^{r-1}u$, where r is the number of prime divisors of $\gcd(m, 4w)$.

PROOF. Let $G = \langle c \mid c^4 = 1 \rangle \times \langle h \mid h^n = 1 \rangle$. Define

$$K = \begin{cases} G/\langle h^w \rangle & \text{if } m \text{ or } w \text{ is odd} \\ G/\langle ch^w \rangle & \text{otherwise.} \end{cases}$$

In the first case $K \cong C_4 \times C_w$; in the second, $K \cong C_{4w}$ (by (iii), $|ch^w| = u$). So the Sylow p-subgroup of K is cyclic for p dividing $\gcd(m, 4w)$. Further, if m is odd then K has exponent $e = 4w/\gcd(4, w)$, whereas $e = 4w$ if m is even. Thus m is self-conjugate modulo e by (i) and (ii).

Let $\rho : G \to K$ be the natural projection, and let χ be a character of K. From (15.9) we get

$$\chi(\rho(D)\rho(D)^{(-1)}) = n\chi\rho(1 - c^2) + nu\chi(K).$$

Since $|c^2| = 2$, $\chi\rho(c^2) = \pm 1$. Thus 2 divides $\chi\rho(1 - c^2)$. Also 2 divides $\chi(K)$, which is 0 or $4w$. Therefore, $2n$ divides $\chi(\rho(D)\rho(D)^{(-1)})$; and, because m^2 divides $2n$,

$$\chi(\rho(D)\rho(D)^{(-1)}) \equiv 0 \pmod{m^2}.$$

So if P is a prime ideal of $\mathbb{Z}[\zeta_e]$ in the factorization of $\langle m \rangle$, then P must divide the product of the ideals $\langle \chi(\rho(D)) \rangle$ and $\langle \overline{\chi(\rho(D))} \rangle$. Since m is self-conjugate modulo e, P divides $\langle \chi(\rho(D)) \rangle$ by Theorem 15.7.6. It follows that

$$\chi(\rho(D)) \equiv 0 \pmod{m}.$$

By (15.10) we now see that

$$\rho(D) = mx_0 + \sum_{i=1}^{r} Q_i x_i$$

for some $x_0, x_1, \ldots, x_r \in \mathbb{Z}[K]$ and where, for each prime p_i dividing $\gcd(m, 4w)$, $Q_i = \langle s_i \rangle$ denotes the subgroup of K of order p_i. Hence

$$\rho(D) \prod_{i=1}^{r} (1 - s_i) = mx$$

for some $x \in \mathbb{Z}[K]$. The left-hand side of this equation cannot be 0, so $x \neq 0$. Then by comparing coefficients on both sides we get $m \leq 2^{r-1}u$ as required. $\square$

The following pretty fact was first proved by Turyn [152].

15.7.8. THEOREM. *If there is a circulant complex Hadamard matrix of order 2^s, then $s \leq 4$.*

PROOF. Let $u = 4$, $w = 2^{s-2}$, and $m = 2^e$, where $e = \lfloor (s+1)/2 \rfloor$. Assuming $s \geq 3$, Theorem 15.7.7 (iii) holds. Since $r = 1$, $2^e \leq 4$. $\square$

No prime congruent to 3 (mod 4) can divide the square-free part of n as in Theorem 15.7.2. Therefore any such prime divisor of n implies a constraint arising from Theorem 15.7.7. The next result (Theorem 2.4 of [3]) provides a constraint of this kind.

15.7.9. THEOREM. *If there exists a circulant complex Hadamard matrix of order $n = p^f u$, where $p \equiv 3 \pmod 4$ is prime and u is coprime to p, then f is even and $p^{f/2} \leq u$.*

PROOF. Let $\rho : \langle c \mid c^4 = 1 \rangle \times C_n \to \langle c \rangle$ be projection onto the first factor such that $\rho(c) = c$, and let χ be the character of $\langle c \rangle$ sending c to i. By (15.9), $a\bar{a} = 2n$ where $a = \chi\rho(D)$. Since -1 is a quadratic non-residue modulo p, p is inertial, i.e., p is a prime element of $\mathbb{Z}[i]$. So p divides both a and $\bar{a}$, implying that the largest power of p dividing $a\bar{a} = 2p^f u$ is even. Since p is self-conjugate modulo $4p^f$, the rest of the theorem follows from Theorem 15.7.7 with $w = p^f$ and $m = p^{f/2}$. $\square$

One may also derive constraints on n from its prime divisors congruent to 1 modulo 4. The following combines Corollaries 2.14 and 2.17 of [**3**].

15.7.10. THEOREM. *Let f be a positive integer. Suppose that $p \equiv 1 \pmod 4$ is prime, and there exists a circulant complex Hadamard matrix of order $p^f u$, or of order $2p^f u$ where u is odd and 2 is self-conjugate modulo p^f. Then $p^e \leq 2u^2 - 2u + 1$ where $e = \lfloor f/2 \rfloor$ if $f \geq 2$ and $e = 1$ if $f = 1$.*

Turyn's other non-existence result for circulant complex Hadamard matrices is a consequence of the preceding two theorems.

15.7.11. THEOREM. *There is no circulant complex Hadamard matrix of order $2p^f$, where $f \geq 1$ and p is an odd prime.*

PROOF. For $p \equiv 3 \pmod 4$, take $u = 2$ in Theorem 15.7.9. For $p \equiv 1 \pmod 4$, take $u = 2$ or $u = 1$ in Theorem 15.7.10. $\square$

15.7.4. Schmidt's descent method. We now come to an important method due to Schmidt [**138**], which concerns the solutions a in $\mathbb{Z}[\zeta_m]$ of the *norm equation*

$$a\bar{a} = k$$

for a positive integer k. Schmidt proved that if m has square divisors, then very often all solutions are of the form $\zeta_m^i b$, where $b \in \mathbb{Z}[\zeta_{F(m,k)}]$, $F(m,k)$ having the same prime divisors as m.

We discuss how $F(m,k)$ is computed. Let $\prod_{i=1}^t p_i^{c_i}$ be the prime factorization of m, and suppose that k has prime divisors $q_1, \ldots, q_s$. For $j = 1, \ldots, s$ let

$$m_j = \begin{cases} \prod_{p_i \neq q_j} p_i & \text{if } m \text{ is odd or } q_j = 2 \\ 4 \prod_{p_i \neq 2, q_j} p_i & \text{otherwise.} \end{cases}$$

Next define the b_i to be the least positive integers such that for all pairs (i,j), $i = 1, \ldots, t$ and $j = 1, \ldots, s$, at least one of the following is satisfied:

- $q_j = p_i$ and $(p_i, b_i) \neq (2,1)$,
- $b_i = c_i$,
- $q_j \neq p_i$ and $q_j^{\text{ord}_{m_j}(q_j)} \not\equiv 1 \pmod{p_i^{b_i+1}}$

where $\text{ord}_{m_j}(q_j)$ is the order of $q_j \pmod{m_j}$. Then

$$F(m,k) = \prod_{i=1}^t p_i^{b_i}.$$

The descent method is most powerful when the b_i are small.

15.7.12. EXAMPLE. We compute $F(200, 10)$. Here $c_1 = 3$, $c_2 = 2$, $q_1 = 2$, $q_2 = 5$, $m_1 = 5$, and $m_2 = 4$. If $b_1 < 3$ then $(2, b_1) \neq (2,1)$ and $5 \not\equiv 1 \pmod{2^{b_1+1}}$. Thus $b_1 = 2$. Similarly $b_2 = 1$, because $(5,1) \neq (2,1)$ and $2^{\text{ord}_5(2)} = 2^4 \not\equiv 1 \pmod{5^2}$. So $F(200, 10) = 20$. In fact $F(2^{3+a}5^{2+b}, 2^c 5^d) = 20$ for all $a, b, c, d \geq 0$.

15.7.13. THEOREM (Schmidt [**140**]). *Let* $a = \sum_{i=0}^{m-1} a_i \zeta_m^i \in \mathbb{Z}[\zeta_m]$ *with* $0 \le a_i \le C$ *for some constant* C, *and suppose that* $a\bar{a}$ *is a positive integer* k. *Then*

$$k \le \frac{C^2 F(m,k)^2}{4\varphi(F(m,k))}$$

where φ *is the Euler totient function.*

Schmidt's theorem gives rise to another necessary condition for existence of a circulant complex Hadamard matrix. Let $G = \langle g \mid g^4 = 1 \rangle \times \langle h \mid h^n = 1 \rangle$, and let χ be a character of G of maximal order mapping g to i. From (15.9) we obtain

(15.11) $$\chi(D)\overline{\chi(D)} = 2n$$

in $\mathbb{Z}[\zeta_m]$, where

$$m = \begin{cases} n & \text{if } n \equiv 0 \pmod{4} \\ 2n & \text{if } n \equiv 2 \pmod{4}. \end{cases}$$

Write

$$D = a + gb + g^2 c + g^3 d$$

where $a, b, c, d \in \mathbb{Z}[\langle h \rangle]$. Then

$$\chi(D) = \chi(a - c) + i\chi(b - d).$$

Since D is a transversal, $a - c$ and $b - d$ have $(1, -1)$-coefficients. When $n \equiv 2 \pmod{4}$, $i \notin \mathbb{Z}[\zeta_n]$. So in this case $\chi(D)$ has $(1, -1)$-coefficients, and we may use Theorem 15.7.13 with $a = \chi(D)$, $k = 2n$, and $C = 2$. When $n \equiv 0 \pmod{4}$, $\chi(D)$ has $(0, \pm 2)$-coefficients, and we may divide both sides of (15.11) by 4. Then we apply Theorem 15.7.13 with $k = n/2$, $a = \chi(D)/2$, and $C = 2$. Thus

15.7.14. THEOREM. *Suppose that there exists a circulant complex Hadamard matrix of order* n. *Then*

$$n \le \begin{cases} 2F(n, n/2)^2/\varphi(F(n, n/2)) & \text{if } n \equiv 0 \pmod{4} \\ F(2n, 2n)^2/2\varphi(F(2n, 2n)) & \text{if } n \equiv 2 \pmod{4}. \end{cases}$$

15.7.15. EXAMPLE. We prove that there is no circulant complex Hadamard matrix of order $n = 2^{3+a} 5^{2+b}$ ($a, b \ge 0$). By Example 15.7.12, $F(n, n/2) = 20$. So

$$2F(n, n/2)^2/\varphi(F(n, n/2)) = 2.(20)^2/8 = 100 < n,$$

and the assertion is justified by Theorem 15.7.14.

This section has shown the relevance to design theory of methods for solving norm equations $a\bar{a} = k$. The paper [**90**] gives such a method in the case that $a \in \mathbb{Z}[\zeta_p]$, p prime. Further research on this problem is needed.

7. RESEARCH PROBLEM. *Devise algorithms for solving norm equations.*

15.8. Composition of cocyclic associates

This final section of the chapter, a sequel to Section 10.8, shows how cocyclic associates may be used to make larger cocyclic designs from smaller ones.

15.8.1. DEFINITION. Suppose that G is a finite group and Z is an abelian group, where $G = G_1 \cdot G_2$ and $Z = Z_1 Z_2$ for subgroups G_1, G_2 of G and Z_1, Z_2 of Z. If $f : G \times G \to Z$ is a cocycle such that $f(G_i \times G_i) \subseteq Z_i$, $i = 1, 2$, then we call f a *composition cocycle* for the restricted cocycles $f_i = f_{|G_i \times G_i} : G_i \times G_i \to Z_i$.

15.8.2. EXAMPLE. We find the composition cocycles that restrict to cocycles of the two small designs in Example 8.1.1. Let $G_1 = \langle \alpha \rangle \cong C_2$ and $G_2 = \langle \beta \rangle \cong C_2$. First note that

$$\begin{bmatrix} 1 & i \\ i & 1 \end{bmatrix} = [f_1(a,b)g_1(ab)]_{a,b \in G_1} \quad \text{and} \quad \begin{bmatrix} x_1 & x_2 \\ x_2 & -x_1 \end{bmatrix} = [f_2(a,b)g_2(ab)]_{a,b \in G_2},$$

where f_1 is the trivial cocycle and

$$[f_2(a,b)]_{a,b \in G_2} = \begin{bmatrix} 1 & 1 \\ 1 & -1 \end{bmatrix}.$$

By definition $G_1 \cap G_2 = 1$, so $G = G_1 \times G_2$. We may also take $Z = Z_1 = \langle i \rangle$ and $Z_2 = \langle -1 \rangle$. All cocycles $f : G \times G \to Z$ are specified by the development table

$$\begin{bmatrix} 1 & 1 & 1 & 1 \\ 1 & A & E^{-1} & AE \\ 1 & B^{-1} & F & BF \\ 1 & AB & EF & ABEF \end{bmatrix}.$$

Here $A, B, E, F \in \langle i \rangle$, $E = \pm B$, and rows and columns are indexed $1, \alpha, \beta, \alpha\beta$. If f is a composition cocycle that restricts to f_1 and f_2, then $A = f(\alpha, \alpha) = f_1(\alpha, \alpha) = 1$ and $F = f(\beta, \beta) = f_2(\beta, \beta) = -1$. Hence

$$(15.12) \qquad \begin{bmatrix} 1 & 1 & 1 & 1 \\ 1 & 1 & E^{-1} & E \\ 1 & B^{-1} & -1 & -B \\ 1 & B & -E & -BE \end{bmatrix}$$

displays all eight possibilities for f. The extension group $E(f)$ has presentation

$$\langle a, b, z \mid a^2 = 1, \ b^2 = z^2, \ z^4 = 1, \ b^a = bz^\epsilon, \ z^a = z^b = z \rangle, \quad \epsilon \in \{0, 2\}.$$

That is, we obtain just two non-isomorphic extension groups for the eight different composition cocycles: $C_4 \times C_2^2$ and $(C_4 \times C_2) \rtimes C_2$.

With Definition 6.3.5 and Section 10.8 (particularly Theorem 10.8.2) in mind, we now define an orthogonality set and ambient ring that may be used in the composition of cocyclic associates. For $i = 1, 2$, let Λ_i be an orthogonality set with alphabet $\mathcal{A}_i$ and ambient ring $\mathcal{R}_i$ containing the central group Z_i. Suppose that $\mathcal{R}$ is an involutory ring containing $\mathcal{R}_1$ and $\mathcal{R}_2$ as involutory subrings that centralize each other. Hence $Z = Z_1 Z_2$ is central in $\mathcal{R}$. Let

$$\text{wt}_i = \{[XX^*]_{11} \mid X \in \Lambda_i\} \quad \text{and} \quad \text{df}_i = \{[XX^*]_{12} \mid X \in \Lambda_i\}.$$

Notice that $Z_i \text{df}_i = \text{df}_i$ $(i = 1, 2)$. The orthogonality set $\Lambda_1 \curlywedge_Z \Lambda_2$ with alphabet $\mathcal{A} = \{a_1 a_2 \mid a_i \in \mathcal{A}_i\}$ (deleting 0 if it appears) consists of all $2 \times v_1 v_2$ $(0, \mathcal{A})$-arrays X with distinct rows such that

$$[XX^*]_{12} \in Z_1 \text{wt}_1 \text{df}_2 \cup Z_2 \text{df}_1 \text{wt}_2 \cup \text{df}_1 \text{df}_2 \quad \text{and} \quad [XX^*]_{11}, [XX^*]_{22} \in \text{wt}_1 \text{wt}_2.$$

To state and prove our composition theorem for cocyclic associates, we will assume that $\mathcal{R}$ is an ambient ring for $\Lambda_1 \curlywedge_Z \Lambda_2$ with central group Z. We also make the usual assumptions that each orthogonality set in the composition has the Gram Property over its ambient ring; and that in each ambient ring the row group, column group, and central group coincide.

15.8.3. THEOREM. *Let $\mathcal{R}_1$, $\mathcal{R}_2$, $\mathcal{R}$, and $\Lambda = \Lambda_1 \curlywedge_Z \Lambda_2$ be as above. Let $f = f_{\iota,\tau}$ be a cocycle of the central extension*

$$1 \longrightarrow Z \xrightarrow{\iota} E \xrightarrow{\pi} G \longrightarrow 1.$$

Suppose that $G = G_1 \cdot G_2$ and $Z = Z_1 Z_2$ where $f(G_i \times G_i) \subseteq Z_i$, $i = 1, 2$. Write

$$\iota_i = \iota_{|Z_i}, \quad \pi_i = \pi_{|E_i}, \quad \tau_i = \tau_{|G_i}, \quad E_i = \langle \tau(G_i), \iota(Z_i) \rangle \le E.$$

Then

    (a) *the restriction $f_i : G_i \times G_i \to Z_i$ of f is the cocycle f_{ι_i,τ_i} of the central extension*

$$1 \longrightarrow Z_i \xrightarrow{\iota_i} E_i \xrightarrow{\pi_i} G_i \longrightarrow 1;$$

    (b) $E = E_1 \diamond_{Z_1 \cap Z_2} E_2.$

Next, put

$$\gamma_i = \sum_{z \in Z_i} z\iota_i(z), \qquad \beta_i = \sum_{x \in G_i} g_i(x)\tau_i(x)$$

and

$$\gamma = \sum_{z \in Z} z\iota(z), \qquad \beta = \sum_{x_1 \in G_1} \sum_{x_2 \in G_2} g(x_1 x_2)\tau(x_1 x_2)$$

where $g(x_1 x_2) = f(x_1, x_2)^{-1} g_1(x_1) g_2(x_2)$. Finally, suppose that

    (1) $\alpha_i = \gamma_i \beta_i \in \mathcal{R}_i[E_i]$ *is an f_i-associate of a* PCD(Λ_i) *for $i = 1, 2$, and*

    (2) α_1 *commutes with $\alpha_2 \alpha_2^{(*)}$ in $\mathcal{R}[E]$.*

Then

    (c) *the array $D = [f(x, y)g(xy)]_{x,y \in G}$ (if it has distinct rows) is a* PCD(Λ).

PROOF. For parts (a) and (b), see Lemma 12.2.2. Now by Theorem 15.2.1,

$$\gamma_i \beta_i \beta_i^{(*)} = \gamma_i \sum_{x \in G_i} h_i(x)\tau_i(x)$$

where

$$[f_i(uy^{-1}, y)^{-1} h_i(uy^{-1})]_{u,y \in G_i} \in \mathrm{Gram}_{\mathcal{R}_i}(\Lambda_i).$$

This implies that

(15.13) $h_i(x) \in \begin{cases} \mathrm{wt}_i & \text{if } x = 1 \\ \mathrm{df}_i & \text{if } x \ne 1. \end{cases}$

Similarly, defining $h : G \to \{0\} \cup \mathcal{A}$ by

$$\gamma \beta \beta^{(*)} = \gamma \sum_{x \in G} h(x)\tau(x),$$

we have that D is a PCD(Λ) if and only if

$$[f(uy^{-1}, y)^{-1} h(uy^{-1})]_{u,y \in G} \in \mathrm{Gram}_{\mathcal{R}}(\Lambda).$$

By the definition of Λ and (15.13), then, it suffices to show that

(15.14) $h(x_1 x_2) = f(x_1, x_2)^{-1} h_1(x_1) h_2(x_2)$

for all $x_1 \in G_1$ and $x_2 \in G_2$.

    Wielding (15.2) we get

$$\gamma\beta = \gamma \sum_{x_1 \in G_1} \sum_{x_2 \in G_2} \iota(f(x_1, x_2)) g_1(x_1) g_2(x_2)\tau(x_1 x_2)$$

$$= \gamma \sum_{x_1 \in G_1} \sum_{x_2 \in G_2} g_1(x_1) g_2(x_2)\tau(x_1)\tau(x_2)$$

$$= \gamma \beta_1 \beta_2.$$

Therefore

$$\gamma\beta\beta^{(*)} = \gamma\beta_1\beta_2\beta_2^{(*)}\beta_1^{(*)} = \frac{1}{|Z_1 \cap Z_2|}\,\gamma_1\gamma_2\beta_1\beta_2\beta_2^{(*)}\beta_1^{(*)}.$$

Since $\alpha_1 = \gamma_1\beta_1$ commutes with $\alpha_2\alpha_2^{(*)} = |Z_2|\gamma_2\beta_2\beta_2^{(*)}$, $\beta_1^{(*)}$ must commute with $\gamma_2\beta_2\beta_2^{(*)}$, and thus

$$\gamma\beta\beta^{(*)} = \frac{1}{|Z_1 \cap Z_2|}\,\gamma_1\sum_{x_1\in G_1} h_1(x)\tau(x_1) \cdot \gamma_2\sum_{x_2\in G_2} h_2(x)\tau(x_2)$$

$$= \gamma\sum_{x_1\in G_1} h_1(x_1)\tau(x_1) \cdot \sum_{x_2\in G_2} h_2(x_2)\tau(x_2)$$

$$= \gamma\sum_{x_1\in G_1} \sum_{x_2\in G_2} f(x_1,x_2)^{-1}h_1(x_1)h_2(x_2)\tau(x_1x_2).$$

It follows that

$$\gamma\sum_{x\in G} h(x)\tau(x) = \gamma\beta\beta^{(*)} = \gamma\sum_{x_1\in G_1} \sum_{x_2\in G_2} f(x_1,x_2)^{-1}h_1(x_1)h_2(x_2)\tau(x_1x_2).$$

Upon comparison of coefficients, (15.14) is proved. □

The following result generalizes Theorem 10.8.4. It is a direct application of Theorem 15.8.3. Note that all of our assumptions about the constituent ambient rings and orthogonality sets are in place for these familiar designs.

15.8.4. THEOREM. *Suppose that $f : G \times G \to Z$ is a cocycle where $G = G_1 \cdot G_2$, $Z = Z_1Z_2$, and $f(G_i \times G_i) \subseteq Z_i$. Let $f_i : G_i \times G_i \to Z_i$ be the cocycle restricted from f. For each triple D_1, D_2, D_3 below, if there are an f_1-developed design D_1 and an f_2-developed design D_2, then there is an f-developed design D_3.*

(1) $H(n_1), H(n_2), H(n_1n_2)$.
(2) $W(n_1, k_1), W(n_2, k_2), W(n_1n_2, k_1k_2)$.
(3) $CH(n_1), CH(n_2), CH(n_1n_2)$.
(4) $CGH(n_1; m_1), CGH(n_2; m_2), CGH(n_1n_2; \mathrm{lcm}(m_1, m_2))$.
(5) $CGW(n_1, k_1; m_1), CGW(n_2, k_2; m_2), CGW(n_1n_2, k_1k_2; \mathrm{lcm}(m_1, m_2))$.
(6) $GH(n_1; H), GH(n_2; H), GH(n_1n_2; H)$.
(7) $GW(n_1, k_1; H), GW(n_2, k_2; H), GW(n_1n_2, k_1k_2; H)$.

PROOF. Since for these designs $\mathrm{wt}(\Lambda_1 \curlywedge_Z \Lambda_2)$ and $\mathrm{df}(\Lambda_1 \curlywedge_Z \Lambda_2)$ are disjoint, $D_3 = D$ as in Theorem 15.8.3 (c) has distinct rows. Also, from Theorem 15.3.1 we see that $\alpha_2\alpha_2^{(*)}$ commutes with α_1 in $\mathcal{R}[E]$. □

Just as in Theorem 10.8.1, we can define the Kronecker product of two cocyclic designs; the product design has indexing group $G_1 \times G_2$ and cocycle $f_1 \times f_2$. However, the construction of Theorem 15.8.3 affords more freedom in composition. We may also compose across different families of designs.

15.8.5. EXAMPLE. Applying part (c) of Theorem 15.8.3 to Example 15.8.2, we get the cocyclic complex orthogonal design

$$\begin{bmatrix} x_1 & \mathrm{i}x_1 & x_2 & -x_2 \\ \mathrm{i}x_1 & x_1 & \mathrm{i}x_2 & \mathrm{i}x_2 \\ x_2 & -\mathrm{i}x_2 & -x_1 & -x_1 \\ -x_2 & -\mathrm{i}x_2 & x_1 & -x_1 \end{bmatrix}.$$

The cocycle here is given by (15.12) with $E = \mathrm{i}$ and $B = -\mathrm{i}$.

CHAPTER 16

Special Classes of Cocyclic Designs

This chapter is a bridge between the more theoretical earlier chapters (such as Chapters 9, 10, 14, 15), and the case studies that occupy a large part of the rest of the book. We review essential theory of cocyclic designs for Hadamard matrices, balanced weighing matrices, and orthogonal designs, in particular. We also prove some specialized results.

The expanded designs in this chapter all have the same form. For a matrix M, let

$$\mathcal{E}(M) = \begin{bmatrix} M & -M \\ -M & M \end{bmatrix},$$

and let A_M be the matrix obtained from $\mathcal{E}(M)$ by changing each entry -1 to 0 (with all other entries unchanged).

16.1. Cocyclic Hadamard matrices

A Hadamard matrix H has ambient ring $\mathbb{Z}$, with row group and column group both equal to $\langle -1 \rangle$. Thus, $\mathrm{Aut}(H)$ is the set of pairs (P, Q) of signed permutation matrices such that $PHQ^\top = H$. The associated design A_H is a $(0, 1)$-matrix, so $\mathrm{Aut}(\mathrm{A}_H) = \mathrm{PermAut}(\mathrm{A}_H)$ is the set of pairs (P, Q) of permutation matrices such that $P\mathrm{A}_H Q^\top = \mathrm{A}_H$.

16.1.1. THEOREM. *A $(1, -1)$-matrix H of order n is a Hadamard matrix if and only if A_H is a $\mathrm{GDD}(2n, n, n/2; 2)$.*

PROOF. See Theorem 9.9.3. □

The next section condenses Sections 9.6, 9.8, and 9.10 for balanced weighing matrices and thus for Hadamard matrices. There we remind ourselves that $\mathrm{Aut}(H)$ is isomorphic to $\mathrm{PermAut}(\mathcal{E}(H)) = \mathrm{Aut}(\mathrm{A}_H)$. With regard to Theorem 16.1.1, this yields a bijection between the rows of H and the point classes of A_H, such that swapping rows i, j of H swaps point classes i, j of A_H. Negating row i of H swaps the points in point class i of A_H.

We use the following handy terminology. Let $f : G \times G \to \langle -1 \rangle$ be a cocycle, and let M be a $(1, -1)$-matrix equivalent to $[f(x, y)g(xy)]_{x,y \in G}$ for some map $g : G \to \{1, -1\}$. We say that f is a cocycle of M. So every cocycle cohomologous to f is a cocycle of M. If M is Hadamard then f is a *Hadamard cocycle*; it is a *pure Hadamard cocycle* if $[f(x, y)]_{x,y \in G}$ is Hadamard. Each Hadamard cocycle is cohomologous to a pure Hadamard cocycle. A *Hadamard group*, according to Ito [100] and Flannery [65], is (any group isomorphic to) the extension group $E(f)$ of a Hadamard cocycle f. We caution the reader that other authors define 'Hadamard group' differently; see, e.g., the survey [60] by Dillon. Also, Horadam [87, p. 121] calls a pure Hadamard cocycle an *orthogonal cocycle*.

The following fact has been noted previously.

16.1.2. LEMMA. *If there exists a Hadamard coboundary $G \times G \to \langle -1 \rangle$ then there exists a G-developed Hadamard matrix, and hence $|G|$ is a square.*

PROOF. This is a consequence of Theorem 13.4.1 (ii) and Theorem 2.3.7.   □

The connection between cocyclic designs and relative difference sets is very tidy in the case of Hadamard matrices. Corollary 15.4.3 gives (i) $\Leftrightarrow$ (ii) in the next theorem, and (i) $\Leftrightarrow$ (iii) is the cocyclic specialization of Theorem 16.1.1 (via centrally regular actions on the expanded design).

16.1.3. THEOREM ([**47**], Theorem 2.4). *Let G be a group of order n, and let $f : G \times G \to \langle -1 \rangle$ be a cocycle. The following are equivalent.*

(i) *f is a Hadamard cocycle.*
(ii) *There is a central relative difference set of size n in $E(f)$ with forbidden subgroup $\langle (1, -1) \rangle$ and index $n/2$.*
(iii) *There is a GDD$(2n, n, n/2; 2)$ whose automorphism group contains $E(f)$ as a regular subgroup, with $\langle (1, -1) \rangle$ acting regularly on each point class.*

16.1.4. REMARK. Explicitly, $[f(x, y)g(xy)]_{x,y \in G}$ is Hadamard if and only if $\{(x, g(x)) \mid x \in G\} \subseteq E(f)$ is a central relative difference set (see the proof of Theorem 15.4.1). When f is a coboundary, [**47**, Theorem 2.6] further states that (i) is equivalent to there being a 'Menon-Hadamard' difference set in G.

Theorem 16.1.5 below is another example of the use of associates in composing cocyclic designs. We need some more notation (cf. Subsection 12.1.3). Let E be a finite group with central subgroup $\langle z \rangle \cong C_2$, and let

(16.1) $$1 \longrightarrow \langle -1 \rangle \overset{\iota}{\longrightarrow} E \overset{\pi}{\longrightarrow} G \longrightarrow 1$$

be a central short exact sequence, where ι maps -1 to z, and π is natural surjection of E onto $G = E/\langle z \rangle$. Denote the cohomology class of cocycles $f_{\iota,\tau}$ corresponding to (16.1) by $\mathcal{C}(E, z)$.

16.1.5. THEOREM. *Let $E = S \diamond_z T$. If there are Hadamard cocycles in $\mathcal{C}(S, z)$ and $\mathcal{C}(T, z)$, then there is a Hadamard cocycle in $\mathcal{C}(E, z)$.*

PROOF. We are supposing that there exist cocyclic associates $\gamma\beta_1$ and $\gamma\beta_2$ as per Theorem 15.3.1 (1), where $\gamma = 1 - z$,

$$\beta_1 = \sum_{x \in \mathcal{T}_1} a_x x, \qquad \text{and} \qquad \beta_2 = \sum_{y \in \mathcal{T}_2} b_y y,$$

for transversals $\mathcal{T}_1, \mathcal{T}_2$ of $\langle z \rangle$ in S, T respectively. Let $\beta_3 = \beta_1 \beta_2 \in \mathbb{Z}[S \diamond_z T]$; then

$$\beta_3 = \sum_{x \in \mathcal{T}_1} \sum_{y \in \mathcal{T}_2} a_x b_y xy = \sum_{x \in \mathcal{T}_3} c_x x,$$

where $\mathcal{T}_3 = \mathcal{T}_1 \mathcal{T}_2$ is a transversal of $\langle z \rangle$ in $S \diamond_z T$. Since $|\mathcal{T}_3| = |\mathcal{T}_1||\mathcal{T}_2|$, β_3 has $(0, \pm 1)$-coefficients. Then for $n_1 = |S|/2$ and $n_2 = |T|/2$,

$$\gamma\beta_3 \beta_3^{(-1)} = \gamma\beta_1\beta_2\beta_2^{(-1)}\beta_1^{(-1)} = \beta_1\gamma\beta_2\beta_2^{(-1)}\beta_1^{(-1)}$$
$$= \beta_1\gamma n_2 \beta_1^{(-1)} = n_2\gamma\beta_1\beta_1^{(-1)} = n_1 n_2 \gamma.$$

Thus β_3 is an associate of a cocyclic Hadamard matrix with indexing group E.   □

16.2. Cocyclic weighing matrices

Let W be a balanced weighing matrix, with parameters v, k, λ. That is, the entrywise product of W with itself is an incidence matrix of an SBIBD(v, k, λ). Hadamard matrices and conference matrices, and therefore all Paley matrices, are balanced weighing matrices: in the former case $v = k = \lambda$ (relaxing the constraint $v > k > \lambda$); and in the latter $k = v - 1$ and $\lambda = v - 2$.

Set

$$\text{(16.2)} \qquad \zeta = \left(\left[\begin{array}{cc} 0_v & I_v \\ I_v & 0_v \end{array} \right], \left[\begin{array}{cc} 0_v & I_v \\ I_v & 0_v \end{array} \right] \right).$$

For any $v \times v$ matrix M, $\zeta \in \mathrm{PermAut}(\mathcal{E}(M))$. A subgroup of $\mathrm{PermAut}(M)$ that contains the central involution ζ is said to *act normally*.

Let W be a weighing matrix. If $(M_1, M_2) \in \mathrm{Aut}(W)$ then $M_i = X_i - Y_i$ for unique disjoint $(0, 1)$-matrices X_i and Y_i. Let

$$\theta(M_i) = \left[\begin{array}{cc} X_i & Y_i \\ Y_i & X_i \end{array} \right].$$

By Theorem 9.6.12, $\Theta : \mathrm{Aut}(W) \to \mathrm{PermAut}(\mathcal{E}(W))$ defined by

$$\Theta((M_1, M_2)) = (\theta(M_1), \theta(M_2))$$

is an isomorphism. The element $\zeta = \Theta((-I_v, -I_v))$ is central in $\mathrm{PermAut}(\mathcal{E}(W))$. Under Θ, an automorphism of W that moves row i to row j corresponds to moving rows i and $v + i$ of $\mathcal{E}(W)$ to rows j and $v + j$ of $\mathcal{E}(W)$, respectively. Negating row i of W corresponds to swapping rows i and $v + i$ of $\mathcal{E}(W)$. Similar remarks apply to the relationship between the effect of $\mathrm{Aut}(W)$ on columns of W and the effect of $\mathrm{PermAut}(\mathcal{E}(W))$ on columns of $\mathcal{E}(W)$.

The map Θ has image in $\mathrm{PermAut}(A_W) = \mathrm{Aut}(A_W)$; hence $\zeta \in \mathrm{Aut}(A_W)$ and is central in $\Theta(\mathrm{Aut}(W))$. In fact $\mathrm{Aut}(W)$ is isomorphic via Θ to the centralizer of ζ in $\mathrm{Aut}(A_W)$. Moreover, if W is balanced then we have by Corollary 9.10.2 that Θ is an isomorphism between $\mathrm{Aut}(W)$ and $\mathrm{PermAut}(\mathcal{E}(W)) = \mathrm{Aut}(A_W)$; hence ζ is a central involution of $\mathrm{Aut}(A_W)$ in that case.

As we saw in Chapter 14, the weighing matrix W is cocyclic if and only if $\mathrm{PermAut}(\mathcal{E}(W))$ contains a centrally regular subgroup (regular subgroup acting normally). In more detail, $f : G \times G \to \langle -1 \rangle$ is a cocycle of W if and only if there is a centrally regular embedding of $E(f)$ in $\mathrm{PermAut}(\mathcal{E}(W))$; i.e., an isomorphism of $E(f)$ onto a regular subgroup that maps $(1, -1)$ to ζ. This characterization of cocyclic development suggests 'cocycle-free' proofs and generalizations of results, such as the following.

16.2.1. THEOREM (Cf. Theorem 11.2.12). *Suppose that H is an $n \times n$ $(1, -1)$-matrix with $\langle -1 \rangle$-distinct rows, such that the action of $\mathrm{PermAut}(\mathcal{E}(H))$ on the row indices of $\mathcal{E}(H)$ is transitive. Then H is Hadamard if and only if its first row and jth row are orthogonal, for all $j \neq 1$.*

PROOF. Suppose that the first row of H is orthogonal to its other rows. Denote the kth row vector of $\mathcal{E}(H)$ by r_k. Given i, $1 \leq i \leq n$, there are $P, Q \in \mathrm{Perm}(2n)$ such that $P\mathcal{E}(H)Q^\top = \mathcal{E}(H)$, and the first row of $P\mathcal{E}(H) = \mathcal{E}(H)Q$ is r_i. For $j \neq i$, $1 \leq j \leq n$, we have $r_j Q^\top = r_l$ for some $l \neq 1, n + 1$. Then

$$r_i r_j^\top = r_1 Q r_j^\top = r_1 r_l^\top = 0.$$

Thus, the ith and jth rows of H are orthogonal. $\qquad\square$

Let $f : G \times G \to \langle -1 \rangle$ be a cocycle. Existence of a cocyclic $\mathrm{BW}(v, k, \lambda)$ with cocycle f is equivalent to existence of a central relative difference set in $E(f)$ of size k and index $\lambda/2$, contained in the transversal $\{(x, 1) \mid x \in G\}$ of the forbidden subgroup $\langle (1, -1) \rangle$. Moreover, if D is any relative difference set in a group E with forbidden central subgroup $\langle z \rangle \cong C_2$, then there is a cocycle $f : K \times K \to \langle z \rangle$ and an isomorphism $\phi : E \to E(f)$ that maps D into $\{(x, 1) \mid x \in K\}$ and z to $(1, -1)$. As noted in Section 16.1, for Hadamard matrices H this equivalence is stronger: H is cocyclic with cocycle f if and only if $\{(x, 1) \mid x \in G\}$ is a relative difference set in $E(f)$ with forbidden subgroup $\langle (1, -1) \rangle$. Remember then that the Hadamard group $E(f)$ acts regularly and normally on $\mathcal{E}(H)$.

16.3. Cocyclic orthogonal designs

Cocyclic orthogonal designs are utilized in Chapters 18, 22, and 23. Here we set out some of the basic theory; more is given in Chapter 22.

Let D be an $\mathrm{OD}(n; a_1, \ldots, a_k)$ with indeterminate set $X = \{x_1, \ldots, x_k\}$. By Theorems 4.4.12 and 4.4.13, $\mathrm{Aut}(D)$ is the set of pairs (P, Q) of signed permutation matrices such that $PDQ^\top = D$. The design is cocyclic, with cocycle $f : G \times G \to \langle -1 \rangle$, if there are a map $h : G \to \{0\} \cup \pm X$ and signed permutation matrices P, Q such that $PDQ^\top = [f(x, y)h(xy)]_{x,y \in G}$.

Section 2.6 and Lemma 12.5.5 imply

16.3.1. THEOREM. *An $\mathrm{OD}(n; a_1, \ldots, a_k)$ with cocycle f is equivalent to a set $\{W_1, \ldots, W_k\}$ of pairwise disjoint anti-amicable cocyclic weighing matrices with respective weights $a_1, \ldots, a_k$ and cocycle f.*

16.3.2. REMARK. Setting each of the indeterminates x_i to either 1 or -1, we see that a cocyclic $\mathrm{OD}(n; a_1, \ldots, a_k)$ with cocycle f is equivalent to a set of 2^k cocyclic weighing matrices $\mathrm{W}(n, a_1 + \cdots + a_k)$ with cocycle f (each matrix is a ± 1-linear combination of the W_i in Theorem 16.3.1). When $a_1 + \cdots + a_k = n$, one obtains Hadamard matrices; so in that case a cocyclic orthogonal design corresponds to a large family of central relative difference sets with forbidden subgroup of size 2.

We provide a direct proof for the following special case of Theorem 14.7.1.

16.3.3. THEOREM. *An $\mathrm{OD}(n; a_1, \ldots, a_k)$, D, is cocyclic with cocycle f if and only if there is a centrally regular embedding of $E(f)$ in $\mathrm{PermAut}(\mathcal{E}(D))$.*

PROOF. Suppose that $D = [f(x, y)h(xy)]_{x,y \in G}$ for some $h : G \to \{0\} \cup \pm X$, and so

$$\mathcal{E}(D) = [abf(x, y)h(xy)]_{(x,a),(y,b) \in E(f)}$$
$$= [g((x, a)(y, b))]_{(x,a),(y,b) \in E(f)}$$

where the map g is defined by $g((x, a)) = ah(x)$. Hence

$$\varphi : (w, d) \mapsto ([\delta^{(x,a)(w,d)}_{(y,b)}]_{(x,a),(y,b) \in E(f)}, [\delta^{(w,d)(y,b)}_{(x,a)}]_{(x,a),(y,b) \in E(f)})$$

is a regular embedding of $E(f)$ in $\mathrm{PermAut}(\mathcal{E}(D))$. Moreover, since $\varphi((1, -1)) = \zeta$ as in (16.2), φ is centrally regular.

Conversely, if there is a centrally regular embedding of $E(f)$ in $\mathrm{PermAut}(\mathcal{E}(D))$, then there is a map $g : E(f) \to \{0\} \cup \pm X$ such that

$$\mathcal{E}(D) = [g((x, a)(y, b))]_{(x,a),(y,b) \in E(f)}$$

for some indexing of rows and (possibly different) indexing of columns of $\mathcal{E}(D)$ by $E(f)$. Since $(1,-1)$ maps to ζ under this embedding, $g((x,-1)) = -g((x,1))$ for all $x \in G$. Furthermore, the top left quadrant of $\mathcal{E}(D)$ must have rows indexed by the elements (x,a) as x runs completely and irredundantly over G; the columns will be indexed similarly. Defining $h : G \to \{0\} \cup \pm X$ by $h(x) = g((x,1))$, we then have

$$g((xy, f(x,y))) = h(xy)f(x,y).$$

Therefore, up to equivalence, $D = [f(x,y)h(xy)]_{x,y \in G}$. $\qquad\square$

We now consider systems of cocyclic orthogonal designs.

16.3.4. DEFINITION. Let $N = [n_{ij}]$ be a $k \times k$ symmetric $(0,1)$-matrix with zero diagonal. Let $\bigcup_{i=1}^{k} X_i$ be a partition of a set X of commuting indeterminates. Suppose that $D_1, \ldots, D_k$ are orthogonal designs of the same order, where D_i has indeterminate set X_i, such that

$$D_i D_j^\top = (-1)^{n_{ij}} D_j D_i^\top, \quad 1 \leq i, j \leq k.$$

Then $\{D_1, \ldots, D_k\}$ is called an N-concordant system of orthogonal designs. The system is cocyclic, with cocycle $f : G \times G \to \langle -1 \rangle$, if there are signed permutation matrices P and Q such that

$$PD_i Q^\top = [f(x,y)g_i(xy)]_{x,y \in G}, \quad 1 \leq i \leq k,$$

for some maps $g_i : G \to \pm X_i$.

Assuming Definition 16.3.4, write

$$D_i = \sum_{x_s \in X_i} x_s W_s.$$

The W_s are weighing matrices such that

$$W_s W_t^\top = (-1)^{q_{st}} W_t W_s^\top,$$

where

$$q_{st} = \begin{cases} 1 & \text{if } x_s, x_t \in X_i \text{ for some } i \\ n_{ij} & \text{if } x_s \in X_i \text{ and } x_t \in X_j \text{ for some } i \neq j. \end{cases}$$

By Lemma 12.5.5, if our system is cocyclic with cocycle f then (up to equivalence) we may simultaneously index the rows and columns of $D_1, \ldots, D_k$ by G so that

$$D_i = \sum_{a \in G} g_i(a) P_a^f.$$

Thus, for some maps $h_s : G \to \{0, \pm 1\}$,

$$W_s = \sum_{a \in G} h_s(a) P_a^f.$$

Using the faithful representation $(a,c) \mapsto P_{(a,c)}^f$ of $E(f)$, we then see that each W_s corresponds to a sum

$$w_s = \sum_{a \in G} h_s(a) \tau(a)$$

of elements of $E(f)$. The group ring elements w_s satisfy

$$w_s w_t^{(-1)} = (-1)^{q_{st}} w_t w_s^{(-1)}.$$

For later reference (in Chapter 22), we summarize the preceding discussion as a theorem.

16.3.5. THEOREM. *In the notation above, a cocyclic N-concordant system of orthogonal designs $\{D_1, \ldots, D_k\}$ with cocycle f corresponds to a set of elements w_s of $\mathbb{Z}[E(f)]$ such that $w_s w_t^{(-1)} = (-1)^{q_{st}} w_t w_s^{(-1)}$. If a_s is the number of non-zero terms in the sum w_s, then the indeterminate x_s appears exactly a_s times in D_i if and only if $x_s \in X_i$.*

16.4. A cocyclic substitution scheme

We show that if the template and plug-in matrices of a certain substitution scheme are cocyclic (i.e., in cocyclic form), then so too is the resulting array. This will have implications in Chapter 18 and elsewhere.

16.4.1. THEOREM. *Let $\mathcal{R}$ be a commutative ring, and let G, H, C be finite groups, where C is abelian and contained in the group of units of $\mathcal{R}$. Suppose that (under a common indexing of rows and columns by the elements of G)*

$$A_i = [f(x,y)g_i(xy)]_{x,y \in G}, \quad 1 \leq i \leq s,$$

where the g_i are maps $G \to \mathcal{R}$ and $f : G \times G \to C$ is a cocycle. Also suppose that

$$D = [h(u,v)g(uv)]_{u,v \in H}$$

where $g : H \to \{ra_1, \ldots, ra_s \mid r \in \mathcal{R}\} \subseteq \mathcal{R}[a_1, \ldots, a_s]$ is a map, and $h : H \times H \to C$ is a cocycle. Define E to be the block matrix obtained by replacing each a_i in D by A_i. Then

$$E = [(f \times h)(\alpha\beta)\bar{g}(\alpha\beta)]_{\alpha,\beta \in G \times H}$$

for some map $\bar{g} : G \times H \to \mathcal{R}$.

PROOF. We have

$$A_i = U^f \sum_{x \in G} g_i(x) P_x^f \quad \text{and} \quad D = U^h \sum_{u \in H} g(u) P_u^h$$

by Lemma 12.5.5. Write $g(u) = \epsilon_u a_{p(u)}$ where $p(u) \in \{1, \ldots, s\}$ and $\epsilon_u \in \mathcal{R}$. Define $\bar{g} : G \times H \to \mathcal{R}$ by $\bar{g}(x \cdot u) = \epsilon_u g_{p(u)}(x)$. Then by Lemma 12.5.6,

$$\begin{aligned}
E &= \sum_{u \in H} \epsilon_u \sum_{x \in G} g_{p(u)} U^f P_x^f \otimes U^h P_u^h \\
&= U^f \otimes U^h \sum_{u \in H} \sum_{x \in G} \epsilon_u g_{p(u)}(x) P_x^f \otimes P_u^h \\
&= U^{f \times h} \sum_{x \cdot u \in G \times H} \bar{g}(x \cdot u) P_{x \cdot u}^{f \times h}.
\end{aligned}$$

Another appeal to Lemma 12.5.5 completes the proof. □

The following is also needed in Chapter 18.

16.4.2. DEFINITION. *Let $X = \{x_1, \ldots, x_k\}$ be a set of commuting indeterminates. A set $\{A_i \mid 1 \leq i \leq m\}$ of $t \times t$ $(0, \pm X)$-matrices is said to be a type $(a_1, \ldots, a_k)$ cocyclic set of suitable matrices, with cocycle $f : G \times G \to \langle -1 \rangle$, if*

   (1) $A_1 A_1^\top + \cdots + A_m A_m^\top = \sum_{i=1}^k a_i x_i^2 I_t$;
   (2) $A_i A_j^\top = A_j A_i^\top$;
   (3) *there are maps $g_i : G \to \{0\} \cup \pm X$ such that (under a common row and column indexing) $A_i = [f(x,y)g_i(xy)]_{x,y \in G}$.*

When $k = 1$ we can replace the lone indeterminate by 1. When $m = 4$ and f is trivial, we say that the A_i are type $(a_1, \ldots, a_k)$ *Williamson-like* matrices developed modulo G.

Here is an example of using a plug-in construction to build a larger cocyclic design from smaller cocyclic designs with simpler structure.

16.4.3. PROPOSITION. *Let G be a group of order t, and suppose that $W, X, Y,$ Z are type $(a_1, \ldots, a_k)$ Williamson-like matrices developed modulo G. Then*

$$\begin{bmatrix} W & X & Y & Z \\ X & -W & Z & -Y \\ Y & -Z & -W & X \\ Z & Y & -X & -W \end{bmatrix}$$

is a cocyclic $\mathrm{OD}(4t; a_1, \ldots, a_k)$ *with extension group* $\mathrm{Q}_8 \times G$.

PROOF. The Williamson array D in the statement of the proposition is well-known to be cocyclic: $D = [f(x,y)g(xy)]_{x,y \in \langle a,b \rangle}$ where $\langle a,b \rangle \cong \mathrm{C}_2^2$, $g(1) = W$, $g(a) = X$, $g(b) = Y$, and $g(ab) = Z$. The extension group is non-abelian, so must be Q_8 (see, e.g., Corollary 15.6.5). By the definitions, plugging in the Williamson-like matrices certainly gives an $\mathrm{OD}(4t; a_1, \ldots, a_k)$. This is cocyclic with extension group $\mathrm{Q}_8 \curlyvee_{\mathrm{C}_2} (\mathrm{C}_2 \times G) = \mathrm{Q}_8 \times G$, by Lemma 12.2.1 and Theorem 16.4.1. $\square$

16.5. Cocyclic complex Hadamard matrices

Let $f : G \times G \to \langle \mathrm{i} \rangle$ be a cocycle, where G is a group of order n. A complex Hadamard matrix H is cocyclic, with cocycle f, if $PHQ^* = [f(x,y)g(xy)]_{x,y \in G}$ for some $P, Q \in \mathrm{Mon}(n, \langle \mathrm{i} \rangle)$ and map $g : G \to \{\pm 1, \pm \mathrm{i}\}$. We say that f is a *binary cocycle* of H if it maps into $\langle -1 \rangle$.

16.5.1. LEMMA. *Suppose that A and B are cocyclic complex Hadamard matrices with cocycles f and h respectively. Then $A \otimes B$ is a cocyclic complex Hadamard matrix with cocycle $f \times h$ and extension group $E(f) \curlyvee E(h)$.*

PROOF. This follows from Theorem 16.4.1 and Lemma 12.2.1. $\square$

The next result is a cocyclic analog of Theorem 2.7.7. It is used at a critical juncture in Chapter 23.

16.5.2. THEOREM. *Let $G_1 = \mathbb{Z}_2 \times G$ where G is a group of order n. Suppose that*

$$H = [f(x,y)g(xy)]_{x,y \in G}$$

where $f : G \times G \to \langle -1 \rangle$ is a cocycle and $g : G \to \{\pm 1, \pm \mathrm{i}\}$ is a set map. Define the cocycle $f_1 : G_1 \times G_1 \to \langle -1 \rangle$ by

$$f_1(ax, by) = (-1)^{ab} f(x,y), \quad a, b \in \mathbb{Z}_2, \ x, y \in G,$$

and define the $(0, \pm 1)$-matrices A and B via the equation $H = A + \mathrm{i}B$. Then

$$D = \begin{bmatrix} A + B & -A + B \\ A - B & A + B \end{bmatrix}$$

is a cocyclic Hadamard matrix of order $2n$ with cocycle f_1 if and only if H is a cocyclic complex Hadamard matrix with binary cocycle f.

PROOF. By Theorem 2.7.7 (and its proof), H is a complex Hadamard matrix if and only if D is a Hadamard matrix. Now

$$A = [f(x,y)g_A(xy)]_{x,y \in G} \qquad \text{and} \qquad B = [f(x,y)g_B(xy)]_{x,y \in G}$$

for maps $g_A, g_B : G \to \{0, \pm 1\}$. So by Theorem 16.4.1, under a common row and column indexing,

$$\begin{bmatrix} A & -A \\ A & A \end{bmatrix} \quad \text{and} \quad \begin{bmatrix} B & B \\ -B & B \end{bmatrix}$$

are both cocyclic with cocycle f_1. Hence the same is true of their sum.   □

CHAPTER 17

The Paley Matrices

This chapter is a study of the symmetries possessed by the Paley Hadamard and conference matrices. Discovered independently by Gilman [**75**] and Paley [**128**] in the 1930s, the Paley matrices form the densest known class of directly constructed Hadamard matrices.

De Launey and Stafford [**56**] determined the automorphism group of the Paley conference matrix by elementary means. Their argument ultimately rests on a result of Carlitz about permutations of a finite field. Kantor [**106**] was the first to obtain a description of the automorphism group of the type I Paley Hadamard matrix. Using the classification of 2-transitive permutation groups (and thereby the classification of finite simple groups), de Launey and Stafford [**57**] determined the automorphism group of the type II Paley Hadamard matrix. In this chapter, we describe these automorphism groups. However, we will not present the lengthy verification (given elsewhere) that all automorphisms have been found.

Each Paley matrix is a *cocyclic* weighing matrix. Knowledge of the centrally regular actions on their expanded designs can be beneficial in constructing relative difference sets and other combinatorial objects. For example, Turyn [**154**] used such an action to construct an infinite family of Williamson matrices; and regular actions for the conference matrix and type I Hadamard matrix enter into the proofs of Chapter 18. The Paley matrices and associated regular actions have been studied by Goethals and Seidel [**76**], Ito [**101**], Yamada [**163**], Yamamoto [**164**], and de Launey and Stafford [**55, 56, 57**].

This chapter is a précis of [**55, 56, 57**]. We refer the reader to those papers for more detailed information and proofs that are omitted here.

17.1. Actions of 2-dimensional linear and semilinear groups

Let $q = p^r$ for an odd prime p, and regard $\mathrm{GF}(q^2)$ as a vector space V over its subfield $\mathrm{GF}(q)$. We fix a $\mathrm{GF}(q)$-basis $\{b_1, b_2\}$ of V.

17.1.1. Two actions of $\mathrm{GL}(2, q)$ on V. Any element $x = x_1 b_1 + x_2 b_2$ of V may be written as a 2×1 matrix $[x_1, x_2]^\top$. For each $A \in \mathrm{GL}(2, q)$, we define $\pi_A \in \mathrm{Sym}(V)$ by

$$\pi_A(x) = Ax.$$

The embedding of $\mathrm{GL}(2, q)$ in $\mathrm{Sym}(V)$ given by $A \mapsto \pi_A$ yields the *standard action* of $\mathrm{GL}(2, q)$ on V (dependent on the choice of basis). Define $\phi_A \in \mathrm{Sym}(V)$ by

$$\phi_A(x) = \det(A)Ax.$$

Then the homomorphism $A \mapsto \phi_A$ yields the *non-standard action* of $\mathrm{GL}(2, q)$ on V.

17.1.2. Two actions of $\Gamma L(2, q)$ on V. Since the Frobenius map $x \mapsto x^p$ generates the automorphism group of $\mathrm{GF}(q)$, the general semilinear group $\Gamma L(2, q)$ is generated by $\mathrm{GL}(2, q)$ and the permutation

$$\sigma : \begin{bmatrix} x_1 \\ x_2 \end{bmatrix} \mapsto \begin{bmatrix} x_1^p \\ x_2^p \end{bmatrix}$$

of V. Indeed, $\Gamma L(2, q) = \mathrm{GL}(2, q) \rtimes \langle \sigma \rangle$. Here we are identifying $\mathrm{GL}(2, q)$ with its image in $\mathrm{Sym}(V)$ under the embedding $A \mapsto \pi_A$. Now for $A = [a_{ij}]_{ij}$, denote the matrix $[a_{ij}^p]_{ij}$ by $A^{(p)}$. Then the conjugation action of $\langle \sigma \rangle$ on $\mathrm{GL}(2, q)$ in $\Gamma L(2, q)$ is given by $\sigma A \sigma^{-1} = \sigma(A) := A^{(p)}\sigma$. Thus, the standard action of $\mathrm{GL}(2, q)$ on V extends to a standard (defining) action of $\Gamma L(2, q)$ on V. The non-standard action of $\mathrm{GL}(2, q)$ on V also extends to $\Gamma L(2, q)$, via the homomorphism $\phi : A\sigma^k \mapsto \phi_A \sigma^k$ from $\Gamma L(2, q)$ into $\mathrm{Sym}(V)$.

17.1.3. Actions on lines and half-lines. Let ν be a non-square element of $\mathrm{GF}(q)^\times$. The set V^* of non-zero elements of $\mathrm{GF}(q^2)$ is partitioned into $q+1$ distinct *lines* $[x] = \{\lambda x \mid \lambda \in \mathrm{GF}(q)^\times\}$. That is, $[x]$ consists of the non-zero elements of the 1-dimensional subspace of V spanned by x. Each line $[x]$ is further partitioned into two *half-lines* $[x]_+ = \{\lambda^2 x \mid \lambda \in \mathrm{GF}(q)^\times\}$ and $[x]_- = \nu[x]_+$.

The scalar groups $Z = \{\lambda I_2 \mid \lambda \in \mathrm{GF}(q)^\times\}$ and $Q = Z^2$ are normal in $\Gamma L(2, q)$. The orbits of Z (respectively, Q) in V^* are the lines (respectively, half-lines). So $\Gamma L(2, q)/Z$ (respectively, $\Gamma L(2, q)/Q$) has an action on the lines (respectively, half-lines) induced by the standard and non-standard actions of $\Gamma L(2, q)$ on V^*. Under the (faithful) action of $\Gamma L(2, q)/Q$ on the half-lines induced by the standard action of $\Gamma L(2, q)$ on V^*,

$$\pi_{A\sigma^k Q}([x]_+) = [A\sigma^k(x)]_+,$$

and under the non-standard induced action,

$$\phi_{A\sigma^k Q}([x]_+) = [\det(A)A\sigma^k(x)]_+.$$

Let $\mathrm{GS}(2, q)$ be the (normal) subgroup of index 2 in $\mathrm{GL}(2, q)$ consisting of the matrices with square determinant. Similarly, $\Gamma S(2, q) = \langle \sigma, \mathrm{GS}(2, q) \rangle$ has index 2 in $\Gamma L(2, q)$. The non-standard and standard actions of $\Gamma L(2, q)/Q$ on half-lines agree on $\Gamma S(2, q)/Q$:

$$\phi_{gQ}([x]_+) = \begin{cases} \pi_{gQ}([x]_+) & \text{if } g \in \Gamma S(2, q) \\ \nu\pi_{gQ}([x]_+) & \text{if } g \notin \Gamma S(2, q). \end{cases}$$

The standard action is 2-transitive on the set of half-lines.

17.1.4. Bilinear forms. Let $f : V \times V \to \mathrm{GF}(q)$ be any alternating bilinear form. For elements $x = x_1 b_1 + x_2 b_2$ and $y = y_1 b_1 + y_2 b_2$ of V,

$$f(x, y) = (x_1 y_2 - x_2 y_1)f(b_1, b_2).$$

This implies (with an abuse of notation) that f is a scalar multiple of the alternating bilinear form det, where

$$\det(x, y) = \begin{vmatrix} x_1 & y_1 \\ x_2 & y_2 \end{vmatrix} = x_1 y_2 - x_2 y_1.$$

It may be checked that

$$(17.1) \quad \det(Ax, Ay) = \det(A)\det(x, y) \quad \text{and} \quad \det(\sigma(x), \sigma(y)) = \det(x, y)^p.$$

17.1.1. LEMMA. *Let $\mathcal{I}$ be a complete irredundant set of line representatives. For each pair $[x]$, $[z]$ of distinct lines,*

(17.2) $$\{\det(x,y)/\det(z,y) \mid y \in \mathcal{I} \text{ and } [y] \neq [x],[z]\} = \mathrm{GF}(q)^{\times}.$$

PROOF. Denote the b_i-component of $a \in V$ by a_i, $i = 1,2$. For any $a \in V^*$, $\det(x,y) = a_1$ and $\det(z,y) = a_2$ if and only if

$$\begin{bmatrix} -x_2 & x_1 \\ -z_2 & z_1 \end{bmatrix} \begin{bmatrix} y_1 \\ y_2 \end{bmatrix} = \begin{bmatrix} a_1 \\ a_2 \end{bmatrix}.$$

The 2×2 matrix above is invertible, so y is determined by a. Now both a_1 and a_2 are non-zero if and only if $y \notin [x] \cup [z] \cup \{0\}$. Thus if $y,u \in V^* \setminus ([x] \cup [z])$ then $\det(x,y)/\det(z,y) = \det(x,u)/\det(z,u)$ if and only if $[y] = [u]$. So the left-hand side of (17.2) is a set of $|\mathcal{I}| - 2 = q - 1$ non-zero elements; i.e., is $\mathrm{GF}(q)^{\times}$. □

We continue with this section's notation for the rest of the chapter.

17.2. The Paley matrices and their automorphism groups

In this section we describe the three classes of Paley matrices. With a few exceptions, the full automorphism group of each matrix is given. These exceptions are the Paley Hadamard matrices of orders 4, 8, and 12. The matrices of orders 4 and 8 are the only Paley matrices that are equivalent to Sylvester Hadamard matrices. The matrix of order 12 is the unique Hadamard matrix that is both a type I and type II Paley matrix; its automorphism group is the unique (up to isomorphism) central extension of C_2 by the Mathieu group M_{12}.

Let $\mathcal{I}$ and $\mathcal{J}$ be complete irredundant sets of representatives for the lines and the half-lines in V^*, respectively.

17.2.1. The Paley conference matrix. Denote the *quadratic character* that maps $x \in \mathrm{GF}(q)$ to $x^{\frac{q-1}{2}} \in \{0,\pm1\}$ by χ. For $x,y \in V$, define

$$c(x,y) = \chi \det(x,y).$$

Then the Paley conference matrix of order $q + 1$ is

$$C = [c(x,y)]_{x,y \in \mathcal{I}}.$$

Since $c(\lambda x, y) = c(x, \lambda y) = \chi(\lambda)c(x,y)$ for any $\lambda \in \mathrm{GF}(q)^{\times}$, different choices of $\mathcal{I}$ lead to equivalent conference matrices. Also, replacing det by another bilinear form does not change the equivalence class of C.

Let $x,y \in \mathcal{I}$. Then $\det(x,y) = 0$ if and only if $x = y$; so the main diagonal of C is zero, and every off-diagonal entry is non-zero. We have

(17.3) $$C^{\top} = (-1)^{(q-1)/2}C,$$

so that C is symmetric if $q \equiv 1 \pmod 4$ and skew-symmetric if $q \equiv 3 \pmod 4$. If $y \neq x$ then $\sum_{z \in \mathcal{I}} c(x,z)c(y,z) = 0$ by (17.2). Thus

(17.4) $$CC^{\top} = qI_{q+1},$$

i.e., C really is a conference matrix.

We now describe the automorphism group $\mathrm{Aut}(C) \cong \mathrm{PermAut}(\mathcal{E}(C))$. Since $c(\nu x, y) = c(x, \nu y) = -c(x,y)$, we have

$$\mathcal{E}(C) = [c(x,y)]_{x,y \in \mathcal{J}}.$$

This definition is independent of the choice of $\mathcal{J}$. For $A \in \mathrm{GL}(2,q)$ and $x,y \in V^*$,

$$
\begin{aligned}
c(Ax, \det(A)Ay) &= \chi \det(Ax, \det(A)Ay) \\
&= \chi(\det(A)\det(Ax, Ay)) \\
&= \chi(\det(A)^2 \det(x,y)) \\
&= \chi(\det(A)^2)\chi \det(x,y) \\
&= c(x,y),
\end{aligned}
$$

using (17.1) in the third line. Also

$$
\begin{aligned}
c(\sigma(x), \sigma(y)) &= \chi \det(\sigma(x), \sigma(y)) \\
&= \chi(\det(x,y)^p) \\
&= \chi \det(x,y) \\
&= c(x,y).
\end{aligned}
$$

The above calculations imply that $\Gamma\mathrm{L}(2,q)/Q$ embeds into $\mathrm{PermAut}(\mathcal{E}(C))$, with the standard action on rows and non-standard action on columns. A proof that there are no other automorphisms of C may be found in [**56**, Section 4.3]. Thus

17.2.1. THEOREM. *The automorphism group of the Paley conference matrix of order $q+1$ is isomorphic to $\Gamma\mathrm{L}(2,q)/Q$.*

17.2.2. The type I Paley Hadamard matrix. The type I Paley matrix $H_1 = I_{q+1} + C$ is defined for $q \equiv 3 \pmod 4$ only. In this case $C^\top = -C$, so

$$
H_1 H_1^\top = I_{q+1} + CC^\top = (q+1)I_{q+1}.
$$

The expanded design is

$$
\mathcal{E}(H_1) = [c^*(x,y)]_{x,y \in \mathcal{J}}
$$

where

$$
c^*(x,y) = \begin{cases} c(x,y) & \text{if } [x] \neq [y] \\ \chi(x/y) & \text{if } [x] = [y]. \end{cases}
$$

Notice that $[A\sigma^k(x)] = [\det(A)A\sigma^k(y)]$ if and only if $[x] = [y]$. So if $[x] \neq [y]$, then

$$
\begin{aligned}
c^*(A\sigma^k(x), \det(A)A\sigma^k(y)) &= c(A\sigma^k(x), \det(A)A\sigma^k(y)) \\
&= c(x,y) \\
&= c^*(x,y).
\end{aligned}
$$

If $[x] = [y]$ then

$$
\begin{aligned}
c^*(A\sigma^k(x), \det(A)A\sigma^k(y)) &= \chi(A\sigma^k(x)/(\det(A)A\sigma^k(y))) \\
&= \chi(\det(A)^{-1})\chi(\sigma^k(x)/\sigma^k(y)) \\
&= \chi(\det(A))\chi((x/y)^{p^k}) \\
&= \chi(\det(A))\chi(x/y) \\
&= \chi(\det(A))c^*(x,y).
\end{aligned}
$$

Therefore

(17.5) $\qquad c^*(A\sigma^k(x), \det(A)A\sigma^k(y)) = \begin{cases} c^*(x,y) & \text{if } [x] \neq [y] \\ \chi(\det(A))c^*(x,y) & \text{if } [x] = [y]. \end{cases}$

It follows that

17.2.2. THEOREM. $\mathrm{Aut}(H_1)$ *contains a subgroup isomorphic to* $\Gamma\mathrm{S}(2, q)/Q$.

If $q > 11$ then $\Gamma\mathrm{S}(2, q)/Q$ is the full automorphism group of H_1: see Theorem 5.5 in [**56**].

17.2.3. The type II Paley Hadamard matrix. Let $q \equiv 1 \pmod 4$. The type II Paley matrix of order $2(q + 1)$ is

$$H_2 = \left[\begin{array}{cc} -I + C & I + C \\ I + C & I - C \end{array} \right].$$

This matrix is Hadamard by (17.3) and (17.4).

We index $\mathcal{E}(H_2)$ by the elements of $\mathcal{J} \times \{\pm 1\}$. A row or a column of $\mathcal{E}(H_2)$ indexed by an element $(x, 1)$ is *even*; the other rows and columns are *odd*. Define $f_1, f_2 : \{\pm 1\} \times \{\pm 1\} \to \{\pm 1\}$ by

$$f_1(a, b) = -1 \iff a = b = 1, \qquad f_2(a, b) = -1 \iff a = b = -1.$$

Then

$$\mathcal{E}(H_2) = [g(u, v)]_{u, v \in \mathcal{J} \times \{\pm 1\}},$$

where

$$g((x, a), (y, b)) = \left\{ \begin{array}{ll} f_1(a, b)\, c^*(x, y) & \text{if } [x] = [y] \\ f_2(a, b)\, c^*(x, y) & \text{if } [x] \neq [y]. \end{array} \right.$$

We now define two actions of $\Gamma\mathrm{L}(2, q)$ on $V \times \{\pm 1\}$. For $l \in \{0, 1\}$ put

$$\pi_{A\sigma^k}(x, (-1)^l) = (\det(A)^l A\sigma^k(x), (-1)^l),$$

$$\phi_{A\sigma^k}(x, a) = (\det(A) A\sigma^k(x), a\chi \det(A)).$$

The corresponding induced actions of $\Gamma\mathrm{L}(2, q)/Q$ on $\{([x]_+, a) \mid x \in \mathcal{J}, a = \pm 1\}$ are then given by

$$\pi_{A\sigma^k Q}([x]_+, (-1)^l) = ([\det(A)^l A\sigma^k(x)]_+, (-1)^l),$$

$$\phi_{A\sigma^k Q}([x]_+, a) = ([\det(A) A\sigma^k(x)]_+, a\chi \det(A)).$$

Note that the first of these induced actions is faithful.

We will show that

(17.6) $$g(\pi_{A\sigma^k}(x, a), \phi_{A\sigma^k}(y, b)) = g((x, a), (y, b)),$$

and hence that $\Gamma\mathrm{L}(2, q)/Q$ embeds into $\mathrm{PermAut}(\mathcal{E}(H_2))$. The next two lemmas are needed; their proofs follow from the definitions and (17.5).

17.2.3. LEMMA. *For* $a, b \in \{\pm 1\}$ *and* $A \in \mathrm{GL}(2, q)$,

$$f_1(a, b\chi \det(A)) = (\chi \det(A))^{(1+a)/2} f_1(a, b),$$

$$f_2(a, b\chi \det(A)) = (\chi \det(A))^{(1-a)/2} f_2(a, b).$$

17.2.4. LEMMA. *For all* $x, y \in V^*$, $\lambda \in \mathrm{GF}(q)^\times$, *and* $A \in \mathrm{GL}(2, q)$,

$$c^*(\lambda x, y) = c^*(x, \lambda y) = \chi(\lambda) c^*(x, y),$$

$$c^*(\sigma(x), \sigma(y)) = c^*(x, y),$$

$$c^*(Ax, Ay) = \left\{ \begin{array}{ll} c^*(x, y) & \text{if } [x] = [y] \\ \chi \det(A) c^*(x, y) & \text{if } [x] \neq [y]. \end{array} \right.$$

We embark upon the proof of (17.6). Lemma 17.2.4 implies that

$$(17.7) \quad c^*(\det(A)^l A\sigma^k(x), \det(A)A\sigma^k(y)) = \begin{cases} \chi \det(A)^{1+l} c^*(x,y) & \text{if } [x] = [y] \\ \chi \det(A)^l c^*(x,y) & \text{if } [x] \neq [y]. \end{cases}$$

Let $a = (-1)^l$. If $[x] = [y]$, then by (17.7) and Lemma 17.2.3 we have

$$g(\pi_{A\sigma^k}(x,a), \phi_{A\sigma^k}(y,b))$$
$$= g((\det(A)^l A\sigma^k(x), a), (\det(A)A\sigma^k(y), b\chi \det(A)))$$
$$= f_1(a, b\chi \det(A)) c^*(\det(A)^l A\sigma^k(x), \det(A)A\sigma^k(y))$$
$$= f_1(a, b)\chi \det(A)^{1-l} \chi \det(A)^{1+l} c^*(x,y)$$
$$= f_1(a, b) c^*(x,y)$$
$$= g((x,a), (y,b)).$$

Similarly, if $[x] \neq [y]$ then

$$g(\pi_{A\sigma^k}(x,a), \phi_{A\sigma^k}(y,b))$$
$$= f_2(a, b\chi \det(A)) c^*(\det(A)^l A\sigma^k(x), \det(A)A\sigma^k(y))$$
$$= f_2(a, b)\chi \det(A)^l \chi \det(A)^l c^*(x,y)$$
$$= f_2(a, b) c^*(x,y)$$
$$= g((x,a), (y,b)).$$

This completes the proof that $\operatorname{Aut}(H_2)$ contains an isomorphic copy of $\Gamma\mathrm{L}(2,q)/Q$.

We next show that $\operatorname{Aut}(H_2)$ contains an additional element ξ of order 4. Define

$$(17.8) \qquad\qquad \pi_\xi(x, (-1)^l) = (\nu^l x, (-1)^{l+1}),$$
$$(17.9) \qquad\qquad \phi_\xi(x, (-1)^l) = (\nu^{l+1} x, (-1)^{l+1}),$$

where as before $\nu \in \mathrm{GF}(q)^\times$ is a non-square. Then for $a = (-1)^k$, $b = (-1)^l$, and $i = 1$ or 2 as appropriate,

$$g(\pi_\xi(x,a), \phi_\xi(y,b)) = g((\nu^k x, (-1)^{k+1}), (\nu^{l+1} y, (-1)^{l+1}))$$
$$= f_i((-1)^{k+1}, (-1)^{l+1}) c^*(\nu^k x, \nu^{l+1} y)$$
$$= f_i((-1)^{k+1}, (-1)^{l+1}) \chi(\nu^{k+l+1}) c^*(x,y)$$
$$= f_i(a, b) c^*(x,y)$$
$$= g((x,a), (y,b)).$$

Except when $q = 5$, we have now seen a full set of generators for $\operatorname{Aut}(H_2)$.

17.2.5. THEOREM. *If $q > 5$ then $\operatorname{Aut}(H_2)$ is generated by a subgroup isomorphic to $\Gamma\mathrm{L}(2,q)/Q$, and an automorphism ξ as defined by (17.8) and (17.9).*

Theorem 17.2.5 is proved in [**57**]. The proof is by contradiction, and divides into three pieces:

- a combinatorial analysis showing that $\operatorname{Aut}(H_2)$ cannot be very large;
- an argument using the classification of 2-transitive permutation groups to prove that $\operatorname{Aut}(H_2)$ cannot be 2-transitive unless it is very large;
- a final step, showing that if $\operatorname{Aut}(H_2)$ is larger than the group specified in Theorem 17.2.5, then $\operatorname{Aut}(H_2)$ is 2-transitive.

Assuming Theorem 17.2.5, and with a certain looseness of notation, we discuss the structure of $\text{Aut}(H_2)$. First,

$$\pi_\xi^2(x, (-1)^l) = (\nu^{2l} \cdot \nu x, (-1)^l) \qquad \text{and} \qquad \phi_\xi^2(x, (-1)^l) = (\nu^{2l+2} \cdot \nu x, (-1)^l).$$

Thus

$$(17.10) \qquad\qquad \xi^2 = \zeta$$

where ζ is the automorphism corresponding to $\nu I_2 Q \in \Gamma\text{L}(2, q)/Q$, i.e., the familiar involution that interchanges each row/column of $\mathcal{E}(H_2)$ with its negation.

Let $a = (-1)^l$. By (17.8) and (17.9),

$$\begin{aligned}
\pi_\xi^{-1} \pi_{A\sigma^k} \pi_\xi(x, a) &= \pi_\xi^{-1} \pi_{A\sigma^k}(\nu^l x, (-1)^{l+1}) \\
&= \pi_\xi^{-1}(\det(A)^{l+1} A\sigma^k(\nu^l x), (-1)^{l+1}) \\
&= (\nu^{-l} \det(A)^{l+1} A\sigma^k(\nu^l x), a) \\
&= (\nu^{(p^k-1)l} \det(A)^{l+1} A\sigma^k(x), a) \\
&= \pi_\xi^{1-\chi \det(A)} \pi_{A\sigma^k}(\lambda^2 x, a)
\end{aligned}$$

for some $\lambda \in \text{GF}(q)^\times$. Similarly

$$\phi_\xi^{-1} \phi_{A\sigma^k} \phi_\xi(x, a) = \phi_\xi^{1-\chi \det(A)} \phi_{A\sigma^k}(\lambda^2 x, a).$$

This proves that

$$(17.11) \qquad\qquad (A\sigma^k Q)^\xi = \zeta^{\frac{1}{2}(1-\chi \det(A))} A\sigma^k Q$$

in $\text{Aut}(H_2)$.

17.2.6. Proposition. *For $q = p^r > 5$,*

(i) *$\text{Aut}(H_2)$ contains a subgroup of index 2 isomorphic to $\Gamma\text{L}(2, q)/Q$,*

(ii) *$\langle \xi \rangle$ is a normal subgroup of $\text{Aut}(H_2)$ of order 4 containing the central involution ζ,*

(iii) *the order of $\text{Aut}(H_2)$ is $4rq(q^2 - 1)$.*

Proof. The first two claims follow from (17.10) and (17.11). So $|\text{Aut}(H_2)| = 2|\Gamma\text{L}(2, q)/Q|$. Since $|\Gamma\text{L}(2, q)| = |\text{GL}(2, q)|r$ and $|Q| = (q - 1)/2$, part (iii) is clear. $\square$

17.3. The regular actions

In this section we survey the centrally regular actions on the expanded designs of Paley matrices (proving incidentally that the Paley matrices are all cocyclic).

17.3.1. Near fields. A *near field* has the axioms of a field, except that its multiplication need not be commutative, and there is a single one-sided distributive law (see, e.g., [**61**, p. 236] for a more formal definition). Most of the regular actions for the Paley matrices are related to near fields corresponding to regular subgroups of $\Gamma\text{L}(2, q)$ containing Q.

Huppert and Blackburn [**91**, XII, §9] discuss the regular subgroups R of $\text{GL}(V)$. Either R is of *type A*, meaning that it contains a normal cyclic irreducible subgroup; or it is of *type B* or *type C*, meaning that it is one of the exceptional near fields of orders $5^2, 7^2, 11^2, 23^2, 29^2, 59^2$. The type of an exceptional near field depends on whether its multiplicative group is solvable.

We will say that a regular group (action) for a Paley matrix is of type A, B, or C according to whether the relevant near field is of that type. At small orders, there are a few *sporadic* regular subgroups not contained in $\Gamma L(2, q)$.

In this section we do not go beyond stating which exceptional near fields give rise to regular actions on the Paley matrices. There are no type B or C regular actions for the type II Hadamard matrix. The type B and C actions for the Paley conference matrix and the type I Hadamard matrix are fully explained in [**56**], from which the following is taken.

17.3.1. EXAMPLE. The multiplicative group of the exceptional near field of order 25 is isomorphic to $SL(2, 3)$. A sample subgroup $G \cong SL(2, 3)$ of $GL(2, 5)$ acting regularly on V^* is generated by

$$\begin{bmatrix} 2 & 0 \\ 0 & 3 \end{bmatrix}, \quad \begin{bmatrix} 0 & 4 \\ 1 & 0 \end{bmatrix}, \quad \begin{bmatrix} 1 & 2 \\ 1 & 3 \end{bmatrix}.$$

The square Q of the scalar subgroup of $GL(2, 5)$ has order 2, so $E = G/Q$ is isomorphic to $PSL(2, 3) \cong \text{Alt}(4)$. The cosets of Q containing each of the matrices below form a $(12, 5, 2, 2)$-relative difference set in E:

$$\begin{bmatrix} 0 & 1 \\ 4 & 0 \end{bmatrix}, \quad \begin{bmatrix} 1 & 2 \\ 1 & 3 \end{bmatrix}, \quad \begin{bmatrix} 4 & 2 \\ 1 & 2 \end{bmatrix}, \quad \begin{bmatrix} 2 & 2 \\ 1 & 4 \end{bmatrix}, \quad \begin{bmatrix} 3 & 2 \\ 1 & 1 \end{bmatrix}.$$

This relative difference set is not central.

A near field may be constructed from a subgroup R of $\Gamma L(2, q)$ acting regularly on V^*, as follows. Fix an element of V^*, say 1. For each $x \in V^*$ there exists a unique $s_x \in R$ such that $s_x x = 1$. We define a binary operation $\odot$ on V by

$$x \odot y = \begin{cases} 0 & \text{if } x = 0 \text{ or } y = 0 \\ s_x^{-1} s_y^{-1} 1 & \text{if } x, y \in V^*. \end{cases}$$

Then $(V^*, \odot)$ is a group isomorphic to R. Moreover, since $s_x \in \Gamma L(2, q)$,

$$x \odot (y + z) = s_x^{-1}(y + z) = s_x^{-1} y + s_x^{-1} z = x \odot y + x \odot z$$

So $(V, +, \odot)$ is a near field.

Conversely, if V is a near field with multiplication $\odot$, and for all $x \in V^*$ the maps $s_x : y \mapsto x \odot y$ are in $\Gamma L(2, q)$, then $R = \{s_x \mid x \in V^*\}$ is a subgroup of $\Gamma L(2, q)$ acting regularly on V^*. Note that by considering V as a $2r$-dimensional vector space W over $GF(p)$, we obtain an embedding of $\Gamma L(2, q)$ into $GL(2r, p)$; so the multiplicative group of every near field order q^2 is isomorphic to a subgroup of $GL(2r, p)$ acting regularly on W^*.

We now look at the multiplicative groups $(V^*, \odot)$ of the type A near fields. An irreducible cyclic subgroup of $GL(2, q)$ is contained in a Singer cycle S of $GL(2, q)$. Define $\alpha \in GL(V)$ by $\alpha : x \mapsto \omega x$, where ω is a primitive element of $GF(q^2)$. Then we recall from Example 3.6.2 that the matrix

$$\begin{bmatrix} 0 & -\omega^{q+1} \\ 1 & \omega + \omega^q \end{bmatrix}$$

of α with respect to the $GF(q)$-basis $\{1, \omega\}$ of V generates a Singer cycle in $GL(2, q)$. It can be shown that S has normalizer $\langle S, \beta \rangle$ in $\Gamma L(2, q)$, where β is the Frobenius map $u \mapsto u^p$. Furthermore,

$$N_{GL(2,q)}(S) = N_{\Gamma L(2,q)}(S) \cap GL(2, q) = S \rtimes \langle \beta^r \rangle \cong C_{q^2-1} \rtimes C_2;$$

note that $\alpha^{\beta^r} = \alpha^q$. Thus, if G is a subgroup of $\mathrm{GL}(2,q)$ with an irreducible cyclic normal subgroup, and G acts regularly on V^*, then $G \leq S \rtimes \langle \beta^r \rangle$ up to conjugacy.

Each Singer cycle contains the scalar subgroup $\langle \alpha^{q+1} \rangle$ of $\mathrm{GL}(2,q)$, which in turn contains $Q = \langle \alpha^{2(q+1)} \rangle$ as a subgroup of index 2. Not all regular subgroups of $\Gamma\mathrm{L}(2,q)$ contain Q. The next proposition states necessary conditions for this to happen (cf. Lemmas 4.5 and 4.7 of [**56**]).

17.3.2. PROPOSITION. *Let m and n be positive integers such that $2r = mn$. For each positive integer t coprime to n, define the regular subgroup $R_{n,t}$ of $\Gamma\mathrm{L}(2,q)$ by*

$$R_{n,t} = \langle a, b \rangle, \quad \text{where } ax = \omega^n x \text{ and } bx = \omega^t x^{p^m} \quad \forall\, x \in V.$$

If $R_{n,t}$ contains Q then one of the following must hold.

(i) *$n = 2$ and p is any odd prime.*

(ii) *$n = 4$ and $p^m \equiv 1 \pmod 4$.*

(iii) *n is odd with prime factorization $p_1^{e_1} \cdots p_k^{e_k}$ and $p^{m/2} \equiv -1 \pmod{p_i}$ for $i = 1, \ldots, k$.*

17.3.2. Regular subgroups for the conference matrix. Theorem 17.2.1 implies that all regular subgroups E for the Paley conference matrix C of order $q + 1$ lie in $\Gamma\mathrm{L}(2,q)/Q$. So E lifts to the multiplicative group of some near field of order q^2. In particular, there are no sporadic regular actions. The type A regular actions correspond to one of the near fields covered by Proposition 17.3.2. The following theorem, which lists the type A actions, is mostly proved as Lemma 4.7 in [**56**].

17.3.3. THEOREM. *Let m and n be positive integers such that $2r = mn$. The following extension groups E and indexing groups G correspond to the type A regular actions on $\mathcal{E}(C)$.*

(1) *$n = 2$, p is any odd prime, and*

$$E \cong \langle a, b \mid a^{q+1} = 1,\ a^b = a^{-1},\ b^2 = a^{(q+1)/2} \rangle.$$

The action is normal, and the indexing group for the cocyclic matrix C is the dihedral group

$$G \cong \langle a, b \mid a^{(q+1)/2} = b^2 = 1,\ a^b = a^{-1} \rangle.$$

(2) *$n = 4$, $m = r/2$, $p^m \equiv 1 \pmod 4$, and*

$$E \cong \langle a, b \mid a^{(q+1)/2} = 1,\ a^b = a^{p^m},\ b^4 = 1 \rangle.$$

The action is not normal.

(3) *n is odd with prime factorization $p_1^{e_1} \cdots p_k^{e_k}$, $m = 2m'$, $p^{m'} \equiv -1 \pmod{p_i}$ for all $i = 1, \ldots, k$, and*

$$E \cong \langle a, b \mid a^{2(q+1)/n} = 1,\ a^b = a^{p^{2m'}},\ b^n = a^{(q+1)/n(p^{m'}+1)} \rangle.$$

The action is normal and the indexing group for the cocyclic matrix C is

$$G \cong \langle a, b \mid a^{(q+1)/n} = 1,\ a^b = a^{p^{2m'}},\ b^n = a^{(q+1)/n(p^{m'}+1)} \rangle.$$

Let ω be a primitive element of $\mathrm{GF}(q^2)$. In each of the above cases, we obtain, for each integer $t < n$ coprime to n, an embedding of E as a subgroup $\langle a, b \rangle / Q$ of $\mathrm{PermAut}(\mathcal{E}(C))$ where

$$ax = \omega^n x \quad \text{and} \quad bx = \omega^t x^{p^m} \quad \forall\, x \in V.$$

Any other regular subgroup of $\mathrm{PermAut}(\mathcal{E}(C))$ *isomorphic to* E *is conjugate by a linear map to* $\langle a, b \rangle / Q$.

The cases $n = 1, 2$ are always allowed. When $n = 1$, E is cyclic and the near field is a finite field.

The following result from [**56**, Section 4.6] lists the exceptional regular actions for the Paley conference matrices.

17.3.4. THEOREM. *In each case listed below, E acts regularly on the expanded design $\mathcal{E}(C)$ of the Paley conference matrix C of order $p + 1$, and there is just one conjugacy class of regular subgroups of $\mathrm{PermAut}(\mathcal{E}(C))$ that are isomorphic to E.*

(1) $p = 5$: $E = \mathrm{PSL}(2, 3) \cong \mathrm{Alt}(4)$ *acts non-normally.*
(2) $p = 11$: $E = \mathrm{SL}(2, 3)$ *acts normally.*
(3) $p = 23$: $E = \mathrm{GL}(2, 3)$ *acts normally.*
(4) $p = 29$: $E = \mathrm{PSL}(2, 5) \cong \mathrm{Alt}(5)$ *acts non-normally.*
(5) $p = 59$: $E = \mathrm{SL}(2, 5)$ *acts normally.*

Example 17.3.1 treats the near field of part (1). The next example illustrates part (2); see [**56**, Subsection 4.6.2].

17.3.5. EXAMPLE. There are two exceptional near fields of order 11^2, but only one of them gives a regular action for the Paley conference matrix C of order 12. One regular embedding of the multiplicative group $(V^*, \odot) \cong \mathrm{SL}(2, 3) \times \mathrm{C}_5$ in $\mathrm{GL}(2, 11)$ has image generated by

$$\begin{bmatrix} 1 & 3 \\ 3 & 10 \end{bmatrix}, \quad \begin{bmatrix} 0 & 10 \\ 1 & 0 \end{bmatrix}, \quad \begin{bmatrix} 6 & 4 \\ 3 & 4 \end{bmatrix}, \quad \begin{bmatrix} 4 & 0 \\ 0 & 4 \end{bmatrix}.$$

The corresponding regular subgroup of $\mathrm{Aut}(\mathrm{A}_C)$ is isomorphic to $\mathrm{SL}(2, 3)$. So C is cocyclic with indexing group $\mathrm{Alt}(4)$. The first three matrices above generate a subgroup $E \cong \mathrm{SL}(2, 3)$ of $\mathrm{GL}(2, 11)$, in which the elements

$$\begin{bmatrix} 8 & 7 \\ 6 & 4 \end{bmatrix}, \quad \begin{bmatrix} 4 & 7 \\ 8 & 6 \end{bmatrix}, \quad \begin{bmatrix} 7 & 7 \\ 6 & 3 \end{bmatrix}, \quad \begin{bmatrix} 0 & 1 \\ 10 & 0 \end{bmatrix}, \quad \begin{bmatrix} 5 & 7 \\ 8 & 7 \end{bmatrix}, \quad \begin{bmatrix} 7 & 6 \\ 7 & 3 \end{bmatrix},$$

$$\begin{bmatrix} 4 & 8 \\ 7 & 6 \end{bmatrix}, \quad \begin{bmatrix} 8 & 6 \\ 7 & 4 \end{bmatrix}, \quad \begin{bmatrix} 5 & 8 \\ 7 & 7 \end{bmatrix}, \quad \begin{bmatrix} 10 & 8 \\ 8 & 1 \end{bmatrix}, \quad \begin{bmatrix} 3 & 10 \\ 10 & 8 \end{bmatrix}$$

form a $(24, 11, 5, 2)$-central relative difference set. By Theorem 17.3.3, the only groups apart from $\mathrm{SL}(2, 3)$ with a normal regular action on A_C are C_{24} and Q_{24}.

17.3.3. Regular subgroups for the type I matrices. Let $q \equiv 3 \pmod 4$. The regular actions for the type I Paley Hadamard matrix H_1 of order $q + 1$ are also determined in [**56**]. There is only one class of type A actions.

17.3.6. THEOREM. *For $q \neq 3, 7, 11, 23, 59$, the dicyclic group*

$$\mathrm{Q}_{2(q+1)} = \langle a, b \mid a^{q+1} = 1, \ a^b = a^{-1}, \ b^2 = a^{(q+1)/2} \rangle$$

is the only group with a centrally regular embedding in $\mathrm{PermAut}(\mathcal{E}(H_1))$.

To obtain a regular embedding of $\mathrm{Q}_{2(q+1)}$ in $\Gamma\mathrm{S}(2, q)/Q$, let ω be any primitive element of $\mathrm{GF}(q^2)$, define $a, b \in \mathrm{Sym}(V)$ by

$$ax = \omega^2 x \quad \text{and} \quad bx = \omega x^q \quad \forall x \in V,$$

and take E to be $\langle a, b\rangle/Q$. It can be shown that any two regular subgroups of $\Gamma\mathrm{S}(2,q)/Q$ isomorphic to $\mathrm{Q}_{2(q+1)}$ are conjugate by an element of $\mathrm{GS}(2,q)/Q$. For $q > 11$, the number of ways in which $\mathrm{Q}_{2(q+1)}$ acts on A_{H_1} is $q(q^2 - 1)\varphi(q + 1)$, where φ is the Euler totient function. The embedding given above is related to the regular action found by Ito [100], Yamada [163], and Yamamoto [164].

The next theorem summarizes the type B and C actions for H_1 (details are in [56, Section 5]).

17.3.7. THEOREM. *For each value of q below, E embeds as a regular subgroup of $\Gamma\mathrm{S}(2,q)/Q$ acting normally on $\mathcal{E}(H_1)$, where H_1 is indexed by G.*

(1) $q = 11$: $E = \mathrm{SL}(2,3)$; $G \cong \mathrm{PSL}(2,3)$.
(2) $q = 23$: $E = \mathrm{GL}(2,3)$; $G \cong \mathrm{PGL}(2,3)$.
(3) $q = 59$: $E = \mathrm{SL}(2,5)$; $G \cong \mathrm{PSL}(2,5)$.

The actions of Theorem 17.3.7 are related to exceptional near field actions for the conference matrix; cf. [56, 99]. Note that a central relative difference set arising from the action for $q = 11$ is obtained by adding I_2 to the relative difference set in Example 17.3.5. This action appeared previously in Example 14.7.3.

Finally we catalog the remaining regular actions for H_1 (see [56, Theorem 5.3]; here we take the opportunity to correct a small error in that result, for $q = 7$).

17.3.8. THEOREM. *The following embed as regular subgroups of $\mathrm{PermAut}(\mathcal{E}(H_1))$ not contained in $\Gamma\mathrm{S}(2,q)/Q$.*

(1) $q = 3$: *all non-cyclic groups of order 8, of which only the dihedral group acts non-normally.*
(2) $q = 7$: *nine of the groups of order 16, which all act normally; three act non-normally as well.*
(3) $q = 11$: $\mathrm{Q}_8 \times \mathrm{C}_3$ *acts normally.*

We remark that this subsection confirms again that the only Hadamard groups of order 24 are $\mathrm{SL}(2,3)$, Q_{24}, and $\mathrm{Q}_8 \times \mathrm{C}_3$ (cf. Example 14.7.3).

17.3.4. Regular subgroups for the type II matrices. The regular actions for the Paley type II Hadamard matrix H_2 of order $2(q+1)$ are worked out in [55]. These are all of type A if $q > 5$. The following lemma adduces the connection.

17.3.9. LEMMA. *Let $q \equiv 1 \pmod 4$ be greater than 5. A regular subgroup E of $\mathrm{PermAut}(\mathcal{E}(H_2))$ must have the form $E = \langle R, \xi\gamma\rangle$ where ξ is defined by (17.8) and (17.9), and*

(i) $R \leq \Gamma\mathrm{L}(2,q)/Q$ *is regular on the even rows,*
(ii) $\gamma \in \Gamma\mathrm{L}(2,q)/Q$,
(iii) $(\xi\gamma)^2 \in R$,
(iv) $\xi\gamma$ *normalizes R.*

Conversely, if R and γ satisfy conditions (i)–(iv), then $E = \langle R, \xi\gamma\rangle$ is a regular subgroup of $\mathrm{PermAut}(\mathcal{E}(H_2))$.

We list the regular actions for H_2 more explicitly. Let ω be a primitive element of $\mathrm{GF}(q^2)$, define $\alpha = \omega I_2 \in \mathrm{GL}(2,q)$, and let β be the Frobenius map.

17.3.10. THEOREM. *Let $q = p^r \equiv 1 \pmod 4$, $q > 5$. A regular subgroup E of $\mathrm{PermAut}(\mathcal{E}(H_2))$ is conjugate in $\mathrm{PermAut}(\mathcal{E}(H_2))$ to one of the following groups.*

(1) $E_{11} = \langle\xi, \alpha Q\rangle$.

(2) $E_{12} = \langle \xi \cdot \beta^r Q, \alpha Q \rangle$.

(3) $E_{21} = \langle \xi, \alpha^2 Q, \alpha \beta^r Q \rangle$.

(4) $E_{22} = \langle \xi \cdot \alpha Q, \alpha^2 Q, \alpha \beta^r Q \rangle$.

(5) $E_n = \langle \xi, \alpha^n Q, \alpha \beta^m Q \rangle$ where $n > 1$ is an odd divisor of r, $m = 2r/n$, and $p^{m/2} \equiv -1 \pmod{p_i}$ for every prime p_i in the prime factorization of n.

In all cases E acts normally.

CHAPTER 18

A Large Family of Cocyclic Hadamard Matrices

In this chapter we prove the following existence result from [**43**].

18.0.1. THEOREM. *Let $q_1, \ldots, q_r \equiv 1 \pmod 4$ and $p_1, \ldots, p_s \equiv 3 \pmod 4$ be prime powers. Let $k_1, \ldots, k_r$ and $m_1, \ldots, m_s$ be non-negative integers. Then there is a cocyclic Hadamard matrix of order*

$$(18.1) \qquad 2^r \left(\prod_{i=1}^{r}(q_i + 1) \right) \left(\prod_{j=1}^{s}(p_j + 1) \right) \left(\prod_{i=1}^{r} q_i^{k_i} \right) \left(\prod_{j=1}^{s} p_j^{m_j} \right).$$

An erroneous stronger version of Theorem 18.0.1 was presented at conferences in 1993 [**41**]; see also [**47**, Theorem 4.2]. This version omitted the exponent r in the factor 2^r at the start of (18.1).

Theorem 18.0.1 gives a large infinite family of cocyclic Hadamard matrices and corresponding central relative difference sets with forbidden subgroup of size 2. Moreover, the theorem implies that many groups are Hadamard groups. In the last section of this chapter, we prove a generalization of Theorem 18.0.1 that yields still more Hadamard groups and relative difference sets.

18.1. On the orders covered

The most powerful existence results for Hadamard matrices are asymptotic ones of the kind: "For all odd positive integers k, there is a Hadamard matrix of order $2^m k$ where $m \leq a_0 + b_0 \log_2 k$." Here a_0 and b_0 are non-negative constants independent of k. The strength of such a result depends first on how small b_0 is, and then on how small a_0 is. The Hadamard Conjecture is equivalent to the result with $a_0 = 2$ and $b_0 = 0$. Seberry [**156**] showed that we can take $a_0 = 0$ and $b_0 = 2$. Later improvements by Craigen, Holzmann, and Kharaghani, culminating in the papers [**21**] and [**23**], allow us to take $b_0 = 3/8$. De Launey and Smith [**54**] showed that $b_0 = 8$ is valid for *cocyclic* Hadamard matrices. A dramatic improvement was announced by de Launey and Kharaghani in [**51**], so that for cocyclic Hadamard matrices we can now take $b_0 = 4/5$. This will be discussed in Chapter 23.

De Launey [**44**] considered the range of orders covered by the Paley Hadamard matrices and their Kronecker products with Sylvester Hadamard matrices. Using number-theoretic arguments based on the distribution of primes, he obtained the following partial asymptotic result.

18.1.1. THEOREM. *Let $\epsilon > 0$. Let $\mathcal{H}(x)$ be the number of odd positive integers $k \leq x$ for which there is a Hadamard matrix of order $2^l k$, where $1 \leq l \leq 2 + \epsilon \log_2 k$. Then there is a constant $c(\epsilon)$, depending only on ϵ, such that $\mathcal{H}(x) > c(\epsilon)x$ for all sufficiently large x.*

Since the Paley and Sylvester Hadamard matrices are cocyclic (see Chapters 17 and 21), this result also holds for cocyclic Hadamard matrices. Theorem 18.1.1 has the advantage that $b_0 = \epsilon$ may be arbitrarily small; but has the disadvantage of applying only for some positive proportion of k. However, it might be possible to use better number-theoretic arguments to show that Theorem 18.0.1 provides so many matrices that $c(\epsilon)$ in Theorem 18.1.1 can be taken very close to $1/2$.

18.2. A construction for prime powers congruent to 3 (mod 4)

In this section, we give a direct construction that proves Theorem 18.0.1 for $r = 0$ and $s = 1$.

Let $p \equiv 3 \pmod 4$ be a prime power. As we observed in Chapter 17, $Q_{2(p+1)}$ has a centrally regular action on the expanded design of the type I Paley Hadamard matrix of order $p + 1$. Here, we will adapt ideas from that chapter to cocyclic orthogonal designs.

18.2.1. THEOREM. *There exists a cocyclic* $\mathrm{OD}(p+1; 1, p)$ *with extension group* $Q_{2(p+1)}$.

PROOF. View $\mathrm{GF}(p^2)$ as a 2-dimensional vector space V over $\mathrm{GF}(p)$. Let χ be the quadratic character on $\mathrm{GF}(p)$, and let $\mathcal{I}$ be a complete irredundant set of representatives for the $p + 1$ lines in V^*. Then for commuting indeterminates a_1, a_2, let
$$D = [\chi \det{}^*(x, y)]_{x,y \in \mathcal{I}} \wedge (a_1 I_{p+1} + a_2 (J_{p+1} - I_{p+1})),$$
where $\det^*$ is defined on V^* by
$$\det{}^*(x, y) = \begin{cases} xy^{-1} & \text{if } xy^{-1} \in \mathrm{GF}(p) \\ \det(x, y) & \text{otherwise,} \end{cases}$$
for any alternating bilinear form $\det$ on V. Remember that if $\gamma \in \mathrm{GL}(V)$ is written as a 2×2 matrix with respect to a basis of V, then $\det(\gamma x, \gamma y) = \det(\gamma) \det(x, y)$ where $\det(\gamma)$ is the ordinary matrix determinant.

We have $D = a_1 I_{p+1} + a_2 C$ where C is a skew-symmetric Paley conference matrix. Hence
$$DD^\top = (a_1{}^2 + p a_2{}^2) I_{p+1},$$
i.e., D is an $\mathrm{OD}(p+1; 1, p)$.

Next, we exhibit a centrally regular action of $Q_{2(p+1)}$ on $\mathcal{E}(D)$. Let $\mathcal{J}$ be a complete irredundant set of representatives for the half-lines in V^*; then
$$\mathcal{E}(D) = [\chi \det{}^*(x, y)]_{x,y \in \mathcal{J}}.$$
For a primitive element ω of $\mathrm{GF}(p^2)$, define the $\mathrm{GF}(p)$-linear maps

(18.2) $\alpha : v \mapsto \omega v, \qquad \sigma : v \mapsto v^p$

on V. Further define
$$a = \alpha^2 \quad \text{and} \quad b = \alpha \circ \sigma.$$
Let t be a square root of -1 in $\mathrm{GF}(p^2)$. Since $p \equiv 3 \pmod 4$, $\{1, t\}$ is a $\mathrm{GF}(p)$-basis of V. With respect to this basis, σ has matrix $\mathrm{diag}(1, -1)$. Thus $\det(\sigma) = -1$. Since $\det(\alpha) = \omega^{p+1}$ (see Example 3.6.2), $\det(a)$ and $\det(b)$ are both squares in $\mathrm{GF}(p)$. So for $xy^{-1} \notin \mathrm{GF}(p)$,
$$\chi \det{}^*(a^i b^j x, a^i b^j y) = \chi(\det(a))^i \chi(\det(b))^j \chi \det{}^*(x, y) = \chi \det{}^*(x, y).$$

If $xy^{-1} \in \mathrm{GF}(p)$ then $(a^i b^j x)(a^i b^j y)^{-1} = xy^{-1} \in \mathrm{GF}(p)$, so $\chi \det^*(a^i b^j x, a^i b^j y) = \chi \det^*(x, y)$ in this case too. Hence, there is a homomorphism from $\langle a, b \rangle$ into $\mathrm{PermAut}(\mathcal{E}(D))$.

Furthermore, $\langle a, b \rangle$ acts regularly on V^*, and we see from (18.2) that this group has presentation

$$\langle a, b \mid a^{(p^2-1)/2} = 1,\ b^2 = a^{(p+1)/2},\ a^b = a^p \rangle.$$

The induced action of $\langle a, b \rangle$ on the half-lines $\{\omega^{2(p+1)i} v \mid 1 \le i \le (p-1)/2\}$ in V^* has kernel $\langle a^{p+1} \rangle$. Factoring out the kernel gives

$$\langle a, b \mid a^{p+1} = 1,\ b^2 = a^{(p+1)/2},\ a^b = a^{-1} \rangle \cong Q_{2(p+1)},$$

which is therefore isomorphic to a regular subgroup E of $\mathrm{PermAut}(\mathcal{E}(D))$. Since $a^{(p+1)/2}$ maps to the automorphism of $\mathcal{E}(D)$ swapping each row and column with its negation, E is centrally regular. $\qquad\square$

18.2.2. EXAMPLE. We walk through some of the calculations in the above proof for $p = 7$.

The polynomial $h(z) = z^2 - z - 1 \in \mathbb{Z}_7[z]$ is irreducible, and $\omega = 2 + z$ is a primitive element of $\mathrm{GF}(49) \cong \mathbb{Z}_7[z]/\langle h(z) \rangle$. Multiplying $a + bz$ by ω yields $2a + b + (a + 3b)z$, and $\sigma(a + bz) = (a + bz)^7 = a + b + 6bz$. Thus

$$\alpha = \begin{bmatrix} 2 & 1 \\ 1 & 3 \end{bmatrix} \quad \text{and} \quad \sigma = \begin{bmatrix} 1 & 1 \\ 0 & 6 \end{bmatrix}.$$

The element α generates a Singer cycle in $\mathrm{GL}(2, 7)$. Then

$$a = \alpha^2 = \begin{bmatrix} 5 & 5 \\ 5 & 3 \end{bmatrix}, \qquad b = \alpha\sigma = \begin{bmatrix} 2 & 1 \\ 1 & 5 \end{bmatrix}$$

satisfy the relations $a^{24} = 1$, $b^2 = a^4$, and $a^b = a^7$. The quotient E of $\langle a, b \rangle$ by its central subgroup $\langle a^8 \rangle$ is isomorphic to Q_{16}.

Line representatives may be written down lexicographically:

$$\{(0, 1), (1, 0), (1, 1), (1, 2), (1, 3), (1, 4), (1, 5), (1, 6)\}.$$

(Alternatively, we could compute the powers $\omega, \ldots, \omega^8$ of the primitive element.) Denote by $(m, n)_+$ any square multiple of (m, n), and by $(m, n)_-$ any non-square multiple of (m, n). That is, the following is a set of representatives for the half-lines in V^*:

$$\{(1, 4)_+, (1, 1)_-, (1, 6)_+, (1, 5)_+, (0, 1)_+, (1, 3)_+, (1, 2)_-, (1, 0)_-,$$
$$(1, 4)_-, (1, 1)_+, (1, 6)_-, (1, 5)_-, (0, 1)_-, (1, 3)_-, (1, 2)_+, (1, 0)_+\}.$$

We have

$$(1,4)_+ \xrightarrow{a} (1,6)_+ \xrightarrow{a} (0,1)_+ \xrightarrow{a} (1,2)_- \xrightarrow{a} (1,4)_- \xrightarrow{a} (1,6)_- \xrightarrow{a} (0,1)_- \xrightarrow{a} (1,2)_+ \xrightarrow{a} (1,4)_+$$
$$b\downarrow \qquad b\downarrow \qquad b\downarrow \qquad b\downarrow \qquad b\downarrow \qquad b\downarrow \qquad b\downarrow \qquad b\downarrow \qquad b\downarrow$$
$$(1,0)_- \xleftarrow{a} (1,3)_+ \xleftarrow{a} (1,5)_+ \xleftarrow{a} (1,1)_- \xleftarrow{a} (1,0)_+ \xleftarrow{a} (1,3)_- \xleftarrow{a} (1,5)_- \xleftarrow{a} (1,1)_+ \xleftarrow{a} (1,0)_-$$

The relation $aba = b$ is reflected in each commuting square in this diagram.

Let D be the orthogonal design as in the proof of Theorem 18.2.1 for $p = 7$, with indeterminates x and y. The expanded design $\mathcal{E}(D)$ is as follows. (Overlines denote negation. We have chosen a row/column indexing that optimally displays the action by E).

	$\binom{1}{4}_+$	$\binom{1}{6}_+$	$\binom{0}{1}_+$	$\binom{1}{2}_-$	$\binom{1}{4}_-$	$\binom{1}{6}_-$	$\binom{0}{1}_-$	$\binom{1}{2}_+$	$\binom{1}{0}_-$	$\binom{1}{3}_+$	$\binom{1}{5}_+$	$\binom{1}{1}_-$	$\binom{1}{0}_+$	$\binom{1}{3}_-$	$\binom{1}{5}_-$	$\binom{1}{1}_+$
$\binom{1}{4}_+$	x	y	y	y	$\bar{x}$	$\bar{y}$	$\bar{y}$	$\bar{y}$	y	$\bar{y}$	y	$\bar{y}$	$\bar{y}$	y	$\bar{y}$	y
$\binom{1}{6}_+$	$\bar{y}$	x	y	y	y	$\bar{x}$	$\bar{y}$	$\bar{y}$	$\bar{y}$	y	$\bar{y}$	$\bar{y}$	y	$\bar{y}$	y	y
$\binom{0}{1}_+$	$\bar{y}$	$\bar{y}$	x	y	y	y	$\bar{x}$	$\bar{y}$	y	$\bar{y}$	$\bar{y}$	y	$\bar{y}$	y	y	$\bar{y}$
$\binom{1}{2}_-$	$\bar{y}$	$\bar{y}$	$\bar{y}$	x	y	y	y	$\bar{x}$	$\bar{y}$	$\bar{y}$	y	$\bar{y}$	y	y	$\bar{y}$	y
$\binom{1}{4}_-$	$\bar{x}$	$\bar{y}$	$\bar{y}$	$\bar{y}$	x	y	y	y	$\bar{y}$	y	$\bar{y}$	y	y	$\bar{y}$	y	$\bar{y}$
$\binom{1}{6}_-$	y	$\bar{x}$	$\bar{y}$	$\bar{y}$	$\bar{y}$	x	y	y	y	$\bar{y}$	y	y	$\bar{y}$	y	$\bar{y}$	$\bar{y}$
$\binom{0}{1}_-$	y	y	$\bar{x}$	$\bar{y}$	$\bar{y}$	$\bar{y}$	x	y	$\bar{y}$	y	y	$\bar{y}$	y	$\bar{y}$	$\bar{y}$	y
$\binom{1}{2}_+$	y	y	y	$\bar{x}$	$\bar{y}$	$\bar{y}$	$\bar{y}$	x	y	y	$\bar{y}$	y	$\bar{y}$	$\bar{y}$	y	$\bar{y}$
$\binom{1}{0}_-$	$\bar{y}$	y	$\bar{y}$	y	y	$\bar{y}$	y	$\bar{y}$	x	y	y	y	$\bar{x}$	$\bar{y}$	$\bar{y}$	$\bar{y}$
$\binom{1}{3}_+$	y	$\bar{y}$	y	y	$\bar{y}$	y	$\bar{y}$	$\bar{y}$	$\bar{y}$	x	y	y	y	$\bar{x}$	$\bar{y}$	$\bar{y}$
$\binom{1}{5}_+$	$\bar{y}$	y	y	$\bar{y}$	y	$\bar{y}$	$\bar{y}$	y	$\bar{y}$	$\bar{y}$	x	y	y	y	$\bar{x}$	$\bar{y}$
$\binom{1}{1}_-$	y	y	$\bar{y}$	y	$\bar{y}$	$\bar{y}$	y	$\bar{y}$	$\bar{y}$	$\bar{y}$	$\bar{y}$	x	y	y	y	$\bar{x}$
$\binom{1}{0}_+$	y	$\bar{y}$	y	$\bar{y}$	$\bar{y}$	y	$\bar{y}$	y	$\bar{x}$	$\bar{y}$	$\bar{y}$	$\bar{y}$	x	y	y	y
$\binom{1}{3}_-$	$\bar{y}$	y	$\bar{y}$	$\bar{y}$	y	$\bar{y}$	y	y	y	$\bar{x}$	$\bar{y}$	$\bar{y}$	$\bar{y}$	x	y	y
$\binom{1}{5}_-$	y	$\bar{y}$	$\bar{y}$	y	$\bar{y}$	y	y	$\bar{y}$	y	y	$\bar{x}$	$\bar{y}$	$\bar{y}$	$\bar{y}$	x	y
$\binom{1}{1}_+$	$\bar{y}$	$\bar{y}$	y	$\bar{y}$	y	y	$\bar{y}$	y	y	y	y	$\bar{x}$	$\bar{y}$	$\bar{y}$	$\bar{y}$	x

The action of $\langle a, b \rangle$ is clearly seen above. Multiplying column indices by a cyclically shifts the first eight columns to the right and cyclically shifts the last eight columns to the left. Multiplying row indices by a cyclically shifts downward the first eight rows, and cyclically shifts upward the last eight rows. Thus all quadrants of $\mathcal{E}(D)$ are preserved by the action of a.

Since $b = a^4 b^{-1}$, b moves the first eight columns to the last eight columns without changing their order, while the last eight columns are moved to the first eight columns and shifted cyclically four times (in either direction). In concert with its row action, the net effect of b is to interchange the upper left and lower right quadrants, and to interchange the other two quadrants. The upper left quadrant is moved without re-ordering of rows and columns, while the other three quadrants have either their rows or columns, or both, cyclically shifted four positions. The upper right and lower left quadrants are negated if their rows or columns are shifted by four positions. Again, inspection confirms that the action of b preserves $\mathcal{E}(D)$.

18.3. A construction for prime powers congruent to 1 (mod 4)

Let $q \equiv 1 \pmod 4$, and recall Definition 16.4.2. We provide a method to construct a cocyclic set of four suitable matrices of order $\frac{q+1}{2}$.

18.3.1. THEOREM. *There exist type $(2, 2q)$ Williamson-like matrices developed modulo* $C_{\frac{1}{2}(q+1)}$.

First we prove a lemma. If ω is a primitive element of $\mathrm{GF}(q^2)$ then

$$(18.3) \qquad \mathcal{I} = \{\omega^{4i} \mid i = 0, 1, \dots, \tfrac{q-1}{2}\} \cup \{\omega^{4i + \frac{q+1}{2}} \mid i = 0, 1, \dots, \tfrac{q-1}{2}\}$$

is a complete set of line representatives in V^*, where V is the $\mathrm{GF}(q)$-space $\mathrm{GF}(q^2)$.

18.3.2. LEMMA. *Let $q \equiv 1 \pmod 4$ be a prime power, and suppose that C is the Paley conference matrix of order $q + 1$ with rows and columns indexed by $\mathcal{I}$ as in (18.3). Then*

$$(18.4) \qquad C = \begin{bmatrix} A & B \\ B & -A \end{bmatrix},$$

where A and B are circulant symmetric matrices satisfying

$$(18.5) \qquad AA^\top + BB^\top = qI_{\frac{1}{2}(q+1)} \quad and \quad AB^\top = BA^\top.$$

PROOF. Define

$$A = [\chi \det(\omega^{4i}, \omega^{4j})]_{ij} \quad \text{and} \quad B = [\chi \det(\omega^{4i}, \omega^{4j + \frac{q+1}{2}})]_{ij},$$

χ as usual denoting the quadratic character on $\mathrm{GF}(q)$. Since $q \equiv 1 \pmod 4$, -1 is a square in $\mathrm{GF}(q)$. Then because det is alternating, $\chi \det(\omega^{4i}, \omega^{4j}) = \chi \det(\omega^{4j}, \omega^{4i})$. Thus A is symmetric.

The $\mathrm{GF}(q)$-linear transformation $\alpha : x \mapsto \omega x$ on V has non-square determinant. So

$$\chi \det(\omega^{4i}, \omega^{4j + \frac{q+1}{2}}) = -\chi \det(\alpha^{\frac{q+1}{2}}) \chi \det(\omega^{4i}, \omega^{4j + \frac{q+1}{2}})$$
$$= -\chi(\omega^{q+1}) \chi \det(\omega^{4i + \frac{q+1}{2}}, \omega^{4j})$$
$$= \chi \det(\omega^{4j}, \omega^{4i + \frac{q+1}{2}}).$$

Thus B is also symmetric. Similarly

$$\chi \det(\omega^{4i}, \omega^{4j}) = \chi \det(\omega^{4(i-j)}, 1)$$

and

$$\chi \det(\omega^{4i}, \omega^{4j + \frac{q+1}{2}}) = \chi \det(\omega^{4(i-j)}, \omega^{\frac{q+1}{2}}),$$

verifying that A and B are circulant.

Finally, since

$$\chi \det(\omega^{4i + \frac{q+1}{2}}, \omega^{4j}) = \chi \det(\omega^{4j}, \omega^{4i + \frac{q+1}{2}}) = \chi \det(\omega^{4i}, \omega^{4j + \frac{q+1}{2}})$$

and

$$\chi \det(\omega^{4i + \frac{q+1}{2}}, \omega^{4j + \frac{q+1}{2}}) = \chi \det(\alpha^{\frac{q+1}{2}}) \chi \det(\omega^{4i}, \omega^{4j})$$
$$= -\chi \det(\omega^{4i}, \omega^{4j}),$$

the Paley conference matrix has the stated format (18.4). $\qquad \square$

PROOF OF THEOREM 18.3.1. We use the matrices A and B defined in the proof of Lemma 18.3.2. Notice that B has all entries equal to ± 1, whereas A has zero diagonal and all off-diagonal entries equal to ± 1. For commuting indeterminates x_1, x_2, define $A_1 = x_1 I + x_2 A$, $A_2 = x_1 I - x_2 A$, and $A_3 = A_4 = x_2 B$. By (18.5) and the fact that they are circulant, the A_i are type $(2, 2q)$ Williamson-like matrices. $\qquad \square$

18.3.3. EXAMPLE. We compute A_1, A_2, A_3, A_4 for $q = 5$. Let ω be the primitive element $z + 1$ of $\mathrm{GF}(25) \cong \mathbb{Z}_5[z]/\langle z^2 - z + 1 \rangle$. Then

$$\mathcal{I} = \{1, \omega^4, \omega^8\} \cup \{\omega^3, \omega^7, \omega^{11}\} = \{(1, 0), (1, 4), (0, 4), (2, 1), (3, 3), (1, 2)\}.$$

Indexing rows and columns by the elements of $\mathcal{I}$ (in the same order of listing), we obtain a Paley conference matrix (18.4) where

$$A = \begin{bmatrix} 0 & 1 & 1 \\ 1 & 0 & 1 \\ 1 & 1 & 0 \end{bmatrix} \quad \text{and} \quad B = \begin{bmatrix} 1 & -1 & -1 \\ -1 & 1 & -1 \\ -1 & -1 & 1 \end{bmatrix}.$$

Then

$$A_1 = \begin{bmatrix} x_1 & x_2 & x_2 \\ x_2 & x_1 & x_2 \\ x_2 & x_2 & x_1 \end{bmatrix}, \quad A_2 = \begin{bmatrix} x_1 & -x_2 & -x_2 \\ -x_2 & x_1 & -x_2 \\ -x_2 & -x_2 & x_1 \end{bmatrix},$$

and

$$A_3 = A_4 = \begin{bmatrix} x_2 & -x_2 & -x_2 \\ -x_2 & x_2 & -x_2 \\ -x_2 & -x_2 & x_2 \end{bmatrix}.$$

18.4. Plug-in matrices

We now use Theorem 16.4.1 to extend Theorems 18.2.1 and 18.3.1. Denote the elementary abelian group of prime power order t by E_t.

18.4.1. THEOREM. *Let* $p \equiv 3 \pmod 4$ *be a prime power. For all* $k \geq 0$, *there is a cocyclic* $\mathrm{OD}((p+1)p^k; p^k, p^{k+1})$ *with extension group* $\mathrm{Q}_{2(p+1)} \times \mathrm{E}_{p^k}$.

PROOF. We begin by proving the result for $k = 1$. Index the Paley conference matrix of order $p + 1$ by the complete set of line representatives

$$\mathcal{I} = \{(0, 1), (1, x) \mid x \in \mathrm{GF}(p)\}$$

for $V = \mathrm{GF}(p^2)$. If we define $\det((a, b), (c, d)) = ad - bc$, then removing the first row and column of the Paley conference matrix produces

$$C = [\chi(y - x)]_{x,y \in \mathrm{GF}(p)}.$$

The matrix C is developed modulo the additive group E_p of $\mathrm{GF}(p)$. We have

$$J_p C^\top = C J_p = 0, \quad C^\top = -C, \quad \text{and} \quad CC^\top = pI_p - J_p.$$

Notice then that $a_2 J_p$ and $a_1 I_p + a_2 C$ are amicable, and that

$$a_2 J_p (a_2 J_p)^\top + p(a_1 I_p + a_2 C)(a_1 I_p + a_2 C)^\top = (p a_1{}^2 + p^2 a_2{}^2) I_p.$$

So if we make the following substitutions

$$a_1 \leftarrow a_2 J_p \quad \text{and} \quad a_2 \leftarrow a_1 I_p + a_2 C$$

into the template cocyclic $\mathrm{OD}(p+1; 1, p)$ provided by Theorem 18.2.1, then by Theorem 16.4.1 we obtain a cocyclic $\mathrm{OD}((p+1)p; p, p^2)$ with extension group $\mathrm{Q}_{2(p+1)} \curlyvee_{\mathrm{C}_2} (\mathrm{C}_2 \times \mathrm{E}_p) = \mathrm{Q}_{2(p+1)} \times \mathrm{E}_p$. Iterating these substitutions then gives the result for $k \geq 1$. $\qquad\square$

18.4.2. THEOREM. *Let* $q \equiv 1 \pmod 4$ *be a prime power. For all* $k \geq 0$, *there are type* $(2q^k, 2q^{k+1})$ *Williamson-like matrices developed modulo* $\mathrm{C}_{\frac{1}{2}(q+1)} \times \mathrm{E}_{q^k}$. *In particular, there is a cocyclic* $\mathrm{OD}(2q^k(q+1); 2q^k, 2q^{k+1})$ *with extension group* $\mathrm{Q}_8 \times \mathrm{C}_{\frac{1}{2}(q+1)} \times \mathrm{E}_{q^k}$.

PROOF. The first statement implies the second, by Proposition 16.4.3. So the theorem is true for $k = 0$ by Theorem 18.3.1. The Williamson-like matrices for $k > 0$ are obtained using a plug-in technique (and manipulations reminiscent of the previous proof).

For the E_q-developed matrix $C = [\chi(y - x)]_{x,y \in \mathrm{GF}(q)}$ we have

$$J_q C^\top = C J_q = 0, \qquad C^\top = C, \qquad CC^\top = qI_q - J_q.$$

Thus bJ_q, dJ_q, $aI_q + bC$, and $cI_q + dC$ are mutually amicable. Also

$$bJ_q(bJ_q)^\top + q(aI_q + bC)(aI_q + bC)^\top + dJ_q(dJ_q)^\top + q(cI_q + dC)(cI_q + dC)^\top$$
$$= (q^2b^2 + q^2d^2 + qa^2 + qc^2)I_q + 2q(ab + cd)C.$$

Now let A and B be the matrices of Lemma 18.3.2, and define

$$A_1 = aI + bA, \quad A_2 = bB, \quad A_3 = cI + dA, \quad A_4 = dB.$$

If we make the substitutions

$$a \leftarrow bJ_q, \quad b \leftarrow aI_q + bC, \quad c \leftarrow dJ_q, \quad d \leftarrow cI_q + dC$$

k times, then set $c = a$ and $d = -b$, we get four type $(2q^k, 2q^{k+1})$ Williamson-like matrices developed modulo $C_{\frac{1}{2}(q+1)} \times E_{q^k}$ as required. $\qquad \square$

18.5. Proof of the main theorem and a generalization

18.5.1. THEOREM. *There exists a Hadamard group isomorphic to*

$$(Q_{2(p_1+1)} \curlyvee Q_{2(p_2+1)} \curlyvee \cdots \curlyvee Q_{2(p_s+1)} \curlyvee Q_8 \curlyvee Q_8 \curlyvee \cdots \curlyvee Q_8)$$
$$\times (C_{\frac{1}{2}(q_1+1)} \times C_{\frac{1}{2}(q_2+1)} \cdots \times C_{\frac{1}{2}(q_r+1)})$$
$$\times (E_{p_1^{m_1}} \times E_{p_2^{m_2}} \times \cdots \times E_{p_s^{m_s}})$$
$$\times (E_{q_1^{k_1}} \times E_{q_2^{k_2}} \times \cdots \times E_{q_r^{k_r}})$$

where the first line has $s + r$ factors. Therefore Theorem 18.0.1 holds.

PROOF. Repeated Kronecker multiplication of the cocyclic Hadamard matrices given by setting all indeterminates to 1 in Theorems 18.4.1 and 18.4.2, and use of Lemma 16.5.1, proves existence. It is easy to calculate the order of the extension group (remembering that the central C_2 in each factor of the central product is amalgamated), and see that it is twice the number (18.1). $\qquad \square$

Composition via associates leads to a wider class of extension groups.

18.5.2. THEOREM. *The extension group in Theorem 18.0.1 can be any group that is the direct product of*

$$(Q_{2(p_1+1)} \diamond_{C_2} \cdots \diamond_{C_2} Q_{2(p_s+1)}) \diamond_{C_2} (Q_8 \diamond_{C_2} \cdots \diamond_{C_2} Q_8)$$

and

$$E_{p_1^{m_1}} \cdot \cdots \cdot E_{p_s^{m_s}} \cdot (C_{\frac{1}{2}(q_1+1)} \times E_{q_1^{k_1}}) \cdot \cdots \cdot (C_{\frac{1}{2}(q_r+1)} \times E_{q_r^{k_r}}).$$

PROOF. First note that for $i = 1, 2$, if A_i is a group and R_i is a group with central involution z_i, then

(18.6) $$(R_1 \diamond_{C_2} R_2) \times (A_1 \cdot A_2) = (R_1 \times A_1) \diamond_{C_2} (R_2 \times A_2)$$

by Proposition 3.1.8. Here the copy of C_2 in R_i is $\langle z_i \rangle$.

By Theorem 16.1.5, if R_1 and R_2 are Hadamard groups, then $R_1 \diamond_{C_2} R_2$ is too. So there is a cocyclic Hadamard matrix with extension group $\Upsilon \diamond_{C_2} \Xi$, where

$$\Upsilon = (Q_{2(p_1+1)} \times E_{p_1^{m_1}}) \diamond_{C_2} \cdots \diamond_{C_2} (Q_{2(p_s+1)} \times E_{p_s^{m_s}})$$

and

$$\Xi = (Q_8 \times C_{\frac{1}{2}(q_1+1)} \times E_{q_1^{k_1}}) \diamond_{C_2} \cdots \diamond_{C_2} (Q_8 \times C_{\frac{1}{2}(q_r+1)} \times E_{q_r^{k_r}}).$$

The distinguished central involution lies in $Q_{2(p_i+1)}$ or Q_8. Now the theorem follows from (18.6). $\qquad\square$

Theorem 18.5.2 shows that many groups are Hadamard groups. Moreover, by Theorem 16.1.3, we have established the existence of many maximal-sized relative difference sets with forbidden central subgroup of size 2.

CHAPTER 19

Substitution Schemes for Cocyclic Hadamard Matrices

The 'Cocyclic Hadamard Conjecture' (de Launey and Horadam [**50**]) states that there is a cocyclic Hadamard matrix of order $4t$ for all $t \geq 1$. Ito conjectured more particularly in [**102**] that the dicyclic group Q_{8t} is always a Hadamard group. Schmidt [**139**] has verified Ito's conjecture for $1 \leq t \leq 46$.

Suppose that there exists a cocyclic Hadamard matrix with indexing group G of order $4t$. If $t = p > 3$ is prime, then $G = K \ltimes N$ where $N \cong C_p$ is a normal 2-complement. The conjugation action of K on N gives rise to a homomorphism from K into $\mathrm{Aut}(C_p) \cong C_{p-1}$. If $K = \langle a, b \rangle \cong C_2^2$ then there are only two actions of K on $N = \langle x \rangle$ corresponding to non-isomorphic G: *trivial action* $x^a = x^b = x$, and *inverting action* $x^a = x^b = x^{-1}$. Similarly, $K = \langle c \rangle \cong C_4$ can act trivially or invertingly on N. For $p \equiv 1 \pmod 4$ we get only one other kind of action: $x^c = x^k$, $|k| = 4 \pmod p$. However, previous results in the book disqualify all actions with $K \cong C_4$. That is, the indexing group of a cocyclic Hadamard matrix of order $4p$ is isomorphic to $C_2^2 \times C_p$ or D_{4p}.

Let us now consider the possible Hadamard groups. If E is an extension of C_2 by $G = K \ltimes N$, then $E = L \ltimes N$ where L is an extension of C_2 by K. Any cocycle $f : G \times G \to \langle -1 \rangle$ arises from a central short exact sequence

$$(19.1) \qquad 1 \longrightarrow \langle -1 \rangle \overset{\iota}{\longrightarrow} E \overset{\pi}{\longrightarrow} G \longrightarrow 1$$

where π is the identity map on N. This chapter is about the cocyclic Hadamard matrices with central short exact sequence (19.1). Our main interest lies in the cases where N is *any* odd order group, and K (of order 4) has trivial or inverting action on N. For each case we determine a substitution scheme with 4×4 template array. These are similar to the Williamson and Goethals-Seidel schemes discussed in Chapter 8. We will see how plug-in matrices for one substitution scheme may often be obtained from plug-in matrices for another. Moreover, the schemes yield group-developed examples of transference between Hadamard matrices, weighing matrices, and complex Hadamard matrices.

The most useful combinatorial equivalent is a (*complex*) *complementary pair*: two $n \times n$ $(\pm 1, \pm i)$-matrices A and B such that

$$AA^* + BB^* = 2nI_n.$$

The pair is *real* if both A and B are $(1, -1)$-matrices. We show how proving the Cocyclic Hadamard Conjecture depends on ascertaining whether complementary pairs of circulant matrices exist for all prime orders; and we give many examples of circulant and other group-developed complementary pairs. Nonetheless, the existence problem for complementary pairs is largely untouched.

19.1. General substitution schemes

19.1.1. PROPOSITION. *Let p be a prime. Suppose that $G = K \ltimes N$ is a finite group with normal p-complement N, and*

$$1 \longrightarrow \langle z \rangle \overset{\iota}{\longrightarrow} E \overset{\pi}{\longrightarrow} G \longrightarrow 1$$

is a central short exact sequence where $|z| = p$ and ι is inclusion. Then $E = L \ltimes M$, where $\pi(L) = K$ and the restriction of π to M is an isomorphism onto N.

PROOF. Let $\bar{N}$ be the normal subgroup $\pi^{-1}(N)$ of E. Since π has kernel $\langle z \rangle \leq \bar{N}$, we have $|\bar{N}| = p|N|$. By the Schur-Zassenhaus theorem, $\bar{N} = M \times \langle z \rangle$ for some subgroup M of $\bar{N}$. Indeed M is characteristic in $\bar{N}$; so is normal in E. It follows that M is a normal p-complement of $\pi^{-1}(K)$ in E. □

Now suppose that $G = K \ltimes N$, $|K| = 4$, and $|N| = t$. Let $f : G \times G \to \langle -1 \rangle$ be a cocycle of a central extension (19.1) where $E = L \ltimes N$, $\pi(L) = K$, and π is the identity on N. In this section, we derive substitution schemes for Hadamard matrices with cocycle f. By Proposition 19.1.1, this accounts for all indexing groups $G = K \ltimes N$ when t is odd.

Let $z = \iota(-1)$. For each K (i.e., $K \cong C_4$ or $K \cong C_2^2$), we break the cocyclic associate $(1 - z)\beta$ up into four pieces, one for each coset of $\langle z, N \rangle$ in E. Then we enforce the group ring equation

$$(19.2) \qquad (1 - z)\beta\beta^{(-1)} = 4t(1 - z)$$

from Theorem 15.3.1 (1). This will give us a substitution scheme consisting of a 4×4 template array whose plug-in matrices are group-developed modulo N and satisfy four constraints.

19.1.1. $K \cong C_4$. Let $K = \langle a \mid a^4 = 1 \rangle$. If $\tilde{a} \in E$ is a preimage of a under π, then $E = L_i \ltimes N$ where

$$(19.3) \qquad L_i = \langle \tilde{a}, z \mid \tilde{a}^4 = z^i, z^2 = 1, z^{\tilde{a}} = z \rangle, \quad i \in \{0, 1\}.$$

Also, $\pi(\tilde{a}) = a$ and $\pi_{|N} = \mathrm{id}_N$ imply that

$$(19.4) \qquad x^{\tilde{a}^j} = x^{a^j} \quad \forall x \in N.$$

A cocyclic Hadamard matrix with indexing group G has cocyclic associate $(1-z)\beta$, where

$$\beta = W + \tilde{a}X + \tilde{a}^2 Y + \tilde{a}^3 Z$$

for some elements

$$W = \sum_{x \in N} g(x)x, \quad X = \sum_{x \in N} g(ax)x, \quad Y = \sum_{x \in N} g(a^2 x)x, \quad Z = \sum_{x \in N} g(a^3 x)x$$

of $\mathbb{Z}[N]$ with (± 1)-coefficients. Using the relations in (19.3), and extending the conjugation action (19.4) of $\langle \tilde{a} \rangle$ on N linearly to all of $\mathbb{Z}[N]$, we get

$$\beta\beta^{(-1)} = WW^{(-1)} + (XX^{(-1)})^{a^3} + (YY^{(-1)})^{a^2} + (ZZ^{(-1)})^a$$

$$+ \tilde{a}(z^i(WZ^{(-1)})^a + XW^{(-1)} + (YX^{(-1)})^{a^3} + (ZY^{(-1)})^{a^2})$$

$$+ \tilde{a}^2(z^i(WY^{(-1)})^{a^2} + z^i(XZ^{(-1)})^a + YW^{(-1)} + (ZX^{(-1)})^{a^3})$$

$$+ \tilde{a}^3(z^i(WX^{(-1)})^{a^3} + z^i(XY^{(-1)})^{a^2} + z^i(YZ^{(-1)})^a + ZW^{(-1)}).$$

Then (19.2) yields

(19.5)

$$WW^{(-1)} + (XX^{(-1)})^{a^3} + (YY^{(-1)})^{a^2} + (ZZ^{(-1)})^a = 4t,$$

$$(-1)^i(WZ^{(-1)})^a + XW^{(-1)} + (YX^{(-1)})^{a^3} + (ZY^{(-1)})^{a^2} = 0,$$

$$(-1)^i(WY^{(-1)})^{a^2} + (-1)^i(XZ^{(-1)})^a + YW^{(-1)} + (ZX^{(-1)})^{a^3} = 0,$$

$$(-1)^i(WX^{(-1)})^{a^3} + (-1)^i(XY^{(-1)})^{a^2} + (-1)^i(YZ^{(-1)})^a + ZW^{(-1)} = 0.$$

So if W, X, Y, Z denote the N-developed $(1, -1)$-matrices that correspond to the group ring elements (N-associates) with the same namesakes, then

$$(19.6) \qquad H_i = \begin{bmatrix} W & X^{a^3} & Y^{a^2} & Z^a \\ X & Y^{a^3} & Z^{a^2} & (-1)^i W^a \\ Y & Z^{a^3} & (-1)^i W^{a^2} & (-1)^i X^a \\ Z & (-1)^i W^{a^3} & (-1)^i X^{a^2} & (-1)^i Y^a \end{bmatrix}$$

is Hadamard. The superscripts a^j in (19.6) now indicate permutation equivalence operations. That is, if U is an N-indexed matrix then U^{a^j} is obtained by permuting rows and columns according to the action of a^j on N. Equations (19.5) and the template (19.6) comprise the substitution scheme for $K \cong C_4$.

19.1.2. $K \cong C_2^2$. Let $K = \langle a, b \mid a^2 = b^2 = 1, a^b = a \rangle$ and choose $\tilde{a}, \tilde{b} \in E$ such that $\pi(\tilde{a}) = a$, $\pi(\tilde{b}) = b$. Then $E = L_{ijk} \ltimes N$ where

$$L_{ijk} = \langle \tilde{a}, \tilde{b}, z \mid \tilde{a}^2 = z^i, \tilde{b}^2 = z^j, z^2 = 1, \tilde{a}^{\tilde{b}} = \tilde{a}z^k, z^{\tilde{a}} = z^{\tilde{b}} = z \rangle, \quad i, j, k \in \{0, 1\}.$$

We have

$$L_{ij0} \cong \begin{cases} C_2^3 & \text{if } i = j = 0 \\ C_4 \times C_2 & \text{otherwise} \end{cases} \qquad \text{and} \qquad L_{ij1} \cong \begin{cases} Q_8 & \text{if } i = j = 1 \\ D_8 & \text{otherwise.} \end{cases}$$

Let $g : N \to \{\pm 1\}$ be a map, and define $\beta = W + \tilde{a}X + \tilde{b}Y + \tilde{a}\tilde{b}Z$ where

$$W = \sum_{x \in N} g(x)x, \quad X = \sum_{x \in N} g(ax)x, \quad Y = \sum_{x \in N} g(bx)x, \quad Z = \sum_{x \in N} g(abx)x.$$

Now $x^{\tilde{a}} = x^a$, $x^{\tilde{b}} = x^b$, and $x^{\tilde{a}\tilde{b}} = x^{ab}$ for $x \in N$. Hence

$$\beta\beta^{(-1)} = WW^{(-1)} + (XX^{(-1)})^a + (YY^{(-1)})^b + (ZZ^{(-1)})^{ab}$$

$$+ \tilde{a}(z^i(WX^{(-1)})^a + XW^{(-1)} + z^i(YZ^{(-1)})^{ab} + (ZY^{(-1)})^b)$$

$$+ \tilde{b}(z^j(WY^{(-1)})^b + z^{j+k}(XZ^{(-1)})^{ab} + YW^{(-1)} + z^k(ZX^{(-1)})^a)$$

$$+ \tilde{a}\tilde{b}(z^{i+j+k}(WZ^{(-1)})^{ab} + z^j(XY^{(-1)})^b + z^{i+k}(YX^{(-1)})^a + ZW^{(-1)}).$$

Then

(19.7)

$$WW^{(-1)} + (XX^{(-1)})^a + (YY^{(-1)})^b + (ZZ^{(-1)})^{ab} = 4t,$$

$$(-1)^i(WX^{(-1)})^a + XW^{(-1)} + (-1)^i(YZ^{(-1)})^{ab} + (ZY^{(-1)})^b = 0,$$

$$(-1)^j(WY^{(-1)})^b + (-1)^{j+k}(XZ^{(-1)})^{ab} + YW^{(-1)} + (-1)^k(ZX^{(-1)})^a = 0,$$

$$(-1)^{i+j+k}(WZ^{(-1)})^{ab} + (-1)^j(XY^{(-1)})^b + (-1)^{i+k}(YX^{(-1)})^a + ZW^{(-1)} = 0$$

by (19.2). Letting W, X, Y, Z denote the matrices corresponding to the N-associates with the same labels, as before, we see that

$$(19.8) \qquad H_{ijk} = \begin{bmatrix} W & X^a & Y^b & Z^{ab} \\ X & (-1)^i W^a & Z^b & (-1)^i Y^{ab} \\ Y & (-1)^k Z^a & (-1)^j W^b & (-1)^{j+k} X^{ab} \\ Z & (-1)^{i+k} Y^a & (-1)^j X^b & (-1)^{i+j+k} W^{ab} \end{bmatrix}$$

is a Hadamard matrix if (19.7) holds.

19.2. Number-theoretic constraints

The plug-in matrices for (19.6) and (19.8) can be any $(1, -1)$-matrices of order t (satisfying (19.5) or (19.7)). However, if they are N-developed then these matrices will be regular, i.e., have constant row and column sums. In this section we suppose that W, X, Y, Z have constant row and column sums w, x, y, z respectively, and state some of the constraints on t that ensue. Note that $w \equiv x \equiv y \equiv z \equiv t \pmod 2$.

19.2.1. $K \cong C_4$. Premultiplying each equation in (19.5) by J_t, we get

$$w^2 + x^2 + y^2 + z^2 = 4t,$$
$$(-1)^i wz + xw + yx + zy = 0,$$
$$(-1)^i wy + (-1)^i xz + yw + zx = 0,$$
$$(-1)^i wx + (-1)^i xy + (-1)^i yz + zw = 0.$$

If $i = 0$ then $wy = -xz$ and $(w + y)(x + z) = 0$. We may suppose that $z = -x$, so $(w + y)^2 = 4t$ and t is a square.

If $i = 1$ then $w(x - z) + y(x + z) = 0$. Thus, if also t is odd then $w, x, y, z \equiv \pm 1 \pmod 4$, and $2 \equiv w(x - z) + y(x + z) \equiv 0 \pmod 4$: a contradiction.

19.2.1. LEMMA. *Suppose that there are regular plug-in matrices for H_i. Then*

  (i) t *is a square if $i = 0$,*
  (ii) t *is even if $i = 1$.*

19.2.2. $K \cong C_2^2$. From (19.7) we obtain

$$w^2 + x^2 + y^2 + z^2 = 4t,$$
$$(1 + (-1)^i)(wx + yz) = 0,$$
$$(1 + (-1)^j)(wy + (-1)^k xz) = 0,$$
$$(1 + (-1)^{i+j+k})(wz + (-1)^j xy) = 0.$$

Reasoning similar to that in the previous subsection applied to these equations yields

19.2.2. LEMMA. *Suppose that there are regular plug-in matrices for H_{ijk}. Then*

  (i) t *is a square if $(i, j, k) = (0, 0, 0)$,*
  (ii) t *is a sum of two squares if $(i, j, k) \in \{(1, 0, 0), (0, 1, 0), (1, 1, 0)\}$,*
  (iii) t *is an even sum of two squares if $(i, j, k) \in \{(0, 0, 1), (1, 0, 1), (0, 1, 1)\}$.*

Note that $2n$ is a sum of two squares if and only if n is a sum of two squares. Also, Lemma 19.2.1 and Lemma 19.2.2 (i) are consequences of Lemma 16.1.2 and Theorem 15.6.1.

19.3. Further results for group-developed plug-in matrices

Set $|N| = t$.

19.3.1. Prime t.

19.3.1. THEOREM. *Let $N \cong \mathrm{C}_p$, p an odd prime.*

   (i) *There are never any N-developed plug-in matrices for H_i.*
   (ii) *There are no N-developed plug-in matrices for H_{ijk} if $(i, j, k) \in \{(0, 0, 0), (0, 0, 1), (1, 0, 1), (0, 1, 1)\}$.*
   (iii) *For $(i, j, k) \in \{(1, 0, 0), (0, 1, 0), (1, 1, 0)\}$, if there are N-developed plug-in matrices for H_{ijk} then $p \equiv 1 \pmod 4$.*

PROOF. An odd prime is a sum of two squares if and only if it is congruent to 1 (mod 4). So the theorem follows from Lemmas 19.2.1 and 19.2.2. □

Thus, if t is an odd prime p, and there is a cocyclic Hadamard matrix with indexing group $K \ltimes N$ of order $4p$, then $K \cong \mathrm{C}_2^2$ and K acts trivially or invertingly on N (this was noted in the chapter preface).

19.3.2. $K \cong \mathrm{C}_4$.

19.3.2. THEOREM. *Let t be odd. If there are N-developed plug-in matrices for H_i, then $i = 0$ and there is a $(\mathrm{C}_4 \ltimes N)$-developed Hadamard matrix of order $4t$.*

PROOF. This follows from Lemma 19.2.1 (ii) and Lemma 16.1.2. □

19.3.3. $K \cong \mathrm{C}_2^2$.

19.3.3. THEOREM. *If there are N-developed plug-in matrices for H_{000} then there is a $(\mathrm{C}_2^2 \ltimes N)$-developed Hadamard matrix. If $(i, j, k) \in \{(0, 0, 1), (1, 0, 1), (0, 1, 1)\}$ and t is odd, then there are no N-developed plug-in matrices for H_{ijk}.*

PROOF. Lemma 16.1.2 takes care of the first claim. For the second we can use Lemma 19.2.2 (iii), or Theorem 15.6.4. □

We next demonstrate transference between cocyclic Hadamard matrices H_{ij0} and group-developed complex Hadamard matrices (cf. Subsection 8.2.2 and Theorem 16.5.2). This is significant in view of the non-existence results for complex Hadamard matrices developed modulo cyclic groups given in Section 15.7.

Let $k = 0$ and let one of i, j be non-zero. The extension group E is $L \ltimes N$ where $L = \langle a, b \mid a^4 = b^2 = 1, \ a^b = a \rangle$ and $a^2 = z$. Suppose that $a \in Z(E)$, so $E = \langle a \rangle \times M$ where $M = \langle b, N \rangle$. Define the homomorphism $\psi : \mathbb{Z}[E] \to \mathbb{Z}[\mathrm{i}][M]$ by

$$\psi : \sum_{j=0}^{3} \sum_{x \in M} \lambda_{j,x} a^j x \mapsto \sum_{j=0}^{3} \sum_{x \in M} \lambda_{j,x} \mathrm{i}^j x.$$

Then

$$\alpha = \tfrac{1}{4}(1+\mathrm{i})\psi((1-z)\beta) = \tfrac{1}{2}(1+\mathrm{i})\psi(W+aX+bY+abZ) = \tfrac{1}{2}(1+\mathrm{i})(W+\mathrm{i}X+b(Y+\mathrm{i}Z))$$

is an element of $\mathbb{Z}[\mathrm{i}][M]$ with $(\pm 1, \pm \mathrm{i})$-coefficients; and

$$\alpha\alpha^{(*)} = \tfrac{1}{16}(1+\mathrm{i})(1-\mathrm{i})\psi(2(1-z)\beta\beta^{(-1)}) = \tfrac{1}{4}\psi(4t(1-z)) = 2t.$$

We have proved

19.3.4. PROPOSITION. *Suppose that* $L_{ijk} = \langle a, b \mid a^4 = 1, b^2 = 1, a^b = a \rangle$, $a^2 = z$, *and* a *centralizes* N. *If* H_{ijk} *is a cocyclic Hadamard matrix then there exists a* $\langle b, N \rangle$-*developed complex Hadamard matrix.*

19.4. Inverting action

Obviously we cannot attempt an exhaustive general treatment of all possible conjugation actions by $G = K \ltimes N$ on N. We will consider inverting and trivial action only.

In this section K acts invertingly on N: so N is abelian.

19.4.1. NOTATION. For matrices X, Y of equal order, define

$$A(X,Y) = \begin{bmatrix} X & Y \\ Y & -X \end{bmatrix} \qquad \text{and} \qquad B(X,Y) = \begin{bmatrix} X & Y \\ Y & X \end{bmatrix}.$$

If X and Y are N-developed then $B(X,Y)$ is $(C_2 \times N)$-developed; $A(X,Y)$ is also in cocyclic form, with non-coboundary cocycle and the same indexing group $C_2 \times N$. Any $(C_2 \times N)$-developed array is permutation equivalent to $B(X,Y)$ for some N-developed X, Y.

19.4.1. $K \cong C_4$.

19.4.2. LEMMA. *Suppose that* $K = \langle a \mid a^4 = 1 \rangle$ *and* $x^a = x^{-1}$ *for all* $x \in N$. *Then there are* N-*developed plug-in matrices for* $H_0^{\mathrm{inv}} = H_0$ *if and only if there is a* $(C_2 \times N)$-*developed* $W(2t, t)$.

PROOF. Here (19.5) becomes

$$WW^\top + XX^\top + YY^\top + ZZ^\top = 4tI_t,$$
$$(X + Z)(W + Y)^\top = 0,$$
$$WY^\top + ZX^\top + YW^\top + XZ^\top = 0.$$

If these equations are satisfied then $\frac{1}{2}B(W + Y, X + Z)$ is a $(C_2 \times N)$-developed $W(2t, t)$.

A $(C_2 \times N)$-developed $W(2t, t)$ may be put into the form $B(P, Q)$ for some N-developed P, Q. Furthermore, $P = \frac{1}{2}(W + Y)$ and $Q = \frac{1}{2}(X + Z)$ where W, X, Y, Z are N-developed plug-in matrices for H_0.   □

Lemma 19.4.2 provides another example of transference. When N is cyclic of odd order (e.g., t is prime), the $W(2t, t)$ is developed modulo C_{2t} and so is equivalent to a circulant matrix.

19.4.3. DEFINITION. Let $C = \langle x \mid x^n = 1 \rangle$. A C-indexed matrix

$$[\omega^{\lfloor (i+j)/n \rfloor} g(x^{i+j})]_{ij}$$

is said to be ω-*cyclic*, or *negacyclic* if $\omega = -1$.

19.4.4. LEMMA.

(i) *If there are* N-*developed plug-in matrices for* H_0^{inv}, *then there is a real complementary pair of* $(C_2 \times N)$-*developed matrices.*

(ii) *If there are* N-*developed plug-in matrices for* $H_1^{\mathrm{inv}} = H_1$, *then there is a real complementary pair of block negacyclic matrices of order* $2t$ *with* N-*developed blocks.*

PROOF. From the constraints listed in the proof of Lemma 19.4.2, we see that $B(Y,W)$, $B(X,Z)$ is a complementary pair as in (i). We similarly check that if H_1^{inv} is a Hadamard matrix then $A(W,Y)$, $A(Z,X)$ is a pair as in (ii). $\square$

19.4.2. $K \cong \mathrm{C}_2^2$. Suppose that $K = \langle a,b \rangle \cong \mathrm{C}_2^2$ and $x^a = x^b = x^{-1}$ for all $x \in N$. The constraints (19.7) become

$$WW^\top + XX^\top + YY^\top + ZZ^\top = 4tI_t,$$
$$(-1)^i XW^\top + XW^\top + (-1)^i YZ^\top + YZ^\top = 0,$$
$$(-1)^j YW^\top + (-1)^{j+k} XZ^\top + YW^\top + (-1)^k XZ^\top = 0,$$
$$(-1)^{i+j+k} WZ^\top + (-1)^j YX^\top + (-1)^{i+k} XY^\top + ZW^\top = 0.$$

19.4.5. LEMMA. *There are N-developed plug-in matrices for H_{110}^{inv} if and only if there is a real complementary pair of $(\mathrm{C}_2 \times N)$-developed matrices.*

PROOF. The complementary pair in question is $B(W,Z)$, $B(-Y,X)$ where $WZ^\top + ZW^\top = XY^\top + YX^\top$. $\square$

Next, let $(i,j,k) = (1,1,1)$. The transpose of H_{111} is equivalent via row and column operations to Ito's array [**98**]:

$$(19.9) \qquad \begin{bmatrix} X & Y & Z & W \\ -Y & X & -W & Z \\ -Z^\top & W^\top & X^\top & -Y^\top \\ -W^\top & -Z^\top & Y^\top & X^\top \end{bmatrix}.$$

If N is cyclic of odd order then the extension group E of H_{111} is $\mathrm{Q}_8 \ltimes \mathrm{C}_t \cong \mathrm{Q}_{8t}$. This case is covered by Ito's conjecture (mentioned at the beginning of the chapter), and was studied by Flannery [**65**] and Schmidt [**139**]. If (19.9) is Hadamard then

$$ZW^\top + XY^\top = WZ^\top + YX^\top,$$

implying that $A(W,Z)$, $A(Y,X)$ and $\frac{1+\mathrm{i}}{2}(W+\mathrm{i}Z)$, $\frac{1+\mathrm{i}}{2}(Y+\mathrm{i}X)$ are complementary pairs. Hence

19.4.6. PROPOSITION. *The following are equivalent.*
   (i) *There are N-developed plug-in matrices for H_{111}^{inv}.*
   (ii) *There are N-developed plug-in matrices for (19.9).*
   (iii) *There is a real complementary pair of 2×2 block negacyclic matrices with N-developed blocks.*
   (iv) *There is a complex complementary pair of N-developed matrices.*

For odd t, Ito's conjecture then reads

6. CONJECTURE. *All objects in Proposition 19.4.6 exist when $N \cong \mathrm{C}_t$.*

We now turn to the remaining values of i,j,k.

19.4.7. LEMMA. *If there are N-developed plug-in matrices for H_{0jk}^{inv} or H_{10k}^{inv} then there are N-developed plug-in matrices for H_{110}^{inv} and H_{111}^{inv}.*

PROOF. If $i = 0$ then $XW^\top + YZ^\top = 0$. Therefore $A(X,W)$, $A(Y,Z)$ and $B(X,W)$, $B(Y,Z)$ are complementary pairs. Similar reasoning works for $j = 0$. Now use Lemma 19.4.5 and Proposition 19.4.6. $\square$

19.5. Trivial action

In this section $G = K \times N$.

19.5.1. When the cocycle is trivial. We have

$$
H_0^{\mathrm{triv}} =
\begin{bmatrix}
W & X & Y & Z \\
X & Y & Z & W \\
Y & Z & W & X \\
Z & W & X & Y
\end{bmatrix}
\quad \text{and} \quad
H_{000}^{\mathrm{triv}} =
\begin{bmatrix}
W & X & Y & Z \\
X & W & Z & Y \\
Y & Z & W & X \\
Z & Y & X & W
\end{bmatrix}.
$$

19.5.1. LEMMA.

(i) *If there are N-developed plug-in matrices for H_0^{triv} then there is a real complementary pair of $(C_2 \times N)$-developed matrices.*

(ii) *There are N-developed plug-in matrices for H_{000}^{triv} if and only if there is an anti-amicable real complementary pair of $(C_2 \times N)$-developed matrices.*

PROOF. The complementary pair is $B(W,Y)$, $B(X,Z)$ in (i), and $B(W,X)$, $B(Y,Z)$ in (ii).   □

If W, X, Y, Z are symmetric and commute with each other (e.g., N is abelian and the matrices are N-developed for suitable indexings) then the last three constraints of (19.7) for $i = j = k = 0$ reduce to

$$
WX + YZ = WY + XZ = WZ + XY = 0.
$$

Such matrices satisfy the last two constraints of (19.5) for $i = 0$, so are plug-in matrices for H_0^{triv} too. Turyn [155] found examples of order $t = 3^{2e}$. He noted that if W_i, X_i, Y_i, Z_i ($i = 1, 2$) are symmetric commuting plug-in matrices for H_{000}^{triv} then

$$
W_3 = \tfrac{1}{2}(W_1 + X_1) \otimes W_2 + \tfrac{1}{2}(W_1 - X_1) \otimes X_2,
$$
$$
X_3 = \tfrac{1}{2}(W_1 + X_1) \otimes Y_2 + \tfrac{1}{2}(W_1 - X_1) \otimes Z_2,
$$
$$
Y_3 = \tfrac{1}{2}(Y_1 + Z_1) \otimes W_2 + \tfrac{1}{2}(Y_1 - Z_1) \otimes X_2,
$$
$$
Z_3 = \tfrac{1}{2}(Y_1 + Z_1) \otimes Y_2 + \tfrac{1}{2}(Y_1 - Z_1) \otimes Z_2
$$

are plug-in matrices for H_{000}^{triv}. Thus we get a group-developed Hadamard matrix of order $4t_1 t_2$ from group-developed Hadamard matrices of orders $4t_1$ and $4t_2$.

A lot more has been accomplished since Turyn's initial contributions. Plug-in matrices have been constructed for various t [158, 161, 162]; e.g., $t = m^2$ where $m = 2^a 3^b 5^{2c_1} 13^{2c_2} 17^{2c_3} q^2$, q a product of primes congruent to 3 (mod 4). Some non-existence results are reported in [118, 119, 133].

19.5.2. $K \cong C_4$ and non-trivial cocycle. Here the substitution scheme is

$$
H_1^{\mathrm{triv}} =
\begin{bmatrix}
W & X & Y & Z \\
X & Y & Z & -W \\
Y & Z & -W & -X \\
Z & -W & -X & -Y
\end{bmatrix}
$$

where

$$
WW^\top + XX^\top + YY^\top + ZZ^\top = 4tI_t,
$$
$$
-WZ^\top + XW^\top + YX^\top + ZY^\top = 0,
$$
$$
-WY^\top - XZ^\top + YW^\top + ZX^\top = 0.
$$

19.5.2. LEMMA. *There are N-developed plug-in matrices for H_1^{triv} if and only if $WZ^\top = XW^\top + YX^\top + ZY^\top$ and $A(W,Y)$, $A(X,Z)$ is a complementary pair.*

19.5.3. $K \cong \mathrm{C}_2^2$ and non-trivial cocycle. We divide the triples (i,j,k) into four sets $\{(0,0,0)\}$, $\{(1,0,0),(0,1,0),(1,1,0)\}$, $\{(0,0,1),(1,0,1),(0,1,1)\}$, and $\{(1,1,1)\}$, such that (i_1,j_1,k_1) and (i_2,j_2,k_2) are in the same set if the substitution schemes that they define are equivalent. For example, we can transform the scheme for $(1,1,0)$ into that for $(1,0,0)$ by relabeling Y as Z and vice versa, then negating Z. Each set of triples corresponds to a single isomorphism type for L_{ijk}.

19.5.3. LEMMA. *The following are equivalent.*

(1) $A(W,X)$, $A(Y,Z)$ *is an amicable complementary pair.*
(2) $A(W,X)$, $A(Z,-Y)$ *is an anti-amicable complementary pair.*
(3) $B(W,Z)$, $B(X,-Y)$ *is an amicable complementary pair.*

PROOF. Let

$$P_1 = \begin{bmatrix} A(W,X) & A(Y,Z) \\ A(Y,Z) & -A(W,X) \end{bmatrix}, \quad P_2 = \begin{bmatrix} A(W,X) & A(Z,-Y) \\ A(Z,-Y) & A(W,X) \end{bmatrix},$$

$$P_3 = \begin{bmatrix} B(W,Z) & B(X,-Y) \\ B(X,-Y) & -B(W,Z) \end{bmatrix}$$

(P_1 is H_{110}^{triv}). We obtain P_2 from P_1 by swapping the last two rows of P_1, then swapping the last two columns and negating the last row and column. Also, $P_3 \approx P_2$: swap rows 2 and 3 of P_2, then swap columns 2 and 3. Now observe that part (α) is the same as P_α being Hadamard. $\square$

19.5.4. COROLLARY. *For $(i,j,k) \in \{(1,0,0),(0,1,0),(1,1,0)\}$, there are N-developed plug-in matrices for H_{ijk}^{triv} if and only if there are*

(i) *an amicable real complementary pair of $(\mathrm{C}_2 \times N)$-developed matrices,*
(ii) *both an amicable real complementary pair and an anti-amicable real complementary pair of block negacyclic matrices of order $2t$ with N-developed blocks.*

Existence of any cocyclic Hadamard matrix as in Corollary 19.5.4 implies a rather surprising fact about cocyclic complex Hadamard matrices.

19.5.5. PROPOSITION. *There is a $(\mathrm{C}_2 \times N)$-developed complex Hadamard matrix if and only if there is a block negacyclic complex Hadamard matrix of order $2t$ with N-developed blocks.*

PROOF. First, it is not hard to verify that the block negacyclic (respectively, $(\mathrm{C}_2 \times N)$-developed) complex Hadamard matrix exists if and only if there is an amicable real complementary pair of block negacyclic (respectively, $(\mathrm{C}_2 \times N)$-developed) matrices of the same form. The result then follows from Lemma 19.5.3. $\square$

The next lemma is immediate from inspection of H_{001}^{triv} and H_{101}^{triv}.

19.5.6. LEMMA. *If there are N-developed plug-in matrices for H_{ijk}^{triv}, where $(i,j,k) \in \{(0,0,1),(1,0,1),(0,1,1)\}$, then there are*

(i) *a real complementary pair of $(\mathrm{C}_2 \times N)$-developed matrices,*
(ii) *a real complementary pair of block negacyclic matrices of order $2t$ with N-developed blocks.*

The final case H_{111}^{triv} is the Williamson array, with constraints

$$WX^\top + YZ^\top = XW^\top + ZY^\top,$$

$$WY^\top + ZX^\top = YW^\top + XZ^\top,$$

$$WZ^\top + XY^\top = ZW^\top + YX^\top.$$

19.5.7. LEMMA. *The N-developed matrices W, X, Y, Z are plug-in matrices for H_{111}^{triv} if and only if* (i) $A(W, X)$, $A(Y, Z)$, (ii) $A(W, Y)$, $A(Z, X)$, (iii) $A(W, Z)$, $A(X, Y)$ *are all complementary pairs.*

Several infinite families of plug-in matrices for the Williamson array are known. Sets of symmetric circulants exist for all odd $t < 60$, excluding $t = 35, 47, 53, 59$: alas, for these t, such plug-in matrices do not exist (this was proved by Holzmann, Kharaghani, and Tayfeh-Rezaie [85]).

19.6. Complementary pairs and the Cocyclic Hadamard Conjecture

We pause to take stock.

19.6.1. THEOREM. *Let G be one of*

(1) $C_4 \times N$,
(2) $C_2^2 \times N$,
(3) $\langle a \mid a^4 = 1 \rangle \ltimes N$ *where N is abelian and $x^a = x^{-1}$ for all $x \in N$,*
(4) $\langle a, b \mid a^2 = b^2 = (ab)^2 = 1 \rangle \ltimes N$ *where N is abelian and $x^a = x^b = x^{-1}$ for all $x \in N$.*

Suppose that there is a cocyclic Hadamard matrix with indexing group G. Then at least one of the following holds:

(i) *there is a real complementary pair of $(C_2 \times N)$-developed matrices;*
(ii) *there is a complex complementary pair of N-developed matrices.*

PROOF. We will be very terse, simply citing the relevant results. Throughout, the equivalence between parts (iii) and (iv) of Proposition 19.4.6 is in play.
1. Lemmas 19.5.1 and 19.5.2.
2. Lemmas 19.5.1, 19.5.6, 19.5.7, and Corollary 19.5.4.
3. Lemma 19.4.4.
4. Lemma 19.4.5, Proposition 19.4.6, and Lemma 19.4.7. $\qquad\square$

If $|N|$ is prime, we get a more succinct statement.

19.6.2. THEOREM. *Let $p > 3$ be prime. There is a cocyclic Hadamard matrix of order $4p$ if and only if at least one of the following is true:*

(i) *there is a real complementary pair of circulant matrices of order $2p$;*
(ii) *there is a complex complementary pair of circulant matrices of order p.*
The former can only occur if $p \equiv 1 \pmod 4$.

PROOF. By Theorem 19.3.1 and Lemma 19.4.7, the substitution array must be H_{110}^{inv} or H_{111}^{inv} or H_{ijk}^{triv} where $(i, j, k) \in \{(1, 0, 0), (0, 1, 0), (1, 1, 0), (1, 1, 1)\}$. Parts (2) and (4) of Theorem 19.6.1 apply. In each case, existence of the plug-in matrices is equivalent to existence of at least one of the complementary pairs. The last claim is Theorem 19.3.1 (iii). $\qquad\square$

19.6.3. REMARK. Only H_{111}^{inv} and H_{111}^{triv} survive when $p \equiv 3 \pmod 4$.

19.7. Existence of group-developed complementary pairs

This section gathers some ideas and results from the wider literature that have an impact on the existence problem for group-developed complementary pairs.

We begin with general facts. The following number-theoretic condition has already been observed.

19.7.1. LEMMA. *If there is a real complementary pair of regular matrices of order n, then $2n$ and hence n is a sum of two squares.*

PROOF. Multiply both sides of $AA^\top + BB^\top = 2nI_n$ by J_n. □

The next lemma is the first composition-type construction in this section.

19.7.2. LEMMA. *If there is an amicable real complementary pair A, B of M-developed matrices, and a complementary pair P, Q of N-developed matrices, then*

$$X = \tfrac{1}{2}((A + B) \otimes P + (A - B) \otimes Q), \quad Y = \tfrac{1}{2}((A + B) \otimes Q - (A - B) \otimes P)$$

is a complementary $(M \times N)$-developed pair.

PROOF. Since $XX^* + YY^* = 2mnI_{mn}$, we see that X, Y are complementary $(\pm 1, \pm i)$-matrices. Also, as sums of Kronecker products of M- and N-developed matrices, X and Y are $(M \times N)$-developed. □

19.7.3. COROLLARY. *If there is an N-developed complementary pair then there is a $(C_2 \times N)$-developed complementary pair.*

PROOF. Let $A = J_2$ and $B = 2I_2 - J_2$ in Lemma 19.7.2. □

19.7.1. Complementary circulants. We now specialize to complementary pairs that are group-developed modulo cyclic groups. Character theory may be used to obtain further order restrictions for such pairs.

19.7.4. THEOREM ([4], Corollary 3.6). *Let $p \equiv 3 \pmod 4$ be a prime. Suppose that there is a real complementary pair of circulant matrices of order $n = up^a$, where u is coprime to p. Then $u \geq 2p^b$, where b is the integer part of $a/2$.*

Thus, of the even sums of two squares less than 50, neither 18 nor 36 can be the order of a real complementary pair of circulants. On the positive side, Golay sequences supply many of the values of n in the following (Golay sequences crop up again in Chapter 23).

19.7.5. THEOREM ([**127**], Proposition 4.1). *There is a real complementary pair of circulant matrices of order $n \leq 50$ if and only if $n \in \{1, 2, 4, 8, 10, 16, 20, 26, 32, 34, 40, 50\}$.*

One might conjecture that circulant real pairs exist for all $n = 2p_1 \cdots p_r$ where the $p_1, \ldots, p_r$ are distinct primes congruent to 1 (mod 4). The smallest open order is $n = 58$. However, these complementary pairs could only represent a vanishingly small fraction of the orders needed to prove the Cocyclic Hadamard Conjecture.

Here is another doubling construction.

19.7.6. LEMMA. *If there is a real (respectively, complex) complementary pair of circulants of order n then there is a real (respectively, complex) complementary pair of circulants of order $2n$.*

PROOF. Let $[a_1, \ldots, a_n]$, $[b_1, \ldots, b_n]$ be the first rows of a complementary pair of circulants; then $[a_1, b_1, \ldots, a_n, b_n]$, $[a_1, -b_1, \ldots, a_n, -b_n]$ are the first rows of complementary circulants.   □

Dropping the (quite restrictive) symmetry hypothesis and two of the constraints for the Williamson array H_{111}^{triv}, we get

19.7.7. LEMMA. *If W, X, Y, Z are circulant $(1, -1)$-matrices such that*

$$(19.10) \quad WW^\top + XX^\top + YY^\top + ZZ^\top = 4nI_n, \quad WX^\top + YZ^\top = XW^\top + ZY^\top$$

then $\frac{1+\mathrm{i}}{2}(W + \mathrm{i}X)$, $\frac{1+\mathrm{i}}{2}(Y + \mathrm{i}Z)$ is a complementary pair of circulants.

By Lemmas 19.7.6 and 19.7.7, we have complex complementary circulants of orders $2^a n$ for $n \leq 57$ and $n \notin \{35, 47, 53\}$.

If there are N-developed $(1, -1)$-matrices W, X, Y, Z such that

$$(19.11) \quad WW^\top + XX^\top + YY^\top + ZZ^\top = 4nI_n, \quad WX^\top + YZ^\top + XW^\top + ZY^\top = 0$$

then $B(W, X)$, $B(Y, Z)$ is a complementary $(\mathrm{C}_2 \times N)$-developed pair. One might seek such complementary circulants from Williamson matrices. However

19.7.8. THEOREM. *If p is an odd prime then there does not exist an amicable real complementary pair of circulant matrices of order $2p^f$, $f \geq 1$. Hence there are no symmetric circulants W, X, Y, Z of order p^f satisfying* (19.11).

PROOF. Suppose that B_1, B_2 are amicable real complementary circulants. Then $C = \frac{1+\mathrm{i}}{2}(B_1 + \mathrm{i}B_2)$ is a circulant complex Hadamard matrix. If W, X, Y, Z exist then $B_1 = B(W, X)$, $B_2 = B(Y, Z)$ are amicable and group-developed modulo $\mathrm{C}_2 \times \mathrm{C}_{p^f} \cong \mathrm{C}_{2p^f}$. By Theorem 15.7.11, this completes the proof.   □

Indeed, by Conjecture 5 (Section 15.7), it is perhaps unlikely that there are symmetric circulants of any odd order satisfying (19.11).

19.7.2. Other group-developed complementary pairs. Now we consider pairs that are developed modulo non-cyclic groups.

19.7.2.1. *Auxiliary matrices.* We say that $(\pm 1, \pm \mathrm{i})$-matrices $A_1, \ldots, A_s$ of order t are *auxiliary matrices* if $\sum_{i=1}^s A_i A_i^* = stI_t$ and $A_i A_j^* = J_t$ for all $i \neq j$. Such matrices are used to construct larger group-developed complementary pairs from smaller pairs.

19.7.9. LEMMA. *Let N, M be finite groups, and set $|N| = n$. If there are n auxiliary M-developed matrices and a complementary pair of N-developed matrices, then there is a complementary pair of $(N \times M)$-developed matrices.*

PROOF. Suppose that $A_1, \ldots, A_n$ are auxiliary M-developed matrices, and $P = [p_{ij}]$, $Q = [q_{ij}]$ is a complementary pair of N-developed matrices. Let $L = [l_{ij}]$ be any Latin square of order n. Then the $n \times n$ block matrices $X = [p_{ij} A_{l_{ij}}]$ and $Y = [q_{ij} A_{l_{ij}}]$ are $(N \times M)$-developed and complementary.   □

Auxiliary matrices tend to exist for square orders.

19.7.10. PROPOSITION. *Let q be a prime.*

   (i) *If $q \equiv 3 \pmod 4$ then there are $(q+1)/2$ complex E_{q^2}-developed auxiliary matrices and $q + 1$ real E_{q^2}-developed auxiliary matrices.*

(ii) *If $q \equiv 1 \pmod 4$ then there are $q + 1$ complex E_{q^2}-developed auxiliary matrices.*

PROOF. For $1 \le k \le q$, let B_k be the $q \times q$ block matrix whose (i,j)th block is $T^{(k-1)(i-j)}$, where T is the $q \times q$ permutation matrix $[\delta^r_{s(1,2,\ldots,q)}]_{rs}$. We check that $B_k B_l^\top = J_{q^2}$ for $k \ne l$, and $B_k B_k^\top$ has (i,j)th block equal to $qT^{(k-1)(i-j)}$.

Suppose that $q \equiv 3 \pmod 4$. The Paley type I Hadamard matrix of order $q+1$ is equivalent to a matrix with an E_q-developed core Q; so $QQ^\top = (q+1)I_q - J_q$ and $QJ_q = -J_q$. Let $A_i = (I_q \otimes Q)B_i$. Then

$$A_i A_j^\top = (I_q \otimes Q)(J_q \otimes J_q)(I_q \otimes Q^\top) = J_{q^2} \quad \forall\, i \ne j.$$

We additionally have

$$
\begin{aligned}
\sum_{i=1}^q A_i A_i^\top &= (I_q \otimes Q) \sum_{i=1}^q B_i B_i^\top (I_q \otimes Q^\top) \\
&= (I_q \otimes Q)(I_q \otimes q^2 I_q + (J_q - I_q) \otimes qJ_q)(I_q \otimes Q^\top) \\
&= q^2(I_q \otimes QQ^\top) + (J_q - I_q) \otimes qQJ_qQ^\top \\
&= q^2(I_q \otimes ((q+1)I_q - J_q)) + (J_q - I_q) \otimes qJ_q \\
&= q^3 I_{q^2} - q^2(I_q \otimes (J_q - I_q)) + (J_q - I_q) \otimes qJ_q.
\end{aligned}
$$

If $A_0 := Q \otimes J_q$ then $A_0 A_i^\top = J_{q^2}$ for $i \ge 1$, and

$$
\begin{aligned}
A_0 A_0^\top &= QQ^\top \otimes qJ_q \\
&= ((q+1)I_q - J_q) \otimes qJ_q \\
&= (qI_q - (J_q - I_q)) \otimes qJ_q \\
&= q^2(I_q \otimes J_q) - (J_q - I_q) \otimes qJ_q \\
&= q^2 I_{q^2} + q^2(I_q \otimes (J_q - I_q)) - (J_q - I_q) \otimes qJ_q.
\end{aligned}
$$

So the A_i are $q + 1$ auxiliary matrices. Furthermore, the $\frac{1+\mathrm{i}}{2}(A_{2i} + \mathrm{i}A_{2i+1})$ are complex auxiliary matrices.

The Paley conference matrix C of order $q + 1$ is symmetric if $q \equiv 1 \pmod 4$. Then $C + \mathrm{i}I_{q+1}$ is a complex Hadamard matrix with E_q-developed core, which we may take as Q. $\qquad\square$

19.7.11. EXAMPLE. Let $q = 3$. We have

$$
B_1 = \begin{bmatrix} I & I & I \\ I & I & I \\ I & I & I \end{bmatrix}, \qquad
B_2 = \begin{bmatrix} I & T^2 & T \\ T & I & T^2 \\ T^2 & T & I \end{bmatrix}, \qquad
B_3 = \begin{bmatrix} I & T & T^2 \\ T^2 & I & T \\ T & T^2 & I \end{bmatrix}
$$

and

$$
A_0 = \left[\begin{array}{ccc|ccc|ccc}
1 & 1 & 1 & -1 & -1 & -1 & -1 & -1 & -1 \\
1 & 1 & 1 & -1 & -1 & -1 & -1 & -1 & -1 \\
1 & 1 & 1 & -1 & -1 & -1 & -1 & -1 & -1 \\
\hline
-1 & -1 & -1 & 1 & 1 & 1 & -1 & -1 & -1 \\
-1 & -1 & -1 & 1 & 1 & 1 & -1 & -1 & -1 \\
-1 & -1 & -1 & 1 & 1 & 1 & -1 & -1 & -1 \\
\hline
-1 & -1 & -1 & -1 & -1 & -1 & 1 & 1 & 1 \\
-1 & -1 & -1 & -1 & -1 & -1 & 1 & 1 & 1 \\
-1 & -1 & -1 & -1 & -1 & -1 & 1 & 1 & 1
\end{array}\right], \quad
A_1 = \left[\begin{array}{ccc|ccc|ccc}
1 & -1 & -1 & 1 & -1 & -1 & 1 & -1 & -1 \\
-1 & 1 & -1 & -1 & 1 & -1 & -1 & 1 & -1 \\
-1 & -1 & 1 & -1 & -1 & 1 & -1 & -1 & 1 \\
\hline
1 & -1 & -1 & 1 & -1 & -1 & 1 & -1 & -1 \\
-1 & 1 & -1 & -1 & 1 & -1 & -1 & 1 & -1 \\
-1 & -1 & 1 & -1 & -1 & 1 & -1 & -1 & 1 \\
\hline
1 & -1 & -1 & 1 & -1 & -1 & 1 & -1 & -1 \\
-1 & 1 & -1 & -1 & 1 & -1 & -1 & 1 & -1 \\
-1 & -1 & 1 & -1 & -1 & 1 & -1 & -1 & 1
\end{array}\right],
$$

$$A_2 = \left[\begin{array}{rrr|rrr|rrr} 1 & -1 & -1 & -1 & -1 & 1 & -1 & 1 & -1 \\ -1 & 1 & -1 & 1 & -1 & -1 & -1 & -1 & 1 \\ -1 & -1 & 1 & -1 & 1 & -1 & 1 & -1 & -1 \\ \hline -1 & 1 & -1 & 1 & -1 & -1 & -1 & -1 & 1 \\ -1 & -1 & 1 & -1 & 1 & -1 & 1 & -1 & -1 \\ 1 & -1 & -1 & -1 & -1 & 1 & -1 & 1 & -1 \\ \hline -1 & -1 & 1 & -1 & 1 & -1 & 1 & -1 & -1 \\ 1 & -1 & -1 & -1 & -1 & 1 & -1 & 1 & -1 \\ -1 & 1 & -1 & 1 & -1 & -1 & -1 & -1 & 1 \end{array}\right], \quad A_3 = \left[\begin{array}{rrr|rrr|rrr} 1 & -1 & -1 & -1 & 1 & -1 & -1 & -1 & 1 \\ -1 & 1 & -1 & -1 & -1 & 1 & 1 & -1 & -1 \\ -1 & -1 & 1 & 1 & -1 & -1 & -1 & 1 & -1 \\ \hline -1 & -1 & 1 & 1 & -1 & -1 & -1 & 1 & -1 \\ 1 & -1 & -1 & -1 & 1 & -1 & -1 & -1 & 1 \\ -1 & 1 & -1 & -1 & -1 & 1 & 1 & -1 & -1 \\ \hline -1 & 1 & -1 & -1 & -1 & 1 & 1 & -1 & -1 \\ -1 & -1 & 1 & 1 & -1 & -1 & -1 & 1 & -1 \\ 1 & -1 & -1 & -1 & 1 & -1 & -1 & -1 & 1 \end{array}\right].$$

Notice that

$$W_1 = \begin{bmatrix} A_0 & A_1 \\ A_1 & A_0 \end{bmatrix}, \qquad W_2 = \begin{bmatrix} A_2 & -A_3 \\ -A_3 & A_2 \end{bmatrix}$$

is an amicable complementary pair. So $\frac{1+i}{2}(W_1 + iW_2)$ is a $(C_2 \times C_3^2)$-developed complex Hadamard matrix, and

$$X_1 = \left[\begin{array}{rrr|rrr|rrr} i & 1 & 1 & -i & -i & -1 & -i & -1 & -i \\ 1 & i & 1 & -1 & -i & -i & -i & -i & -1 \\ 1 & 1 & i & -i & -1 & -i & -1 & -i & -i \\ \hline -i & -1 & -i & i & 1 & 1 & -i & -i & -1 \\ -i & -i & -1 & 1 & i & 1 & -1 & -i & -i \\ -1 & -i & -i & 1 & 1 & i & -i & -1 & -i \\ \hline -i & -i & -1 & -i & -1 & -i & i & 1 & 1 \\ -1 & -i & -i & -i & -i & -1 & 1 & i & 1 \\ -i & -1 & -i & -1 & -i & -i & 1 & 1 & i \end{array}\right], \quad X_2 = \left[\begin{array}{rrr|rrr|rrr} 1 & -1 & -1 & i & -i & -1 & i & -1 & -i \\ -1 & 1 & -1 & -1 & i & -i & -i & i & -1 \\ -1 & -1 & 1 & -i & -1 & i & -1 & -i & i \\ \hline i & -1 & -i & 1 & -1 & -1 & i & -i & -1 \\ -i & i & -1 & -1 & 1 & -1 & -1 & i & -i \\ -1 & -i & i & -1 & -1 & 1 & -i & -1 & i \\ \hline i & -i & -1 & i & -1 & -i & 1 & -1 & -1 \\ -1 & i & -i & -i & i & -1 & -1 & 1 & -1 \\ -i & -1 & i & -1 & -i & i & -1 & -1 & 1 \end{array}\right]$$

is a C_3^2-developed complementary pair.

19.7.12. EXAMPLE. There would be an E_{90}-developed complementary pair by Lemma 19.7.2 and the example above, if there was an amicable real complementary pair of circulants of order 10. But the latter pair does not exist by Theorem 19.7.8.

Auxiliary matrices can be composed.

19.7.13. LEMMA. *Suppose that* $A_k = [a_{ij}^{(k)}]$ *for* $1 \leq k \leq r$ *are auxiliary matrices of order* s. *For any auxiliary matrices* $B_1, \ldots, B_{rs}$, *let* C_k *be the block matrix with* (i,j)th *block* $a_{ij}^{(k)} B_{(k-1)s+i-j+1}$, *reading subscripts modulo* rs. *Then* $C_1, \ldots, C_r$ *are auxiliary matrices.*

19.7.14. EXAMPLE. There are two auxiliary matrices of order 9, and 18 auxiliary matrices of order 17^2. So there are two auxiliary matrices of order $3^2 17^2$. These form a C_{51}^2-developed complementary pair.

19.7.15. EXAMPLE. Since there are eight real auxiliary E_{49}-developed matrices, there are amicable real $E_{2^2 7^2}$- and $(C_4 \times E_{49})$-developed complementary pairs.

19.7.2.2. *Weak special plug-in matrices.* Call $(1,-1)$-matrices W, X, Y, Z such that

$$\text{(19.12)} \qquad WW^\top + XX^\top + YY^\top + ZZ^\top = 4nI_n,$$
$$\text{(19.13)} \qquad WX^\top + YZ^\top = WY^\top + XZ^\top = WZ^\top + XY^\top = 0$$

special plug-in matrices. As noted in Subsection 19.5.1, E_n-developed special plug-in matrices exist for all $n = (2^a 3^b 5^{2c_1} 13^{2c_2} 17^{2c_3} q^2)^2$ where q is a product of primes congruent to 3 (mod 4). If (19.12) and

$$(19.14) \qquad\qquad WX^\top + YZ^\top = 0$$

hold, then W, X, Y, Z are *weak special plug-in matrices*. Since (19.12) and (19.14) together imply (19.10) and (19.11), weak special plug-in matrices of order n yield both real and complex complementary pairs of orders $2n$ and n respectively. Turyn's weak special plug-in matrices of order 9 then give real pairs developed modulo E_{18} and E_{36}. By Theorem 19.7.5, this proves

19.7.16. THEOREM. *Group-developed real complementary pairs exist for all even orders $n \leq 50$ such that n is a sum of two squares.*

We have one more composition result.

19.7.17. LEMMA. *If there are weak special plug-in matrices of order m and special plug-in matrices of order n, then there are weak special plug-in matrices of order mn.*

PROOF. (Cf. Lemma 19.7.2.) Let A, B, C, D be the special plug-in matrices, and let P, Q, R, S be the weak special plug-in matrices. Then

$$W = \tfrac{1}{2}((P+Q)\otimes A + (P-Q)\otimes B), \quad Y = \tfrac{1}{2}((R+S)\otimes A + (R-S)\otimes B),$$
$$X = \tfrac{1}{2}((P+Q)\otimes C + (P-Q)\otimes D), \quad Z = \tfrac{1}{2}((R+S)\otimes C + (R-S)\otimes D)$$

are $(1,-1)$-matrices satisfying (19.12) and (19.14). $\qquad\qquad\square$

We might suspect that E_n-developed weak special plug-in matrices exist for all square orders. The smallest open order is 25.

19.7.3. Future work. A proper study of group-developed complementary pairs has yet to be done; this section just scratched the surface. We should aim to prove that there are E_n-developed matrices satisfying

- (19.10) for all n;
- (19.11) for each n that is a sum of two squares;
- (19.12) and (19.13) for all fourth powers n;
- (19.12) and (19.14) for all square n.

CHAPTER 20

Calculating Cocyclic Development Rules

Ever since Bose introduced his 'method of differences' for BIBDs [**13**], various development rules have become known to design theorists. Prior to the discovery of cocyclic development, there was no unifying theory for these rules.

A *development table* encapsulates all cocyclic development rules for a finite group G. We will describe how to compute development tables when G is given by a polycyclic presentation. Our method is independent of the (finitely generated abelian) coefficient group, and thus is applicable to many classes of designs.

This chapter is long, but has the virtues of being elementary and self-contained. References for standard theory in this chapter are [**12, 108**].

20.1. Introduction to development tables

In plainest terms, calculating cocycles is just linear algebra over $\mathbb{Z}$. The cocycle identity for a group G of order v and abelian coefficient group C can be viewed as a homogeneous system of v^3 linear equations

$$(20.1) \qquad x(a,b) + x(ab,c) - x(b,c) - x(a,bc) = 0, \quad a,b,c \in G.$$

Let M be the $v^3 \times v^2$ $(0,\pm1)$-matrix of the system (20.1). That is, if x is the row vector of variables $x(a,b)$, then $Mx^\top = 0$. We may put M into Smith normal form: there are $\mathbb{Z}$-matrices P, Q invertible over $\mathbb{Z}$, and a uniquely determined matrix D, such that $M = PDQ$. Here D has entry d_i in position (i,i), $1 \le i \le m$, and all other entries zero, where $d_1, \ldots, d_m$ are positive integers such that $d_i | d_{i+1}$. Put $d_{m+1} = \cdots = d_{v^2} = 0$.

Denote the ith component of $y = xQ^\top$ by y_i. Since P is invertible over $\mathbb{Z}$, $Mx^\top = 0$ if and only if $d_i y_i = 0$, $1 \le i \le m$. Thus

$$x(a,b) = \sum_{i=1}^{v^2} \lambda(i,a,b)c_i$$

where $\lambda(i,a,b)$ is the ith entry in column (a,b) of $(Q^\top)^{-1}$, and $c_i \in C$ has order dividing d_i if $d_i \ne 0$. We present solutions of the system (20.1) in an array.

20.1.1. DEFINITION. Let $X_1, \ldots, X_n$ be commuting indeterminates satisfying $X_i^{e_i} = 1$ where $e_1, \ldots, e_n \in \mathbb{N}$. For $1 \le i \le n$ and $a, b \in G$, let $\lambda(i,a,b) \in \mathbb{Z}$. Then

$$(20.2) \qquad [\textstyle\prod_{i=1}^n X_i^{\lambda(i,a,b)}]_{a,b \in G}$$

is a *development table* if, for each finitely generated abelian group C, the set of cocycles $f : G \times G \to C$ is the set of maps f such that $f(a,b) = \prod_{i=1}^n c_i^{\lambda(i,a,b)}$ for some $c_i \in C$ satisfying $c_i^{e_i} = 1$.

In other words, if we replace X_i in the development table by any element c_i of C such that $c_i^{e_i} = 1$ $(1 \leq i \leq t)$, then we obtain a cocyclic matrix $[f(a,b)]_{a,b \in G}$; and every cocycle $f : G \times G \to C$ is so obtained. It is sensible to require that our tables are *minimal*, insofar as no X_i is redundant.

20.2. Development tables for abelian groups

In Section 11.3, we determined all cocycles with a cyclic ground group. A more general argument, taken from de Launey and Horadam's paper [**50**], may be used to calculate a development table for any abelian ground group G.

Suppose that G is in torsion-invariant form $\langle x_1 \rangle \oplus \cdots \oplus \langle x_m \rangle$, where $|x_i| = n_i$ and $n_1 | \cdots | n_m$. For $a \in G$, let a_i be the integers such that $0 \leq a_i < n_i$ and $a = \sum_{i=1}^m a_i x_i$. Denote the kth partial sum $\sum_{i=1}^k a_i x_i$ of a by A_k. Let $f : G \times G \to C$ be a map, and define

$$R(a, b, c) = f(a, b) + f(a + b, c) - f(b, c) - f(a, b + c).$$

We assume that $R(a, b, c) = 0$ for all $a, b, c \in G$; i.e., f satisfies the cocycle identity. Next define

$$S(a, b) = f(a, b) - f(b, a).$$

Since

$$R(a, c, b) - R(c, a, b) + R(c, b, a) = S(a, b) + S(a, c) - S(a, b + c),$$

$S : G \times G \to C$ is an alternating bilinear form. Also,

$$R(A_{k-1}, a_k x_k, B_k) - R(A_{k-1}, B_{k-1}, (a_k + b_k)x_k) - R(a_k x_k, b_k x_k, B_{k-1})$$
$$= f(A_k, B_k) + f(A_{k-1}, a_k x_k) - f(A_{k-1}, B_{k-1}) - f(A_{k-1} + B_{k-1}, (a_k + b_k)x_k)$$
$$- f(a_k x_k, b_k x_k) + S(B_{k-1}, (a_k + b_k)x_k) + f(b_k x_k, B_{k-1})$$

implies that

(20.3)
$$f(A_k, B_k) = f(A_{k-1}, B_{k-1}) + f(a_k x_k, b_k x_k) + f(A_{k-1} + B_{k-1}, (a_k + b_k)x_k)$$
$$- f(A_{k-1}, a_k x_k) - f(B_{k-1}, b_k x_k) - S(B_{k-1}, a_k x_k).$$

Let $s_{ij} = S(x_i, x_j)$; so $s_{ii} = 0$, $s_{ij} = -s_{ji}$, and $n_i s_{ij} = n_j s_{ij} = 0$. Iterating (20.3) and using Theorem 11.3.2, we get

(20.4)
$$f(a, b) = \sum_{i=1}^m \left\{ f(0, 0) + \sum_{j=0}^{b_i - 1} [f((a_i + j)x_i, x_i) - f(jx_i, x_i)] \right\}$$
$$+ \sum_{i=2}^m \left\{ f(A_{i-1} + B_{i-1}, (a_i + b_i)x_i) - f(A_{i-1}, a_i x_i) - f(B_{i-1}, b_i x_i) \right\}$$
$$- \sum_{i=1}^{m-1} \sum_{j=i+1}^m b_i a_j s_{ij}.$$

20.2.1. THEOREM. *The map f satisfies the cocycle identity if and only if (20.4) holds for all $a, b \in G$.*

PROOF. We have proved the 'only if' part. Proving the 'if' part is left as an exercise. □

Let $P_1 = \{f(0,0)\} \cup P_1'$ where

$$P_1' = \{f(y_k, x_k), f(z_k, y_k) \mid y_k \in \langle x_k \rangle, \; z_k \in \langle x_1, \ldots, x_{k-1} \rangle, \; y_k, z_k \neq 0, \; 1 \le k \le m\},$$

and let $P_2 = \{s_{ij} \mid 1 \le i < j \le m\}$. We can write down a development table for the abelian group G, whose indeterminates correspond to the elements of $P_1 \cup P_2$, by Theorem 20.2.1. There are precisely

$$1 + \sum_{i=1}^{m}(n_i - 1) + \sum_{i=1}^{m}(n_1 n_2 \cdots n_{i-1} - 1)(n_i - 1) = 1 + n_1 n_2 \cdots n_m - 1 = |G|$$

elements in P_1, all unconstrained. The $\binom{m}{2}$ elements s_{ij} in P_2 are all constrained: $n_i s_{ij} = 0$.

20.2.2. EXAMPLE. Let $G = \langle x, y, z \rangle \cong \mathbb{Z}_2^3$. Then

$$P_1' = \{f(x,x), f(y,y), f(z,z), f(x,y), f(x,z), f(y,z), f(x+y,z)\}.$$

We label the elements of P_1' (in the same order of listing) X, Y, Z, D, E, F, H. Set $s_{12} = K$, $s_{13} = L$, and $s_{23} = M$, so that $2K = 2L = 2M = 0$. Index rows and columns as $0, x, y, x+y, z, x+z, y+z, x+y+z$. Allowing only normalized cocycles (i.e., $f(0,0) = 0$), and using multiplicative notation to save space, the Hadamard product of

$$\begin{bmatrix}
1 & 1 & 1 & 1 & 1 & 1 & 1 & 1 \\
1 & X & D & Xd & E & Xe & DfH & XdFh \\
1 & D & Y & Yd & F & DeH & Yf & YdEh \\
1 & Xd & Yd & XYd^2 & H & XdeF & YdEf & XYd^2h \\
1 & E & F & H & Z & Ze & Zf & Zh \\
1 & Xe & DeH & XdeF & Ze & XZe^2 & ZDef & XZdeh \\
1 & DfH & Yf & YdEf & Zf & ZDef & YZf^2 & YZdfh \\
1 & XdFh & YdEh & XYd^2h & Zh & XZdeh & YZdfh & XYZd^2h^2
\end{bmatrix}$$

and

$$\begin{bmatrix}
1 & 1 & 1 & 1 & 1 & 1 & 1 & 1 \\
1 & 1 & 1 & 1 & 1 & 1 & 1 & 1 \\
1 & K & 1 & K & 1 & K & 1 & K \\
1 & K & 1 & K & 1 & K & 1 & K \\
1 & L & M & LM & 1 & L & M & LM \\
1 & L & M & LM & 1 & L & M & LM \\
1 & KL & M & KLM & 1 & KL & M & KLM \\
1 & KL & M & KLM & 1 & KL & M & KLM
\end{bmatrix}$$

is a development table for G. A lower case letter stands for its upper case version with exponent -1. Notice that the first array is symmetric.

20.3. Development tables revisited

By Subsection 3.1.15, there is a bijection between the solution set of (20.1) over C and $\mathrm{Hom}(R(G), C)$, where

$$R(G) = \mathrm{Ab}\langle (a,b), \, a, b \in G \mid (a,b) + (ab,c) - (b,c) - (a,bc) = 0 \;\; \forall a, b, c \in G \rangle$$
$$= \mathrm{Ab}\langle u_1, \ldots, u_n \mid e_1 u_1 = \cdots = e_n u_n = 0 \rangle$$

say, for $e_i \in \mathbb{N}$. So there are integers $\lambda(i, a, b)$ such that $(a, b) = \sum_{i=1}^{n} \lambda(i, a, b) u_i$. Then $[\prod_{i=1}^{n} X_i^{\lambda(i,a,b)}]_{a,b \in G}$ is a development table for G. Thus, a development table for a given group G is not canonically defined. Even the number of indeterminates in two tables for the same group may differ. For example, we could take the torsion part of $R(G)$ to be in primary-invariant or torsion-invariant form.

20.3.1. EXAMPLE. The array

$$\begin{bmatrix} Z & Z & Z & Z \\ Z & AZ & dKZ & ADKZ \\ Z & dZ & BZ & BDZ \\ Z & ADZ & BDKZ & ABD^2KZ \end{bmatrix}$$

is a development table for $G = \langle a, b \rangle \cong \mathbb{Z}_2^2$. Here Z, A, B, D are free, and K has order 2 in $R(G)$. Indeed $R(G) \cong \mathbb{Z}_2 \oplus \mathbb{Z}^4$.

Let $\mathrm{DT}(G)$ be a development table (20.2). Then $\mathrm{DT}(G) = T \wedge F$ where

$$T = [\textstyle\prod_{i:e_i>0} X_i^{\lambda(i,a,b)}]_{a,b\in G} \quad \text{and} \quad F = [\textstyle\prod_{i:e_i=0} X_i^{\lambda(i,a,b)}]_{a,b\in G}.$$

By definition, F displays all cocycles $f : G \times G \to C$ for free abelian C. We denote such a table $\mathrm{DT}_{\mathrm{free}}(G)$, and say that it is a *free development table*. The table T is related to the torsion subgroup $H(G)$ of $R(G)$. Pick a complement N of $H(G)$ in $R(G)$. (There are several choices for the free abelian group N.) Bi-additivity of Hom implies that

$$\mathrm{Hom}(R(G), C) = \mathrm{Hom}(N, C) \oplus \mathrm{Hom}(H(G), C).$$

A table $\mathrm{DT}_{\mathrm{tor}}(G)$ is a *torsion development table* for G if it displays all cocycles corresponding to the elements of $\{\phi \in \mathrm{Hom}(R(G), C) \mid \phi(x) = 0 \ \forall x \in N\}$.

20.3.2. REMARK. Let G be abelian. To obtain $\mathrm{DT}_{\mathrm{free}}(G)$ and $\mathrm{DT}_{\mathrm{tor}}(G)$, we set the indeterminates in P_2 to 1, or set the indeterminates in P_1 to 1, respectively (cf. Example 20.2.2). It follows from (20.4) and symmetry of cocycles with a cyclic ground group that $\mathrm{DT}_{\mathrm{free}}(G)$ is symmetric.

We will give a construction of $\mathrm{DT}_{\mathrm{free}}(G)$ for any G that uses our solution of the problem for abelian G in Section 20.2. Then we show how to construct $\mathrm{DT}_{\mathrm{tor}}(G)$ when G is solvable.

20.4. Group cohomology

For maps f_1 and f_2 from the n-fold Cartesian product $G^n = G \times \cdots \times G$ to the finitely generated abelian group C, define $(f_1 f_2)(x) = f_1(x) f_2(x)$. The set of maps $f : G^n \to C$ under this pointwise product is an abelian group $\mathrm{Fun}^n(G, C)$, where $\mathrm{Fun}^0(G, C) = C$. Define $d_{n+1} f \in \mathrm{Fun}^{n+1}(G, C)$ for $f \in \mathrm{Fun}^n(G, C)$ by

$$(d_{n+1}f)(x_1, x_2, \ldots, x_{n+1}) = f(x_2, \ldots, x_{n+1}) f(x_1, \ldots, x_n)^{(-1)^{n+1}}$$

$$\times \prod_{i=1}^{n} f(x_1, \ldots, x_{i-1}, x_i x_{i+1}, x_{i+2}, \ldots, x_{n+1})^{(-1)^i}.$$

Since C is abelian, $d_{n+1} : \mathrm{Fun}^n(G, C) \to \mathrm{Fun}^{n+1}(G, C)$ is a homomorphism. Denote the kernel of d_{n+1} by $Z^n(G, C)$, and the image of d_n by $B^n(G, C)$. The elements of $Z^n(G, C)$ are *n-cocycles*, and those of $B^n(G, C)$ are *n-coboundaries*. It may be checked that the $d_{n+1} d_n$ are all trivial; thus $B^n(G, C) \leq Z^n(G, C)$. The quotient

$$H^n(G, C) = Z^n(G, C) / B^n(G, C)$$

is the *nth cohomology group of G with coefficients in C*.

If f is a 1-cocycle then $(d_2 f)(x, y) = f(y)f(x)f(xy)^{-1} = 1$ for all $x, y \in G$. Since $(d_1 f)(x) = f(\ )f(\ )^{-1} = 1$, the image of d_1 is the identity of $\mathrm{Fun}^1(G, C)$. Thus $H^1(G, C) = \mathrm{Hom}(G, C) \cong \mathrm{Hom}(G/G', C)$.

An element of $Z^2(G, C)$ is a map $f \in \mathrm{Fun}^2(G, C)$ such that $f(x, y)f(xy, z) = f(y, z)f(x, yz)$ for all x, y, z; so the elements of $Z^2(G, C)$ are 2-cocycles, as they were defined from the beginning. Moreover, f is a 2-coboundary if and only if $f(x, y) = \rho(x)\rho(y)\rho(xy)^{-1}$ for some map $\rho : G \to C$. We now see that $H^2(G, C)$ is the set $\{[f] \mid f \in Z^2(G, C)\}$ of cohomology classes of cocycles, in the notation of Subsection 12.1.4.

20.5. Constructing a free table

In this section, C is a finitely generated free abelian group. We show how to calculate $Z^2(G, C)$.

20.5.1. Reduction to abelian ground groups. Let H be a subgroup of finite index n in a group K, and choose a transversal $\{g_1, \ldots, g_n\}$ of H in K. Fix $g \in K$. For each g_i there is a unique g_j such that $g_j^{-1} g g_i \in H$. Define $h_i = g_j^{-1} g g_i$ and $\gamma(g) = h_1 h_2 \cdots h_n H'$. Then $\gamma : K \to H/H'$ is a homomorphism (well-known in group theory as 'transfer'). Since γ maps K into an abelian group, its kernel contains K'. If H is central in K, then $\gamma(h) = h^n$ for $h \in H$. Hence

20.5.1. LEMMA. *If H is a central subgroup of finite index n in a group K, then every element of $H \cap K'$ has order dividing n.*

20.5.2. COROLLARY. *For any central extension*

$$(20.5) \qquad\qquad 1 \longrightarrow C \xrightarrow{\iota} E \xrightarrow{\pi} G \longrightarrow 1$$

of the torsion-free group C by the finite group G, $\iota(C) \cap E' = 1$.

Let $N \trianglelefteq G$. For each $f \in Z^2(G/N, C)$, we define $f' \in Z^2(G, C)$ by $f'(x, y) = f(xN, yN)$. The map $\mathrm{inf} : f \mapsto f'$ is a homomorphism, called *inflation*. Inflation on $Z^2(G/N, C)$ induces a homomorphism from $H^2(G/N, C)$ to $H^2(G, C)$, also denoted inf.

20.5.3. THEOREM. $\mathrm{inf} : H^2(G/G', C) \to H^2(G, C)$ *is an isomorphism.*

PROOF. Suppose that $\mathrm{inf}(f)$ is a coboundary for some (normalized) cocycle $f \in Z^2(G/G', C)$; say $f(xG', yG') = \rho(x)\rho(y)\rho(xy)^{-1}$. Then $\rho(x)\rho(y)\rho(xy)^{-1} = 1$ if x or y is in G', which means that the restriction of ρ to G' is a homomorphism. Since G' is finite and C is torsion-free, ρ must be the identity on G'. So $\tilde\rho : G/G' \to C$ defined by $\tilde\rho : xG' \mapsto \rho(x)$ is well-defined, and we have $f(xG', yG') = \tilde\rho(xG')\tilde\rho(yG')\tilde\rho(xG'yG')^{-1}$. Thus inf is injective on $H^2(G/G', C)$.

It remains to prove that inf is surjective. Let $f = f_{\iota,\tau}$ be a cocycle of (20.5). The restriction π' of π to E' is an isomorphism onto G' by Corollary 20.5.2. Let $x_1 = 1, x_2, \ldots, x_n$ be a transversal of G' in G, and for each i select $y_i \in E$ such that $\pi(y_i) = x_i$. We extend the inverse $\mu : G' \to E'$ of π' to a normalized transversal map $\mu : G \to E$ by setting $\mu(x_i g) = y_i \mu(g)$ for $g \in G'$. The cocycle $f_\mu = f_{\iota,\mu}$ is therefore cohomologous to f. If $g, h \in G'$ we then have

$$\iota f_\mu(x_i h, g) = \mu(x_i h)\mu(g)\mu(x_i hg)^{-1}$$
$$= \mu(x_i)\mu(h)\mu(g)\mu(g)^{-1}\mu(h)^{-1}\mu(x_i)^{-1}$$
$$= 1$$

and

$$\iota f_\mu(g, x_i h) = \mu(g)\mu(x_i h)\mu(gx_i h)^{-1}$$
$$= \mu(g)\mu(x_i)\mu(h)\mu(x_i g^{x_i} h)^{-1}$$
$$= \mu(g)\mu(x_i)\mu(h)\mu(h)^{-1}\mu(g^{x_i})^{-1}\mu(x_i)^{-1}$$
$$= \mu(g)\mu(x_i)\mu(g^{x_i})^{-1}\mu(x_i)^{-1}$$
$$= 1,$$

because $\mu(g)\mu(x_i)\mu(g^{x_i})^{-1}\mu(x_i)^{-1} \in E'(E')^{\mu(x_i)^{-1}} = E'$, and $\iota(C) \cap E' = 1$. This shows that $f_\mu(x, y) = 1$ if x or y is in G'. By the cocycle identity,

$$f_\mu(x, y)f_\mu(xy, g) = f_\mu(y, g)f_\mu(x, yg) \quad \text{and} \quad f_\mu(x, g)f_\mu(xg, y) = f_\mu(g, y)f_\mu(x, gy).$$

Thus, if $x, y \in G$ and $g \in G'$ then $f_\mu(x, yg) = f_\mu(x, y)$; also

$$f_\mu(xg, y) = f_\mu(x, gy) = f_\mu(x, yg^y) = f_\mu(x, y).$$

Consequently the map $\tilde{f}$ such that $\tilde{f}(xG', yG') = f_\mu(x, y)$ is a (well-defined) element of $Z^2(G/G', C)$. Since $\inf(\tilde{f}) = f_\mu$, and f is cohomologous to f_μ, inf is surjective on $H^2(G/G', C)$. $\qquad\square$

20.5.2. A free development table for any G.

20.5.4. THEOREM. *Let $\mathcal{T} = \{x_1 = 1, x_2, \ldots, x_n\}$ be a transversal of G' in G. For $x \in G \setminus \mathcal{T}$, let F_x be a free indeterminate, and set $F_x = 1$ if $x \in \mathcal{T}$. Then*

$$(\mathrm{DT}_{\mathrm{free}}(G/G') \otimes J_{|G'|}) \wedge [F_x F_y F_{xy}^{-1}]_{x,y \in G}$$

is a free development table for G.

PROOF. If $f \in Z^2(G, C)$ then $f(x, y) = h(x, y)\rho(x)\rho(y)\rho(xy)^{-1}$ for some map ρ and cocycle h inflated from $Z^2(G/G', C)$, by Theorem 20.5.3. Thus $h(x, y) = h(x_i, x_j)$ where $x \in x_i G'$ and $y \in x_j G'$. So h is displayed by $\mathrm{DT}_{\mathrm{free}}(G/G') \otimes J_{|G'|}$. Now for each $x \in G$, define $\bar{x} \in \mathcal{T}$ by $xG' = \bar{x}G'$, and let $F_x = \rho(x)\rho(\bar{x})^{-1}$. Then $\rho(x)\rho(y)\rho(xy)^{-1} = \inf(f_1)(x, y) F_x F_y F_{xy}^{-1}$ where $f_1 \in B^2(G/G', C)$. This completes the proof. $\qquad\square$

Therefore, $\mathrm{DT}_{\mathrm{free}}(G)$ can be constructed by Theorem 20.5.4 and Section 20.2. We proceed to address the problem of constructing $\mathrm{DT}_{\mathrm{tor}}(G)$.

20.6. Group homology

Each table $\mathrm{DT}_{\mathrm{tor}}(G)$ corresponds to a complement N in $R(G)$ of the torsion part $H(G)$. We will shortly identify $H(G)$ as the second homology group of G (our reasoning is along the lines of [**88**, Section 11]).

In this section, $v := |G|$.

20.6.1. Homology groups. Let $M_0(G) = \mathbb{Z}$, and let $M_m(G)$ for $m \geq 1$ be the finitely generated free abelian group $\mathrm{Ab}\langle(x_1, \ldots, x_m) : x_1, \ldots, x_m \in G\rangle$. We define a homomorphism $\partial_{m+1} : M_{m+1}(G) \to M_m(G)$ by setting $\partial_1(x_1) = 0$ and

$$\partial_{m+1}(x_1, x_2, \ldots, x_{m+1}) = (-1)^{m+1}(x_1, x_2, \ldots, x_m) + (x_2, x_3, \ldots, x_{m+1})$$

$$+ \sum_{i=1}^m (-1)^i (x_1, \ldots, x_i x_{i+1}, \ldots, x_{m+1})$$

for $m \geq 1$. Next, define $L_m(G) = \partial_{m+1}(M_{m+1}(G))$ and $K_{m+1}(G) = \ker(\partial_{m+1})$. Since $\partial_m \partial_{m+1} = 0$, $L_m(G) \leq K_m(G)$. The quotient $H_m(G) = K_m(G)/L_m(G)$ is the *mth homology group of G*. We have $H_1(G) \cong G/G'$. The second homology group $H_2(G)$ is the *Schur multiplier* of G.

Let $R_m(G) = M_m(G)/L_m(G)$. Then $H_m(G) \leq R_m(G)$, $R_1(G) = H_1(G)$, and $R_2(G) = R(G)$.

20.6.2. The isomorphism type of $R_2(G)$. In this subsection we show that $H_2(G)$ is the torsion part of $R_2(G)$; the exponent of $H_2(G)$ divides v; and $R_2(G)/H_2(G)$ has rank v. For abelian G, all of this is evident in Section 20.2.

Let $[x] := (\sum_{y \in G}(x,y)) + L_2(G) \in R_2(G)$, and let $S_2(G)$ be the subgroup of $R_2(G)$ generated by these elements. The homomorphism $R_2(G) \to M_1(G)$ induced by ∂_2 maps $[x]$ to $v(x)$; so $S_2(G) \cong M_1(G) \cong \mathbb{Z}^v$.

20.6.1. THEOREM. $R_2(G) \cong \mathbb{Z}^v \oplus T$, *where the torsion subgroup T of $R_2(G)$ is finite of exponent dividing v.*

PROOF. Write $R_2(G) = \hat{T} \oplus N$ where $\hat{T}$ is the torsion subgroup. Now

$$v(a,b) = \sum_{c \in G}(a,b) = \sum_{c \in G}(a,bc) + \sum_{c \in G}(b,c) - \sum_{c \in G}(ab,c) = [a] + [b] - [ab]$$

modulo $L_2(G)$. Thus, multiplication by v is a homomorphism $R_2(G) \to S_2(G)$, whose kernel T has exponent dividing v. As a finitely generated abelian group of finite exponent, T is finite. Since $T \leq \hat{T}$ and $R_2(G)/T$ is torsion-free, $T = \hat{T}$. Also $N \cong vN \leq S_2(G)$ implies that $N \cong S_2(G)$. $\square$

20.6.2. THEOREM. $H_2(G)$ *is the torsion subgroup of $R_2(G)$.*

PROOF. Denote by ∂_2^* the injective homomorphism on $M_2(G)/K_2(G)$ induced by ∂_2. Since

$$M_2(G)/K_2(G) \cong \partial_2^*(M_2(G)/K_2(G)) \leq M_1(G),$$

$M_2(G)/K_2(G)$ is free abelian of rank at most v. For $a \in G$ define

$$\varsigma(a) = \sum_{i=0}^{|a|-1}(a,a^i) + K_2(G) \in M_2(G)/K_2(G).$$

We have $\partial_2^*(\varsigma(a)) = |a|(a)$, and if $a \neq b$ then $|a|(a) \neq |b|(b)$. Thus $M_2(G)/K_2(G)$ has rank v. Then

$$R_2(G)/H_2(G) = (M_2(G)/L_2(G))/(K_2(G)/L_2(G)) \cong M_2(G)/K_2(G),$$

so $R_2(G)/H_2(G)$ is free abelian of rank v, and every torsion element of $R_2(G)$ is in $H_2(G)$. By Theorem 20.6.1, $R_2(G) = T \oplus N$ where $N \cong \mathbb{Z}^v$ and T is the torsion subgroup of $R_2(G)$. Since $T \leq H_2(G)$, it follows that $H_2(G) = T \oplus (H_2(G) \cap N)$. If $H_2(G) \cap N$ were non-trivial then $R_2(G)/H_2(G)$ would not have rank v. Thus $T = H_2(G)$ as claimed. $\square$

20.6.3. Finiteness of the second cohomology group. Let C be finitely generated. For each cocycle $f \in Z^2(G,C)$ there is a homomorphism $\phi_f : R_2(G) \to C$ such that $\phi_f((x,y)) = f(x,y)$. In fact

(20.6) $$\Phi : f \mapsto \phi_f$$

defines an isomorphism of $Z^2(G,C)$ onto $\mathrm{Hom}(R_2(G), C)$.

If C is finite then clearly $Z^2(G, C)$ and thus $H^2(G, C)$ are finite. More generally we have

20.6.3. LEMMA. $H^2(G, C)$ *is finite of exponent dividing* v.

PROOF. Let $f \in Z^2(G, C)$. Define $\phi : G \to C$ by $\phi(x) = \Phi(f)([x])$, where $[x]$ is as in Subsection 20.6.2. Applying $\Phi(f)$ to the equation $v(x, y) = [x] + [y] - [xy]$ modulo $L_2(G)$ gives $vf(x, y) = \phi(x) + \phi(y) - \phi(xy)$. Thus vf is a coboundary. Since $H^2(G, C)$ is a finitely generated abelian group, this completes the proof. $\square$

20.6.4. Torsion and free parts of a cocycle. If N is a complement of $H_2(G)$ in $R_2(G)$, then $\text{Hom}(R_2(G), C) = A \oplus B$ where

$$A = \{\phi \in \text{Hom}(R_2(G), C) \mid \phi(N) = 0\} \cong \text{Hom}(H_2(G), C)$$

and

$$B = \{\phi \in \text{Hom}(R_2(G), C) \mid \phi(H_2(G)) = 0\} \cong \text{Hom}(N, C).$$

Let Φ be the isomorphism (20.6). Then

$$Z^2(G, C) = Z^2_{\text{tor}}(G, C) \oplus Z^2_{\text{free}}(G, C)$$

where

$$Z^2_{\text{tor}}(G, C) = \Phi^{-1}(A) \qquad \text{and} \qquad Z^2_{\text{free}}(G, C) = \Phi^{-1}(B).$$

The cocycles in $Z^2_{\text{tor}}(G, C)$ are *fully torsion*, and those in $Z^2_{\text{free}}(G, C)$ are *fully free*. Although $Z^2_{\text{free}}(G, C)$ is canonically defined, $Z^2_{\text{tor}}(G, C)$ is not, because it depends on the choice of N. We assume an arbitrary but fixed choice of N.

This decomposition of cocycles mirrors the separation of a development table into torsion and free tables. If $f \in Z^2_{\text{free}}(G, C)$ then $\Phi(f) = \psi(f) \circ \chi$ for some $\chi \in \text{Hom}(R_2(G), \mathbb{Z}^v)$ and $\psi(f) \in \text{Hom}(\mathbb{Z}^v, C)$. Therefore each element of $Z^2_{\text{free}}(G, C)$ is a cocycle with coefficient group $\mathbb{Z}^v$ composed with a homomorphism $\mathbb{Z}^v \to C$ (note that all coboundaries have this property). So we obtain the fully free cocycles by Subsection 20.5.2. By the end of the next two sections, we will be able to calculate all fully torsion cocycles.

20.7. Presentations and the Schur multiplier

The main subject of this section is *transgression*. Given a presentation of the finite group G, transgression is a device that we use to obtain an embedding of $\text{Hom}(H_2(G), C)$ into $Z^2(G, C)$ that is amenable to calculating $Z^2_{\text{tor}}(G, C)$.

20.7.1. Cocycle decomposition and cocycle equivalence.

20.7.1. LEMMA. *If* $|G| = v$ *then* $Z^2_{\text{free}}(G, C)$ *consists precisely of the cocycles* f *that factor through* $\mathbb{Z}^v$, *i.e., such that* $f = \psi \circ \bar{f}$ *for some* $\psi \in \text{Hom}(\mathbb{Z}^v, C)$ *and* $\bar{f} \in Z^2(G, \mathbb{Z}^v)$.

PROOF. Let Φ_A be the usual isomorphism of $Z^2(G, A)$ onto $\text{Hom}(R_2(G), A)$. At the end of Subsection 20.6.4 we observed that each element of $Z^2_{\text{free}}(G, C)$ factors through $\mathbb{Z}^v$. Suppose that $f \in Z^2(G, C)$ and $f = \psi \circ \bar{f}$ for $\bar{f}$ and ψ as stated. Then $\Phi_C(f) = \psi \circ \Phi_{\mathbb{Z}^v}(\bar{f})$. Since the only homomorphism from the finite group $H_2(G)$ to $\mathbb{Z}^v$ is trivial, $\Phi_C(f)$ is zero on $H_2(G) \leq R_2(G)$. This means that $f \in Z^2_{\text{free}}(G, C)$ by definition. $\square$

20.7.2. COROLLARY. $B^2(G, C) \leq Z^2_{\text{free}}(G, C)$.

Hence the cocycles in $Z^2_{\mathrm{tor}}(G,C)$ are pairwise non-cohomologous; so $\mathrm{DT}_{\mathrm{tor}}(G)$ displays all fully torsion cocycles without repetition of cohomology class. Furthermore,

$$H^2(G,C) = H^2_{\mathrm{tor}}(G,C) \oplus H^2_{\mathrm{free}}(G,C)$$

where $H^2_{\mathrm{tor}}(G,C) \cong Z^2_{\mathrm{tor}}(G,C) \cong \mathrm{Hom}(H_2(G),C)$ and

$$H^2_{\mathrm{free}}(G,C) = Z^2_{\mathrm{free}}(G,C)/B^2(G,C)$$
$$= \{[\psi \circ \bar{f}] \mid \psi \in \mathrm{Hom}(\mathbb{Z}^v,C),\ [\bar{f}] \in H^2(G,\mathbb{Z}^v)\}.$$

20.7.2. Transgression. Let f be a cocycle of the central extension

$$(20.7) \qquad\qquad 1 \longrightarrow A \xrightarrow{\ \iota\ } E \xrightarrow{\ \pi\ } G \longrightarrow 1.$$

For any $\phi \in \mathrm{Hom}(A,C)$, $\phi \circ f \in Z^2(G,C)$; and if h is another cocycle of (20.7) then $[\phi \circ f] = [\phi \circ h]$. The map $\phi \mapsto [\phi \circ f]$ associated to (the equivalence class of) the extension (20.7) is a homomorphism $\mathrm{Hom}(A,C) \to H^2(G,C)$, known as *transgression*.

Suppose that for any central extension

$$(20.8) \qquad\qquad 1 \longrightarrow C \xrightarrow{\ \iota'\ } B \xrightarrow{\ \pi'\ } G \longrightarrow 1$$

there are $\phi \in \mathrm{Hom}(A,C)$ and $\alpha \in \mathrm{Hom}(E,B)$ such that

$$
\begin{array}{ccccccccc}
1 & \longrightarrow & A & \xrightarrow{\ \iota\ } & E & \xrightarrow{\ \pi\ } & G & \longrightarrow & 1 \\
 & & \downarrow{\scriptstyle \phi} & & \downarrow{\scriptstyle \alpha} & & \| & & \\
1 & \longrightarrow & C & \xrightarrow{\ \iota'\ } & B & \xrightarrow{\ \pi'\ } & G & \longrightarrow & 1
\end{array}
$$

commutes. Then $\phi \circ f$ is a cocycle of (20.8), and the transgression associated to (20.7) is therefore surjective. Existence of such a central extension (20.7) is proved below.

20.7.3. Presentations and transgression. We identify G with F/R, where F is a finitely generated free group and R is the normal closure in F of a finite subset. We say that F/R is a presentation of G.

The subgroup $[R,F]$ of R is normal in F, and $R/[R,F] \le Z(F/[R,F])$. So

$$(20.9) \qquad 1 \longrightarrow R/[R,F] \xrightarrow{\ \mathrm{inc}\ } F/[R,F] \xrightarrow{\ \mathrm{proj}\ } F/R \longrightarrow 1$$

is a central short exact sequence, where inc denotes inclusion and proj is natural projection $F/[R,F] \to (F/[R,F])/(R/[R,F]) \cong F/R$.

20.7.3. LEMMA. *If*

$$1 \longrightarrow C \xrightarrow{\ \iota\ } E \xrightarrow{\ \pi\ } G \longrightarrow 1$$

is any central extension of a (finitely generated) abelian group C by G, then there are homomorphisms ϕ and α such that the diagram

$$(20.10)$$
$$
\begin{array}{ccccccccc}
1 & \longrightarrow & R/[R,F] & \xrightarrow{\ \mathrm{inc}\ } & F/[R,F] & \xrightarrow{\ \mathrm{proj}\ } & F/R & \longrightarrow & 1 \\
 & & \downarrow{\scriptstyle \phi} & & \downarrow{\scriptstyle \alpha} & & \| & & \\
1 & \longrightarrow & C & \xrightarrow{\ \iota\ } & E & \xrightarrow{\ \pi\ } & G & \longrightarrow & 1
\end{array}
$$

commutes. Hence the transgression $\mathrm{tra} : \mathrm{Hom}(R/[R,F],C) \to H^2(G,C)$ *associated to (20.9) is surjective.*

PROOF. Let $F = \langle x_1, \ldots, x_k \rangle$, choose a transversal map $\tau : G \to E$, and define $\bar{\alpha} : F \to E$ by $\bar{\alpha}(x_i) = \tau(x_i R)$, $1 \le i \le k$. Then $\pi\bar{\alpha}$ is the natural projection of F onto G. Thus $\bar{\alpha}(R) \le \ker \pi = \iota(C)$ is central in E. For $r \in R$ and $x \in F$ we have $\bar{\alpha}([r, x]) = [\bar{\alpha}(r), \bar{\alpha}(x)] = 1$, implying that $[R, F] \le \ker \bar{\alpha}$. So $\bar{\alpha}$ induces a homomorphism $\alpha : F/[R, F] \to E$ such that $\pi\alpha = \text{proj}$. Since $\alpha(R/[R, F]) \le \iota(C)$, if we let ϕ be $\iota^{-1}\alpha_{|R/[R,F]}$ then we get the commuting diagram (20.10). $\square$

20.7.4. $H^2_{\text{tor}}(G, C)$ as the image of transgression. As before $G = F/R$.

20.7.4. LEMMA. *The finitely generated abelian group $R/[R, F]$ has torsion part $(R \cap F')/[R, F]$.*

PROOF. Since F/F' is torsion-free and
$$(R/[R, F])/((R \cap F')/[R, F]) \cong R/(R \cap F') \cong RF'/F' \le F/F',$$
$(R \cap F')/[R, F]$ contains the torsion part of $R/[R, F]$. Conversely, $(R \cap F')/[R, F]$ is finite by Lemma 20.5.1. $\square$

Thus
$$R/[R, F] = (R \cap F')/[R, F] \times S/[R, F]$$
for some *Schur complement* $S/[R, F]$. Note that $S/[R, F]$ is a free abelian group of rank at most $v = |G|$. By Lemma 20.7.3,

$$(20.11) \quad H^2(G, C) = \text{tra}(\text{Hom}((R \cap F')/[R, F], C)) + \text{tra}(\text{Hom}(S/[R, F], C)).$$

Taking $C = \mathbb{Z}^v$ we get
$$H^2(G, \mathbb{Z}^v) = \text{tra}(\text{Hom}(S/[R, F], \mathbb{Z}^v)) = \{[\psi \circ f] \mid \psi \in \text{Hom}(S/[R, F], \mathbb{Z}^v)\}$$
for a fixed $f \in Z^2(G, R/[R, F])$. Since every homomorphism $S/[R, F] \to C$ can be factored through $\mathbb{Z}^v$, this implies that

$$(20.12) \qquad\qquad H^2_{\text{free}}(G, C) = \text{tra}(\text{Hom}(S/[R, F], C)).$$

Our next objective is to prove that (20.11) is a direct sum.

20.7.5. LEMMA. *Each element of $\ker \text{tra}$ may be extended to a homomorphism $F/[R, F] \to C$.*

PROOF. We use Lemma 20.7.3 and its notation. If $\phi \in \ker \text{tra}$ then E splits over C. Let μ be a (projection) homomorphism from $E \cong C \times G$ onto C such that $\mu\iota = \text{id}_C$. Then $\mu\alpha$ extends ϕ. $\square$

20.7.6. COROLLARY. *The transgression* $\text{tra} : \text{Hom}(R/[R, F], C) \to H^2(G, C)$ *is injective on* $\text{Hom}((R \cap F')/[R, F], C)$.

PROOF. Define $\phi \in \text{Hom}((R \cap F')/[R, F], C)$ as a homomorphism on all of $R/[R, F]$ by setting $\phi(S/[R, F]) = 1$. Suppose that $\phi \in \ker \text{tra}$, and extend ϕ to a homomorphism ψ on $F/[R, F]$ by Lemma 20.7.5. Now $\psi([x, y][R, F]) = 1$ because C is abelian. Hence ϕ is the identity on $(R \cap F')/[R, F]$, as required. $\square$

20.7.7. LEMMA. $\text{tra}(\text{Hom}((R \cap F')/[R, F], C)) \cap \text{tra}(\text{Hom}(S/[R, F], C)) = 1$.

PROOF. Suppose that $\phi \in \text{Hom}((R \cap F')/[R, F], C)$, $\mu \in \text{Hom}(S/[R, F], C)$, and $\text{tra}(\phi) = \text{tra}(\mu)$. We extend ϕ and μ to all of $R/[R, F]$, and then $\phi - \mu$ is trivial on $(R \cap F')/[R, F]$ by Corollary 20.7.6. As μ is trivial on $(R \cap F')/[R, F]$ too, the result follows. $\square$

Lemma 20.7.7, together with (20.11) and (20.12), yield

$$H^2(G, C) = H^2_{\text{free}}(G, C) \oplus \text{tra}(\text{Hom}((R \cap F')/[R, F], C)).$$

Thus

(20.13) $\text{tra}(\text{Hom}((R \cap F')/[R, F], C)) \cong H^2_{\text{tor}}(G, C) \cong \text{Hom}(H_2(G), C).$

By Corollary 20.7.6 and (20.13), if e_1, e_2 are the respective exponents of $H_2(G)$ and $(R \cap F')/[R, F]$, then

$$H_2(G) \cong \text{Hom}(H_2(G), \mathbb{Z}_{e_1 e_2}) \cong \text{Hom}((R \cap F')/[R, F], \mathbb{Z}_{e_1 e_2}) \cong (R \cap F')/[R, F].$$

This proves *Hopf's formula*:

$$H_2(G) \cong (R \cap F')/[R, F].$$

20.8. Constructing a torsion table

Let G be given by the presentation F/R, and suppose that $S/[R, F]$ is a Schur complement. Since $[S, F] \leq [R, F] \leq S$, we see that $S \triangleleft F$. Furthermore $R/S \leq Z(F/S)$ and

$$R/S \cong (R/[R, F])/(S/[R, F]) \cong (R \cap F')/[R, F] \cong H_2(G).$$

Hence the finite group F/S is a central extension

(20.14) $1 \longrightarrow R/S \xrightarrow{\text{inc}} F/S \xrightarrow{\text{proj}} F/R \longrightarrow 1$

of $H_2(G)$ by G, called a *Schur cover* (or *covering group*) of G. The following picture may be helpful.

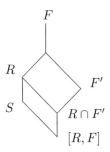

FIGURE 20.1. Hasse diagram for torsion cocycles calculation

Each vertex in this diagram is a normal subgroup of F. Meets of edges represent intersections and products. For example, $S \cap F' = [R, F]$, and $SF' = RF'$ is the vertex adjacent to F. Parallel edges represent isomorphic quotients; in particular $S/[R, F] \cong RF'/F'$.

20.8.1. LEMMA. *$S/[R, F]$ is a free abelian group of the same rank as F.*

PROOF. Since $S/[R, F]$ is isomorphic to the finite index subgroup RF'/F' of the free abelian group F/F', we must have $S/[R, F] \cong F/F'$.   □

20.8.2. REMARK. As Lemma 20.8.1 shows, (the isomorphism type of) a Schur complement is not uniquely determined by G.

Let ϱ denote the natural surjection $F/[R, F] \to F/S$. The restriction $\tilde{\varrho}$ of ϱ to $(R \cap F')/[R, F]$ is an isomorphism onto R/S. Each homomorphism $\tilde{\phi} : R/S \to C$ is of the form $\phi \circ \tilde{\varrho}^{-1}$, where ϕ is a homomorphism $R/[R, F] \to C$ such that $\phi(S/[R, F]) = 1$. If f is a cocycle of (20.9), and $\hat{\varrho}$ denotes the restriction of ϱ to $R/[R, F]$, then $\hat{f} = \hat{\varrho} \circ f$ is a cocycle of (20.14). Next we note that if $x \in R$ then there are unique $y \in R \cap F'$ and $z \in S$ such that $x[R, F] = yz[R, F]$, so

$$\phi \circ \tilde{\varrho}^{-1} \circ \hat{\varrho}(x[R, F]) = \phi \circ \tilde{\varrho}^{-1}(yS)$$
$$= \phi(y[R, F])$$
$$= \phi(y[R, F])\phi(z[R, F])$$
$$= \phi(x[R, F]).$$

Hence

$$\tilde{\phi} \circ \hat{f}(xR, yR) = \phi \circ \tilde{\varrho}^{-1} \circ \hat{\varrho} \circ f(xR, yR) = \phi \circ f(xR, yR).$$

This shows that $\mathrm{tra}(\phi) = \widetilde{\mathrm{tra}}(\tilde{\phi})$ where $\widetilde{\mathrm{tra}}$ is the transgression $\mathrm{Hom}(R/S, C) \to H^2(G, C)$ associated to (20.14).

We are now ready to state our algorithm for calculating $H^2_{\mathrm{tor}}(G, C)$, given an input presentation F/R for G.

(1) Determine the Schur multiplier as $(R \cap F')/[R, F]$, say

$$(R \cap F')/[R, F] = \langle a_1[R, F], \ldots, a_h[R, F]\rangle$$

in torsion-invariant form.
(2) Find a Schur complement $S/[R, F]$. Then $R/S = \langle a_1 S, \ldots, a_h S\rangle$.
(3) Calculate $\mathrm{Hom}(R/S, C)$.
(4) Let f be a cocycle of (20.14); i.e., select a transversal map $\tau : F/R \to F/S$ and set $f(x, y) = \tau(x)\tau(y)\tau(xy)^{-1}$.
(5) For each $\phi \in \mathrm{Hom}(R/S, C)$, compose ϕ with f.

By (20.13) and the preceding discussion, the set of outputs $[\phi \circ f]$ is $H^2_{\mathrm{tor}}(G, C)$.

Steps (3)–(5) are straightforward, and there are implemented algorithms that may be used to carry out steps (1) and (2). The algorithm of Flannery and O'Brien [66] relies on Holt's method [82] for computing the Schur multiplier of a finite permutation group. We next outline a procedure due to Nickel [125] for computing the Schur multiplier and a Schur cover of a finite solvable group.

Suppose that G is given by a consistent polycyclic presentation F/R, where F is free on $X = \{x_1, \ldots, x_n\}$. We seek a polycyclic presentation of $F/[R, F]$.

Say the set of power relations (12.12) is $\{x_i^{r_i} = u_i \mid i \in I\}$, and the set of conjugate relations (12.13), (12.14) is

$$\{x_j^{-1}x_ix_j = v_{ij}, \ x_jx_ix_j^{-1} = w_{ij} \mid 1 \le j < i \le n\}.$$

Define a new presentation $F^\star/R^\star$ as follows. The new generating set is $X \cup X^\star$, where

$$X^\star = \{x_{ii} \mid i \in I\} \cup \{x_{ij} \mid 1 \le j < i \le n\} \cup \{\bar{x}_{ij} \mid 1 \le j < i \le n, i \notin I\}.$$

The new relation set is

$$x_i^{r_i} = u_ix_{ii}, \quad i \in I,$$
$$x_j^{-1}x_ix_j = v_{ij}x_{ij}, \quad 1 \le j < i \le n,$$
$$x_jx_ix_j^{-1} = w_{ij}\bar{x}_{ij}, \quad 1 \le j < i \le n, \ i \notin I,$$

together with conjugate relations ensuring that the generators in $X^\star$ are central.

20.8.3. THEOREM. $F^\star/R^\star$ is a presentation of $F/[R, F]$, and $R/[R, F]$ is the subgroup generated by $X^\star$.

For a proof of Theorem 20.8.3 we refer to [**125**, pp. 42–48]. Without proof, the claims should make sense intuitively: setting everything in $X^\star$ to 1 reduces to the original presentation of G, a new generator has been added to $X^\star$ for each generator of $R/[R, F]$, and the subgroup generated by $X^\star$ is central (so $F/[R, F]$ is a central extension of $R/[R, F]$ by G).

The remainder of the procedure is to turn $F^\star/R^\star$ into a consistent polycyclic presentation for $F/[R, F]$, which yields a consistent polycyclic presentation for $R/[R, F]$. Then $H_2(G)$ is apparent from the generators in this presentation with power relations. A Schur complement is similarly apparent from the generators with no power relations.

20.8.1. Examples.

20.8.1.1. *Cyclic G.* If G is cyclic then $G = F/R$ where F is infinite cyclic. Since $F' = 1$, $H_2(G)$ is trivial by Hopf's formula. Thus $\mathrm{DT}_{\mathrm{tor}}(G)$ is trivial. This is in accord with Theorem 11.3.2, which places no order constraints on table entries.

20.8.1.2. *Abelian G.* Let R be the normal closure in $F = \langle x_1, \ldots, x_m \rangle$ of

$$\langle [x_i, x_j], x_i^{n_i} : 1 \le i < j \le m \rangle$$

where $n_i | n_{i+1}$. Section 20.2 dealt with abelian G. We now give a more rigorous construction of $\mathrm{DT}_{\mathrm{tor}}(G)$ that uses transgression. Note that $H_2(G)$ is well-known as the *exterior square* of G (see [**134**, pp. 347–348]).

The relations

$$[xy, z][R, F] = [x, z]^y [y, z][R, F] = [x, z][y, z][R, F]$$

in $F/[R, F]$, and $[y, x] = [x, y]^{-1}$ in F, imply that

$$(R \cap F')/[R, F] = F'/[R, F] = \langle [x_i, x_j][R, F] : 1 \le i < j \le m \rangle.$$

Since

$$([x_i, x_j][R, F])^{n_i} = [x_i^{n_i}, x_j][R, F] = 1,$$

$F'/[R, F]$ has order dividing $\prod_{i=1}^m n_i^{m-i}$. Clearly $\langle x_i^{n_i}[R, F] : 1 \le i \le m \rangle$ is a Schur complement $S/[R, F]$.

Define a transversal map $\tau : F/R \to F/S$ by

$$\tau(x_1^{k_1} \cdots x_m^{k_m} R) = x_1^{k_1} \cdots x_m^{k_m} S.$$

Let f be the cocycle of (20.14) for this τ. That is, if $x = x_1^{k_1} \cdots x_m^{k_m}$, $y = x_1^{l_1} \cdots x_m^{l_m}$, and denoting reduction modulo n_i of exponents by a subscript n_i, then

$$f(xR, yR) = x_1^{k_1} \cdots x_m^{k_m} x_1^{l_1} \cdots x_m^{l_m} x_m^{-\{l_m + k_m\}_{n_m}} \cdots x_1^{-\{l_1 + k_1\}_{n_1}} S$$

$$= x_1^{k_1} \cdots x_{m-1}^{k_{m-1}} x_1^{l_1} \cdots x_{m-2}^{l_{m-2}} x_{m-1}^{-k_{m-1}} \cdots x_1^{-l_1 - k_1} \prod_{i=1}^{m-1} [x_m, x_i]^{l_i k_m} S$$

$$= \prod_{i<j} [x_j, x_i]^{l_i k_j} S$$

$$= \prod_{i<j} [x_i, x_j]^{-l_i k_j} S,$$

since $x_i^{n_i} \in S$ and $F'S/S \leq Z(F/S)$. The exponent $l_i k_j$ counts the number of times x_j is moved past x_i, $j > i$, in reducing $x_1^{k_1} \cdots x_m^{k_m} \cdot x_1^{l_1} \cdots x_m^{l_m}$ to collected form.

Note that $\phi \in \mathrm{Hom}(R/S, C)$ is defined by setting its value $\phi_{ij} \in C$ on each of the $\binom{m}{2}$ generators $[x_i, x_j]S$ of R/S, and that ϕ_{ij} has order e_{ij} dividing n_i. Thus, if we replace each ϕ_{ij} that appears in $[f(x, y)]_{x, y \in G}$ by an indeterminate X_{ij} with the constraint $X_{ij}^{e_{ij}} = 1$, then we get a torsion development table for G.

20.8.1.3. *Metacyclic* G. The Schur multiplier of any finite metacyclic group is cyclic: see [**108**, Theorem 2.11.3, p. 98]. Let

(20.15) $$G = \langle a, b \mid a^r = b^s = 1, a^b = a^{-1} \rangle,$$

so $F = \langle a, b \rangle$ and R is the normal closure in F of $\langle a^r, b^s, a^b a \rangle$.

Suppose first that $r = 2m > 2$. Denote $x[R, F] \in F/[R, F]$ by $\bar{x}$. Set $x_1 = \overline{a^r}$, $x_2 = \overline{b^s}$, and $x_3 = \overline{a^b a}$. The x_i are central in $F/[R, F]$, and $\bar{b}^{-1} \bar{a}^t \bar{b} = \bar{a}^{-t} x_3^t$ for all $t \geq 1$. Then

$$1 = [x_1, \bar{b}] = \bar{a}^{-r} \bar{b}^{-1} \bar{a}^r \bar{b} = \bar{a}^{-r} \bar{a}^{-r} x_3^r = \bar{a}^{-2r} x_3^r$$

implies the relation $x_3^r = x_1^2$ in $R/[R, F] = \langle x_1, x_2, x_3 \rangle$. So $R/[R, F] = \langle y, x_2, x_3 \rangle$ where $y = x_1 x_3^{-m}$ has order dividing 2. Since

$$\langle b^s F', a^b a F' \rangle = \langle b^s F', b^{-1} a b a^{-1} a^2 F' \rangle = \langle b^s F', a^2 F' \rangle = RF'/F',$$

we may take $S/[R, F] = \langle x_2, x_3 \rangle$ by Lemma 20.8.1. Thus $(R \cap F')/[R, F] = \langle y \rangle$, and $R/S = \langle a^r (a^b a)^{-m} S \rangle = \langle a^r S \rangle$.

We now calculate a cocycle f of the Schur cover F/S. Define a transversal map $\tau : F/R \to F/S$ by $\tau : a^i b^j R \mapsto a^i b^j S$ for $0 \leq i \leq r - 1$ and $0 \leq j \leq s - 1$. Using $baS = a^{-1}bS$ and $ba^{-1}S = abS$, we get

$$b^j a^k S = a^{(-1)^j k} b^j S \quad \text{and} \quad b^j a^k R = a^{(-1)^j k} b^j R.$$

Hence

$$
\begin{aligned}
f(a^i b^j R, a^k b^l R) &= \tau(a^i b^j R) \tau(a^k b^l R) \tau(a^i b^j a^k b^l R)^{-1} \\
&= a^i b^j a^k b^l S \cdot \tau(a^{\{i + (-1)^j k\}_r} b^{\{j + l\}_s} R)^{-1} \\
&= a^{i + (-1)^j k} b^{j + l} S (a^{\{i + (-1)^j k\}_r} b^{\{j + l\}_s} S)^{-1} \\
&= a^{i + (-1)^j k - \{i + (-1)^j k\}_r} S
\end{aligned}
$$

where the exponent subscripts denote reduction modulo r or s. So, for even j,

(20.16) $$f(a^i b^j R, a^k b^l R) = \begin{cases} S & \text{if } i + k < r \\ a^r S & \text{if } i + k \geq r, \end{cases}$$

and for odd j,

(20.17) $$f(a^i b^j R, a^k b^l R) = \begin{cases} S & \text{if } i \geq k \\ a^r S & \text{if } i < k. \end{cases}$$

It would be clear that $|H_2(G)| = 2$ if we had followed the Nickel procedure. We give a separate justification. Suppose that $|H_2(G)| \neq 2$. Then $H_2(G)$ is trivial, and so $Z^2(G, C) = Z^2_{\mathrm{free}}(G, C)$. For any G, it is a consequence of Theorem 20.5.3, Lemma 20.7.1, and symmetry of fully free cocycles with abelian ground group, that each $h \in Z^2_{\mathrm{free}}(G, C)$ is *almost symmetric*; i.e., $h(a, b) = h(b, a)$ whenever $[a, b] = 1$. However, let f be the cocycle defined by (20.16) and (20.17), with S replaced by 1 and $a^r S$ replaced by -1. Then f is not almost symmetric: $f(a^m R, bR) = 1$ but $f(bR, a^m R) = -1$. This contradiction proves that $|H_2(G)| = 2$.

In summary, $|H^2_{\text{tor}}(G,C)| = |\text{Hom}(\mathbb{Z}_2, C)|$ is the size of the largest elementary abelian 2-subgroup of C. If we index a torsion development table $\text{DT}_{\text{tor}}(G)$ by the elements of G under the ordering

$$a^i b^j < a^k b^l \iff j < l, \text{ or both } j = l \text{ and } i < k,$$

then

$$\text{DT}_{\text{tor}}(G) = \begin{bmatrix} X & X & \cdots & X \\ Y & Y & \cdots & Y \\ X & X & \cdots & X \\ \vdots & \vdots & \vdots & \vdots \\ Y & Y & \cdots & Y \end{bmatrix},$$

where the $r \times r$ blocks X and Y are defined according to (20.16) and (20.17), replacing S by 1 and $a^r S$ by an indeterminate A coupled with the order constraint $A^2 = 1$.

Now suppose that

$$G = \langle a, b \mid a^r = b^s = 1, \ a^b = a^k \rangle$$

where $1 < k < r$, r divides k^s, and $\gcd(r, 1-k) = 1$ (if r is odd then (20.15) is a group of this kind). As before, we denote images under the natural surjection $F \to F/[R, F]$ by overlining. Let $x_1 = \overline{a^r}$, $x_2 = \overline{b^s}$, and $x_3 = \overline{a^b a^{-k}}$. In this case $x_3^r = x_1^{1-k}$. If l and m are integers such that $rl + (1-k)m = 1$ then let $y = x_3^m x_1^l$. We have

$$y^{1-k} = x_3^{(1-k)m}(x_1^{1-k})^l = x_3^{1-rl} x_3^{rl} = x_3$$

and

$$y^r = (x_3^r)^m x_1^{lr} = x_1^{(1-k)m} x_1^{1-(1-k)m} = x_1,$$

proving that

$$R/[R, F] = \langle x_1, x_2, x_3 \rangle = \langle y, x_2 \rangle.$$

Therefore $H_2(G) = 1$: otherwise $R/[R, F] \cong \mathbb{Z}^2 \oplus H_2(G)$ could not be generated by fewer than three elements.

20.9. Listing the elements of the second cohomology group

By Theorem 13.4.1, if we are classifying cocyclic designs via indexing group then we would like to have a complete and irredundant list of cocycles; i.e., a list containing a single explicit representative from each relevant cohomology class. We can compile such a list (which is finite by Lemma 20.6.3) using the theory of this chapter.

20.9.1. The fully torsion case. Distinct fully torsion cocycles are non-cohomologous by Corollary 20.7.2. We showed in Section 20.8 how to list these cocycles.

20.9.2. The fully free case. Calculation of $Z^2_{\text{free}}(G, C)$ reduces to the case of abelian G. However, it is unclear how to sort the fully free cocycles by cohomology class using Theorem 20.2.1 alone (cf. also Section 12.4). In this subsection we take a different tack to calculate $H^2_{\text{free}}(G, C)$. The method is very simple.

20.9.1. LEMMA. *If $A_1, \ldots, A_k$ are finite abelian groups then*

$$H^2_{\text{free}}(A_1 \times \cdots \times A_k, C) \cong H^2_{\text{free}}(A_1, C) \oplus \cdots \oplus H^2_{\text{free}}(A_k, C).$$

PROOF. It suffices to prove the lemma for $k = 2$. Let $f \in Z^2_{\text{free}}(A_1 \times A_2, C)$; remember that f is symmetric. Define $\rho : A_1 \times A_2 \to C$ by $\rho(ab) = f(a, b)^{-1}$. We have

$$f(a, a'b) = f(a, a')\rho(a'b)\rho(aa'b)^{-1} \quad \forall\, a, a' \in A_1,\ b \in A_2$$

and

$$f(ab', b) = f(b', b)\rho(ab')\rho(abb')^{-1} \quad \forall\, a \in A_1,\ b, b' \in A_2.$$

Now f restricts to cocycles $f_{A_1} \in Z^2_{\text{free}}(A_1, C)$ and $f_{A_2} \in Z^2_{\text{free}}(A_2, C)$. Then

$$\begin{aligned}
f(ab, a'b') &= f(a, a'bb')f(b, a'b')f(a, b)^{-1} \\
&= f(a, a')f(b, b')\rho(a'bb')\rho(aa'bb')^{-1}\rho(a'b')\rho(a'bb')^{-1}\rho(ab) \\
&= f_{A_1}(a, a')f_{A_2}(b, b')\rho(ab)\rho(a'b')\rho(aba'b')^{-1},
\end{aligned}$$

proving that f is cohomologous to the product cocycle $f_{A_1} \times f_{A_2}$.

For $f \in Z^2_{\text{free}}(A_1 \times A_2, C)$ and $(f_1, f_2) \in Z^2_{\text{free}}(A_1, C) \oplus Z^2_{\text{free}}(A_2, C)$, define

$$\theta : f \mapsto (f_{A_1}, f_{A_2}) \in Z^2_{\text{free}}(A_1, C) \oplus Z^2_{\text{free}}(A_2, C)$$

and

$$\eta : (f_1, f_2) \mapsto f_1 \times f_2 \in Z^2_{\text{free}}(A_1 \times A_2, C).$$

It may be readily checked that the maps θ and η induce homomorphisms between $H^2_{\text{free}}(A_1 \times A_2, C)$ and $H^2_{\text{free}}(A_1, C) \oplus H^2_{\text{free}}(A_2, C)$, which are mutually inverse by the previous paragraph. □

The required description of the second cohomology of finite cyclic groups may be sifted from Theorem 11.3.2.

20.9.2. LEMMA. *Let A be a cyclic group of order n, and denote the subgroup $\langle c^n : c \in C \rangle$ of nth powers in C by $C(n)$. Then*

$$H^2(A, C) = H^2_{\text{free}}(A, C) \cong C/C(n).$$

PROOF. Let $A = \langle a \rangle$. Each (normalized) cocycle $f : A \times A \to C$ is defined by choosing values A_i for the $f(a, a^i)$, and setting

$$f(a^i, a^j) = \prod_{k=0}^{j-1} f(a^{i+k}, a)f(a^k, a)^{-1}.$$

Hence f is a coboundary if and only if there is a normalized map $\rho : A \to C$ such that $A_i = \rho(a)\rho(a^i)\rho(a^{i+1})^{-1}$, $1 \leq i \leq n - 1$.

Supposing that $n \geq 3$, let $\rho(a) = \rho(a^2) = A_1$ and $\rho(a^{i+1}) = A_1 A_i^{-1}\rho(a^i)$ for $2 \leq i \leq n - 2$. Then $h \in Z^2(A, C)$ such that $h(a, a^i) = A_i$ for $1 \leq i \leq n - 2$ and $h(a, a^{n-1}) = A_1^{n-1}A_2^{-1}\cdots A_{n-2}^{-1}$ is a coboundary. Let h be the trivial cocycle if $n = 2$. Next, define $f_c \in Z^2(A, C)$ by

$$(20.18) \qquad\qquad f_c(a^i, a^j) = c^{\lfloor (i+j)/n \rfloor}.$$

We see that $f = hf_c$ where $c = A_{n-1}h(a, a^{n-1})^{-1}$. Thus each element of $H^2(A, C)$ contains a cocycle f_c as in (20.18).

If $f_d(a^i, a^j) = \mu(a^i)\mu(a^j)\mu(a^{i+j})^{-1}$ then $\mu(a)\mu(a^i) = \mu(a^{i+1})$ for $i \leq n - 2$; so $d = \mu(a)^n \in C(n)$. Conversely, $f_{c^n}(a^i, a^j) = \mu(a^i)\mu(a^j)\mu(a^{i+j})^{-1}$ where $\mu(a^i) = c^i$. The map $c \mapsto f_c$ therefore induces a homomorphism of C onto $H^2(A, C)$ with kernel $C(n)$. □

20.9.3. EXAMPLE. The Kronecker product

$$\begin{bmatrix} 1 & 1 \\ 1 & X_1 \end{bmatrix} \otimes \cdots \otimes \begin{bmatrix} 1 & 1 \\ 1 & X_n \end{bmatrix}$$

displays representatives of all fully free cocyles for $\mathbb{Z}_2^n$. Such a table can be seen in Example 20.2.2: set the coboundary indeterminates D, E, F, H to 1.

Lemmas 20.9.1 and 20.9.2 underpin the following algorithm for computing $H^2_{\text{free}}(G, C)$. Here G need not be solvable.

(1) Find G/G' in torsion-invariant form $C_{n_1} \times \cdots \times C_{n_m}$.
(2) Let $D_k := [X_k^{\lfloor (i+j)/n_k \rfloor}]_{0 \le i,j \le n_k - 1}$.
(3) Let $D := D_1 \otimes \cdots \otimes D_m \otimes J_{|G'|}$.

As each indeterminate X_k ranges over a transversal for $C(n_k)$ in C, the output table D displays $H^2_{\text{free}}(G, C)$ irredundantly. Briefly, if f is a fully free cocycle, then any f-developed matrix is equivalent to a Kronecker product of ω-cyclic matrices.

It is worth mentioning that the results of this subsection enable us to recognize the decomposition

$$H^2(G, C) = H^2_{\text{free}}(G, C) \oplus H^2_{\text{tor}}(G, C)$$

as the 'Universal Coefficient Theorem' of second cohomology [**134**, 11.4.18, p. 349]. The summand $H^2_{\text{free}}(G, C)$ is better known as $\text{Ext}(G/G', C)$, where

$$\text{Ext}(A, C) = \{[f] \in H^2(A, C) \mid f \in Z^2(A, C) \text{ is symmetric}\}$$

for any finite abelian group A. The central extensions of C by A corresponding to the elements of $\text{Ext}(A, C)$ are precisely those which are abelian.

20.10. Another look at the Cocyclic Hadamard Conjecture

In this coda to the chapter, we note a ring-theoretic statement of de Launey and Horadam's conjecture. We define quotients of a certain abelian group ring that determine the existence of cocyclic (generalized) Hadamard matrices.

Write

$$R_2(G) = \text{Ab}\langle (x, y) : x, y \in G \mid (x, y)(xy, z)(x, yz)^{-1}(y, z)^{-1} : x, y, z \in G \rangle$$

multiplicatively. Let $\mathcal{I}$ be the ideal of $\mathbb{C}[R_2(G)]$ generated by

$$(1, 1) - 1, \quad (x, y)^2 - 1, \quad \sum_{y \in G}(z, y), \quad x, y, z \in G, \ z \ne 1.$$

20.10.1. THEOREM. *There is a non-trivial homomorphism from $\mathbb{C}[R_2(G)]$ to $\mathbb{C}$ whose kernel contains $\mathcal{I}$ if and only if there exists a cocyclic Hadamard matrix with indexing group G.*

PROOF. Suppose that $[f(x, y)]_{x,y \in G}$ is a Hadamard matrix for some normalized cocycle f. Let ϕ_f be the homomorphism from $R_2(G)$ into $\langle -1 \rangle$ as in (20.6). If we extend ϕ_f linearly to $\mathbb{C}[R_2(G)]$ then we obtain a ring homomorphism from $\mathbb{C}[R_2(G)]$ to $\mathbb{C}$ with kernel containing $\mathcal{I}$.

Conversely, suppose that there is a non-trivial homomorphism $\varphi : \mathbb{C}[R_2(G)] \to \mathbb{C}$ with kernel containing $\mathcal{I}$. Then $\varphi((x, y)) \in \{\pm 1\}$ for all x, y, and $[\varphi((x, y))]_{x,y \in G}$ is a cocyclic Hadamard matrix by Theorem 11.2.12. $\square$

Theorem 20.10.1 also follows from Hilbert's Nullstellensatz. We close with two problems suggested by the theorem.

8. RESEARCH PROBLEM. *Study the quotient of $\mathbb{C}[R_2(G)]$ modulo*

(i) *$\mathcal{I}$, or*

(ii) *the ideal generated by $(1,1) - 1$ and the elements $\sum_{y \in G}(z, y)$, $z \neq 1$.*

A reasonable choice of G in part (i) would be D_{4p}, p an odd prime.

CHAPTER 21

Cocyclic Hadamard Matrices Indexed by Elementary Abelian Groups

We may begin a search for cocyclic PCD(Λ)s by calculating non-cohomologous cocycles for each possible indexing group. The next step is determining whether a coboundary can modify one of these cocycles to obtain a PCD(Λ). Algebra is sometimes brought to bear on this combinatorial part of the process as well, but the arguments there are usually ad hoc and less conventional.

Elementary abelian 2-groups are obvious candidates for indexing groups of cocyclic Hadamard matrices. We will see that for such groups the combinatorial problem is utterly routine: each Hadamard cocycle is a cocycle of a Sylvester matrix. Apart from being an interesting case study in its own right, this chapter is required reading for Chapter 22. Both chapters draw heavily on the paper [**54**] by de Launey and Smith.

21.1. Motivation: indexing groups for the Sylvester matrices

The Sylvester Hadamard matrices

$$H_0 = [1], \qquad H_k = \begin{bmatrix} H_{k-1} & H_{k-1} \\ H_{k-1} & -H_{k-1} \end{bmatrix} = H_1 \otimes H_{k-1}$$

are pure cocyclic.

21.1.1. LEMMA. H_k is cocyclic over C_2^k.

PROOF. We have $H_1 = [f_1(x, y)]_{x,y \in \mathrm{C}_2}$ where f_1 is the non-trivial cocycle $\mathrm{C}_2 \times \mathrm{C}_2 \to \langle -1 \rangle$. By induction, $H_{k+1} = H_1 \otimes H_k$ has cocycle $f_1 \times f_1^k = f_1^{k+1}$. □

In the proof of Lemma 21.1.1, $f = f_1^k$ is a non-coboundary (fully free) collection cocycle. Let us next find an extension group of H_k. View

$$\langle a_1, \ldots, a_k \mid a_i^2 = 1 \, (1 \leq i \leq k), \, [a_j, a_i] = 1 \, (1 \leq i < j \leq k) \rangle \cong \mathrm{C}_2^k$$

as the GF(2)-vector space V_k of dimension k. For $0 \leq i \leq 2^k - 1$, index the ith row and ith column of H_k by $(r_1, \ldots, r_k) \in V_k$ where $i = \sum_{j=1}^k 2^{j-1} r_j$. Then

$$H_k = [(-1)^{r \cdot s}]_{r,s \in V_k} = [f_c(a^r, a^s)]_{r,s \in V_k}$$

where $f_c = f$ is the collection cocycle as in (12.22) with $Q = I_k$. By Lemma 12.4.6,

$$E(f) = \langle b_1, \ldots, b_k, z \mid b_i^2 = z, \, z^2 = 1, \, b_i b_j = b_j b_i, \, b_i z = z b_i, \, 1 \leq i, j \leq k \rangle$$
$$= \langle b_1 \rangle \times \langle b_1 b_2 \rangle \times \cdots \times \langle b_{k-1} b_k \rangle$$
$$\cong \mathrm{C}_4 \times \mathrm{C}_2^{k-1}.$$

Since $E(f_1) \cong \mathrm{C}_4$ and $\mathrm{C}_4 \curlyvee_{\mathrm{C}_2} \mathrm{C}_4 \cong \mathrm{C}_4 \times \mathrm{C}_2$, this agrees with Lemma 12.2.1.

In this chapter, we will determine *all* cocycles and Hadamard groups for the Sylvester matrix when the indexing group is elementary abelian. Essentially we are going to carry out part of the agenda in Section 13.5, giving comprehensive answers to three natural questions: what are the Hadamard cocycles, cocyclic Hadamard matrices, and Hadamard groups, for an elementary abelian indexing group?

In Section 21.2 we describe the extension groups. Merging this algebra with the design theory, we prove subsequently that every Hadamard group that is an extension of C_2 by C_2^k has a centrally regular action on the expanded design of the Sylvester matrix H_k. Moreover, if $k > 1$ is odd then each cohomology class contains a (pure) Hadamard cocycle; whereas only the trivial class is ruled out if $k > 2$ is even. In short, nearly all elements of $H^2(C_2^k, \langle -1 \rangle)$ contain Hadamard cocycles: a remarkable fact.

Of course, there are possibilities other than just C_2^k for an indexing group of H_k. For example, H_{16} is cocyclic over every group of order 16 apart from the cyclic and dihedral groups. The following problem may be challenging.

9. RESEARCH PROBLEM. *Determine all indexing groups, Hadamard cocycles, and Hadamard groups for the Sylvester matrix H_k.*

Section 9.2 tells us that

$$\mathrm{Aut}(H_k) \cong Z(\mathrm{Aut}(H_k)) \times C_2^k \rtimes \mathrm{AGL}(k, 2) \quad \text{and} \quad Z(\mathrm{Aut}(H_k)) \cong C_2$$

for $k > 1$. So an indexing group over which H_k is cocyclic is a subgroup of order 2^k in the central quotient $C_2^{2k} \rtimes \mathrm{GL}(k, 2)$ of $\mathrm{Aut}(H_k)$.

21.2. The extension problem

Our aim in this section is to work out the full range of isomorphism types of extensions of C_2 by an elementary abelian 2-group. We use two approaches. One of these is solely group-theoretic; the other relates the isomorphism type of each extension to its cohomology class. Information garnered from both approaches will be needed in Chapter 23.

21.2.1. Group-theoretic approach.

21.2.1. LEMMA. *Let G be an abelian extension of $C \cong C_2$ by C_2^k. Then either $G \cong C_2^{k+1}$, or $G \cong A \times C_2^{k-1}$ where $A \cong C_4$ and $C \le A$.*

PROOF. Suppose that G is not elementary abelian. Then $C = \langle a^2 \rangle$ for some $a \in G$. We can select a normalized transversal for $A = \langle a \rangle$ in G whose non-identity elements all have order 2. The subgroup that these elements generate contains a complement of A in G. □

It turns out that a non-abelian extension is a direct product of an elementary abelian 2-group with one of the following (we adhere to notation from [54]):

(21.1)
$$A_m = \underbrace{D_8 \curlyvee \cdots \curlyvee D_8}_{m \text{ times}}$$

(21.2)
$$B_m = \underbrace{D_8 \curlyvee \cdots \curlyvee D_8}_{m-1 \text{ times}} \curlyvee Q_8$$

(21.3)
$$C_m = \underbrace{D_8 \curlyvee \cdots \curlyvee D_8}_{m \text{ times}} \curlyvee C_4.$$

In each central product, the amalgamated subgroup is a C_2. The groups A_m and B_m are the two (non-isomorphic) extraspecial 2-groups of order 2^{2m+1}. Also, $|C_m| = 2^{2m+2}$ and $C_m \cong A_m \curlyvee C_4 \cong B_m \curlyvee C_4$.

Let $\xi(K)$ denote the number of elements of order 4 in a group K. Together with $|Z(K)|$, this invariant determines the isomorphism type of an extension K of C_2 by C_2^k.

21.2.2. LEMMA. $\xi(A_m) = 2^m(2^m - 1)$, $\xi(B_m) = 2^m(2^m + 1)$, and $\xi(C_m) = 2^{2m+1}$.

PROOF. We first prove the formula for $\xi(A_m)$. Certainly $\xi(A_1) = 2(2 - 1)$, so assume that $m \geq 2$. Let $D = \langle x, y \mid x^4 = y^2 = 1, x^y = x^{-1} \rangle \cong D_8$. Any element of $A_m \cong A_{m-1} \curlyvee_{\langle x^2 \rangle} D$ is uniquely expressible as a, ax, ay, or axy, where $a \in A_{m-1}$. Counting the number of elements of order 4 in each case, we get

$$\xi(A_m) = \xi(A_{m-1}) + (|A_{m-1}| - \xi(A_{m-1})) + \xi(A_{m-1}) + \xi(A_{m-1})$$
$$= 2\xi(A_{m-1}) + 2^{2m-1}.$$

Solving the recurrence relation $a_m = 2a_{m-1} + 2^{2m-1}$ gives the formula for $\xi(A_m)$. Similar analyses of $B_m \cong A_{m-1} \curlyvee Q_8$ and $C_m \cong A_m \curlyvee C_4$ then give the other two formulae. $\square$

21.2.3. THEOREM. *Let E be a non-abelian extension of $C \cong C_2$ by C_2^k, $k \geq 2$. Then E is isomorphic to one and only one of the following groups:*

  (i) $A_m \times C_2^{j-1}$,
  (ii) $B_m \times C_2^{j-1}$,
  (iii) $C_m \times C_2^{j-2}$,

where $|Z(E)| = 2^j$, $1 \leq j \leq k - 1$, $k - j$ is odd, and $m = \frac{1}{2}(k - j + 1) \geq 1$.

PROOF. Let $Z = Z(E)$; of course $C \leq Z$. Since a group with cyclic central quotient is abelian, $|Z| = 2^j \leq 2^{k-1}$. By Lemma 21.2.1, either $Z \cong C_2^j$ or $Z \cong C_4 \times C_2^{j-2}$.

The subspace Z/C of the GF(2)-vector space E/C is complemented: $E/C = D/C \times Z/C$. Thus $E = D \curlyvee_C Z$. Indeed $Z(D) = C$, because $Z(D) \leq Z(E)$ implies that $Z(D) \leq D \cap Z = C$. So D is extraspecial, i.e., $D \cong A_m$ or $D \cong B_m$ for some m. Since $2^{k+1} = |E| = |D||Z|/2 = 2^{2m+j}$, $k - j$ is odd.

As we have seen, there are two possible isomorphism types of D, and two possible isomorphism types of Z. If $Z \cong C_2^j$ then E is as in (i) or (ii), with C splitting off Z and being absorbed into D. If $Z \cong C_4 \times C_2^{j-2}$ then both possibilities for the isomorphism type of D give the same isomorphism type (iii) of E.

If two listed groups are isomorphic then their centers have the same order, hence the same value of j. Fixing j fixes m. However, the groups in (i)–(iii) for the same j and m have different numbers of elements of order 4: $2^{j-1}\xi(A_m)$, $2^{j-1}\xi(B_m)$, and $2^{j-2}\xi(C_m)$ respectively, by Lemma 21.2.2. $\square$

21.2.4. COROLLARY. *Extensions E_1, E_2 of C_2 by C_2^k are isomorphic if and only if $|Z(E_1)| = |Z(E_2)|$ and $\xi(E_1) = \xi(E_2)$.*

21.2.2. Cocyclic approach. In Chapter 20 we gave a general technique for calculating cocycles, that is independent of the coefficient group. However, for our present purposes it is simpler to use the machinery of Section 12.4.

Let
$$G = \langle a_1, \ldots, a_k \mid a_i^2 = 1, [a_j, a_i] = 1, 1 \le i, j \le k \rangle \cong C_2^k.$$

For a $k \times k$ upper triangular $(0,1)$-matrix $Q = [q_{ij}]$, define the (collection) cocycle $f_Q : G \times G \to \langle -1 \rangle$ by

(21.4) $$f_Q(a^r, a^s) = (-1)^{sQr^\top}$$

using exponent vector notation. Treating Q as a 'relation matrix', the canonical extension group $E(Q) := E(f_Q)$ has presentation

$$\langle b_1, \ldots, b_k, z \mid b_i^2 = z^{q_{ii}}, z^2 = 1, [b_j, b_i] = z^{q_{ij}} \ (i < j), [z, b_i] = 1, 1 \le i, j \le k \rangle.$$

Each element of $E(Q)$ is of the form b^r or $b^r z$. Note that

(21.5) $$b^r b^s = b^{\overline{r+s}} z^{(1 - f_Q(a^r, a^s))/2}$$

where overlining denotes reduction modulo 2. The cocycles f_Q as Q runs over all $k \times k$ upper triangular binary matrices constitute a complete and irredundant set of representatives for the elements of $H^2(G, C)$.

According to Corollary 21.2.4, we must find $|Z(E(Q))|$ and $\xi(E(Q))$. Now $b^r z$ has order 4 if and only if b^r has order 4, which by (21.5) is true if and only if

(21.6) $$rQr^\top \equiv 1 \qquad (\text{mod } 2).$$

That is, the order 4 elements of $E(Q)$ are of the form b^r and $b^r z$ where r satisfies (21.6). Such vectors r are said to be *Q-anisotropic*. In the examples of Chapter 22, it is particularly easy to calculate $\xi(E(Q))$ by counting anisotropic vectors.

It is also easy to determine $|Z(E(Q))|$, as follows. The element $b^r z$ of $E(Q)$ is central if and only if b^r is central. By (21.5), b^r commutes with b^s if and only if $f_Q(a^r, a^s) = f_Q(a^s, a^r)$, which by (21.4) is equivalent to $rPs^\top \equiv 0$, where $P = Q + Q^\top$. Therefore

(21.7) $$b^r \in Z(E(Q)) \quad \Longleftrightarrow \quad r^\top \in \ker P.$$

Our discussion implies

21.2.5. PROPOSITION.

(1) $\xi(E(Q))$ *is twice the number of Q-anisotropic vectors.*
(2) $|Z(E)| = 2^j$ *where* $j = \dim(\ker P) + 1$.

We summarize the results of this section as an algorithm.

ALGORITHM 1

Input: Q, a $k \times k$ upper triangular binary matrix.
Output: the isomorphism type of $E(Q)$.

(1) If Q is not a diagonal matrix then go to step 2. Else $E(Q)$ is the split extension C_2^{k+1} if $Q = 0$, and $E(Q) \cong C_4 \times C_2^{k-1}$ otherwise.
(2) Count the number $\xi/2$ of Q-anisotropic length k binary vectors.
(3) Calculate $j = \dim(\ker(Q + Q^\top)) + 1$, and set $m = \frac{1}{2}(k - j + 1)$.
(4) If $\xi/2^{j-1} = 2^m(2^m - 1)$ then $E(Q) \cong A_m \times C_2^{j-1}$; if $\xi/2^{j-1} = 2^m(2^m + 1)$ then $E(Q) \cong B_m \times C_2^{j-1}$; else $E(Q) \cong C_m \times C_2^{j-2}$.

21.3. Pure Hadamard collection cocycles

We first characterize the pure Hadamard collection cocycles with an abelian indexing group. Then we show that each such cocycle is a cocycle of a Sylvester matrix.

21.3.1. THEOREM. *Let G be a finite abelian group.*

(i) *Suppose that $[f(x,y)]_{x,y\in G}$ is Hadamard, where f is a collection cocycle. Then G is an elementary abelian 2-group, and the extension group $E(f)$ is generated by elements of order 4.*

(ii) *Let f be the collection cocycle f_Q defined by (21.4) for $G \cong C_2^k$ and Q a $k \times k$ upper triangular $(0,1)$-matrix. Then $[f(x,y)]_{x,y\in G}$ is Hadamard if and only if Q is invertible, i.e., upper unitriangular.*

PROOF.

i. We refer again to Section 12.4, taking f to be the collection cocycle (12.18) arising from the extension (12.15) of $C = \langle -1 \rangle$ by G, where G has presentation (12.16) and E has presentation (12.17). Denote $\iota^{-1}(z_{ij})$ for $i \le j$ by c_{ij}.

Each non-initial row of the Hadamard matrix $[f(x,y)]_{x,y\in G}$ has zero sum. The sum of row a_1 is

$$\sum_{s_1,\ldots,s_k} f(a_1, a_1^{s_1} \cdots a_k^{s_k}) = \sum_{s_1,\ldots,s_k} \iota^{-1}(b_1^{1+s_1-\overline{1+s_1}}),$$

so that $\sum_{s_1=0}^{m_1-1} \iota^{-1}(b_1^{1+s_1-\overline{1+s_1}}) = m_1 - 1 + c_{11} = 0$. Thus $m_1 = 2$ and $c_{11} = -1$. Now assume that $m_i = 2$ and $c_{ii} = -1$ for all $i < l$ and some fixed $l \le k$. By Proposition 12.4.1,

$$f(a_1^{r_1} \cdots a_{l-1}^{r_{l-1}} a_l, a_1^{s_1} \cdots a_k^{s_k}) = f(a_l, a_l^{s_l}) \prod_{i=1}^{l-1} \prod_{j \ge i}^{l} c_{ij}^{s_i r_j}.$$

Setting $c_{ij} = (-1)^{q_{ij}}$, $q_{ij} \in \{0,1\}$, the sum of row $a_1^{r_1} \cdots a_{l-1}^{r_{l-1}} a_l$ is therefore

$$(21.8) \qquad \left(\prod_{i=1}^{l-1} \sum_{s_i=0}^{m_i-1} (-1)^{s_i \sum_{j=i}^{l} q_{ij} r_j} \right) \sum_{s_l=0}^{m_l-1} f(a_l, a_l^{s_l}).$$

The condition $\sum_{j=i}^{l} q_{ij} r_j \equiv 0 \pmod 2$ for all $i = 1, \ldots, l-1$ is equivalent to the system

$$\hat{Q}(r_1, \ldots, r_{l-1})^\top = (q_{1l}, \ldots, q_{l-1,l})^\top$$

having a solution over GF(2), where $\hat{Q}$ is the $(l-1) \times (l-1)$ upper triangular matrix with (i,j)th entry q_{ij}. By assumption, $q_{ii} = 1$, $1 \le i < l$; therefore $\hat{Q}$ is invertible, and the above system has a solution. Since (21.8) is zero, we then have

$$\sum_{s_l=0}^{m_l-1} f(a_l, a_l^{s_l}) = m_l - 1 + (-1)^{q_{ll}} = 0,$$

so $m_l = 2$ and $q_{ll} = 1$. This proves by induction that $m_i = 2$ and $q_{ii} = 1$ for all i. Thus G is an elementary abelian 2-group, and the extension group E is generated by elements of order 4.

ii. We may assume that $k > 1$. If $[f(x,y)]_{x,y\in G}$ is a Hadamard matrix then, as we saw in the proof of part (i), the upper unitriangular matrix $Q = [q_{ij}]$ is invertible. Conversely, suppose that Q is invertible. Let $r \in V_k$ be non-zero, and denote by U the subspace of V_k consisting of all s such that $sQr^\top = 0$. Then $V_k = U \oplus W$ where W is 1-dimensional, spanned by w say. Thus

$$\sum_{s \in V_k} f(a^r, a^s) = \sum_{s \in U} ((-1)^{sQr^\top} + (-1)^{(s+w)Qr^\top}) = |U|(1-1) = 0.$$

By Theorem 11.2.12, this completes the proof. $\qquad\qquad\square$

21.3.2. PROPOSITION. *Let $G \cong C_2^k$, and let $f : G \times G \to \langle -1 \rangle$ be a collection cocycle. If $[f(x,y)]_{x,y \in G}$ is Hadamard then it is equivalent to the Sylvester matrix of order 2^k.*

PROOF. We know that $f(a^r, a^s) = (-1)^{sQr^\top}$ for some upper unitriangular $(0,1)$-matrix Q. Define a permutation ρ of the elements of G by $\rho : a^s \mapsto a^t$ if and only if $s = tQ$. Postmultiplying $[f(x,y)]_{x,y \in G}$ by $[\delta^x_{\rho(y)}]_{x,y \in G}$ produces the matrix whose entry in position (a^r, a^s) is $(-1)^{sQ^{-1}Qr^\top} = (-1)^{r \cdot s}$. $\qquad\square$

The Hadamard group for a given pure Hadamard collection cocycle can be found via Algorithm 1 in Section 21.2. As we will see, all non-abelian extensions of C_2 by C_2^k occur, except for $D_8 \times C_2^{k-2}$.

21.4. Bilinearity and Hadamard cocycles

Let G be a finite solvable group, and let C be a finite abelian group. We denote by $\mathrm{Coll}(G, C)$ the set of collection cocycles defined with respect to a fixed (polycyclic) presentation of G. In fact $\mathrm{Coll}(G, C)$ is a subgroup of $Z^2(G, C)$. By Corollary 12.4.3, if G and C are both elementary abelian p-groups for the same prime p, then $Z^2(G, C) = \mathrm{Coll}(G, C) \oplus B^2(G, C)$. As a consequence $\mathrm{Coll}(G, C) \cong H^2(G, C)$.

Let $G \cong C_2^k$ and $C = \langle -1 \rangle$, so $Z^2(G, C)$ is likewise an elementary abelian 2-group. Counting the number of $k \times k$ upper triangular $(0,1)$-matrices, we see that $\mathrm{Coll}(G, C)$ has rank $\binom{k+1}{2}$. Now for each $k \times k$ $(0,1)$-matrix M, we have a bilinear cocycle $f : G \times G \to C$ defined by

(21.9) $$f(a^r, a^s) = (-1)^{sMr^\top}.$$

The set $\mathrm{Bil}(G, C)$ of all such cocycles is a subgroup of $Z^2(G, C)$ of rank k^2, that contains $\mathrm{Coll}(G, C)$. By Theorem 21.3.1, there are $2^{\binom{k}{2}}$ pure Hadamard cocycles in $\mathrm{Coll}(G, C)$. The next lemma characterizes the pure Hadamard cocycles in $\mathrm{Bil}(G, C)$.

21.4.1. LEMMA. *Let f be a cocycle as defined by (21.9).*

 (i) *f is a coboundary if and only if M is symmetric with zero diagonal.*
 (ii) *f is a pure Hadamard cocycle if and only if M is invertible.*
 (iii) *If the diagonal of M is all 1s then f is cohomologous to a pure Hadamard cocycle.*
 (iv) *If f is pure Hadamard then it is a cocycle of the Sylvester matrix of order 2^k.*

PROOF.
i. See Lemma 12.4.7.

ii. Suppose that f is pure Hadamard. If M is singular then $Mr^\top = 0$ for some non-zero r; but this then means that row $a^r \neq 1$ of $[f(x,y)]_{x,y \in G}$ consists entirely of 1s. For the other direction, cf. the proof of Theorem 21.3.1 (ii).

iii. Let N be the symmetric matrix with zero diagonal whose (i,j)th entry for $i > j$ is the (i,j)th entry of M. Then let f' be the coboundary defined by (21.9) with N in place of M. We have

$$(ff')(a^r, a^s) = (-1)^{sMr^\top}(-1)^{sNr^\top} = (-1)^{s(M+N)r^\top}.$$

Since $M + N$ is upper unitriangular, hence invertible, ff' is pure Hadamard.

iv. The proof of Proposition 21.3.2 carries over, *mutatis mutandis*. □

21.5. Solution of the Hadamard cocycle problem

In this section we prove that almost every element of $Z^2(C_2^k, \langle -1 \rangle)$ is a cocycle of the Sylvester matrix of order 2^k.

The next lemma is a converse of Theorem 21.3.1 (i).

21.5.1. LEMMA. *Let $G \cong C_2^k$ and let $f : G \times G \to \langle -1 \rangle$ be a cocycle. Suppose that $E(f)$ is generated by elements of order 4. Then f is cohomologous to a pure Hadamard collection cocycle.*

PROOF. We have $E(f) = \langle b_1, \dots, b_k \rangle$ where $b_i^2 = (1, -1)$ for all i, and then G has basis $\{a_1, \dots, a_k\}$ where a_i is the image of b_i under the canonical surjection $E(f) \to G$. By Proposition 12.4.1, f is cohomologous to a collection cocycle f_Q defined as in (21.4) with respect to the basis $\{a_1, \dots, a_k\}$. Since $b_i^2 = (1, -1)$, the diagonal entries of Q are all 1s. So f_Q is pure Hadamard by Lemma 21.4.1 (ii). □

21.5.2. REMARK. (Cf. Remark 12.4.5.) If $f = f_Q$ is Hadamard then a cocycle f' obtained from f via change of basis is Hadamard too. However $f \in \mathrm{Coll}(G, C)$, whereas f' is in the larger set $\mathrm{Bil}(G, C)$.

There is only one non-abelian extension of C_2 by C_2^k that cannot be generated by elements of order 4.

21.5.3. THEOREM. *A cocycle $f : C_2^k \times C_2^k \to \langle -1 \rangle$ that is not a coboundary is cohomologous to a pure Hadamard collection cocycle, unless $E(f) \cong D_8 \times C_2^{k-2}$ and f is cohomologous to f_Q defined by (21.4) where*

$$(21.10) \qquad Q = \begin{bmatrix} 1 & 0 & \cdots & 0 & 1 \\ 0 & 1 & \cdots & 0 & 1 \\ \vdots & \vdots & \ddots & \vdots & \vdots \\ 0 & 0 & \cdots & 1 & 1 \\ 0 & 0 & \cdots & 0 & 0 \end{bmatrix}.$$

PROOF. Let $m > 0$ be the maximum number of generators of order 4 in any generating set for $E(f)$. Then $E(f)$ has presentation

$$\langle b_1, \dots, b_m, b'_1, \dots, b'_{k-m}, z \mid b_i^2 = z, \ b_j b_i = b_i b_j z^{r_{ij}}, \ 1 \le i, j \le m,$$
$$b_i'^2 = 1, \ b'_j b'_i = b'_i b'_j z^{s_{ij}}, \ 1 \le i, j \le k - m,$$
$$b'_j b_i = b_i b'_j z^{t_{ij}}, \ 1 \le i \le m, \ 1 \le j \le k - m \rangle,$$

where the relations specifying $z = (1, -1)$ as a central involution are omitted. By Lemma 21.5.1, we assume that $m < k$. If any t_{ij} is 0 then $b_i b'_j$ has order 4 and can replace b'_j as a generator, violating maximality of m. So every t_{ij} is 1. Similarly, $s_{ij} = 0$ (else we can replace b'_i by $b'_i b'_j$) and $r_{ij} = 0$ (else we can replace any b'_l by

$b_i b_j b'_l$). Finally, $m = k - 1$; for if not then $(b_i b'_1 b'_2)^2 = z$ and we can replace b'_1 by $b_i b'_1 b'_2$ to increase m. Hence $E(f)$ has presentation

$$\langle b_1, \ldots, b_{k-1}, b, z \mid b_i^2 = z, \; b^2 = 1, \; b_j b_i = b_i b_j, \; b b_i = b_i b z, \; 1 \leq i, j \leq k \rangle,$$

so is the direct product of $\langle b_1 b_j : 2 \leq j \leq k - 1 \rangle \cong C_2^{k-2}$ and $\langle b_1, b \rangle \cong D_8$. With the help of Lemma 12.4.6, we see that this extension has collection cocycle (21.4) with Q as in (21.10). □

Now we show that for $k > 2$, f is cohomologous to a pure Hadamard cocycle even when $E(f) \cong D_8 \times C_2^{k-2}$.

21.5.4. THEOREM. *Let $f : C_2^k \times C_2^k \to \langle -1 \rangle$ be a cocycle.*
   (i) *If $k > 2$ is even then f is Hadamard.*
   (ii) *If $k \geq 1$ is odd, then f is Hadamard if and only if it is not a coboundary.*
   (iii) *If $k = 2$, then f is Hadamard if and only if $E(f) \not\cong D_8$.*
   (iv) *If f is Hadamard then it is a cocycle of the Sylvester matrix of order 2^k.*

PROOF. A Hadamard coboundary $C_2^k \times C_2^k \to \langle -1 \rangle$ can exist only for even k. If k is even then let M be the permutation matrix with repeated diagonal block

$$\begin{bmatrix} 0 & 1 \\ 1 & 0 \end{bmatrix}.$$

The cocycle (21.9) with this M is a coboundary by Lemma 21.4.1 (i), and it is Hadamard by Lemma 21.4.1 (ii).

Suppose that $k \geq 2$ and f is not a coboundary. By Theorem 21.5.3, f is a Hadamard cocycle unless $E(f) \cong D_8 \times C_2^{k-2}$ and f is cohomologous to (21.4) with Q as in (21.10). Suppose that $k \geq 3$, and let $S = [s_{ij}]$ be the $k \times k$ matrix such that $s_{k-2,k-1} = s_{k-1,k-2} = s_{k-1,k} = s_{k,k-1} = 1$ and $s_{ij} = 0$ otherwise. Then the cocycle defined by (21.9) with $M = Q + S$ is pure Hadamard and cohomologous to f. If $k = 2$ then none of the cocycles f such that $E(f) \cong D_8$ are Hadamard, by Corollary 15.6.5.

We have proved parts (i)–(iii). Now recall that if f is Hadamard, then it is cohomologous to a bilinear pure Hadamard cocycle. By Lemma 21.4.1 (iv), the proof is complete. □

21.5.5. COROLLARY. *An extension E of C_2 by C_2^k is a Hadamard group if and only if E is not D_8 nor C_2^{k+1}, k odd. Such a Hadamard group E acts centrally regularly on the expanded design of the Sylvester matrix of order 2^k.*

We again provide a summary algorithm. For a binary $k \times k$ matrix N, define $g_N : C_2^k \to \{\pm 1\}$ by $g_N(a^r) = (-1)^{r N r^\top}$.

ALGORITHM 2

Input: $G \cong C_2^k$ and a $k \times k$ upper triangular $(0, 1)$-matrix Q.
Output: a map $g : G \to \{\pm 1\}$ such that $[f_Q(x, y) g(xy)]_{x,y \in G}$ is Hadamard, or a message that no such map g exists.

   (1) If $Q = 0$ and k is odd then g does not exist; if $Q = 0$ and k is even then return g_N where N has 1 in position $(2i - 1, 2i)$, $1 \leq i \leq k/2$, and zeros elsewhere.
   (2) If Q is unitriangular then return $g = 1$, i.e., return g_N where $N = 0$.

(3) If $k = 2$, then g does not exist if Q has 1 in position $(1, 2)$. Else return g_N where N has 1 in position $(1, 2)$ and zeros elsewhere.
(4) If $k \geq 3$, find a symmetric $(0, 1)$-matrix S with zero diagonal such that $Q + S$ is invertible. Then return g_N where $S = N + N^\top$.

CHAPTER 22

Cocyclic Concordant Systems of Orthogonal Designs

Let $M = [m_{ij}]$ be an $m \times m$ symmetric $(0,1)$-matrix with zero diagonal, and let n be a positive integer. Recall (from Definition 16.3.4) that a set $\{D_1, \ldots, D_m\}$ of orthogonal designs of order n on disjoint indeterminate sets is an M-concordant system if $D_i D_j^\top = (-1)^{m_{ij}} D_j D_i^\top$ for all i, j. The system is cocyclic, with cocycle $f : G \times G \to \langle -1 \rangle$, if there are signed permutation matrices P, Q and maps g_i such that $PD_iQ = [f(x,y)g_i(xy)]_{x,y \in G}$ for $1 \le i \le m$. In Chapter 23 we will see how these systems lend themselves to powerful constructions for cocyclic Hadamard matrices.

As usual, the most basic question to be answered is the existence one: for which M, n, $a_{i1}, \ldots, a_{ik_i}$, and f, does there exist a cocyclic M-concordant system of $\mathrm{OD}(n; a_{i1}, \ldots, a_{ik_i})$ with cocycle f? We will answer this question in an important special case; namely, when each indeterminate occurs precisely once in each row and column of the design containing it.

22.1. Existence and uniqueness of cocyclic systems of $\mathrm{OD}(n; 1^k)$

We introduce some notation required for our main theorem.

Let $k_1, \ldots, k_m$ be positive integers. Define $l_0 = 0$ and $l_i = k_1 + \cdots + k_i$ for $i \ge 1$. If $M = [m_{ij}]$ is a symmetric $(0,1)$-matrix of order m with zero diagonal, let $N(M, k_1, \ldots, k_m) = [n_{rs}]$ be the $l_m \times l_m$ symmetric $(0,1)$-matrix such that

$$(22.1) \qquad n_{rs} = \begin{cases} 0 & \text{if } r = s \\ 1 & \text{if } l_{i-1} < r, s \le l_i, \ r \ne s \\ m_{ij} & \text{if } l_{i-1} < r \le l_i \text{ and } l_{j-1} < s \le l_j \text{ with } i \ne j. \end{cases}$$

Next, for any $k \times k$ binary matrix $N = [n_{ij}]_{0 \le i,j \le k-1}$, define $Q_N = [q_{ij}]_{1 \le i,j \le k-1}$ by

$$(22.2) \qquad q_{ij} = \begin{cases} n_{0i} & \text{if } i = j \\ n_{0i} + n_{j0} + n_{ij} \pmod 2 & \text{if } j > i \\ 0 & \text{otherwise.} \end{cases}$$

The map $N \mapsto Q_N$ is a bijection from the set of binary $k \times k$ symmetric matrices with zero diagonal to the set of binary $(k-1) \times (k-1)$ upper triangular matrices.

Let Q be an upper triangular binary matrix. In Chapter 21, we saw that an extension $E = E(Q)$ of $\langle z \rangle \cong C_2$ by an elementary abelian 2-group has the form $\mathrm{Hub}(E) \times S$, where the 'hub group' $\mathrm{Hub}(E)$ contains z and is isomorphic to one of A_k, B_k, or C_k as in (21.1)–(21.3). Note that $\mathrm{Hub}(E)$ is abelian if and only if $k = 0$, where $A_0 = B_0 = \langle z \rangle$ and C_0 is cyclic of order 4 containing z. The center of E is $\langle z \rangle \times S$.

Section 22.4 is devoted to a proof of the following result, which appears as Theorem 7.1 in de Launey and Smith's paper [**54**].

22.1.1. THEOREM. *Let M be an $m \times m$ binary symmetric matrix with zero diagonal. Let $N = N(M, k_1, \ldots, k_m)$ and $Q = Q_N$ as per* (22.1) *and* (22.2).

(i) *There exists a cocyclic M-concordant system of $\mathrm{OD}(n; 1^{k_i})$ with cocycle f if and only if $E(f)$ has a subgroup T isomorphic to $\mathrm{Hub}(E(Q))$ containing the central involution $(1, -1)$ of $E(f)$.*

(ii) *There is essentially just one cocyclic M-concordant system of $\mathrm{OD}(n; 1^{k_i})$ if $|\mathrm{Hub}(E(Q))| = 2n$.*

Regarding (i), note that $\mathrm{Hub}(E(Q))$ has a unique central involution. Further, $E(Q) = E(f_Q)$ where f_Q is the collection cocycle defined by (21.4). Thus, a cocyclic system as stated certainly exists when $f = f_Q$. Part (ii) says that minimizing the order of the extension group of the cocyclic system (i.e., minimizing n) guarantees essential uniqueness of the system, for given M and k_i.

22.2. A reduction

Suppose that $\{D_1, \ldots, D_m\}$ is a cocyclic M-concordant system of $\mathrm{OD}(n; 1^{k_i})$, with cocycle $f : G \times G \to \langle -1 \rangle$. Put $l_0 = 0$ and $l_i = k_1 + \cdots + k_i$ if $i \geq 1$. Then

$$D_i = U \sum_{a \in G} g(a) P_a^f$$

by Lemma 12.5.5, for some map $g : G \to \{0, \pm x_{l_{i-1}+1}, \ldots, \pm x_{l_i}\}$ and P_a^f, U as in (12.23) and Definition 12.5.3. Since the P_a are monomial and disjoint,

$$D_i = U \sum_{r=l_{i-1}+1}^{l_i} x_r P_{a_r}$$

for some $a_r \in E(f)$. In Section 16.3, we saw that existence of the cocyclic system is equivalent to a set of relations between the elements a_r and a_s in the extension group of the system. In particular, Theorem 16.3.5 implies the following lemma. This is our first large step toward the proof of Theorem 22.1.1, and reduces the existence question to a problem in the extension group.

22.2.1. LEMMA. *Let M be an $m \times m$ binary symmetric matrix with zero main diagonal, and let $N = [n_{rs}]$ be the matrix $N(M, k_1, \ldots, k_m)$ defined as in* (22.1) *for integers $k_1, \ldots, k_m > 0$. Given a cocycle $f : G \times G \to \langle -1 \rangle$ and $a_1, \ldots, a_{l_m} \in E(f)$, define $D_i = \sum_{r=1+l_{i-1}}^{l_i} x_r P_{a_r}^f$. Denote the central involution $(1, -1) \in E(f)$ by z. Then $\{D_1, \ldots, D_m\}$ is a cocyclic M-concordant system of $\mathrm{OD}(|G|; 1^{k_i})$ with cocycle f if and only if $(a_r a_s^{-1})^2 = z^{n_{rs}}$ for all $r, s \in \{1, \ldots, l_m\}$.*

Let N be any binary $(k + 1) \times (k + 1)$ symmetric matrix with zero diagonal. A sequence of elements $a_0, \ldots, a_k$ of a group with central involution z such that

$$(22.3) \qquad\qquad (a_i a_j^{-1})^2 = z^{n_{ij}}, \qquad 0 \leq i, j \leq k,$$

is N-*concordant with respect to z*. Since $a_0, \ldots, a_k$ is N-concordant if and only if the normalized sequence $1, a_1 a_0^{-1}, \ldots, a_k a_0^{-1}$ is too, Lemma 22.2.1 may be expressed as follows.

22.2.2. LEMMA. *Let M be an $m \times m$ binary symmetric matrix with zero main diagonal. There exists a cocyclic M-concordant system of $\mathrm{OD}(n; 1^{k_i})$ with cocycle f if and only if $E(f)$ contains a normalized $N(M, k_1, \ldots, k_m)$-concordant sequence with respect to $(1, -1)$.*

22.3. Solution of the reduced problem

22.3.1. LEMMA. *Let $Q = [q_{ij}]$ be a $k \times k$ upper triangular binary matrix, and let E be a group with central involution z. There exist $b_1, \ldots, b_k \in E$ such that*

$$(22.4) \qquad b_i^2 = z^{q_{ii}}, \qquad [b_j, b_i] = z^{q_{ij}}, \qquad 1 \le i < j \le k,$$

if and only if E contains an isomorphic copy T of $\mathrm{Hub}(E(Q))$ with $z \in T$.

PROOF. Since $E(Q)$ has presentation

$$(22.5) \quad \langle e_1, \ldots, e_k, y \mid e_i^2 = y^{q_{ii}}, \ y^2 = 1, \ [e_j, e_i] = y^{q_{ij}}, \ [y, e_i] = 1, \ 1 \le i < j \le k \rangle,$$

if there exist $b_i \in E$ satisfying the relations (22.4) then $\tilde{E} = \langle b_1, \ldots, b_k, z \rangle$ is the image of $E(Q)$ under the homomorphism $\theta : E(Q) \to E$ defined by $e_i \mapsto b_i$, $y \mapsto z$. Let K denote the kernel of this homomorphism. Since $[E(Q), E(Q)] \le \langle y \rangle$, we get that $[K, E(Q)] \le K \cap \langle y \rangle = 1$, i.e., K is central in $E(Q)$. Then as y is the unique central involution of $\mathrm{Hub}(E(Q))$, this implies that $\mathrm{Hub}(E(Q)) \cap K = 1$. Therefore θ is an isomorphism from $\mathrm{Hub}(E(Q))$ onto a subgroup of E containing z.

Now suppose that E has a subgroup $T \cong \mathrm{Hub}(E(Q))$, and $z \in T$. As we know, $E(Q) = \mathrm{Hub}(E(Q)) \times S$ for some S. Denote the image of $e \in E(Q)$ under the projection $E(Q) \to \mathrm{Hub}(E(Q))$ by $\bar{e}$; then $\bar{e_i}^2 = y^{q_{ii}}$ and $[\bar{e_j}, \bar{e_i}] = y^{q_{ij}}$. Since any isomorphism from $\mathrm{Hub}(E(Q))$ onto T must map y to z, we are done. $\square$

22.3.2. COROLLARY. *Let E be a group with central involution z, and let N be a binary symmetric matrix with zero diagonal. There exists an N-concordant sequence of elements in E with respect to z if and only if E contains an isomorphic copy T of $\mathrm{Hub}(E(Q_N))$ with $z \in T$.*

PROOF. If (22.3) holds for elements a_i of E, then the $b_i = a_i a_0^{-1}$ satisfy (22.4). Conversely, if there are $b_i \in E$ satisfying (22.4), then (22.3) holds for $a_0 = 1$ and $a_i = b_i$. $\square$

22.3.3. REMARK. Suppose that the group E has a subgroup $T \cong \mathrm{Hub}(E(Q))$, where $Q = Q_N$. We recapitulate from the proof of Lemma 22.3.1 how to find an N-concordant sequence in E with respect to the central involution of T, working with (22.5). (Our method actually locates the sequence in T.) First, by (21.7) we calculate $\ker(Q + Q^\top)$, which tells us the center $\langle y \rangle \times S$ of $E(Q)$. We then find a generating set for a complement $\mathrm{Hub}(E(Q))$ of S in $E(Q)$. Armed with this, it is straightforward to obtain the $\bar{e_i} \in \mathrm{Hub}(E(Q))$ and $s_i \in S$ such that $e_i = \bar{e_i} s_i$, $1 \le i \le k$. Then $1, \bar{e_1}, \ldots, \bar{e_k}$ is an N-concordant sequence. (Note that the elements in this sequence need not be distinct.)

22.3.4. LEMMA. *Let N be a binary symmetric matrix. Suppose that z is a central involution of $E \cong \mathrm{Hub}(E(Q_N))$. Then $\mathrm{Aut}(E)$ acts transitively on the set $\mathcal{N}$ of all normalized N-concordant sequences in E with respect to z.*

PROOF. Set $Q = Q_N$. If $\phi \in \mathrm{Aut}(E)$ and $(1, a_1, \ldots, a_k) \in \mathcal{N}$ then it is clear that $(1, \phi(a_1), \ldots, \phi(a_k)) \in \mathcal{N}$, because ϕ must map the unique central involution z of E to z. Thus $\mathrm{Aut}(E)$ acts on $\mathcal{N}$.

As seen in the proof of Lemma 22.3.1 (and by the proof of Corollary 22.3.2), the map θ defined by $e_i \mapsto a_i$, $y \mapsto z$ is a surjective homomorphism from $E(Q)$ with presentation (22.5) onto $\tilde{E} := \langle a_1, \ldots, a_k, z \rangle$, that embeds $\mathrm{Hub}(E(Q))$ in $\tilde{E}$. Thus $\tilde{E} = E$ and θ restricted to $\mathrm{Hub}(E(Q))$ is an isomorphism onto E. It follows that if $(1, a'_1, \ldots, a'_k)$ is any other sequence in $\mathcal{N}$, then there is $\phi \in \text{Aut}(E)$ such that $\phi(a_i) = a'_i$ for all i. □

For the $\text{Aut}(E)$-action in Lemma 22.3.4 to be transitive, the condition that E be as small as possible (i.e., a hub group) is unavoidable: some concordant sequences generate a hub group, whereas others generate all of E.

22.4. Proof of Theorem 22.1.1

We come now to the proof of Theorem 22.1.1. Part (i) of the theorem follows from Lemma 22.2.2 and Corollary 22.3.2. So we only have to confirm uniqueness. Before doing that, we discuss what uniqueness of a cocyclic system of orthogonal designs means.

The input data for a system is the concordance matrix M and the positive integers $k_1, \ldots, k_m$. This input determines $N = N(M, k_1, \ldots, k_m)$ and $Q = Q_N$. The minimality condition $2n = |\text{Hub}(E(Q))|$ then determines the extension group E and the indexing group G for a cocyclic M-concordant system of $\text{OD}(n; 1^{k_i})$. That is, $E \cong \text{Hub}(E(Q))$ and $G \cong E/Z(E)$. The next matter is to decide which cocycles $f : G \times G \to \langle -1 \rangle$ are cocycles of the system. Strictly speaking, f is not uniquely determined: a system satisfying $2n = |\text{Hub}(E(Q))|$ for given input can possess several cocycles, and moreover these will be distributed across several cohomology classes. However, by allowing variability in N-concordant sequences, we arrive at a criterion for essential uniqueness. We proceed by considering separately the effect of varying the cocycle while keeping all other parameters fixed; and of varying other parameters while the cocycle is fixed.

Let

$$\{D_i = \textstyle\sum_{j=1+l_{i-1}}^{l_i} x_j P_{a_j}^f \mid 1 \leq i \leq m\}$$

be an M-concordant system of $\text{OD}(n; 1^{k_i})$ with cocycle f, where $(a_1, \ldots, a_{l_m})$ is N-concordant with respect to $(1, -1)$ in $E(f)$. Negating x_1 if necessary, we may take $a_1 = 1$. For fixed f, our only freedom of choice when making up the D_i lies in the labeling of rows and columns by the elements of G, and in the selection of normalized N-concordant sequence. We will regard two concordant sequences in $E(f)$, and their corresponding systems, as equivalent if the sequences are in the same orbit under the action by $\text{Aut}(E(f))$. By Lemma 22.3.4, then, we can move on to considering the effect of changing cocycles. Let f, f' be cocycles of the same system, so that $E(f) \cong E(f')$. Indeed (see Subsection 12.1.8 of Chapter 12), $f' = f \circ \alpha$ for some $\alpha \in \text{Aut}(G)$, and the map $\phi : (x, c) \mapsto (\alpha(x), c)$ is an isomorphism $\phi : E(f') \to E(f)$. Also, ϕ maps concordant sequences to concordant sequences. We have

$$P_{(x,c)}^{f'} = c[\delta_s^{rx} f(\alpha(r), \alpha(x))]_{r,s \in G}$$
$$= [\delta_r^{\alpha(a)}]_{a,r} \, c[\delta_s^{r\alpha(x)} f(r, \alpha(x))]_{r,s} \, [\delta_{\alpha(b)}^s]_{s,b}$$
$$= [\delta_r^{\alpha(a)}]_{a,r} \, P_{(\alpha(x),c)}^f \, [\delta_{\alpha(b)}^s]_{s,b}.$$

Thus, a system with cocycle f' and concordant sequence $(a_1, \ldots, a_{l_m})$ is equivalent to a system with cocycle f and sequence $(\phi(a_1), \ldots, \phi(a_{l_m}))$. As we have decreed that changing concordant sequences while keeping the cocycle fixed is an equivalence operation (in this situation where the extension group has minimal order), the two systems are equivalent.

22.4.1. REMARK. Suppose that f is a cocycle of the system $\{D_i \mid 1 \leq i \leq m\}$ of orthogonal designs such that $E(f)$ has minimal order, and let $\phi \in \mathrm{Aut}(E(f))$. The regular representations of $E(f)$ defined by $a \mapsto P_a^f$ and $a \mapsto P_{\phi(a)}^f$ are similar, and it is conceivable that they are similar via conjugation by a signed permutation matrix (cf. Lemma 12.5.2). Two systems of cocyclic orthogonal designs obtained for different choices of concordant sequence in $E(f)$ would then be equivalent in the usual sense. However, we cannot expect this to be the case generally. For if the k_i are distinct then $D_i = [f(x,y)g_i(xy)]_{x,y \in G}$ must be equivalent to $D_i = [f(\alpha(x), \alpha(y))g_i(xy)]_{x,y \in G}$ for an automorphism α of G, and this just need not be true. If we want to classify minimal systems as essentially unique, our only option is to expand the notion of equivalence to include $\mathrm{Aut}(E(f))$-action on concordant sequences.

22.5. Removing the zeros

Plug-in techniques for constructing Hadamard matrices from orthogonal designs require that the designs have no zeros. Here is an inductive existence result for cocyclic systems of such designs (this was used by de Launey and Smith [54] to prove their Theorem 1.1).

22.5.1. THEOREM. *Let $h : C_2^t \times C_2^t \to \{1\}$ be the trivial map. If there exist an M-concordant system of $\mathrm{OD}(n; 1^{2^t})$ with cocycle f, and a Hadamard matrix of order n with cocycle f, then there exists an M-concordant system of $\mathrm{OD}(n2^t; n^{2^t})$ with cocycle $f \times h$.*

PROOF. Suppose that $\{D_1, \ldots, D_m\}$ is an M-concordant system of $\mathrm{OD}(n; 1^{2^t})$ with cocycle $f : G \times G \to \langle -1 \rangle$. Set $l_i = 2^t i$. We have

$$D_i = \sum_{s=1+l_{i-1}}^{l_i} y_s P_{a_s}^f$$

for some $a_1, \ldots, a_{l_m} \in E(f)$, by Lemma 22.2.1. If $a_s = (x, -1)$ then we may negate y_s and replace a_s by $(x, 1)$; so we assume that $a_s = (x_s, 1)$, $1 \leq s \leq l_m$.

For each i, $1 \leq i \leq m$, let $\alpha_{l_{i-1}+1}, \ldots, \alpha_{l_i}$ be a listing of the elements of $K \cong C_2^t$. The assignment $\alpha \mapsto P_\alpha^h$ is a regular representation of K as a group of $2^t \times 2^t$ permutation matrices. Let

$$E_i = \sum_{s=l_{i-1}+1}^{l_i} \left(y_s H P_{x_s}^f \otimes P_{\alpha_s}^h \right)$$

where H is a Hadamard matrix with cocycle f. We will show that $f' := f \times h$ is a cocycle of E_i. By Lemma 12.5.5, $H = U^f \sum_{x \in G} g(x) P_x^f$ for some map $g : G \to \{\pm 1\}$. Define $g_s(x) = g(xx_s^{-1})$, so that

$$\sum_{x \in G} g(x) P_x^f \, P_{x_s}^f = \sum_{x \in G} g_s(x) P_x^f.$$

Then

$$E_i = (U^f \otimes I_{2^t}) \sum_{x \in G} \sum_{s=l_{i-1}+1}^{l_i} y_s g_s(x) (P_x^f \otimes P_{\alpha_s}^h)$$

$$= U^{f'} \sum_{x \in G} \sum_{s=l_{i-1}+1}^{l_i} y_s g_s(x) P_{(x_s \alpha, 1)}^{f'}$$

by Lemma 12.5.6. So

$$E_i = U^{f'} \sum_{x \in G} \sum_{\alpha \in K} \hat{g}(x\alpha) P_{x\alpha}^{f'}$$

where $\hat{g}(x\alpha_s) = g_s(x)y_s$, $x\alpha_s \in G \times K$. Hence E_i is cocyclic with cocycle f', by Lemma 12.5.5 again. Note that all entries of E_i are non-zero.

Next we verify that E_i is an $\mathrm{OD}(2^t n; n^{2^t})$:

$$
\begin{aligned}
E_i E_i^\top &= (\textstyle\sum_{s=l_{i-1}+1}^{l_i} y_s H P_{x_s}^f \otimes P_{\alpha_s}^h)(\textstyle\sum_{u=l_{i-1}+1}^{l_i} y_u P_{x_u^{-1}}^f H^\top \otimes P_{\alpha_u}^h) \\
&= \textstyle\sum_s y_s^2 H P_{x_s}^f P_{x_s^{-1}}^f H^\top \otimes P_{\alpha_s}^h P_{\alpha_s}^h \\
&\quad + \textstyle\sum_{s \neq u} y_s y_u H (P_{x_s}^f P_{x_u^{-1}}^f + P_{x_u}^f P_{x_s^{-1}}^f) H^\top \otimes P_{\alpha_s}^h P_{\alpha_u}^h \\
&= \textstyle\sum_s y_s^2 H H^\top \otimes I_{2^t} \\
&= n I_{n2^t} \textstyle\sum_{s=l_{i-1}+1}^{l_i} y_s^2.
\end{aligned}
$$

We relied on D_i being an orthogonal design to conclude that the summation in the third line is zero. Similarly we may prove that

$$
D_j D_i^\top = (-1)^{m_{ij}} D_i D_j^\top \quad \Longleftrightarrow \quad E_j E_i^\top = (-1)^{m_{ij}} E_i E_j^\top;
$$

so $\{E_i \mid 1 \le i \le m\}$ is a cocyclic system of orthogonal designs of the desired kind. $\qquad\square$

22.6. Examples

22.6.1. THEOREM. *For $k \ge 1$, let*

$$
T_k = \begin{cases}
A_{(k-2)/2} & k \equiv 0 \pmod{8} \\
A_{(k-1)/2} & k \equiv \pm 1 \pmod{8} \\
C_{(k-2)/2} & k \equiv \pm 2 \pmod{8} \\
B_{(k-1)/2} & k \equiv \pm 3 \pmod{8} \\
B_{(k-2)/2} & k \equiv 4 \pmod{8},
\end{cases}
$$

using the notation defined in (21.1)–(21.3). Denote the central involution of T_k by z. Then there exists a cocyclic $\mathrm{OD}(n; 1^k)$ with cocycle f if and only if there is an embedding of T_k into $E(f)$ that maps z to $(1, -1)$. If $2n = |T_k|$ then the cocyclic $\mathrm{OD}(n; 1^k)$ is essentially unique.

PROOF. We apply Theorem 22.1.1. Here $m = 1$, $M = [0]$, $N = J_k - I_k$, and Q is the $(k-1) \times (k-1)$ upper triangular matrix with 1s everywhere on and above the main diagonal. The isomorphism class of $E(Q)$, and thus of $T_k = \mathrm{Hub}(E(Q))$, can be found using Algorithm 1 of Chapter 21 (see Subsection 21.2.2).

Set $P = Q + Q^\top \equiv J_{k-1} + I_{k-1} \pmod 2$. An element of $\ker P$ has all entries equal. If k is odd then P is self-inverse modulo 2. Thus

$$
\ker P = \begin{cases}
\langle (1, \ldots, 1) \rangle & k \equiv 0 \pmod 2 \\
0 & k \equiv 1 \pmod 2.
\end{cases}
$$

If x is a length $k-1$ binary vector of weight w, then

$$
x Q x^\top = \tfrac{1}{2} x (Q + Q^\top) x^\top = \tfrac{1}{2}(x J x^\top + x x^\top) = \tfrac{1}{2} w(w+1).
$$

A Q-anisotropic length $k-1$ binary vector must therefore have $w \equiv 1$ or $2 \pmod 4$, so that the number $\xi/2$ of such vectors is $\sigma_1 + \sigma_2$, where $\sigma_s = \sum_{w \equiv s(4)} \binom{k-1}{w}$. Now

$$
\begin{aligned}
(1+1)^{k-1} &= \sigma_0 + \sigma_1 + \sigma_2 + \sigma_3 \\
(1+\mathrm{i})^{k-1} &= \sigma_0 + \mathrm{i}\sigma_1 - \sigma_2 - \mathrm{i}\sigma_3 \\
(1-\mathrm{i})^{k-1} &= \sigma_0 - \mathrm{i}\sigma_1 - \sigma_2 + \mathrm{i}\sigma_3,
\end{aligned}
$$

and so

$$\xi = 2(\sigma_1 + \sigma_2)$$
$$= (1+1)^{k-1} - \tfrac{1}{2}((1+\mathrm{i})(1+\mathrm{i})^{k-1} + (1-\mathrm{i})(1-\mathrm{i})^{k-1})$$
$$= 2^{k-1} - 2^{k/2}(e^{\mathrm{i}\pi k/4} + e^{-\mathrm{i}\pi k/4})/2$$
$$= 2^{k-1} - 2^{k/2}\cos(\pi k/4).$$

By Algorithm 1, we get $j := \dim(\ker P) + 1 = 2$ if $k \equiv 0 \pmod 2$ and $j = 1$ if $k \equiv 1 \pmod 2$; then we compare $\xi/2^{j-1}$ with $2^{k-j} \pm 2^{(k-j)/2}$ to find the isomorphism type of T_k. This undemanding exercise is left to the reader. $\square$

We have a cocyclic $\mathrm{OD}(2^{s+t}; (2^s)^{2^t})$ for $s, t > 0$ by Theorems 22.5.1 and 22.6.1, and Corollary 21.5.5. So the number of indeterminates that can appear in some cocyclic orthogonal design with no zero entries is unbounded.

Note also that Theorem 22.6.1 gives a dependence of the minimal possible order of the design on the number of its indeterminates. However, that bound grows much more rapidly than Radon's bound [**71**, Corollary 1.4, p. 4].

22.6.2. EXAMPLE. We construct the cocyclic $\mathrm{OD}(4; 1^4)$ with smallest possible extension group. Here

$$N = \begin{bmatrix} 0 & 1 & 1 & 1 \\ 1 & 0 & 1 & 1 \\ 1 & 1 & 0 & 1 \\ 1 & 1 & 1 & 0 \end{bmatrix}, \qquad Q = \begin{bmatrix} 1 & 1 & 1 \\ 0 & 1 & 1 \\ 0 & 0 & 1 \end{bmatrix}, \qquad P = \begin{bmatrix} 0 & 1 & 1 \\ 1 & 0 & 1 \\ 1 & 1 & 0 \end{bmatrix},$$

and

$$E(Q) = \langle b_1, b_2, b_3, z \mid b_1^2 = b_2^2 = b_3^2 = z,\ b_2 b_1 = b_1 b_2 z,\ b_3 b_2 = b_2 b_3 z,\ b_3 b_1 = b_1 b_3 z \rangle.$$

Since $\ker P = \langle (1,1,1) \rangle$, the center of $E(Q)$ is $\langle b_1 b_2 b_3, z \rangle \cong \mathrm{C}_2^2$ by (21.7). In the notation of Remark 22.3.3, $S = \langle b_1 b_2 b_3 \rangle$. We can easily pick out a complement $\mathrm{Hub}(E(Q))$ of S in $E(Q)$:

$$\mathrm{Hub}(E(Q)) = \langle b_1, b_2, z \mid b_1^2 = b_2^2 = z,\ b_2 b_1 = b_1 b_2 z \rangle \cong \mathrm{Q}_8.$$

Factoring out $\langle z \rangle$, we get the indexing group $G = \langle a_1, a_2 \rangle \cong \mathrm{C}_2^2$. The method of Remark 22.3.3 yields that $1, b_1, b_2, b_1 b_2$ is an N-concordant sequence in $\mathrm{Hub}(E(Q))$ with respect to z.

We may choose any cocycle $f : G \times G \to \langle z \rangle$ such that $E(f) \cong \mathrm{Hub}(E(Q))$ as a cocycle of the design. So let τ be the transversal map that sends a^r to b^r, and define $f = f_\tau$ as usual (see (12.5)). With rows and columns indexed $1, a_1, a_2, a_1 a_2$, the matrix U of Definition 12.5.3 is

$$\begin{bmatrix} 1 & 0 & 0 & 0 \\ 0 & - & 0 & 0 \\ 0 & 0 & - & 0 \\ 0 & 0 & 0 & - \end{bmatrix}.$$

By Lemma 22.2.1,

$$U(aP_1 + bP_{b_1} + cP_{b_2} + dP_{b_1 b_2}) = \begin{bmatrix} a & b & c & d \\ b & -a & d & -c \\ c & -d & -a & b \\ d & c & -b & -a \end{bmatrix}$$

is the resulting orthogonal design.

Theorem 22.6.4 below is an accompanying result on cocyclic amicable pairs.

22.6.3. LEMMA. *For* $s, t \geq 1$, *let*

$$T_{s,t} = \begin{cases} A_{(s+t-2)/2} & s-t \equiv 0 \pmod 8 \\ A_{(s+t-1)/2} & s-t \equiv \pm 1 \pmod 8 \\ C_{(s+t-2)/2} & s-t \equiv \pm 2 \pmod 8 \\ B_{(s+t-1)/2} & s-t \equiv \pm 3 \pmod 8 \\ B_{(s+t-2)/2} & s-t \equiv 4 \pmod 8. \end{cases}$$

Define Q *to be the* $(s+t-1) \times (s+t-1)$ *matrix with* 1*s everywhere above the main diagonal and in the first* $s-1$ *main diagonal positions, and* 0*s everywhere else. Then*

(i) $\mathrm{Hub}(E(Q)) \cong T_{s,t}$,
(ii) $E(Q) \cong T_{s,t}$ *if* $s-t \equiv \pm 2 \pmod 8$ *or* $s-t$ *is odd, and* $E(Q) \cong T_{s,t} \times \mathrm{C}_2$ *otherwise.*

PROOF. Cf. the proof of Theorem 22.6.1. We have $P = Q + Q^\top \equiv J_{s+t-1} + I_{s+t-1} \pmod 2$, and so

$$\ker P = \begin{cases} \langle (1, \ldots, 1) \rangle & s-t \equiv 0 \pmod 2 \\ 0 & s-t \equiv 1 \pmod 2. \end{cases}$$

For a $(0,1)$-vector x of length $s+t-1$, let w_1 denote the number of 1s in the first $s-1$ positions of x. Then $xQx^\top = w(w+1)/2 - w_2$, where w is the weight of x and $w_2 = w - w_1$. Hence, we need to find the number of all binary vectors x such that

$$(22.6) \qquad\qquad w(w+1) - 2w_2 \equiv 2 \pmod 4.$$

Since

$$w(w+1) - 2w_2 \equiv 2 \pmod 4 \quad \Longleftrightarrow \quad (w_1 - w_2)(w_1 - w_2 + 1) \equiv 2 \pmod 4,$$

the pairs (w_1, w_2) satisfying (22.6) are precisely those such that $w_1 - w_2 \equiv 1$ or $2 \pmod 4$. The required count is thus $\sigma_1 + \sigma_2$, where

$$\sigma_u := \sum_{w_1 - w_2 \equiv u(4)} \binom{s-1}{w_1} \binom{t}{w_2}.$$

Since

$$\sigma_1 + \sigma_2 + \sigma_0 + \sigma_3 = 2^{s+t-1}$$

and

$$2(\sigma_1 + \sigma_2 - \sigma_0 - \sigma_3) = -(1+\mathrm{i})^{s-1}(1-\mathrm{i})^t - (1-\mathrm{i})^{s-1}(1+\mathrm{i})^t \\ - \mathrm{i}(1+\mathrm{i})^{s-1}(1-\mathrm{i})^t + \mathrm{i}(1-\mathrm{i})^{s-1}(1+\mathrm{i})^t,$$

we have

$$4(\sigma_1 + \sigma_2) = 2^{s+t} - (1+\mathrm{i})^s(1-\mathrm{i})^t - (1-\mathrm{i})^s(1+\mathrm{i})^t \\ = 2^{s+t} - ((\sqrt{2}e^{\mathrm{i}\pi/4})^s(\sqrt{2}e^{-\mathrm{i}\pi/4})^t + (\sqrt{2}e^{-\mathrm{i}\pi/4})^s(\sqrt{2}e^{\mathrm{i}\pi/4})^t) \\ = 2^{s+t} - 2^{(s+t)/2}(e^{(s-t)\mathrm{i}\pi/4} + e^{-(s-t)\mathrm{i}\pi/4}) \\ = 2^{s+t} - 2^{(s+t+2)/2}\cos((s-t)\pi/4).$$

Now Algorithm 1 of Subsection 21.2.2 may be used to find the isomorphism types of $E(Q)$ and $\mathrm{Hub}(E(Q))$. $\qquad\square$

22.6.4. THEOREM. *Assume the notation of Lemma 22.6.3, and let z be the unique central involution in $T_{s,t}$. There exists a pair of cocyclic amicable $\mathrm{OD}(n; 1^s)$ and $\mathrm{OD}(n; 1^t)$ with cocycle f if and only if there is an embedding of $T_{s,t}$ into $E(f)$ that maps z to $(1, -1)$. If $2n = |T_{s,t}|$ then the cocyclic amicable pair is essentially unique.*

PROOF. Let $M = \begin{bmatrix} 0 & 0 \\ 0 & 0 \end{bmatrix}$. Then the symmetric matrix $N = N(M, s, t)$ has zero diagonal, a $t \times t$ block of zeros in the upper right-hand corner, and 1s everywhere else above the diagonal. The theorem is proved by invoking Lemma 22.6.3 with $Q = Q_N$, and Theorem 22.1.1. $\qquad\square$

22.6.5. EXAMPLE. We construct the minimal cocyclic system of two amicable $\mathrm{OD}(4; 1^2)$. This time

$$N = \begin{bmatrix} 0 & 1 & 0 & 0 \\ 1 & 0 & 0 & 0 \\ 0 & 0 & 0 & 1 \\ 0 & 0 & 1 & 0 \end{bmatrix}, \qquad Q = \begin{bmatrix} 1 & 1 & 1 \\ 0 & 0 & 1 \\ 0 & 0 & 0 \end{bmatrix}, \qquad P = \begin{bmatrix} 0 & 1 & 1 \\ 1 & 0 & 1 \\ 1 & 1 & 0 \end{bmatrix}.$$

Thus $E(Q)$ is

$$\langle b_1, b_2, b_3, z \mid b_1^2 = z,\ b_2^2 = b_3^2 = z^2 = 1,\ b_2 b_1 = b_1 b_2 z,\ b_3 b_2 = b_2 b_3 z,\ b_3 b_1 = b_1 b_3 z \rangle,$$

and $Z(E(Q)) = \langle b_1 b_2 b_3, z \rangle$. Furthermore,

$$\mathrm{Hub}(E(Q)) = \langle b_1, b_2, z \mid b_1^2 = z,\ b_2^2 = z^2 = 1,\ b_1 b_2 = b_2 b_1 z \rangle \cong \mathrm{D}_8.$$

The two designs will correspond to the pairs $1, b_1$ and $b_2, b_1 b_2$ making up the N-concordant sequence $1, b_1, b_2, b_1 b_2$ in $E(Q)$. Our indexing group is $G = \langle a_1, a_2 \rangle \cong \mathrm{C}_2^2$. With rows and columns indexed $1, a_1, a_2, a_1 a_2$, a cocycle of the system is given by

$$\begin{bmatrix} 1 & 1 & 1 & 1 \\ 1 & - & 1 & - \\ 1 & - & 1 & - \\ 1 & 1 & 1 & 1 \end{bmatrix}.$$

Under the same indexing,

$$\begin{bmatrix} a & b & 0 & 0 \\ b & -a & 0 & 0 \\ 0 & 0 & a & -b \\ 0 & 0 & b & a \end{bmatrix}, \qquad \begin{bmatrix} 0 & 0 & c & d \\ 0 & 0 & d & -c \\ c & -d & 0 & 0 \\ d & c & 0 & 0 \end{bmatrix}$$

is the stated system.

In the next example we construct a cocyclic pair of amicable $\mathrm{OD}(16; 8^2)$. An interesting unanswered question is whether a cocyclic pair of amicable $\mathrm{OD}(8; 4^2)$, or better still a cocyclic pair of amicable $\mathrm{OD}(4; 2^2)$, exists. (There is no cocyclic amicable pair of $\mathrm{OD}(2; 1^2)$. Also, we cannot use Theorem 22.5.1 and the $\mathrm{OD}(4; 1^2)$ of Example 22.6.5, because D_8 is not a Hadamard group.) Existence in either case would allow us to replace the integer 8 in [**54**, Theorem 1.1 (2)] by 6 or 4 respectively.

22.6.6. EXAMPLE. Using Q as in Example 22.6.5, but with concordant sequence $1, b_1, b_2, b_3$ and $E(Q) \cong D_8 \times C_2$ rather than $\text{Hub}(E(Q))$ as extension group, we construct a (non-minimal) system of two amicable OD(8; 1^2).

The indexing group is $G = \langle a_1, a_2, a_3 \rangle \cong C_2^3$, and rows and columns are indexed $1, a_1, a_2, a_1a_2, a_3, a_1a_3, a_2a_3, a_1a_2a_3$. We take the cocycle of the system to be f_Q, for which

$$
U = \begin{bmatrix}
1 & 0 & 0 & 0 & 0 & 0 & 0 & 0 \\
0 & - & 0 & 0 & 0 & 0 & 0 & 0 \\
0 & 0 & 1 & 0 & 0 & 0 & 0 & 0 \\
0 & 0 & 0 & 1 & 0 & 0 & 0 & 0 \\
0 & 0 & 0 & 0 & 1 & 0 & 0 & 0 \\
0 & 0 & 0 & 0 & 0 & 1 & 0 & 0 \\
0 & 0 & 0 & 0 & 0 & 0 & - & 0 \\
0 & 0 & 0 & 0 & 0 & 0 & 0 & 1
\end{bmatrix}, \quad
P_{b_1} = \begin{bmatrix}
0 & 1 & 0 & 0 & 0 & 0 & 0 & 0 \\
- & 0 & 0 & 0 & 0 & 0 & 0 & 0 \\
0 & 0 & 0 & - & 0 & 0 & 0 & 0 \\
0 & 0 & 1 & 0 & 0 & 0 & 0 & 0 \\
0 & 0 & 0 & 0 & 0 & - & 0 & 0 \\
0 & 0 & 0 & 0 & 1 & 0 & 0 & 0 \\
0 & 0 & 0 & 0 & 0 & 0 & 0 & 1 \\
0 & 0 & 0 & 0 & 0 & 0 & - & 0
\end{bmatrix},
$$

$$
P_{b_2} = \begin{bmatrix}
0 & 0 & 1 & 0 & 0 & 0 & 0 & 0 \\
0 & 0 & 0 & 1 & 0 & 0 & 0 & 0 \\
1 & 0 & 0 & 0 & 0 & 0 & 0 & 0 \\
0 & 1 & 0 & 0 & 0 & 0 & 0 & 0 \\
0 & 0 & 0 & 0 & 0 & 0 & - & 0 \\
0 & 0 & 0 & 0 & 0 & 0 & 0 & - \\
0 & 0 & 0 & 0 & - & 0 & 0 & 0 \\
0 & 0 & 0 & 0 & 0 & - & 0 & 0
\end{bmatrix}, \quad
P_{b_3} = \begin{bmatrix}
0 & 0 & 0 & 0 & 1 & 0 & 0 & 0 \\
0 & 0 & 0 & 0 & 0 & 1 & 0 & 0 \\
0 & 0 & 0 & 0 & 0 & 0 & 1 & 0 \\
0 & 0 & 0 & 0 & 0 & 0 & 0 & 1 \\
1 & 0 & 0 & 0 & 0 & 0 & 0 & 0 \\
0 & 1 & 0 & 0 & 0 & 0 & 0 & 0 \\
0 & 0 & 1 & 0 & 0 & 0 & 0 & 0 \\
0 & 0 & 0 & 1 & 0 & 0 & 0 & 0
\end{bmatrix}.
$$

Then

$$
\begin{bmatrix}
a & b & 0 & 0 & 0 & 0 & 0 & 0 \\
b & -a & 0 & 0 & 0 & 0 & 0 & 0 \\
0 & 0 & a & -b & 0 & 0 & 0 & 0 \\
0 & 0 & b & a & 0 & 0 & 0 & 0 \\
0 & 0 & 0 & 0 & a & -b & 0 & 0 \\
0 & 0 & 0 & 0 & b & a & 0 & 0 \\
0 & 0 & 0 & 0 & 0 & 0 & -a & -b \\
0 & 0 & 0 & 0 & 0 & 0 & -b & a
\end{bmatrix}, \quad
\begin{bmatrix}
0 & 0 & c & 0 & d & 0 & 0 & 0 \\
0 & 0 & 0 & -c & 0 & -d & 0 & 0 \\
c & 0 & 0 & 0 & 0 & 0 & d & 0 \\
0 & c & 0 & 0 & 0 & 0 & 0 & d \\
d & 0 & 0 & 0 & 0 & 0 & -c & 0 \\
0 & d & 0 & 0 & 0 & 0 & 0 & -c \\
0 & 0 & -d & 0 & c & 0 & 0 & 0 \\
0 & 0 & 0 & d & 0 & -c & 0 & 0
\end{bmatrix}
$$

is our system of OD(8; 1^2).

By Corollary 21.5.5, f_Q is also a cocycle of the Sylvester matrix of order 8 (up to equivalence, there is only one choice for H in this example anyway). To find g such that

$$H = [f_Q(a^r, a^s) g(a^r a^s)]_{a^r, a^s \in G}$$

is Hadamard, we avail of Algorithm 2 at the end of Chapter 21. For the 3×3 $(0, 1)$-matrix L with a single non-zero entry, in position $(1, 3)$, $Q + L + L^\top$ is invertible. So $g(a^r) = (-1)^{r L r^\top}$, i.e., $g(a_1 a_3) = g(a_1 a_2 a_3) = -1$ and $g(a^r) = 1$ otherwise. Then

$$
H = \begin{bmatrix}
1 & 1 & 1 & 1 & 1 & - & 1 & - \\
1 & - & 1 & - & - & - & - & - \\
1 & - & 1 & - & 1 & 1 & 1 & 1 \\
1 & 1 & 1 & 1 & - & 1 & - & 1 \\
1 & 1 & - & - & 1 & - & - & 1 \\
- & 1 & 1 & - & 1 & 1 & - & - \\
1 & - & - & 1 & 1 & 1 & - & - \\
- & - & 1 & 1 & 1 & - & - & 1
\end{bmatrix}.
$$

The pair of amicable $OD(16; 8^2)$ that results from application of Theorem 22.5.1 with $f = f_Q$, this H, and the above pair of $OD(8; 1^2)$, is shown below (an upper case letter represents the negation of its lower case version).

$$
\begin{bmatrix}
v & v & v & v & v & V & v & V & U & u & u & U & U & U & u & u \\
v & V & v & V & V & V & V & V & u & u & U & U & U & u & u & U \\
v & V & v & V & v & v & v & v & u & u & U & U & u & U & U & u \\
v & v & v & v & V & v & V & v & U & u & u & U & u & u & U & U \\
v & v & V & V & v & V & V & v & U & u & U & u & U & U & U & U \\
V & v & v & V & v & v & V & V & U & U & U & U & u & U & u & U \\
v & V & V & v & v & v & V & V & u & u & u & u & u & U & u & U \\
V & V & v & v & v & V & V & v & u & U & u & U & U & U & U & U \\
U & u & u & U & U & U & u & u & v & v & v & v & v & V & v & V \\
u & u & U & U & U & u & u & U & v & V & v & V & V & V & V & V \\
u & u & U & U & u & U & U & u & v & V & v & V & v & v & v & v \\
U & u & u & U & u & u & U & U & v & v & v & v & V & v & V & v \\
U & u & U & u & U & U & U & U & v & v & V & V & v & V & V & v \\
U & U & U & U & u & U & u & U & V & v & v & V & v & v & V & V \\
u & u & u & u & u & U & u & U & v & V & V & v & v & v & V & V \\
u & U & u & U & U & U & U & U & V & V & v & v & v & V & V & v
\end{bmatrix}
$$

$$
\begin{bmatrix}
y & y & y & y & Y & y & Y & y & x & X & x & X & x & x & x & x \\
y & Y & y & Y & y & y & y & y & X & X & X & X & x & X & x & X \\
y & Y & y & Y & Y & Y & Y & Y & x & x & x & x & x & X & x & X \\
y & y & y & y & y & Y & y & Y & X & x & X & x & x & x & x & x \\
Y & Y & y & y & y & Y & Y & y & x & X & X & x & x & x & X & X \\
y & Y & Y & y & y & y & Y & Y & x & x & X & X & X & x & x & X \\
Y & y & y & Y & y & y & Y & Y & x & x & X & X & x & X & X & x \\
y & y & Y & Y & y & Y & Y & y & x & X & X & x & X & X & x & x \\
x & X & x & X & x & x & x & x & y & y & y & y & Y & y & Y & y \\
X & X & X & X & x & X & x & X & y & Y & y & Y & y & y & y & y \\
x & x & x & x & x & X & x & X & y & Y & y & Y & Y & Y & Y & Y \\
X & x & X & x & x & x & x & x & y & y & y & y & y & Y & y & Y \\
x & X & X & x & x & x & X & X & Y & Y & y & y & y & Y & Y & y \\
x & x & X & X & X & x & x & X & y & Y & Y & y & y & y & Y & Y \\
x & x & X & X & x & X & X & x & Y & y & y & Y & y & y & Y & Y \\
x & X & X & x & X & X & x & x & y & y & Y & Y & y & Y & Y & y
\end{bmatrix}
$$

CHAPTER 23

Asymptotic Existence of Cocyclic Hadamard Matrices

The centerpiece of this final chapter is the following theorem of de Launey and Kharaghani [**51**].

THEOREM. *If $q > 1$ is odd and*

$$k \geq 10 + 8 \left\lfloor \frac{\log_2(q-1)}{10} \right\rfloor,$$

then there exists a cocyclic Hadamard matrix of order $2^k q$.

This result provides further support for the Cocyclic Hadamard Conjecture. Our proof of the theorem uses circulant Hermitian and skew-Hermitian matrices, and cocyclic monomial matrices derived from a concordant system of orthogonal designs (for which we turn to Chapter 22). The circulant complex matrices are supplied by work of Craigen, Holzmann, and Kharaghani [**23**]. Composing these and the monomial matrices in a rather intricate way (adapted from [**23**]), we are able to manufacture a certain cocyclic complex Hadamard matrix, which implies existence of the required cocyclic Hadamard matrix.

The paper [**54**] contains an earlier asymptotic existence result for cocyclic Hadamard matrices. De Launey and Smith verified existence for orders $2^k q$ where $q > 1$ is odd and $k \geq \lfloor 8 \log_2 q \rfloor$. The above theorem is a substantial improvement on their result. It is even an improvement on Seberry's bound $2 \log_2(q-3)$ in the exponent of 2, for general Hadamard matrices [**156**]; but it is about twice the bound in [**23**] for general Hadamard matrices.

Much of the material in this chapter first appeared in [**51**].

23.1. Complex sequences with zero aperiodic autocorrelation

Let

$$a_1 = (a_{1,1}, \ldots, a_{1,n_1}), \ldots, a_r = (a_{r,1}, \ldots, a_{r,n_r})$$

be r sequences of complex fourth roots of unity. The *total length* of these sequences is $\sum_{i=1}^{r} n_i$. The sequences have *zero aperiodic autocorrelation* if

$$\sum_{i=1}^{r} \sum_{j=1}^{n_i-c} a_{i,j} \overline{a_{i,j+c}} = 0 \quad \forall \, c > 0.$$

In this chapter, sequences are complex unless otherwise stated, and autocorrelation is always aperiodic. If the a_i have zero autocorrelation, then each $n_i > 1$ occurs an even number of times in the list $n_1, \ldots, n_r$. Hence, if $n_i > 1$ for all i then we may arrange the sequences into pairs so that $n_{2i} = n_{2i-1}$ for $1 \leq i \leq r/2$.

Golay complementary sequences are a special case of the above definition, with $a_{i,j} \in \{\pm 1\}$, $r = 2$, and $n_1 = n_2$. They are known to exist for all total lengths $2^a 10^b 26^c$.

23.1.1. LEMMA. *If* $(a_1, a_2, \ldots, a_n)$, $(b_1, b_2, \ldots, b_n)$ *are Golay complementary sequences then*

$$(a_1, b_1, a_2, b_2, \ldots, a_n, b_n), \quad (a_1, -b_1, a_2, -b_2, \ldots, a_n, -b_n)$$

are Golay complementary sequences.

PROOF. We have autocorrelation

$$b_{t+1}a_1 + a_{t+2}b_1 + b_{t+2}a_2 + \cdots + a_n b_{n-t-1} + b_n a_{n-t}$$
$$-(b_{t+1}a_1 + a_{t+2}b_1 + b_{t+2}a_2 + \cdots + a_n b_{n-t-1} + b_n a_{n-t}) = 0$$

for offsets $2t + 1$, and

$$2(a_{t+1}a_1 + b_{t+1}b_1 + a_{t+2}a_2 + \cdots + a_n a_{n-t} + b_n b_{n-t}) = 0$$

for offsets $2t$.   □

23.1.2. EXAMPLE. Starting with the sequences (1), (1), by Lemma 23.1.1 we get Golay sequences of total lengths 2^a for $a = 1, 2, 3$:

$(1, 1)$ $(1, -1)$

$(1, 1, 1, -1)$ $(1, -1, 1, 1)$

$(1, 1, 1, -1, 1, 1, -1, 1)$ $(1, -1, 1, 1, 1, -1, -1, -1)$.

23.1.3. THEOREM. *Let* $p \geq 1$. *If the binary expansion of* p *has* d 1s, *then there are* $2d$ $(1, -1)$*-sequences with zero aperiodic autocorrelation and total length* $2p$.

PROOF. There is a pair of Golay sequences for each length 2^a. By hypothesis, $p = 2^{e_1} + \cdots + 2^{e_d}$ where $0 \leq e_1 < \cdots < e_d$. Then we just collect together the d pairs of Golay sequences of lengths 2^{e_i}, $1 \leq i \leq d$.   □

Let $L(p)$ denote the least number of sequences of total length $2p$ with zero autocorrelation. Theorem 23.1.3 gives an upper bound on $L(p)$. The following bounds are from [23] (there is a small misprint in Corollary 4.2 of that paper).

23.1.4. THEOREM. *If* $1 \leq p \leq 1024$ *then* $L(p) \leq 4$; *except possibly when* $p \in \{799, 927, 967, 959\}$, *in which cases* $L(2p) \leq 4$.

Combining these sequences gives

23.1.5. THEOREM ([23, Section 4]). *Let* $p > 1$ *and* $t = \lfloor (1 + \log_2 p)/10 \rfloor$. *For some* $l \leq 4t + 4$ *there are* l *complex sequences with zero aperiodic autocorrelation and total length* $2p$. *Consequently,* $L(p) \leq 4 + 4t$.

It would be desirable to have more qualitative information about the behavior of the function $L(p)$. We make the following conjecture.

7. CONJECTURE. *If* $\epsilon > 0$ *then there is* $m_\epsilon \in \mathbb{N}$ *such that for all* $p > m_\epsilon$,

$$L(p) \leq \epsilon \left\lfloor \frac{1 + \log_2 p}{10} \right\rfloor.$$

23.2. Sets of Hermitian and skew-Hermitian circulant matrices

Guided by [**23**], we now explain how the sequences of the previous section may be used to make special sets of circulant Hermitian and skew-Hermitian complex matrices.

Suppose that

$$a_1 = (a_{1,1}, \ldots, a_{1,n_1}), \ldots, a_{2r} = (a_{2r,1}, \ldots, a_{2r,n_{2r}})$$

are sequences with $n_{2i} = n_{2i-1}$, zero autocorrelation, and total length $2p$.

23.2.1. LEMMA. *If $m > n_i$ for $i = 1, \ldots, r$ then the $m \times m$ circulant matrices M_i with initial rows $[a_{i,1}, \ldots, a_{i,n_i}, 0, 0, \ldots, 0]$ satisfy*

$$M_1 M_1^* + \cdots + M_{2r} M_{2r}^* = 2p I_m.$$

PROOF. Let $k > l$. The (k, l)th entry of $M_i M_i^*$ is either the autocorrelation at offset $k - l$, or the autocorrelation at offset $k - l$ plus the autocorrelation at offset $l - k + m$.   □

We begin the construction. Let $q = 2p + 1$, and for $i = 1, \ldots, r$ define X_i, Y_i to be the $q \times q$ circulant matrices with initial rows

$$[0^{1+p-n_1-n_3-\cdots-n_{2i-1}}, a_{2i-1,1}, a_{2i-1,2}, \ldots, a_{2i-1,n_{2i-1}}, 0^{n_1+n_3+\cdots+n_{2i-3}},$$
$$0^{n_2+n_4+\cdots+n_{2i-2}}, a_{2i,1}, a_{2i,2}, \ldots, a_{2i,n_{2i}}, 0^{p-n_2-n_4-\cdots-n_{2i}}]$$

and

$$[0^{1+p-n_1-n_3-\cdots-n_{2i-1}}, a_{2i-1,1}, a_{2i-1,2}, \ldots, a_{2i-1,n_{2i-1}}, 0^{n_1+n_3+\cdots+n_{2i-3}},$$
$$0^{n_2+n_4+\cdots+n_{2i-2}}, -a_{2i,1}, -a_{2i,2}, \ldots, -a_{2i,n_{2i}}, 0^{p-n_2-n_4-\cdots-n_{2i}}]$$

respectively.

23.2.2. EXAMPLE. Let $p = 2$. Using the pair $(1, 1)$, $(1, -1)$, we have that X_1, Y_1 are circulant matrices with respective first rows $[0, 1, 1, 1, -1]$ and $[0, 1, 1, -1, 1]$. Using the pair $(1, \mathrm{i})$, $(1, -\mathrm{i})$, we obtain X_1, Y_1 with initial rows $[0, 1, \mathrm{i}, 1, -\mathrm{i}]$ and $[0, 1, \mathrm{i}, -1, \mathrm{i}]$.

Let $p = 3$. From the four real sequences (1), (1), $(1, 1)$, $(1, -1)$ we obtain X_1, Y_1, X_2, Y_2 with initial rows $[0, 0, 0, 1, 1, 0, 0]$, $[0, 0, 0, 1, -1, 0, 0]$, $[0, 1, 1, 0, 0, 1, -1]$, and $[0, 1, 1, 0, 0, -1, 1]$.

The next two lemmas are needed to prove Theorem 23.2.6 below.

23.2.3. LEMMA. *The matrices $X_1, \ldots, X_r$ have disjoint supports that cover all but the diagonal entries. Indeed, $I_q + \sum_{i=1}^{r} X_i$ is a $(\pm 1, \pm \mathrm{i})$-matrix. Analogous statements hold for the Y_i.*

PROOF. Immediate from the definitions of X_i and Y_i.   □

23.2.4. LEMMA. $\sum_{i=1}^{r} (X_i X_i^* + Y_i Y_i^*) = 4p I_q.$

PROOF. Apply Lemma 23.2.1.   □

Our Hermitian and skew-Hermitian circulants are built out of the X_i and Y_i. Define

$$S_i = \tfrac{1}{2}(X_i + X_i^*), \quad T_i = \tfrac{1}{2}(Y_i + Y_i^*), \quad U_i = \tfrac{1}{2}(X_i - X_i^*), \quad V_i = \tfrac{1}{2}(Y_i - Y_i^*).$$

23.2.5. EXAMPLE. We continue with Example 23.2.2. For the real sequences $(1, 1), (1, -1)$, the circulants S_1, T_1, U_1, V_1 have first rows $[0, 0, 1, 1, 0]$, $[0, 1, 0, 0, 1]$, $[0, 1, 0, 0, -1]$, $[0, 0, 1, -1, 0]$. For the pair $(1, i), (1, -i)$ we get

$$S_1 = \tfrac{1}{2}\text{circ}[0, 1 + i, 1 + i, 1 - i, 1 - i] \qquad U_1 = \tfrac{1}{2}\text{circ}[0, 1 - i, -1 + i, 1 + i, -1 - i]$$
$$T_1 = \tfrac{1}{2}\text{circ}[0, 1 - i, -1 + i, -1 - i, 1 + i] \quad V_1 = \tfrac{1}{2}\text{circ}[0, 1 + i, 1 + i, -1 + i, -1 + i].$$

23.2.6. THEOREM. *Let $q = 2p + 1$. If there are $2r$ paired complex sequences with zero aperiodic autocorrelation and total length $2p$, then there are circulant $q \times q$ matrices $S_1, \ldots, S_r, T_1, \ldots, T_r, U_1, \ldots, U_r, V_1, \ldots, V_r$ satisfying the following conditions.*

(1) S_i, T_i are Hermitian, and U_i, V_i are skew-Hermitian.

(2) $I_q + \sum_{i=1}^{r}(S_i \pm U_i)$ and $I_q + \sum_{i=1}^{r}(T_i \pm V_i)$ are $(\pm 1, \pm i)$-matrices.

(3) $\sum_{i=1}^{r}(S_i S_i^* + T_i T_i^* + U_i U_i^* + V_i V_i^*) = 4p I_q$.

PROOF. Part (1) is clear, so we proceed to part (2). Note that $X_i^* = S_i - U_i$ and $X_i = S_i + U_i$ both have the same support. Thus $S_i \pm U_i$ is a $(0, \pm 1, \pm i)$-matrix whose support is the same as that of X_i; and so $I_q + \sum_{i=1}^{r}(S_i \pm U_i)$ is a $(\pm 1, \pm i)$-matrix by Lemma 23.2.3. The same argument with Y_i replacing X_i proves the rest of part (2).

For part (3), note that (because circulant matrices commute)

$$S_i S_i^* + U_i U_i^* = \tfrac{1}{4}(X_i + X_i^*)^2 - \tfrac{1}{4}(X_i - X_i^*)^2 = X_i X_i^*,$$
$$T_i T_i^* + V_i V_i^* = \tfrac{1}{4}(Y_i + Y_i^*)^2 - \tfrac{1}{4}(Y_i - Y_i^*)^2 = Y_i Y_i^*.$$

Then we are done by Lemma 23.2.4. $\square$

23.3. Sets of cocyclic signed permutation matrices

23.3.1. LEMMA. *Let Q be the $(4r + 1) \times (4r + 1)$ matrix with 1s everywhere above the main diagonal and in the first $2r$ diagonal entries, and 0s elsewhere. Let Q' be the $4r \times 4r$ matrix obtained by deleting the last row and last column of Q. Then there are cocyclic signed permutation matrices $P_{1,0}, \ldots, P_{1,2r}, P_{2,0}, \ldots, P_{2,2r}$, all with the same cocycle $f := f_{Q'} : C_2^{4r} \times C_2^{4r} \to \langle -1 \rangle$, such that*

- *the $P_{1,i}$s are anti-amicable;*
- *the $P_{2,i}$s are anti-amicable;*
- *every pair $\{P_{1,i}, P_{2,j}\}$ is amicable.*

Moreover $E(f) \cong A_{2r}$ (the central product of $2r$ copies of D_8), and there is a map $g : C_2^{4r} \to \{\pm 1\}$ such that

$$H := \sum_{x \in C_2^{4r}} g(x) P_x^f$$

is equivalent to the Sylvester matrix of order 2^{4r}.

PROOF. We first use the theory of cocyclic orthogonal designs from the previous chapter. By Lemma 22.6.3, $\text{Hub}(E(Q)) \cong E(Q') \cong A_{2r}$ and $E(Q) \cong A_{2r} \times C_2$. By Theorem 22.6.4, we deduce that there exists a pair of amicable $OD(2^{4r}; 1^{2r+1})$ with cocycle f. Let N be the matrix $N(M, 2r + 1, 2r + 1)$ as defined in (22.1), where M is the 2×2 zero matrix. Lemma 22.2.2 shows that there is an N-concordant sequence $x_0, \ldots, x_{2t}, y_0, \ldots, y_{2t}$ in $E(f)$ with respect to $(1, -1)$. That is, $(ab^{-1})^2 =$

$(1, (-1)^{n_{ij}})$ for any two elements a, b of the sequence. Then we can take $P_{1,i} = P_{x_i}^f$ and $P_{2,i} = P_{y_i}^f$, by Lemma 12.5.1 (2).

The last assertion follows from Lemma 12.5.5 and Corollary 21.5.5.   □

All ingredients for our asymptotic existence result have now been assembled. We prove this result in the next section.

23.4. Existence of cocyclic complex Hadamard matrices

23.4.1. THEOREM. *Let $q = 2p+1 > 1$. If there are $2r$ paired $(\pm 1, \pm i)$-sequences with zero aperiodic autocorrelation and total length $2p$, then there is a cocyclic complex Hadamard matrix of order $q2^{4r+1}$ with indexing group $C_q \times C_2^{4r+1}$ and extension group $C_{2q} \times D_8 \curlyvee \cdots \curlyvee D_8$.*

PROOF. Let $T = \begin{bmatrix} 0 & 1 \\ 1 & 0 \end{bmatrix}$. Define

$$K = I_2 \otimes \left(I_q \otimes P_{1,0}H + \sum_{i=1}^{r}(S_i \otimes P_{1,2i}H + U_i \otimes P_{2,2i-1}H) \right)$$

$$+ iT \otimes \left(I_q \otimes P_{2,0}H + \sum_{i=1}^{r}(T_i \otimes P_{2,2i}H + V_i \otimes P_{1,2i-1}H) \right),$$

where $P_{1,i}, P_{2,i}, H$ and S_i, T_i, U_i, V_i are as in Lemma 23.3.1 Theorem 23.2.6. We claim that K is a complex Hadamard matrix of the required kind. Our proof of this claim is split into three parts.

1. We begin by proving that K is a $(\pm 1, \pm i)$-matrix. As I_2 and T are disjoint and sum to J_2, it suffices to verify that

$$A = I_q \otimes P_{1,0}H + \sum_{i=1}^{r}(S_i \otimes P_{1,2i}H + U_i \otimes P_{2,2i-1}H)$$

and

$$B = I_q \otimes P_{2,0}H + \sum_{i=1}^{r}(T_i \otimes P_{2,2i}H + V_i \otimes P_{1,2i-1}H)$$

are $(\pm 1, \pm i)$-matrices. But this is a consequence of Lemma 23.2.6 (2), because the $P_{1,i}H$ and $P_{2,i}H$ are $(1, -1)$-matrices.

2. There are cocycles f_1, f_2, f_3 with respective extension groups $C_2 \times C_2, C_{2q}$, A_{2r}, such that K is a sum of matrices each of which is the Kronecker product of three cocyclic matrices with cocycles f_1, f_2, f_3 respectively. Hence K is a sum of matrices which all have cocycle $f_1 \times f_2 \times f_3$ and extension group $C_{2q} \times A_{2r}$. So K is cocyclic with extension group $C_{2q} \times A_{2r}$ and (by Theorem 21.2.3) indexing group $C_q \times C_2^{4r+1}$.

3. The last thing we must do is prove that K has the correct Grammian. Now the S_i, T_i, U_i, V_i are commuting matrices such that $S_i^* = S_i$, $T_i^* = T_i$, $U_i^* = -U_i$, and $V_i^* = -V_i$. By Lemma 23.3.1 and since $HH^\top = 2^{4r}I_{2^{4r}}$ commutes with the $P_{1,i}$ and $P_{2,i}$,

(†) $I_q \otimes P_{1,0}H, \quad S_i \otimes P_{1,2i}H, \quad U_i \otimes P_{2,2i-1}H \qquad (1 \leq i \leq r)$

are pairwise anti-amicable, and

(‡) $I_q \otimes P_{2,0}H, \quad T_i \otimes P_{2,2i}H, \quad V_i \otimes P_{1,2i-1}H \qquad (1 \leq i \leq r)$

are pairwise anti-amicable. Thus

$$
\begin{aligned}
AA^* &= (I_q \otimes P_{1,0}H)(I_q \otimes P_{1,0}H)^* + \sum_{i=1}^{r}(S_i \otimes P_{1,2i}H)(S_i \otimes P_{1,2i}H)^* \\
&\quad + (U_i \otimes P_{2,2i-1}H)(U_i \otimes P_{2,2i-1}H)^* \\
&= \Big(I_q + \sum_{i=1}^{r}(S_i S_i^* + U_i U_i^*)\Big) \otimes 2^{4r} I_{2^{4r}}.
\end{aligned}
$$

Similarly,

$$
BB^* = \Big(I_q + \sum_{i=1}^{r}(T_i T_i^* + V_i V_i^*)\Big) \otimes 2^{4r} I_{2^{4r}}.
$$

Since the matrices listed in (†) are amicable with the matrices listed in (‡), A and B are amicable. Therefore, KK^* is

$$
\begin{aligned}
I_2 \otimes (AA^* + BB^*) &= I_2 \otimes \Big(2I_q + \sum_{i=1}^{r}(S_i S_i^* + T_i T_i^* + U_i U_i^* + V_i V_i^*)\Big) \otimes 2^{4r} I_{2^{4r}} \\
&= q 2^{4r+1} I_{q 2^{4r+1}}
\end{aligned}
$$

by Theorem 23.2.6 (3). This completes the proof.   □

23.4.2. COROLLARY. *If there are $2r$ paired complex sequences with zero auto-correlation and total length $q - 1$, then there is a cocyclic Hadamard matrix with indexing group $C_q \times C_2^{4r+2}$ and extension group $C_{2q} \times D_8 \curlyvee \cdots \curlyvee D_8 \curlyvee C_4$.*

PROOF. By Theorem 16.5.2, a cocyclic complex Hadamard matrix with cocycle f and indexing group G yields a cocyclic Hadamard matrix with cocycle $f' \times f$ and indexing group $\mathbb{Z}_2 \times G$, where $f'(x, y) = (-1)^{xy}$ for $x, y \in \mathbb{Z}_2$. Since $E(f') \cong C_4$, the result follows from Theorem 23.4.1 and Lemma 16.5.1.   □

23.4.3. COROLLARY. *Let $q > 1$ be odd. If the binary expansion of q has d 1s, then there exists a cocyclic Hadamard matrix of order $q 2^{4d-2}$.*

PROOF. The number of 1s in the binary expansion of $p = (q - 1)/2$ is $d - 1$. The result follows from Corollary 23.4.2 and Theorem 23.1.3.   □

Applying [**23**], we get an even better result, as stated at the beginning of the chapter.

23.4.4. COROLLARY. *For any odd $q > 1$, and $k \geq 10 + 8 \lfloor \log_2(q - 1)/10 \rfloor$, there is a cocyclic Hadamard matrix of order $q 2^k$.*

PROOF. Write $q = 2p + 1$. By Corollary 23.4.2, there is a cocyclic Hadamard matrix H of order $q 2^{2L(p)+2}$. So there is a cocyclic Hadamard matrix of order $q 2^k$, by Theorem 23.1.5: namely the Kronecker product of H and the Sylvester matrix of order $2^{k-2L(p)-2}$.   □

23.5. Concluding remarks

The number $L(p)$ determines the number of matrices $P_{1,i}$, $P_{2,i}$ used in proving Theorem 23.4.1, which in turn determines the order of these matrices. We can minimize this order given any $L(p)$. Hence, our asymptotic existence result can be bettered only by minimizing $L(p)$. If Conjecture 7 is true, then the following holds.

23.5.1. PROPOSITION. *For each $\epsilon > 0$ there is an integer n_ϵ such that, for any odd $q > n_\epsilon$, there exists a cocyclic Hadamard matrix of order $q 2^{\epsilon \log q}$.*

Proposition 23.5.1 would be a major advance. Note that if Conjecture 7 is true, then current methods [**23**] give us nothing stronger than Proposition 23.5.1, even if we do not insist that the $P_{1,i}$ and $P_{2,i}$ are cocyclic. So this conjecture seems to be the key issue to settle.

10. RESEARCH PROBLEM. *Prove Conjecture 7, or an analog for a workable generalization of zero aperiodic autocorrelation sequences.*

Bibliography

1. V. Alvarez, J. A. Armario, M. D. Frau, and P. Real, *The homological reduction method for computing cocyclic Hadamard matrices*, J. Symbolic Comput. **44** (2009), no. 5, 558–570.
2. K. T. Arasu and W. de Launey, *Two-dimensional perfect quaternary arrays*, IEEE Trans. Inform. Theory **47** (2001), no. 4, 1482–1493.
3. K. T. Arasu, W. de Launey, and S. L. Ma, *On circulant complex Hadamard matrices*, Des. Codes Cryptogr. **25** (2002), no. 2, 123–142.
4. K. T. Arasu and Q. Xiang, *On the existence of periodic complementary binary sequences*, Des. Codes Cryptogr. **2** (1992), no. 3, 257–262.
5. E. F. Assmus, Jr. and C. J. Salwach, *The (16, 6, 2) designs*, Internat. J. Math. Math. Sci. **2** (1979), no. 2, 261–281.
6. R. D. Baker, *An elliptic semiplane*, J. Combin. Theory Ser. A **25** (1978), no. 2, 193–195.
7. A. Baliga and K. J. Horadam, *Cocyclic Hadamard matrices over $Z_t \times Z_2^2$*, Australas. J. Combin. **11** (1995), 123–134.
8. L. D. Baumert, *Cyclic difference sets*, Lecture Notes in Math., vol. 182, Springer-Verlag, Berlin, 1971.
9. G. Berman, *Weighing matrices and group divisible designs determined by* $EG(t, p^r)$, $p > 2$, Utilitas Math. **12** (1977), 183–191.
10. ———, *Families of generalized weighing matrices*, Canad. J. Math. **30** (1978), no. 5, 1016–1028.
11. T. Beth, D. Jungnickel, and H. Lenz, *Design theory. Vol. I*, Encyclopedia of Mathematics and its Applications, vol. 69, Cambridge University Press, Cambridge, 1999.
12. F. R. Beyl and J. Tappe, *Group extensions, representations, and the Schur multiplicator*, Lecture Notes in Math., Springer-Verlag, Berlin, 1982.
13. R. C. Bose, *On the construction of balanced incomplete block designs*, Ann. Eugenics **9** (1939), 353–399.
14. W. Bosma, J. Cannon, and C. Playout, *The Magma algebra system. I. The user language*, J. Symbolic Comput. **24** (1997), no. 3-4, 235–265.
15. B. W. Brock, *Hermitian congruence and the existence and completion of generalized Hadamard matrices*, J. Combin. Theory Ser. A **49** (1988), no. 2, 233–261.
16. A. T. Butson, *Generalized Hadamard matrices*, Proc. Amer. Math. Soc. **13** (1962), 894–898.
17. ———, *Relations among generalized Hadamard matrices, relative difference sets, and maximal length linear recurring sequences*, Canad. J. Math. **15** (1963), 42–48.
18. G. Cohen, D. Rubie, J. Seberry, C. Koukouvinos, S. Kounias, and M. Yamada, *A survey of base sequences, disjoint complementary sequences and* $OD(4t; t, t, t, t)$, J. Combin. Math. Combin. Comput. **5** (1989), 69–103.
19. C. J. Colbourn and W. de Launey, *Difference matrices*, The CRC handbook of combinatorial designs, CRC Press, Boca Raton, 1996, pp. 287–296.
20. R. Compton, R. Craigen, and W. de Launey, *Unreal* $BH(n, 6)s$ *and Hadamard matrices*, preprint.
21. R. Craigen, *Signed groups, sequences, and the asymptotic existence of Hadamard matrices*, J. Combin. Theory Ser. A **71** (1995), no. 2, 241–254.
22. R. Craigen and W. de Launey, *Generalized Hadamard matrices whose transposes are not generalized Hadamard matrices*, J. Combin. Des. **17** (2009), no. 6, 456–458.
23. R. Craigen, W. H. Holzmann, and H. Kharaghani, *On the asymptotic existence of complex Hadamard matrices*, J. Combin. Des. **5** (1997), no. 5, 319–327.
24. R. Craigen and H. Kharaghani, *On the nonexistence of Hermitian circulant complex Hadamard matrices*, Australas. J. Combin. **7** (1993), 225–227.

25. _____, *A combined approach to the construction of Hadamard matrices*, Australas. J. Combin. **13** (1996), 89–107.

26. _____, *Weaving Hadamard matrices with maximum excess and classes with small excess*, J. Combin. Des. **12** (2004), no. 4, 233–255.

27. R. Craigen, J. Seberry, and X. M. Zhang, *Product of four Hadamard matrices*, J. Combin. Theory Ser. A **59** (1992), no. 2, 318–320.

28. T. Czerwinski, *On finite projective planes with a single (P, l) transitivity*, J. Combin. Theory Ser. A **48** (1988), no. 1, 136–138.

29. J. E. Dawson, *A construction for generalized Hadamard matrices $\mathrm{GH}(4q, \mathrm{EA}(q))$*, J. Statist. Plann. Inference **11** (1985), no. 1, 103–110.

30. D. de Caen, D. A. Gregory, and D. Pritikin, *Minimum biclique partitions of the complete multigraph and related designs*, Graphs, matrices, and designs, Lecture Notes in Pure and Appl. Math., vol. 139, Dekker, New York, 1993, pp. 93–119.

31. D. de Caen, R. Mathon, and G. E. Moorhouse, *A family of antipodal distance-regular graphs related to the classical Preparata codes*, J. Algebraic Combin. **4** (1995), no. 4, 317–327.

32. W. de Launey, *Generalised Hadamard matrices whose rows and columns form a group*, Combinatorial mathematics, X (Adelaide, 1982), Lecture Notes in Math., vol. 1036, pp. 154–176.

33. _____, *On the nonexistence of generalised Hadamard matrices*, J. Statist. Plann. Inference **10** (1984), no. 3, 385–396.

34. _____, *On the nonexistence of generalised weighing matrices*, Ars Combin. **17** (1984), no. A, 117–132.

35. _____, *A survey of generalised Hadamard matrices and difference matrices $\mathrm{D}(k, \lambda; G)$ with large k*, Utilitas Math. **30** (1986), 5–29.

36. _____, *$(0, G)$-designs with applications*, Ph.D. thesis, University of Sydney, 1987.

37. _____, *On difference matrices, transversal designs, resolvable transversal designs and large sets of mutually orthogonal F-squares*, J. Statist. Plann. Inference **16** (1987), no. 1, 107–125.

38. _____, *GBRDs: some new constructions for difference matrices, generalised Hadamard matrices and balanced generalised weighing matrices*, Graphs Combin. **5** (1989), no. 2, 125–135.

39. _____, *Square GBRDs over nonabelian groups*, Ars Combin. **27** (1989), 40–49.

40. _____, *On the construction of n-dimensional designs from 2-dimensional designs*, Australas. J. Combin. **1** (1990), 67–81, Combinatorial Mathematics and Combinatorial Computing, Vol. 1 (Brisbane, 1989).

41. _____, *Cocyclic Hadamard matrices and relative difference sets*, Ohio State Conference on Groups and Difference Sets; The Hadamard Centenary Conference, University of Wollongong, 1993.

42. _____, *On the asymptotic existence of partial complex Hadamard matrices and related combinatorial objects*, Discrete Appl. Math. **102** (2000), no. 1-2, 37–45, Coding, cryptography and computer security (Lethbridge, AB, 1998).

43. _____, *On a family of cocyclic Hadamard matrices*, Codes and designs (Columbus, OH, 2000), Ohio State Univ. Math. Res. Inst. Publ., vol. 10, de Gruyter, Berlin, 2002, pp. 187–205.

44. _____, *On the asymptotic existence of Hadamard matrices*, J. Combin. Theory Ser. A **116** (2009), no. 4, 1002–1008.

45. W. de Launey and J. E. Dawson, *A note on the construction of $\mathrm{GH}(4tq; \mathrm{EA}(q))$ for $t = 1, 2$*, Australas. J. Combin. **6** (1992), 177–186.

46. _____, *An asymptotic result on the existence of generalised Hadamard matrices*, J. Combin. Theory Ser. A **65** (1994), no. 1, 158–163.

47. W. de Launey, D. L. Flannery, and K. J. Horadam, *Cocyclic Hadamard matrices and difference sets*, Discrete Appl. Math. **102** (2000), no. 1-2, 47–61, Coding, cryptography and computer security (Lethbridge, AB, 1998).

48. W. de Launey and D. M. Gordon, *A comment on the Hadamard conjecture*, J. Combin. Theory Ser. A **95** (2001), no. 1, 180–184.

49. _____, *A remark on Plotkin's bound*, IEEE Trans. Inf. Th. **47** (2001), no. 1.

50. W. de Launey and K. J. Horadam, *A weak difference set construction for higher-dimensional designs*, Des. Codes Cryptogr. **3** (1993), no. 1, 75–87.

51. W. de Launey and H. Kharaghani, *On the asymptotic existence of cocyclic Hadamard matrices*, J. Combin. Theory Ser. A **116** (2009), no. 6, 1140–1153.

52. W. de Launey and D. Levin, *(1, −1)-matrices with near-extremal properties*, SIAM J. Discrete Math. **23** (2009), no. 3, 1422–1440.

53. W. de Launey and J. Seberry, *The strong Kronecker product*, J. Combin. Theory Ser. A **66** (1994), no. 2, 192–213.

54. W. de Launey and M. J. Smith, *Cocyclic orthogonal designs and the asymptotic existence of cocyclic Hadamard matrices and maximal size relative difference sets with forbidden subgroup of size 2*, J. Combin. Theory Ser. A **93** (2001), no. 1, 37–92.

55. W. de Launey and R. M. Stafford, *The regular subgroups of the Paley type II Hadamard matrix*, preprint.

56. _____, *On cocyclic weighing matrices and the regular group actions of certain Paley matrices*, Discrete Appl. Math. **102** (2000), no. 1-2, 63–101, Coding, cryptography and computer security (Lethbridge, AB, 1998).

57. _____, *On the automorphisms of Paley's type II Hadamard matrix*, Discrete Math. **308** (2008), no. 13, 2910–2924.

58. P. Delsarte and J.-M. Goethals, *Tri-weight codes and generalized Hadamard matrices*, Information and Control **15** (1969), 196–206.

59. J. F. Dillon, *Variations on a scheme of McFarland for noncyclic difference sets*, J. Combin. Theory Ser. A **40** (1985), no. 1, 9–21.

60. _____, *Some REALLY beautiful Hadamard matrices*, Cryptogr. Commun. **2** (2010), no. 2, 271–292.

61. J. D. Dixon and B. Mortimer, *Permutation groups*, Graduate Texts in Mathematics, vol. 163, Springer-Verlag, New York, 1996.

62. D. A. Drake, *Partial λ-geometries and generalized Hadamard matrices over groups*, Canad. J. Math. **31** (1979), no. 3, 617–627.

63. P. Eades, *Integral quadratic forms and orthogonal designs*, J. Austral. Math. Soc. Ser. A **30** (1980/81), no. 3, 297–306.

64. D. L. Flannery, *Calculation of cocyclic matrices*, J. Pure Appl. Algebra **112** (1996), no. 2, 181–190.

65. _____, *Cocyclic Hadamard matrices and Hadamard groups are equivalent*, J. Algebra **192** (1997), no. 2, 749–779.

66. D. L. Flannery and E. A. O'Brien, *Computing 2-cocycles for central extensions and relative difference sets*, Comm. Algebra **28** (2000), no. 4, 1939–1955.

67. J. C. Galati, *A group extensions approach to relative difference sets*, J. Combin. Des. **12** (2004), no. 4, 279–298.

68. H. M. Gastineau-Hills, *Quasi-Clifford algebras and systems of orthogonal designs*, J. Austral. Math. Soc. Ser. A **32** (1982), no. 1, 1–23.

69. A. V. Geramita and J. M. Geramita, *Complex orthogonal designs*, J. Combin. Theory Ser. A **25** (1978), no. 3, 211–225.

70. A. V. Geramita, J. M. Geramita, and J. S. Wallis, *Orthogonal designs*, Linear and Multilinear Algebra **3** (1975/76), no. 4, 281–306.

71. A. V. Geramita and J. Seberry, *Orthogonal designs*, Lecture Notes in Pure and Applied Mathematics, vol. 45, Marcel Dekker Inc., New York, 1979, Quadratic forms and Hadamard matrices.

72. P. B. Gibbons and R. Mathon, *Construction methods for Bhaskar Rao and related designs*, J. Austral. Math. Soc. Ser. A **42** (1987), no. 1, 5–30.

73. _____, *Signings of group divisible designs and projective planes*, Australas. J. Combin. **11** (1995), 79–104.

74. P. B. Gibbons and R. A. Mathon, *Group signings of symmetric balanced incomplete block designs*, Proceedings of the Singapore conference on combinatorial mathematics and computing (Singapore, 1986), vol. 23, 1987, pp. 123–134.

75. R. E. Gilman, *On the Hadamard determinant theorem and orthogonal determinants*, Bull. Amer. Math. Soc. **37** (1931), 30–31.

76. J.-M. Goethals and J. J. Seidel, *Orthogonal matrices with zero diagonal*, Canad. J. Math. **19** (1967), 1001–1010.

77. D. Gorenstein, *Finite groups*, Chelsea Publishing Company, New York, 1980.

78. S. W. Graham and I. E. Shparlinski, *On RSA moduli with almost half of the bits prescribed*, Discrete Appl. Math. **156** (2008), no. 16, 3150–3154.

79. The GAP group, GAP - *Groups, Algorithms, and Programming*, Version 4.4.9 (2006), http://www.gap-system.org.

80. J. Hadamard, *Resolution d'une question relative aux determinants*, Bull. des Sci. Math. **17** (1893), 240–246.

81. J. Hammer and J. R. Seberry, *Higher-dimensional orthogonal designs and applications*, IEEE Trans. Inform. Theory **27** (1981), no. 6, 772–779.

82. D. F. Holt, *The calculation of the Schur multiplier of a permutation group*, Computational group theory (Durham, 1982), Academic Press, London, 1984, pp. 307–319.

83. D. F. Holt, B. Eick, and E. A. O'Brien, *Handbook of computational group theory*, Chapman & Hall/CRC Press, Boca Raton, London, New York, Washington, 2005.

84. W. H. Holzmann and H. Kharaghani, *On the Plotkin arrays*, Australas. J. Combin. **22** (2000), 287–299.

85. W. H. Holzmann, H. Kharaghani, and B. Tayfeh-Rezaie, *Williamson matrices up to order 59*, Des. Codes Cryptogr. **46** (2008), no. 3, 343–352.

86. K. J. Horadam, *An introduction to cocyclic generalised Hadamard matrices*, Discrete Appl. Math. **102** (2000), no. 1-2, 115–131, Coding, cryptography and computer security (Lethbridge, AB, 1998).

87. ———, *Hadamard matrices and their applications*, Princeton University Press, Princeton, NJ, 2007.

88. K. J. Horadam and W. de Launey, *Cocyclic development of designs*, J. Algebraic Combin. **2** (1993), no. 3, 267–290.

89. H. Hotelling, *Some improvements in weighing and other experimental techniques*, Ann. Math. Statistics **15** (1944), 297–306.

90. N. Howgrave-Graham and M. Szydlo, *A method to solve cyclotomic norm equations $f * \overline{f}$*, Algorithmic number theory, Lecture Notes in Comput. Sci., vol. 3076, Springer, Berlin, 2004, pp. 272–279.

91. B. Huppert and N. Blackburn, *Finite groups. III*, Grundlehren der Mathematischen Wissenschaften, vol. 243, Springer-Verlag, Berlin, 1982.

92. Y. J. Ionin, *New symmetric designs from regular Hadamard matrices*, Electron. J. Combin. **5** (1998), Research Paper 1, 8 pp. (electronic).

93. ———, *A technique for constructing symmetric designs*, Des. Codes Cryptogr. **14** (1998), no. 2, 147–158.

94. ———, *Building symmetric designs with building sets*, Des. Codes Cryptogr. **17** (1999), no. 1-3, 159–175.

95. ———, *Symmetric subdesigns of symmetric designs*, J. Combin. Math. Combin. Comput. **29** (1999), 65–78.

96. ———, *Applying balanced generalized weighing matrices to construct block designs*, Electron. J. Combin. **8** (2001), no. 1, Research Paper 12, 15 pp. (electronic).

97. I. M. Isaacs, *Algebra: a graduate course*, Brooks/Cole, Pacific Grove, 1994.

98. N. Ito, *Note on Hadamard matrices of type Q*, Studia Sci. Math. Hungar. **16** (1981), no. 3-4, 389–393.

99. ———, *Note on Hadamard groups of quadratic residue type*, Hokkaido Math. J. **22** (1993), no. 3, 373–378.

100. ———, *On Hadamard groups*, J. Algebra **168** (1994), no. 3, 981–987.

101. ———, *On Hadamard groups, II*, J. Algebra **169** (1994), no. 3, 936–942.

102. ———, *On Hadamard groups III*, Kyushu J. Math. **51** (1997), no. 3, 369–379.

103. Z. Janko, H. Kharaghani, and V. D. Tonchev, *Bush-type Hadamard matrices and symmetric designs*, J. Combin. Des. **9** (2001), no. 1, 72–78.

104. ———, *The existence of a Bush-type Hadamard matrix of order 324 and two new infinite classes of symmetric designs*, Des. Codes Cryptogr. **24** (2001), no. 2, 225–232.

105. D. Jungnickel, *On difference matrices, resolvable transversal designs and generalized Hadamard matrices*, Math. Z. **167** (1979), no. 1, 49–60.

106. W. M. Kantor, *Automorphism groups of Hadamard matrices*, J. Combinatorial Theory **6** (1969), 279–281.

107. ———, *Symplectic groups, symmetric designs, and line ovals*, J. Algebra **33** (1975), 43–58.

108. G. Karpilovsky, *The Schur multiplier*, London Mathematical Society Monographs. New Series, vol. 2, The Clarendon Press Oxford University Press, New York, 1987.

109. H. Kharaghani, *An asymptotic existence result for orthogonal designs*, Combinatorics advances (Tehran, 1994), Math. Appl., vol. 329, Kluwer Acad. Publ., Dordrecht, 1995, pp. 225–233.

110. _____, *On the twin designs with the Ionin-type parameters*, Electron. J. Combin. **7** (2000), Research Paper 1, 11 pp. (electronic).

111. H. Kharaghani and J. Seberry, *Regular complex Hadamard matrices*, Proceedings of the Nineteenth Manitoba Conference on Numerical Mathematics and Computing (Winnipeg, MB, 1989), vol. 75, 1990, pp. 187–201.

112. H. Koch, *Number theory. Algebraic numbers and functions*, Graduate Studies in Mathematics, vol. 24, American Mathematical Society, Providence, RI, 2000.

113. T. Y. Lam and K. H. Leung, *On vanishing sums of roots of unity*, J. Algebra **224** (2000), no. 1, 91–109.

114. J. S. Leon, *An algorithm for computing the automorphism group of a Hadamard matrix*, J. Comb. Theory, Ser. A **27** (1979), no. 3, 289–306.

115. C. Mackenzie and J. Seberry, *Maximal ternary codes and Plotkin's bound*, Ars Combin. **17** (1984), no. A, 251–270.

116. V. C. Mavron, T. P. McDonough, and C. A. Pallikaros, *A difference matrix construction and a class of balanced generalized weighing matrices*, Arch. Math. (Basel) **76** (2001), no. 4, 259–264.

117. V. C. Mavron and V. D. Tonchev, *On symmetric nets and generalized Hadamard matrices from affine designs*, J. Geom. **67** (2000), no. 1-2, 180–187, Second Pythagorean Conference (Pythagoreion, 1999).

118. R. L. McFarland, *Hadamard difference sets in abelian groups of order $4p^2$*, Mitt. Math. Sem. Giessen. **192** (1989), 1–70.

119. _____, *Sub-difference sets of Hadamard difference sets*, J. Combin. Theory Ser. A **54** (1990), no. 1, 112–122.

120. B. McKay, *Practical graph isomorphism*, Congressus Numerantium **30** (1981), 45–87.

121. A. C. Mukhopadhyay, *Generalized weighing matrices,* SGDD*s possessing dual property and related configurations*, Sankhyā Ser. A **54** (1992), no. Special Issue, 291–298, Combinatorial mathematics and applications (Calcutta, 1988).

122. R. C. Mullin, *A note on balanced weighing matrices*, Combinatorial mathematics, III (Proc. Third Australian Conf., Univ. Queensland, St. Lucia, 1974), Springer, Berlin, 1975, pp. 28–41. Lecture Notes in Math., Vol. 452.

123. R. C. Mullin and R. G. Stanton, *Balanced weighing matrices and group divisible designs*, Utilitas Math. **8** (1975), 303–310.

124. _____, *Group matrices and balanced weighing designs*, Utilitas Math. **8** (1975), 277–301.

125. W. Nickel, *Central extensions of polycyclic groups*, Ph.D. thesis, Australian National University, 1993.

126. P. Ó Catháin and M. Röder, *The cocyclic Hadamard matrices of order less than* 40, Des. Codes Cryptogr. **58** (2011), no. 1, 73–88.

127. D. Ž. Đoković, *Periodic complementary sets of binary sequences*, Int. Math. Forum **4** (2009), no. 13-16, 717–725.

128. R. E. A. C. Paley, *On orthogonal matrices*, J. Math. Phys. **12** (1933), 311–320.

129. A. A. I. Perera and K. J. Horadam, *Cocyclic generalised Hadamard matrices and central relative difference sets*, Des. Codes Cryptogr. **15** (1998), no. 2, 187–200.

130. A. Pott, *Finite geometry and character theory*, Lecture Notes in Math., vol. 1601, Springer-Verlag, Berlin, 1995.

131. D. P. Rajkundlia, *Some techniques for constructing new infinite families of incomplete block designs*, Ph.D. thesis, Queens University, Kingston, Canada, 1978.

132. _____, *Some techniques for constructing infinite families of BIBDs*, Discrete Math. **44** (1983), no. 1, 61–96.

133. D. K. Ray-Chaudhuri and Q. Xiang, *New necessary conditions for abelian Hadamard difference sets*, J. Statist. Plann. Inference **62** (1997), 69–79.

134. D. J. S. Robinson, *A course in the theory of groups*, second ed., Graduate Texts in Mathematics, vol. 80, Springer-Verlag, New York, 1996.

135. D. G. Sarvate and J. Seberry, *Group divisible designs, GBRSDS and generalized weighing matrices*, Util. Math. **54** (1998), 157–174.

136. P. J. Schellenberg, *A computer construction for balanced orthogonal matrices*, Proceedings of the Sixth Southeastern Conference on Combinatorics, Graph Theory and Computing (Florida Atlantic Univ., Boca Raton, Fla., 1975) (Winnipeg), Utilitas Math., 1975, pp. 513–522. Congressus Numerantium, No. XIV.

137. S. T. Schibell and R. M. Stafford, private communications, 1993, 2011.

138. B. Schmidt, *Cyclotomic integers and finite geometry*, J. Amer. Math. Soc. **12** (1999), no. 4, 920–952.

139. _____, *Williamson matrices and a conjecture of Ito's*, Des. Codes Cryptogr. **17** (1999), no. 1-3, 61–68.

140. _____, *Towards Ryser's conjecture*, European Congress of Mathematics, Vol. I (Barcelona, 2000), Progr. Math., vol. 201, Birkhäuser, Basel, 2001, pp. 533–541.

141. _____, *Characters and cyclotomic fields in finite geometry*, Lecture Notes in Math., vol. 1797, Springer-Verlag, Berlin, 2002.

142. J. Seberry, *Some remarks on generalised Hadamard matrices and theorems of Rajkundlia on SBIBDs*, Combinatorial mathematics, VI (Proc. Sixth Austral. Conf., Univ. New England, Armidale, 1978), Lecture Notes in Math., vol. 748, Springer, Berlin, 1979, pp. 154–164.

143. _____, *A construction for generalized Hadamard matrices*, J. Statist. Plann. Inference **4** (1980), no. 4, 365–368.

144. J. Shawe-Taylor, *Coverings of complete bipartite graphs and associated structures*, Discrete Math. **134** (1994), no. 1-3, 151–160, Algebraic and topological methods in graph theory (Lake Bled, 1991).

145. P. J. Shlichta, *Higher dimensional Hadamard matrices*, IEEE Trans. Inform. Theory **25** (1979), no. 5, 566–572.

146. S. S. Shrikhande, *Generalized Hadamard matrices and orthogonal arrays of strength two*, Canad. J. Math. **16** (1964), 736–740.

147. C. C. Sims, *Computation with finitely presented groups*, Encyclopedia of Mathematics and its Applications, vol. 48, Cambridge University Press, Cambridge, 1994.

148. D. J. Street, *Generalized Hadamard matrices, orthogonal arrays and F-squares*, Ars Combin. **8** (1979), 131–141.

149. J. J. Sylvester, *Thoughts on inverse orthogonal matrices, simultaneous sign successions, and tesselated pavements in two or more colours, with applications to Newton's rule, ornamental tile-work, and the theory of numbers*, Phil. Mag. **34** (1867), 461–475.

150. V. Tarokh, H. Jafarkhani, and A. R. Calderbank, *Space-time block codes from orthogonal designs*, IEEE Trans. Inform. Theory **45** (1999), no. 5, 1456–1467.

151. R. J. Turyn, *Sequences with small correlation*, Error Correcting Codes (Proc. Sympos. Math. Res. Center, Madison, Wis., 1968), John Wiley, New York, 1968, pp. 195–228.

152. _____, *Complex Hadamard matrices*, Combinatorial Structures and their Applications (Proc. Calgary Internat. Conf., Calgary, Alta., 1969), Gordon and Breach, New York, 1970, pp. 435–437.

153. _____, *On C-matrices of arbitrary powers*, Canad. J. Math. **23** (1971), 531–535.

154. _____, *An infinite class of Williamson matrices*, J. Combinatorial Theory Ser. A **12** (1972), 319–321.

155. _____, *A special class of Williamson matrices and difference sets*, J. Combin. Theory Ser. A **36** (1984), no. 1, 111–115.

156. Jennifer Seberry Wallis, *On the existence of Hadamard matrices*, J. Combinatorial Theory Ser. A **21** (1976), no. 2, 188–195.

157. J. Williamson, *Hadamard's determinant theorem and the sum of four squares*, Duke Math. J. **11** (1944), 65–81.

158. R. M. Wilson and Q. Xiang, *Constructions of Hadamard difference sets*, J. Combin. Theory Ser. A **77** (1997), no. 1, 148–160.

159. W. Wolfe, *Rational quadratic forms and orthogonal designs*, Number theory and algebra, Academic Press, New York, 1977, pp. 339–348.

160. _____, *Limits on pairwise amicable orthogonal designs*, Canad. J. Math. **33** (1981), no. 5, 1043–1054.

161. M. Y. Xia, *Some infinite classes of special Williamson matrices and difference sets*, J. Combin. Theory Ser. A **61** (1992), no. 2, 230–242.

162. Q. Xiang and Y. Q. Chen, *On Xia's construction of Hadamard difference sets*, Finite Fields and Their Applications **2** (1996), 87–95.

163. M. Yamada, *Hadamard matrices of generalized quaternion type*, Discrete Math. **87** (1991), no. 2, 187–196.

164. K. Yamamoto, *On a generalized Williamson equation*, Finite and infinite sets, Vol. I, II (Eger, 1981), Colloq. Math. Soc. János Bolyai, vol. 37, North-Holland, Amsterdam, 1984, pp. 839–850.

165. Y. X. Yang, X. X. Niu, and C. Q. Xu, *Theory and applications of higher-dimensional Hadamard matrices*, 2nd ed., CRC Press, Boca Raton, 2010.

Index

Titles in This Series

TITLES IN THIS SERIES

For a complete list of titles in this series, visit the
AMS Bookstore at **www.ams.org/bookstore/**.